# Innovative Mobile Learning:
## Techniques and Technologies

Hokyoung Ryu
*Massey University, New Zealand*

David Parsons
*Massey University, New Zealand*

Information Science
REFERENCE

INFORMATION SCIENCE REFERENCE
Hershey · New York

Director of Editorial Content: Kristin Klinger
Director of Production: Jennifer Neidig
Managing Editor: Jamie Snavely
Assistant Managing Editor: Carole Coulson
Typesetter: Larissa Vinci
Cover Design: Lisa Tosheff
Printed at: Yurchak Printing Inc.

Published in the United States of America by
Information Science Reference (an imprint of IGI Global)
701 E. Chocolate Avenue, Suite 200
Hershey PA 17033
Tel: 717-533-8845
Fax: 717-533-8661
E-mail: cust@igi-global.com
Web site: http://www.igi-global.com

and in the United Kingdom by
Information Science Reference (an imprint of IGI Global)
3 Henrietta Street
Covent Garden
London WC2E 8LU
Tel: 44 20 7240 0856
Fax: 44 20 7379 0609
Web site: http://www.eurospanbookstore.com



Library of Congress Cataloging-in-Publication Data

Innovative mobile learning : techniques and technologies / Hokyoung Ryu and David Parsons, editor.

p. cm.

Includes bibliographical references and index.

Summary: "This book includes the challenges and practical experience of the design of M-Learning environments, covering current developments in M-learning experiences in both academia and industry"--Provided by publisher.

ISBN 978-1-60566-062-2 (hardcover) -- ISBN 978-1-60566-063-9 (ebook)

1. Mobile communication systems in education. I. Ryu, Hokyoung. II. Parsons, David, 1959 Oct. 13-

LB1044.84.I56 2009

658.3'1240402854678--dc22

2008010308

British Cataloguing in Publication Data
A Cataloguing in Publication record for this book is available from the British Library.

# Table of Contents

# Detailed Table of Contents

## Section II
## Enhancing Individual Learning Experiences

**Section III**
**Enhancing Collaborative Learning Experiences**

In this chapter, the authors use the scaffolding technique to design an interactive and collaborative participatory simulation. Evalution reveals that this participatory simulation helps both the individual learner and groups of learners to gain a deeper understanding of a sorting algorithm, and encourages more active participation in group work.

**Section IV**
**Enhancing Situated Learning Experiences**



This chapter presents the SketchMap system, which supports children's situated learning by creating maps. Its use is to integrate outdoor and classroom activities, and share the children's experiences through the maps in order to promote situated learning activities.



This chapter presents an approach to designing a mobile application for contextual lifelong learning. It assists learners to access, compose and manage their learning in a range of institutional, informal and work-based settings by keeping them connected with content that is relevant to their studies, and its use is demonstrated in three lifelong learning scenarios.



This chapter discusses a location-aware learning organizer that helps university students manage their learning activities and supports situated learning experiences.



This chapter discusses a professional mobile learning experience for those who are in need of more contextual understanding of different work procedures in a specific learning environment.

**Section V**
**Challenges and Future Mobile Learning**

# Foreword

I have the pleasure of writing some words for this timely and thought-provoking collection. The field of mobile learning is somewhat different from many other technological advancements of the past. The rapid growth of mobile learning has its roots as much in the educational and pedagogical needs of flexibility and the anytime anywhere demand for learning as it does in the readily available mass market of tiny devices that almost everyone seems to have. "Technology first, pedagogy later" is not new when it comes to technology adoption in education. However, effective mobile learning has evolved not so much from the over-enthusiasm of early technology adopters but rather due to the high penetration of technology already available for application penetration. This is perhaps also the reason why it did not take long before researchers started to focus on getting pedagogy right, rather than letting technology dominate it (which has all too often been the case in the past).

The danger of such readily available technology is the possibility of succumbing to the desire of using it as a fancy replacement for every possible existing solution. On the other hand, with tiny screens, low and unreliable bandwidth, small memory, and other similar limitations, one could easily argue against using such technology where far richer technological solutions already exist. Therefore, caution needs to be exercised when deciding where mobile technology would be most effective in the learning process and whom it would best serve.

With the marriage of multimedia-rich content on mobile devices and the design of appropriate learning tasks, mobile learning has the potential to provide a learning experience that is of a far higher quality than present day classrooms are able to provide. Mix it with the latest generation of technologies, such as iPods, Wii, and smart chips, add all those multiple input and output capabilities that are available on most mobile devices in the market, such as freehand annotations, audio, photo snapshots and video, and you have learning experience that knows no bounds!

It is very refreshing to see these issues appropriately discussed in this collection. The editors have done a superb job of eliciting chapters from pioneering authors and put them together in a sequence that moves the reader to increasing levels of understanding. Both editors come from Massey University where I worked for seven years and it is ironic that I first knew about them not by a meeting of any sort but through their research papers! It is an honor for me to be able to contribute to this important volume. The list of authors boasts the same tradition of high quality researchers who are engaged in state-of-the-art yet down-to-earth research on mobile learning.

The book contains a good balance between the enthusiastic uses of emerging innovations and finding appropriate pedagogy that would ensure that benefits are in fact achieved in terms of improved and omnipresent flexibility, interactivity, learner engagement, learning opportunities, and in-context learning through collaboration. Technology is not used just for technology's sake but appropriate assessment methods are used to ensure the educational effectiveness of these technologies. Readers, whether they are new to the mobile learning world or are seasoned players, would find the book equally interesting.

Individual chapters of this collection would guide today's educators in deciding how to use different mobile technologies in the most effective ways. The whole book would also serve as a referential archive for future generations of researchers by helping them to better understand the systematic uptake of mobile technology that is guided by the educational paradigms.

*Kinshuk*
*Professor, Athabasca University, Canada*

# Preface

The aim of *Innovative Mobile Learning: Techniques and Technologies* is to introduce the reader to the current directions of mobile learning (a.k.a., m-learning). More precisely, it is about providing a comprehensive survey of mobile learning research and projects that both academics and educational practitioners may utilize in their work.

It is simply not possible to give a neat definition of mobile learning, specifying necessary and sufficient characteristics of all those activities that have ever been called "mobile learning." It is a new phenomenon that has developed through the recent spread of mobile ICT (information and communication technology), both a separate endeavor to traditional pedagogy and a complementary approach to it. Hence, in order to keep the book within manageable bounds, some difficult decisions had to be made about what to include and what to exclude. In making these decisions we were guided by our own experience and the recommendations of our reviewers who contributed significantly to the book's development.

In approaching this novel learning mode, this book argues that a holistic approach for encompassing diverse mobile learning themes is necessary to design new kinds of learning activities with mobile technologies, where there is still a lack of well-defined characteristics and features of mobile learning spaces. It integrates concerns about tools, methods, and technologies for mobile learning development with concerns about validating the learning experience that each project produces. This integration is believed to offer a pragmatic solution to critical challenges in both technical design and learning outcomes.

The contribution of this book is thus to bring together a range of approaches to technologies and techniques, presenting them in a common format and at a detailed level. The goal is to provide detailed information about each project, and to make primary sources more accessible.

The rest of this preface introduces the scope of the book and the approach that is employed for its level of description.

## THE CHALLENGES

The increasing spread of mobile devices is dramatically affecting people's daily lives. They not only increase the pace and efficiency of everyday life, but also allow more flexibility at business and professional levels.

Arguably, mobile technologies, particularly the increasingly sophisticated mobile phone (blurring the boundaries between communication and computation), combine both ubiquity and utility. This phenomenon has given rise to opportunities to employ mobile technologies more broadly than just as communication tools.

Learning design with mobile technologies has been, to some extent, a successful technological and scientific undertaking, helping to broaden the arenas of the educational sector in ways that no one could have anticipated a decade ago. Of course, we cannot predict its future; however, the progress of the past decade highlights specific current challenges.

To the extent that the success of mobile learning is due to its ambitiously multidisciplinary application, an m-learning textbook should span its multidisciplinary scientific and technical foundations. Future m-learning systems will only succeed if they can continuously synthesize coherent learning experiences from these foundations. For instance, the Ambient Wood project carried out by Sussex University (UK) has demonstrated how a mobile and ambient environment can provide a more effective and situated learning experience. In another learning context, MIT (USA) has developed a collaborative m-learning tool to help students understand complex and dynamic epidemic phenomena, simulating them with learners' wearable mobile devices, demonstrating significant advantages over non-mobile e-learning applications. Even Nintendo's DS Lite™ handheld console has been used to provide an enjoyable way to improve Japanese students' individual English skills. M-learning is not simply limited to delivering teaching materials onto student's mobile handsets, which the term 'learning' implicitly points out, but also encompasses public information or even commercial information on lifestyle choices and health promotion. Many other domains (e.g., health practitioners) see the advantages of m-learning environments in providing personalized content (e.g., dietary information, quit smoking programs) via the mobile phones that have become so pervasive in recent years. These projects are targeted at specific objectives but they are also designed to extract the critical success factors that can be used to generalize findings to other m-learning environments. In this way, we can develop a better understanding of how mobile technologies can be used to enhance various user experiences, empower the user with the knowledge and ability to self-manage, and learn how these technologies can improve quality of life across a spectrum of contexts whilst containing costs and stimulating demand for services.

As the technologies that may support m-learning continue to evolve, this field will become increasingly more challenging as new opportunities emerge, and academics and practitioners need to learn from one another's experience. For instance, how to effectively take the user (i.e., learner) into account within emerging m-learning environments has formed a persistent theme in the academic field. In contrast, much of the practitioner's perspective on m-learning applications has been what kind of learning products and content can facilitate the uptake of this new learning environment. To achieve significant outcomes from this research that both deliver technological solutions and enhance the usability and sustainability of the technologies, this book aims to draw together expertise from a range of international academic and industrial contributors.

We also believe that simply being driven by technical initiatives, with a narrow focus only on the quality of mobile technologies, does not capture the potential variety and emergent aspects of mobile learning activities. Practitioners, as well as researchers, should instead embrace the notion of learning experiences, for a better understanding of the important values that mobile learning can provide. Although this 'experience' or 'learning theme' has been widely discussed, there are still few available empirical, exploratory or large-scale success cases.

## OVERVIEW OF THE BOOK

This edited book is intended to discuss the latest mobile learning environments beyond the desktop learning environment, an area of research that is increasingly seeing new developments and techniques in both the academic and commercial fields. It comprises articles from leading researchers and practitioners in the field of mobile learning. One purpose of the book is to disseminate writings about the challenges and practical experience of the design of mobile learning environments, current developments in mobile learning experiences in both academia and industry, current methods and approaches to mobile learning development, the current economic and social context of m-learning development, and empirical research into deployed m-learning environments. More importantly, a key aim of this book is to explore the technical aspects of m-learning development, where we need to systematically take into account learner interactions, learning activities and the completely renewed social and cultural environments that m-learning environments can integrate with and that technologies are now capable of delivering.

Thus, this volume is organized around wide-ranging mobile learning projects, briefly describing each project, and how they address different learning activities with mobile technologies. It then provides a more detailed description, emphasizing what sort of learning outcomes or benefits are produced. Finally, each chapter briefly comments on future research directions, opportunities, or additional ideas offered by the authors of each chapter, and issues that may be important in the next decade.

The chapters are organized in the book along some general dimensions of learning activities: individual, collaborative, and situated. Before this level of description, some theoretical foundations for learning experiences are firstly described in Section I. This will help the reader to understand the structure of this volume. We then focus on individual learning activities with mobile technologies in Section II, collaborative learning activities in Section III, situated learning activities in Section IV; and finally, Section V addresses challenges in developing mobile learning applications. This organization will hopefully assist the reader in seeing various perspectives of current mobile learning projects, but may be regarded as somewhat idealized. In practice, the applications can differ in many dimensions, and many of them target more than one learning activity, so you may consider that they have been somewhat arbitrarily placed in the book. However, we see the important connecting factor between the chapters is their focus on common themes and arguments.

In detail, we have organized these research efforts into four sections and 17 chapters. A brief description of each of the chapters follows:

Section I provides an overview of theoretical approaches, and describes a way of understanding mobile learning experiences. In Chapter I, Hokyoung Ryu and David Parsons focus on the development of a theoretical framework, setting out three distinct learning spaces that are markedly differently, and considered throughout the book: individual, collaborative, and situated learning. This framework provides systematic support for mobile learning experience design, and it is used to analyze three mobile learning environments. Extending this approach, in Chapter II, Patrick Danaher, Raj Gururajan and Abdul Hafeez-Baig deploy mobile learning experiences in conjunction with three key educational principles: engagement, presence, and flexibility. Each principle is accompanied by an elicitation of practical strategies that have proved effective in implementing the principles sustainably within particular courses and programs of study, as well as factors that inhibit that implementation.

In Section II, we include four mobile learning projects as practical examples of how individual learners may have mobile learning experiences that lead to specific learning outcomes. Firstly, in chapter III, Eusebio Scornavacca, Sid Huff, and Stephen Marshall describe the development of a SMS-based classroom interaction system and explore the impact that the TXT-2-LRN system can have on students' learning experience. Their findings indicate that instructors and students perceive a number of benefits from the additional channel of communication in the classroom. With a more sophisticated mobile technology, Christian Kittl, Francika Edegger, and Otto Petrovic (in Chapter IV) demonstrate how mobile game-based learning can be used for an efficient transfer of knowledge in learning processes, revealing its long-term learning outcomes and individual learning efficiency. The empirical results also imply game-based learning leads individual learners to higher energetic activation, more positive emotions, more positive attitudes towards learning content and more efficient knowledge transfer than other instructional formats. In contrast, in Chapter V, Peter Dolittle, Danielle Lusk, C. Noel Byrd, and Gina Mariano explore the use of the iPod™ as an educational platform and report on a study designed to examine individual differences in iPod™ use. It empirically demonstrates an important factor for the success of mobile-based individual learning activities, that is, working memory capacity (WMC). Finally, Chapter VI by Martin Owen, surveys diverse mobile learning projects, including a simple game-based learning system, a complex multi-role simulation and an environmental tagging and hypermedia project. It explicitly shows how mobile learning projects have been evolving from individual learning support to located and contextual activity-based learning experiences, themes that led us into the remaining sections of the book (Sections III and IV).

Section III delivers empirical data and case studies on collaborative learning experiences with mobile technologies, where the themes in mobile learning are markedly different from traditional e-learning environments. Firstly, Martha Grabowski, Greg Lepak, and George Kulick, working collaboratively between the United States and Poland, empirically examine the impacts of new collaborative technologies (including mobile technologies) on distributed learners. They also introduce a technology-independent framework for taking into account collaborative mobile technologies, relating expected technology impacts to user preferences. The following two chapters (VIII and IX) have been invited from a Pan-European mobile learning research initiative. In Chapter VIII, María Felisa Verdejo, Carlos Celorrio, Emilio Julio Lorenzo, Marta Millán, Sergio Prades, and Javier Vélez present a broad overview of the approach, design and implementation of a collaborative mobile learning infrastructure (i.e., the ENLACE project). Following this, in Chapter IX, Daniel Spikol, Arianit Kurti, and Marcelo Milrad describe the AMULETS (advanced mobile and ubiquitous learning environments for teachers and students) project. In the last chapter of Section III, Chapter X, Chengjiu Yin, Hiroaki Ogata, and Yoneo Yano provide a further example of the collaborative learning experience, that is, participatory simulation that helps both the individual learner and a group of learners to understand sorting algorithms by enacting collaborative processes with mobile devices.

The primary focus of Section IV is how situated learning can be shaped with mobile technologies. Sosuke Miura, Pamela Ravasio, and Masanori Sugimoto, in Chapter XI, present the SketchMap system that supports children's situated learning by creating maps. The goal of the SketchMap system is to investigate whether integrating outdoor and classroom activities and sharing of the children's experiences through the maps can actually promote situated learning. From a somewhat different perspective, Dionisios Dimakopoulos and George Magoulas, in Chapter XII, respond to the ever-increasing need of individuals and organizations for lifelong learning, presenting an approach to designing a mobile application for contextual lifelong learning. It assists learners to access, compose and manage their learning in a range of institutional, informal and work-based settings by keeping them connected with content that is relevant to their studies, and its use is demonstrated in three lifelong learning scenarios. In Chapter XIII, Hokyoung Ryu designs and evaluates a location-aware learning organizer that helps university students to manage their learning activities at campus. Finally in this section, Ana Dzartevska (Chapter XIV) extends this mobile learning experience to professionals who are in need of more contextual understanding of different work procedures.

By way of conclusion, we return to some of the issues and challenges raised at the beginning of this Preface and look at how they may be addressed by the work described in Chapters XV and XVI. Yanjie Song (Chapter XV) reviews and discusses research on applications of handheld devices in education. She classifies these mobile learning applications into six categories based on their functions: educational communication, managing, multimedia access, games and simulations, data collection, and context-aware applications. From a software engineering perspective, Ajax (asynchronous JavaScript and XML) is explored in Chapter XVI to increase the mobile Web page's interactivity, speed, functionality, and usability, which seem to be essential qualities in designing mobile learning contents.

In addition to the main body of this book, we also provide a list of recommended readings and resources to help the reader. The final chapter (Chapter XVII) compiles a list of recommended books, articles, scholarly journals and conferences, to offer one possible source of reading guidance on mobile learning research. With this list, the editors seek to serve both academics and practitioners who want to find out the basic details of mobile learning or disseminate their latest findings through the research network. At the end of this volume there is also a comprehensive glossary, covering most of the terms that may be new to the reader or that are being used in an unfamiliar way.

## TOWARDS A SOLUTION

Mobile learning is a relatively new research area. There is an increasing demand for tools and techniques but perhaps less enthusiasm or support for researchers to have the opportunity to fully articulate the relationships among these tools, techniques and underlying pedagogical theory. Therefore, a comprehensive volume of articles covering current trends, technologies and techniques in mobile learning is necessary. In this sense, we believe that this book will be a timely publication for both academics and practitioners who are interested in the design and development of future learning environments. However, this is of course a collection of readings on related topics, not an extended narrative with a beginning, middle, and end. Readers of the book should not feel constrained by the order of the chapters and the structure of the book. Obviously, we have arranged the material in an order that makes sense to us, trying wherever possible to locate readings that speak to the same or closely related issues, but many different arrangements are possible, and these reinterpretations may suggest other solutions to the future challenges of mobile learning.

*Hokyoung Ryu and David Parsons (Editors)*
*Auckland, New Zealand*

# Acknowledgment

Many people have helped us during the writing and preparation of this book. We are especially indebted to the reviewers who commented at length on earlier versions of the manuscript. As these reviewers can now verify, their constructive suggestions have played a major role in shaping the book's final form. They are Ken Hawick, Kinshuk, Stephen Downes, Matthias Lampe, Siobhan Thomas, Yunhi Chang, Jongbae Kim, Demetrios Sampson, Friderich W. Hesse, Irma Becerra-Fernandez, Deniz Eseryel, Jose Bravo, Daniel Wessel, Alain Derycke, and Peter Doolittle. In addition, we are in indebted to both Marcelo Milrad and Hiroaki Ogata, whose efforts in gathering together an excellent set of submissions have guaranteed the very high quality of the work presented in this volume. We are also grateful to Professor Tony Norris for his leadership of the Centre for Mobile Computing at Massey University, within which our research into mobile learning is based. In closing, we would like to warmly thank all the authors for their insights and valuable contributions to this book.

Special thanks also go to the publishing team at IGI Global, whose contributions throughout the whole process from inception of the initial idea to final publication have been invaluable.

Finally, there are our children, Jihoon, Youngji, Jenny, Kate, and Abbie, to whom any formal expression of thanks seems inadequate.

*Hokyoung Ryu and David Parsons (Editors)*
*Auckland, New Zealand*
*June 2008*

# Section I
# Theoretical Foundations of Mobile Learning Experiences

# Chapter I
# Designing Learning Activities with Mobile Technologies

**Hokyoung Ryu**
*Massey University, New Zealand*

**David Parsons**
*Massey University, New Zealand*

## ABSTRACT

*This introductory chapter focuses not so much on mobile learning technologies per se, but rather on a theoretical foundation and its pragmatic application to designing learning activities with mobile technologies. It sets out three learning spaces that are explicitly considered in the book: individual, collaborative, and situated learning. On these differing learning spaces, we begin by proposing the essential factors in effective mobile learning experience design that should be addressed by different features or functions of the relevant learning activities. In turn, derived is a conceptual framework to provide systematic support for mobile learning expererience design. This chapter concludes by surveying the mobile learning systems included in this book, reviewing their differing learning activities within context of the framework. We hope that this analysis will help to expose the key qualities and features that can support the future development of increasingly effective mobile learning applications.*

## INTRODUCTION

There is little doubt that information and communication technologies (ICT) are among the defining technological transformations of the late 20$^{th}$ and early 21$^{st}$ centuries. The changes in society brought about by ICT advances equally lead to new types of education systems that are not restricted to traditional education providers. They seemingly heighten both the effects and expectations of the advent of new pedagogies. We may witness such changes as the rise of new knowledge source providers, including the entertainment industry, defining themselves as knowledge producers and educators.

Of course, institutional pedagogy has its special and important frames of reference for what constitutes learning. However, as learning activities begin to escape these frames, as a consequence partly of mobile technologies, there are some important questions that need to be addressed about how traditional educational institutions can meet these rising challenges and opportunities. One consequence of this is an increasing interest in the potential of mobile learning.

In this chapter we intend to interpret various learning spaces with mobile technologies in order to lay out some important concepts and themes that will provide a context for reading the remaining chapters of this book. For the purposes of our discussion, a theoretical framework is introduced that may help us to understand how appropriate learning activities can be supported by particular aspects of mobile technologies. In the following section, we begin by introducing the three major learning activities that have been identified as significant, and how these differing learning activities have been dealt with by new mobile technologies. Within this frame of reference we lay out the themes and scope of this book.

## THREE PILLARS OF LEARNING

Educators have made considerable efforts to exploit the unique capabilities and characteristics of mobile technologies to enable new and engaging forms of learning activity. For instance, Korean commuters, using third generation (3G) mobile handsets, can access a multimedia-based English language learning tool supported by location aware services. This attractive pilot service holds out the promise of unlimited access to educational resources beyond the traditional institutional boundaries, amalgamating currently separated learning activities into one with an integrated technical platform. Yet, of course, such ambitious goals are still beyond reach, partly because most

*Figure 1. Three learning spaces, extended from De Jong (2001)*

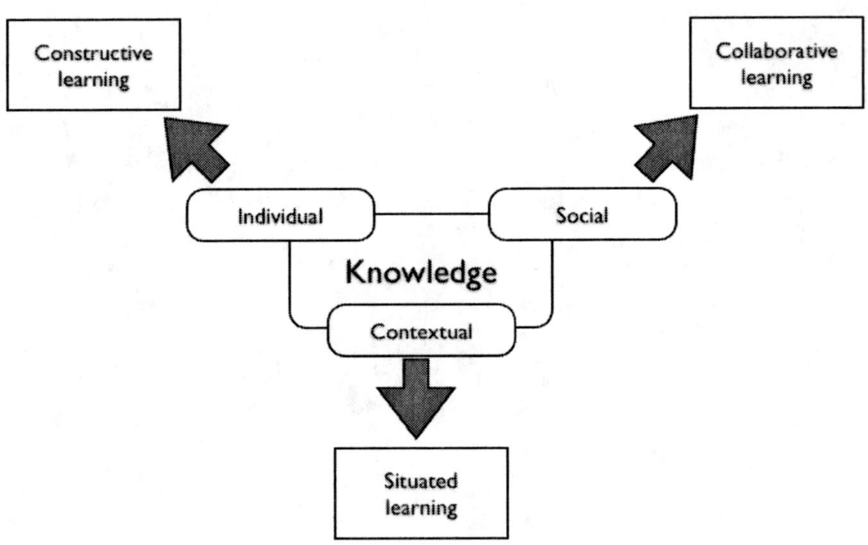

current mobile learning activities have focussed less on utilitarian education needs, but mostly because it is still not entirely clear what advances or benefits mobile technologies actually bring to learning activities. As a consequence, it seems that neither learners nor educators have as yet seen significantly persuasive forms of learning experience from the new mobile learning environment.

The concept of mobile learning promises users new and/or advanced user experiences, which are quite often markedly different from those afforded by conventional desktop computer-based learning systems (e-learning). However, simply providing another format for online courses using mobile devices does not address the potential learning outcomes that mobile learning provides. In fact, this technology-oriented approach overlooks the possibilities for embodying the three pillars of learning activities: *individual, collaborative* and *situated* learning, as shown in Figure 1.

In the literature, many education theorists have identified three differing learning spaces. Firstly, the individual learning space relies mostly on individual learners who are constructing their own knowledge by learning from books, online resources or teachers. This has long been considered the major learning activity in institutional pedagogy. To ensure success, techniques such as '*scaffolding*' (Hogan & Pressley, 1997) and the '*training wheel*' (Carroll & Carrithers, 1984) have been used, providing a process by which a teacher provides temporary support to learners to "help bridge the gap between what [the learner] know[s] and can do and what [he or she] need[s] to accomplish in order to succeed at a particular learning task" (p. 169, ibid.). Such techniques prove useful by allowing complex learning activities to be customized for individual learners (see Chapter X for more detail of this technique). In contrast, both situated learning (creating authentic learning in life-like settings), and collaborative learning (encouragement of pair and group work to address complex problems and tasks) have been paid more recent pedagogical attention, aiming to empower learners in different ways. In particular, the close relationship between learning and the context and situation in which the learning needs arise has been acknowledged by many authors (Brown, Collins, & Duguid, 1989; Brown & Duguid, 2000; Lave & Wenger, 1991). These various approaches to learning can have a profound impact on both the learner and the teacher in terms of their practices, roles, and values and how they may evolve to meet new learning requirements and opportunities. The details of these three learning spaces are further discussed later in this chapter but are also seen throughout the rest of the book.

*"Learning is mobile in terms of space, i.e., it happens at the workplace, at home, and at places of leisure; it is mobile between different areas of life, i.e., it may relate to work demands, self-improvement, or leisure; and it is mobile with respect to time, i.e., it happens at different times during the day, on working days or on weekends"* (p. 152, Vavoula & Sharples, 2002)

With regard to the three learning spaces, Vavoula and Sharples (2002) rightly point out that mobile learning should be viewed in particular settings, and integrated into our lives, which implicitly represents a seamless flow of learning experiences. They thus conclude that it is less useful to focus on the technical factors, but rather there is a need to understand the ways in which learning activities are embedded within the technology. In this respect, the rest of this section explores how mobile technologies have offered new possibilities for the three learning spaces or experiences, by which the reader may see the benefits and limitations of the mobile learning environment and the choices we must make when designing learning experiences within it.

## Individual Learning and its Mobile Applications

In the literature the definition of individual learning depends on the author's point of view, but it is often expressed as individually obtaining some knowledge or information to enhance one's own knowledge space. The contemporary theoretical foundation to support this individual learning activity is '*constructivism*' developed during the 1960s and 70s, inspired by the rise in cognitive theories of learning (Bruner, 1966; 1986; 1996). It depicts that learners are constructing their own knowledge by testing ideas and approaches based on their prior knowledge and experiences, actively applying these to a new situation and constructively integrating the new knowledge gained with pre-existing intellectual constructs. Therefore, the commitments and practices that the teacher or other educational resources should support would be the creation of suitable learning situations and settings, rather than the organized content to be delivered (though the predefined learning contents are also necessary for individual learning activities).

Individual and constructive learning activities have been dramatically changed by the advent of personal home computers (PCs) in the 1980s. These offered tremendous advances in terms of interaction with, and access to, educational material, particularly with the advent of the World Wide Web in the 1990s, enabling the user to gain an active locus of control in their own learning activities. The intensity of interest in computer technologies has also embraced the potential of virtual reality, where teachers can give learners an artificial environment in which to participate in the learning process, and the appropriate settings to work with that knowledge. Recent years have also seen an increasing interest in pervasive learning environments (Benford, Magerkurth, & Ljungstrand, 2005) that support individual learning outcomes with related mobile technologies (e.g., mobile phones, satellite positioning systems, sensing technologies, RFID tags, and so forth). Increasingly, these virtual and pervasive worlds are drawing together in the field of augmented reality, with mobile learning tools such as augmented reality goggles being used to support individual constructive learning (Doswell, 2006).

Mobile, game-based learning provides an illustrative case of an individual mobile learning activity. This combination of computer games and learning content provides an opportunity to provide students with an engaging pedagogical medium. Chapter IV addresses the individual benefits that a mobile game based learning environment can produce. In short, it demonstrates that learning with mobile devices itself has prominent benefits that other educational media cannot present, such as personal engagement, satisfaction, and a highly motivated learning process.

In another example, the first author of this chapter developed a mobile quiz system that supports students' reflective learning activities beyond the classroom environment. This project was actually aimed at developing a useful and entertaining addition to any kind of classroom learning activity, removing some of the formality from the learning experiences (Ryu & Wong, 2008). It concluded that the porting of current learning experiences to mobile devices might enhance the autonomy of individual learners and boost their constructive learning experience, and suggested that sophisticated technologies are not necessarily required to build an effective mobile learning environment.

There is an important point of detail in a general account of *constructivism*, that leads us into *situated learning*. Educational psychologists distinguish *individual constructivism* from *social constructivism*. A dominant influence on the latter type is Vygotsky's work (1926) on socio-cultural learning, describing how interactions with adults, more capable peers (or friends or colleagues), and cognitive tools are internalized to form mental constructs. Vygotsky's version of constructivist theory has led to the view that behavior, skills, attitudes and beliefs are inherently situated, that is, bound to a specific socio-cultural setting which is linked to *situated learning activities* (described later in this section, as well as in Section IV of this book.)

## Collaborative Learning and its Mobile Applications

Apart from the individual learning activities discussed previously, researchers (e.g., Beckman, 1990; Collier, 1980; Slavin, 1983a, b) have also identified that, regardless of the subject matter, students working in groups tend to learn more of what is taught and retain it longer than when the same content is presented in other instructional formats. That is, learning can be even more effective when learners can converse with each other, by interrogating and sharing their descriptions of the world.

Both the capabilities of mobile devices and their wide context of use contribute to their propensity to foster collaborative learning activities. Mobile devices can easily communicate with other devices, enabling learners to share data, files and messages in groups. For instance, one digitally augmented classroom environment (Moher, Hussain, Halter, & Kilb, 2005) employed Pocket PCs to provide dynamic readings of simulated earthquakes. Students had to monitor these readings and share the state of the simulation through distributed media positioned around the classroom, collecting evidences to solve problems or answer questions related to the earthquakes. Furthermore, it asked them to build a physical model of the phenomena with other peers, thereby sharing their understanding of the simulated earthquakes.

Many current mobile learning environments fall into this category, drawing upon the essential benefits of the collaborative learning space provided by mobile technologies. It is argued that participants in a collaborative learning environment can learn equally well from each other as from the teacher or course materials, and that learning is not static subject matter but the process of participating itself. Recent years have seen a growth in the social networking capability of Web-based services, known as 'Web 2.0.' This refers to online collaboration tools, such as photo and video-sharing services, Weblogs, wikis, social bookmarking, and syndication of site contents, which facilitate the sharing of content by users. Laine and Suhonen (2008) introduced the idea that these social networking tools, that is, blogs and wikis, can also be used by learners in the new mobile paradigm. This seems to be particularly important in mobile learning where interaction otherwise often occurs only between the user and the mobile device. This enhanced learning style in mobile learning supports a new breed of users, best understood as a loose grouping of participants who share a set of common aims and practices around user-led content creation communities, rather than passive information consumers, that is, new wines (new learning activities) in new bottles (mobile or Web 2.0 technologies), for generation Y and beyond.

Creating collaborative learning practices can obviously be a rewarding opportunity but it is also full of challenges and dilemmas (see for example Chapters VII, VIII, and IX). In particular, designing group work is demanding, and teaching in collaborative settings brings to the fore a tension between the process of student learning and content coverage. Nonetheless, this learning experience is widely believed to give students an insight into a sharing learning community, with all the joys and difficulties that attend, via their own mobile devices.

## Situated Learning and its Mobile Applications

A number of mobile learning projects have also explored the use of mobile and sensor technologies for learning in a range of outdoor contexts. Some of these projects are based on field trips (e.g., The Ambient Wood project, Rogers et al., 2005), using systems for tagging information (physically and/or digitally) that allow access to location-specific information attached to the tags, for learning in outdoor spaces (e.g., Chapter IX). The idea behind these projects is that learning may be supported in new ways by combining the real world with digitally re-presented information. In other words, by linking learning activities to contextually relevant or associated information in the form of digital representations, where learners can experience phenomena and explore concepts and relationships through combined physical and digital artefacts (Rogers et al., 2005; Sharples, Taylor, & Vavoula, 2005).

Lave and Wenger's (1991) *situated learning theory* differentiates this situated learning space from the other learning spaces, acknowledging that learning needs to be presented in an authentic context, that is, settings and applications that would normally involve that knowledge. For example, on a recent visit to Korea we were interested to see many people using their mobile phones to learn English on the train, with contextually-managed English learning courses (e.g., location-aware vocabulary or environment-aware sentences). In this case, location awareness means gathering information from the surroundings to provide the relevant content based on what is currently going on around the user and the device. Learning activities and content that are particularly relevant to that environment can then be made available. This authentic environment in which learners actually practice speaking and understanding the new language is an illustrative example of situated and contextual learning activities in life-like settings.

As such, mobile devices are especially well suited to context-aware applications simply because they are available in different contexts, and so can draw on those contexts to enhance the learning activity. Both Chapter XI and XIII in this book are such cases. In particular, Chapter XI displays a technological innovation characterized by new types of interaction devices and communication technologies that facilitate human-to-human communication in cooperative work scenarios as well as interactive and situated learning activities. When it comes to contextual learning situations in the tertiary or life-long learning environment, Chapter XIII provides some practical assessment of using situated learning support for continuous personal time and task management. Still, mobile learning tools for situated learning activities are still some way from working seamlessly with learning outcomes, giving us many challenges for future research.

In defining the pedagogies for mobile learning within the three pillars of learning spaces as discussed previously, it is important to be clear about what exactly mobile learning contributes that is new and different from previous technologies of learning. Therefore the following section describes how it is designed, interpreted, employed, constructed, and reconstructed, perhaps showing a way forward for future mobile learning design.

## DESIGNING MOBILE LEARNING EXPERIENCES

The potential for mobile technologies to offer opportunities for new ways of learning have clearly been demonstrated, but currently there are few theoretical frameworks looking specifically at the role of mobile learning (Chapter II supports this perspective) as per the different learning experiences of the three learning spaces previously discussed. Therefore, we understand little about the underlying mechanisms of how mobile learning actually works for the learner. To do so, further research is needed to better understand the kinds of learning tasks, activities and experiences it most benefits.

In this section, a practical approach to developing mobile learning environments is presented, by which a designer can build learning activities with mobile technologies. The proposed framework allows the designer to make explicit both the technical perspectives in designing mobile learning systems and the learning aspects, to generate the best possible learning experiences. In this way one can describe the intimate connection between the generic mobile environment, the mobile learning context, and the learning activities and objectives.

Of course there have been a number of learning design frameworks from which we can borrow when designing mobile learning experiences, such as the curricular or conversational design frameworks. Patten et al.'s *curricular approach* (2006), in particular, proposes mobile learning systems design by the functional outcomes such as administrative, data collection, location awareness, and collaborative outcomes. It is a first attempt to encompass current principles of meaningful and situated teaching and learning in relation to mobile devices from the simple (administration) to the complex (location aware, collaborative, situational, motivated). Whilst the functional framework defined a set of curricular uses in the fast growing mobile market, it focused only on the functional or content approach of mobile learning design without specific concern for learning experiences. In contrast, the *conversational design framework* (Laurillard, 2002) is of great value for developing new experiences common to all forms of learning; conventional, distance, mobile, and blended. The form of this framework defines a dialogic process between 'teacher' and 'student' on two levels; the discursive level, where the focus is theory, and the experiential level, where the focus is on practice, activity and procedure-building. Laurillard claimed that the conversational framework could provide a powerful way of reviewing both traditional and digital learning designs, illustrating in a reasonably formal way why digital forms offer a better integrated and more motivating learning environment. However, no pragmatic case studies confirm this framework for designing mobile learning design, perhaps because of its lack of a technical perspective.

This section proposes, instead, a practical framework for mobile learning systems design that also provides systematic support for mobile learning experience design. It is believed to be a pragmatic two-tiered approach (i.e., technical perspective and learning perspective) by which a mobile learning designer can develop learning experiences along with the technical platform. To demonstrate the feasibility of using the design framework proposed in this chapter, a set of mobile learning projects from this book are chosen and analyzed. This section firstly explores what factors and design requirements are crucial to the three pillars of the current mobile learning activities, and suggests how future mobile learning applications can be designed with an understanding of these factors and requirements, as shown in Figure 2.

## Technical Perspectives: Generic Mobile Environment

Our framework is designed to describe the minimal requirements for developing mobile learning environments. It can be interpreted as saying that, on the basis of a range of findings from research on other mobile learning-related studies and teaching methods to motivate the learner, it should be able to act as a framework for designing learning experiences with mobile technologies. From a broader perspective, we note that the early frameworks have given little consideration to technical aspects, and that this lack seems to create difficulties in applying the frameworks to the design of mobile learning systems. The symmetry and relation of all these factors is illustrated in Figure 2, showing how the two perspectives (technical and learning issues) constitute a complete and simple set of requirements in designing mobile learning spaces.

The generic mobile environment issues, which are located in the first column of Figure 2, are

*Figure 2. A framework for designing mobile learning spaces*

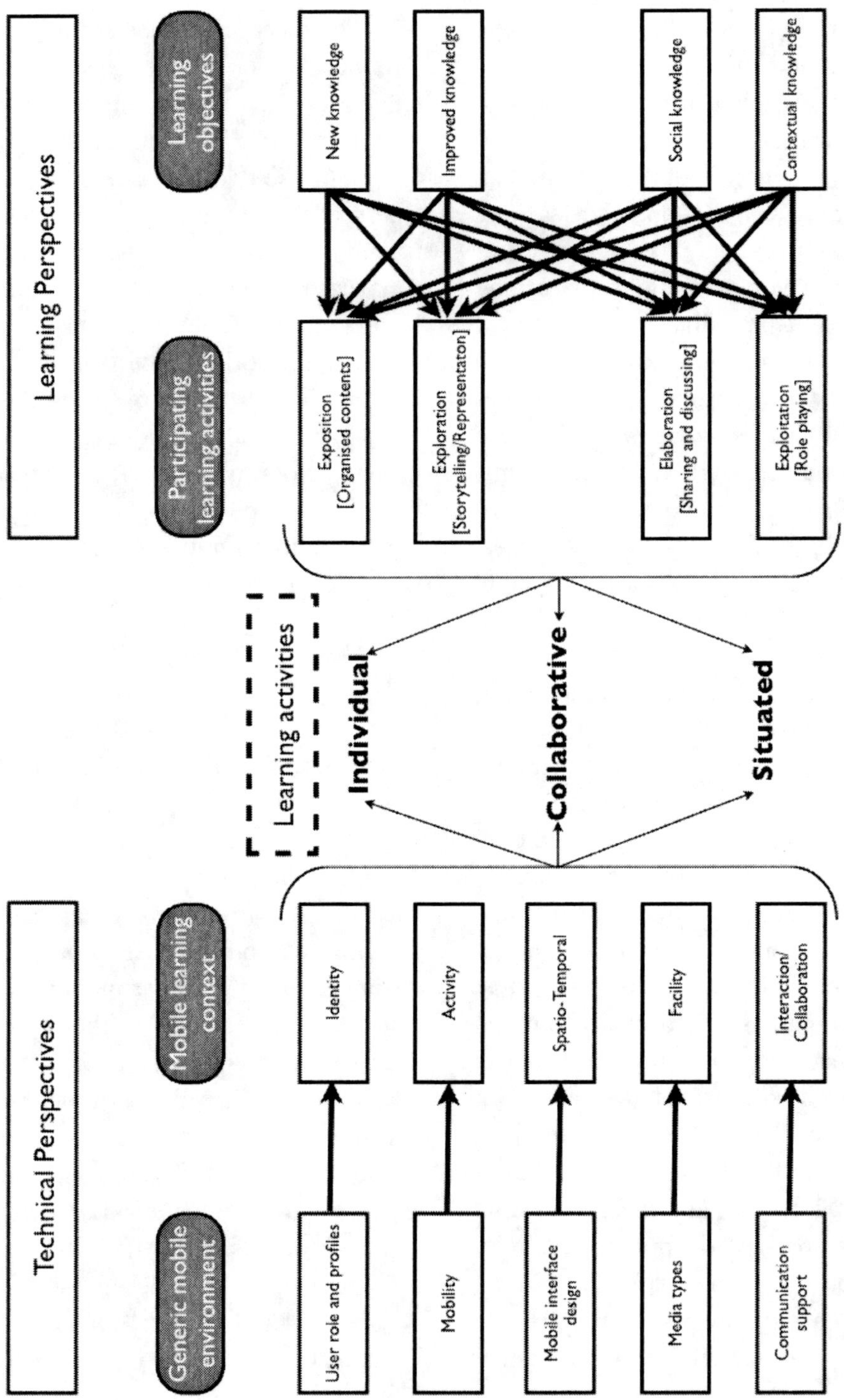

related to how a given mobile environment would dictate the related technical challenges for mobile learning systems. The most important feature in the mobile environment is *mobility* itself. It enables us to be in contact while we are outside the reach of conventional communication spaces. Unlike other learning environments, mobility allows the learner access to knowledge resources where and when they want them. This convenience would of course contribute positively to students' autonomy. *Mobility* can be conceptualized in different ways, that is, mobility of the user, mobility of the device, and mobility of services, and these three aspects should be addressed both technically and contextually in the design process. In many cases, this mobility issue should be related to the contextual activities that the mobile learning system is intended to support.

We should also understand that each user employs their own mobile device in a different way (even the devices themselves are not the same). For instance, teenagers frequently use SMS (short message service) text messaging to communicate, while professionals may be more likely to use corporate email. In this sense, the feature of different *user profiles* and their *roles* in the use of mobile technologies becomes even more important in designing mobile learning environments. The users' roles and profiles are used to determine their different identities in the learning context. For instance, the use of SMS in mobile learning has been identified as an effective tool to enhance both student's learning experience and tutor's instructing experience, yet the requirements of the two parties may be quite different in terms of their previous experience of SMS (Scornavacca, Huff, & Marshall, 2007; Stone, Briggs, & Smith, 2002).

Awareness of the constraints of the *user interface* is also vital. Mobile devices suffer from small screens, constrained input devices, and limited battery life. Therefore, the interface design for mobile learning services must meet users' needs without overloading them with unnecessary complexity or operating too slowly, aspects that are closely related to spatial and temporal relationships in mobile learning context design. That is, users need to be able to interact comfortably and freely with application content, having pages load quickly and animations play smoothly and seamlessly. Addressing these issues early on, when developing mobile learning programs, helps ensure success. Chapter XVI provides one technical approach to this problem, leveraging the power of Asynchronous JavaScript and XML (Ajax) to improve the performance of a mobile learning system, enhancing the user's mobile learning experience.

Successful mobile applications tend to employ many rich *media objects*. One of the commonly stated characteristics of mobile learning content is that it can be seen to be delivered in short 'nuggets' rather than large units of information. These media types should support content appropriately. For instance, Luchini et al. (2004) stress the importance of the user participating in and learning about underlying concepts and processes, perhaps using simulation tools (Colella, 2000). However, the available media types are dependent on the facilities of the mobile device being used. For instance, in Miura et al.'s participatory simulation in Chapter XI, TabletPCs were employed for content delivery, since the system required more powerful processing and data management facilities than could be supported by conventional mobile devices such as PDAs or mobile phones. In Dzartevska's professional mobile learning system (Chapter XIV), ruggedized smart phones with bar code readers were used to support mobile learning in a warehouse environment.

One widely noted feature of mobile technologies is that they afford the possibility of perpetual contact for collaboration or interaction. This sense of *communication support* tends to contribute to the other possibilities of mobile learning in use. As an illustrative example, communication support has helped to meet the collective learning objectives of a field trip in Spikol et al.'s project in Chapter IX of this book. In particular, mobile devices with communication support have been mainly used for collaborative learning activities, such as gathering responses from students (Chapter IV in this book), participatory simulation (Chapter X), and for collaborative data gathering (Chapters VIII and IX).

## Technical Perspectives: Mobile Learning Context

Each of the components within our framework plays its part in the design of learning activities, creating a close relationship between them. The designer may take the generic mobile environment issues to simply map to appropriate mobile learning activities, given the means to do it. In practice, however, mobile learning has frequently been used in conjunction with other types of learning (Colazzo, Molinari, Ronchetti, & Trifonova, 2003; Rogers et al., 2004), complementing previous learning experiences. This is one reason why we look to the traditional learning contexts (identity, activity, spatio-temporal, facility and interaction, as shown in the second column of Figure 2), which are based on Wang's (2004) six dimensions of the mobile learning context. The contextual issues in mobile learning are closely related to the generic mobile environment issues described above, so the framework provides a way of checking that a mobile learning design ensures the minimal technical requirements, and in particular, provides a way of analyzing what technical tools would ensure the related mobile learning context.

*Identity* of mobile learning users (e.g., learners or teachers) is a necessary contextual factor. Depending on the identity of the user, a mobile learning system can provide the right content at the right time. For instance, Chapter XII by Dimakpolous et al. shows how to manage the different user identities to deliver the different learning materials or data in the life-long learning environment. Their research has been geared towards the provision of systems that will better support those who pursue education at a number of different stages in their life, moving between institutions while pursuing a course of study, or making use of resources and facilities from more than one institution or organization. The mobile system thus helps users to access, compose, and manage their activities by keeping them connected with content that is relevant to their identity and its use in three learning scenarios. This identity issue is also related to the learner's cognitive attitude or style, which means that learning experiences can be more individually targeted and thus more effective.

Several considerations of the *learner* should also be made, because each learner has a different psychological temperament that relates to their learning experience. For instance, highly-motivated learners would probably improve their learning experience by using all the features that the system can provide. Dolittle et al.'s Chapter V shows how mobile learning can have different learning effects based on students' cognitive resources (i.e., working memory capacity) and their dependence on the context where the learning takes place. There has been limited work on the technical tools to address this mechanism, but Al-Dujaily's work (2007) and Brusilovsky (2001) on the different ways that people use e-learning systems may suggest some practical ways to include the learner's context from the technical perspective.

The most important feature of the mobile learning context is *learning activities*. Indeed, compelling individual learning activities have been addressed in many mobile learning systems (see Section II of this book). Also, collaborative learning experiences hold much promise for new mobile learning activities (see Section III). Several empirical studies from this book indicate that mobile learning can support collaborative activities in differing ways by strengthening the organization of the learning material and information, supporting communication among group members, and helping the coordination between the learning activities. Such collaboration can take many forms. It may take place in a classroom or be a remote connection to a live tutor (Chapter IV in this book). The last section (Section IV) of the book pays attention to social learning activities that take place in the situated learning context. To ensure appropriate technical support for the differing learning activities in consideration, many projects in this book employed a variety of technical solutions, so the reader may draw their own conclusions about the technical feasibility and requirements of the various learning activities.

The *spatio-temporal* dimension means an awareness of time and/or location. Spatial location has been acknowledged as the underlying environmental context in many applications, while the temporal context can deal with issues like scheduling or collaborative interactions. Sensing or tagging technologies have been widely used to measure the spatio-temporal dimension. For instance, Chapter XIII shows how to link the spatial and temporal context using the global positioning system (GPS) as an appropriate mobile sensing technology and a scheduling database for learning activities. With regard to the user experience, these spatial and temporal issues are reported to foster greater student engagement, more effective learning experiences and increasingly social practices built around simple-to-use but functional and reliable technology (as GPS has proven to be).

The context of *facility* can impact on the design of mobile learning interfaces. Normally, the type of facility employed in the learning experience dictates the type of interactions, and vice versa. Most recent mobile learning environments tend to employ sophisticated mobile devices to enable a richer facility context, as the cost of such devices has significantly reduced, but nevertheless an understanding of the demographics of technology uptake by the target learner audience is important. Some studies therefore focus on lowest common denominator technology, in order to be inclusive (e.g., the SMS-based system described in Chapter III), while others are based on specific tools known to be available to the mobile learner, such as the professional mobile learning platform described in Chapter XIV. Where it is feasible to use them, more innovative technological tools (see Chapter IX for more detail) can enable a richer *facility* context for the mobile learning space being developed.

## Learning Perspectives: Learning Objectives and Participatory Learning Activities

Most of the literature regarding mobile learning design tends to simply link available technologies with the learning activities in mind. However, we may gain a greater understanding by setting mobile learning designs against a background of pedagogical requirements, so the focus of design should turn towards providing learning experiences, *exposition, exploration, elaboration* and *exploitation,* that are markedly different from those available in other learning contexts.

### Expositional Participation

Expositional learning experience is commonly presented as a sequential module of knowledge with predefined learning content. The learner's task is then to follow the order of the organized content, which is closely related to the acquisition of new knowledge. In mobile learning design, unlike static computer-based learning systems, it should be structured in such a way that the user can easily stop and re-start their learning in an episodic fashion, since much mobile learning takes place in 'down-time' or as part of other time constrained activities. This special kind of use requires the organization of the learning content into small "information nuggets," which generally fit into one or two views within the limited screen estate. For this reason, still photo images, or short audio or video files, are often used for expositional learning experiences (see Chapter XIV).

## Exploratory Participation

Many consider that mobile learning should be more than just expositional learning activities. *Exploratory learning* is such an approach to teaching and training that encourages the learner to explore a learning space in their own learning process, with much less of a focus on expositional learning activities. However, it does not necessarily mean an unguided or unconstrained learning environment, rather it means that learners may discover unexpected lessons and reach conclusions following various paths of participation.

For successful mobile learning programs, learning content must be reliable and relevant and it must engage users through more diverse personalized learning experiences. The value of exploratory learning has been studied in many learning scenarios (e.g., the Ambient Wood project). However, due to the less controlled learning path, exploratory learning seems to be more suitable for those who already have some basic knowledge and the skills for learning on their own, that is, for improving knowledge rather than for new knowledge acquisition.

There are many differing ways of supporting this exploratory learning experience. For example the representation of content may be embedded in a narrative or story element that the learners can easily follow. The narrative in learning experiences can be seen as an important structural factor in developing improved skills, which places it at the center of learning activities, helping learners reflect on what they have learnt. In effect, it enables a sort of meta-learning and illuminates the very processes by which they learn and provides an organizing structure for knowledge. For instance, the Savannah project described in Chapter VI has focused more on exploring the potential of digital overlay of physical environments where there is little or no direct correspondence with the particular physical environment and the learning domain. Instead, a new story is generated in an augmented reality environment. In contrast, both Chapters VIII and IX demonstrate the benefits of this exploratory learning experience in digitally-facilitated site-specific activities.

## Elaboration and Sharing Participation

With increasing frequency, students are working with each other, and alongside their teachers, to grasp subject matter or elaborate their understanding of it. There have been social dimensions to this learning process, but only recently have specially designed collaborative learning experiences been applied as an innovative alternative to the lecture-centered approach. In this process, learners are developing their social, as well as intellectual, knowledge, mutually working toward a common goal and elaborating a unique intellectual and social synergy.

The key of an elaboration or sharing of learning is the formation of communities of practice. For instance, learners may link and connect with others in similar fields from around the world and begin to share and improve on each other's work. These communities can foster improved excellence and quality, the wide nature of which is well represented in Chapter VII by Grabowski et al. and Spikol et al.'s work (Chapter IX) shows how one group of students collecting data, working remotely with another group using that data to locate knowledge, together create collaborative learning experiences that would not otherwise be possible. This social interaction process to elaborate their learning experiences provides group experiences that establish relevant collective learning. This helps the learner to develop appropriate social skills and assists team building in reaching joint learning objectives.

### Exploitation and Experiential Participation

Some projects have emphasized the active nature of student involvement, associated with "hands-on" learning and experiential or activity-based learning. For instance, Chapter IV shows how mobile game-based learning can help people have learning experiences based on active involvement, which reinforces the learning process and allows them to construct their own knowledge through experimentation in constrained models of real world domains. The nature of this participation can be described as "learning-by-doing." Experiential and discovery learning may be supported in new ways by combining real-world physical interaction with digital representation. For example, actions can be linked to contextually relevant or associated information in the form of digital representations, or digitally activated experiences of scientific phenomena can be provided that are not normally encountered in everyday situations. Using these techniques, learners can experience phenomena and explore concepts and relationships through combined physical and digital artefacts. Chapter XI by Miura et al. demonstrates how children can enhance their understanding of the surrounding environment from an experiential learning activity, supported by location-awareness, to contextualize learning activities by enabling them to interact appropriately with their environment.

The mobile learning design framework can therefore provide an explicit but simple way of illustrating how these four mobile learning design requirements interact to create effective learning activities. As an example of these interactions, consider mobile learning for dynamic complex situations, such as rescue services or intensive care. These require collective learning objectives (social and contextual knowledge), particularly developing team skills. These learning objectives would be supported by the appropriate learning experiences, participating in social interaction and/or collaboration. These learning activities would require the corresponding mobile learning contexts, including activity, spatio-temporal, facility and communication support, which in turn would map in a context specific way to generic mobile design requirements.

To validate this framework in detail, we apply it to the three mobile learning environments described in Chapters IV, IX and XIII. For each example, we have provided a figure that maps the specific details of the mobile learning environment to the components of the framework.

## APPLYING THE FRAMEWORK FOR UNDERSTANDING INDIVIDUAL MOBILE LEARNING DESIGN

Several of the research projects in this book demonstrate how mobile technology can be used to augment everyday interaction with the world, and illustrate ways in which they can be used to enhance learning experiences in different contexts. For instance, Chapter IX employs sensor technology in the environment to trigger contextually relevant information that is not otherwise available in the physical world. Learner's information and data can also be dynamically integrated over time and space, facilitating new forms of collaborative learning as well as broadening and connecting student's understandings and reflections both in the physical world and in classroom settings.

In this section we apply the two-tiered mobile learning design framework for three projects in this book that have differing learning activities. These are the projects in Chapter IV (individual exploratory learning activities), Chapter IX (collaborative exploratory learning activities), and Chapter XIII (situated expositive learning activities). The mobile game-based exploratory learning activities described in Chapter IV enhance the problem solving techniques using game-based contents embedded in the physical location information, whilst Chapter IX is based on children exploring a physical environment (i.e., a forest) that included a number of cues and learning contents embedded in the physical space to collaboratively find the answer from the field study group and the students in the classroom. The mobile learning organizer from Chapter XIII provides a mobile location-aware information system for university students with predefined organized contents along with their location. Whereas in the first two examples we use the framework for a post-hoc analysis of mobile learning experiences, for the last example we see how the framework may be used for the analysis of new mobile learning systems design requirements. Despite the rather diverse nature of these experiments, we show in the following sections how they can all be analyzed within our framework.

## Applying the Framework for Understanding Individual Exploratory Learning Activities

Kittl et al. in Chapter IV describe a mobile learning project designed to assist students' game-based learning activities. The project used simple technologies, short message service (SMS) and digital pictures, to enable students to create a learning community with their fellow players on a mobile phone communication channel in order to set up a new and improved learning environment for themselves. Even though this project proposed the collaborative use of the mobile learning system, its prominent learning outcomes described in the chapter were mostly from individual learning activities.

This game-based mobile learning system asks students via SMS to find and analyze special everyday situations in the area of the "Digital Economy." The students work individually and must document the specified situation (i.e., take a photo) and submit a proposal for improvement as quickly as possible by means of still picture messages and texts on their own mobile phones. Their contributions are then displayed on a Website in a mobile blog. Text messaging was used for information, supervision, and feedback. An important aspect of the images was that the database enabled explanatory text to be included for each picture.

A part of the analysis of the mobile game-based learning project with our framework is shown in Figure 3, showing the path from learning perspectives through to mobile design issues. Though it is not a comprehensive analysis (please refer to the chapter for more detail), it depicts how both the technical and learning perspectives work together to establish the individual exploratory learning activities.

## Applying the Framework for Understanding Collaborative Exploratory Learning Activities

In the AMULETS project described in Chapter IX, a collaborative learning experience was developed where children could combine outdoor and indoor learning activities that used mobile computing technologies together with stationary computers. The main collaboration modes they explored were peer-to-peer, individual-to-group and individual-to-teacher collaboration. In particular, their collaboration happened while the learners were solving the tasks and encouraged to collaborate with the other groups outside or inside, thus bridging the location contexts of the same activities. This was technology mediated and relied upon text (mobile instant messaging) and content (audio/video/picture messages).

One of the learning scenarios they assessed with these mobile technologies was learning about a forest. In this scenario, a group of seven children took part in an exploration of the physical environment (i.e., the forest), identifying different types of tree and measuring their height and age. Part of the children's task was to record still images and video clips using smartphones, detailing how

*Figure 3. Reviewing the learning space of a game-based mobile learning system*

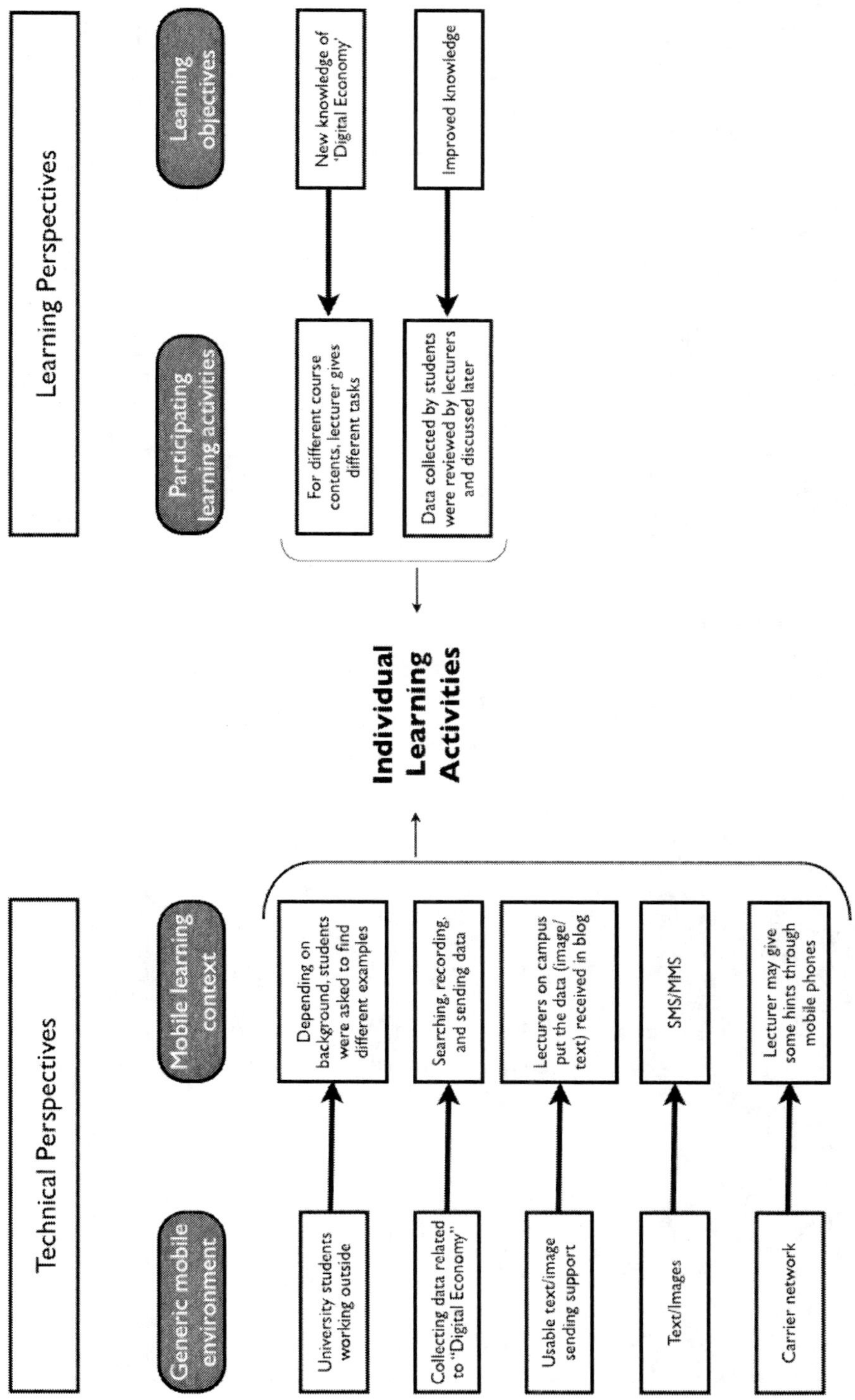

*Figure 4. Reviewing the learning space of AMULETS*

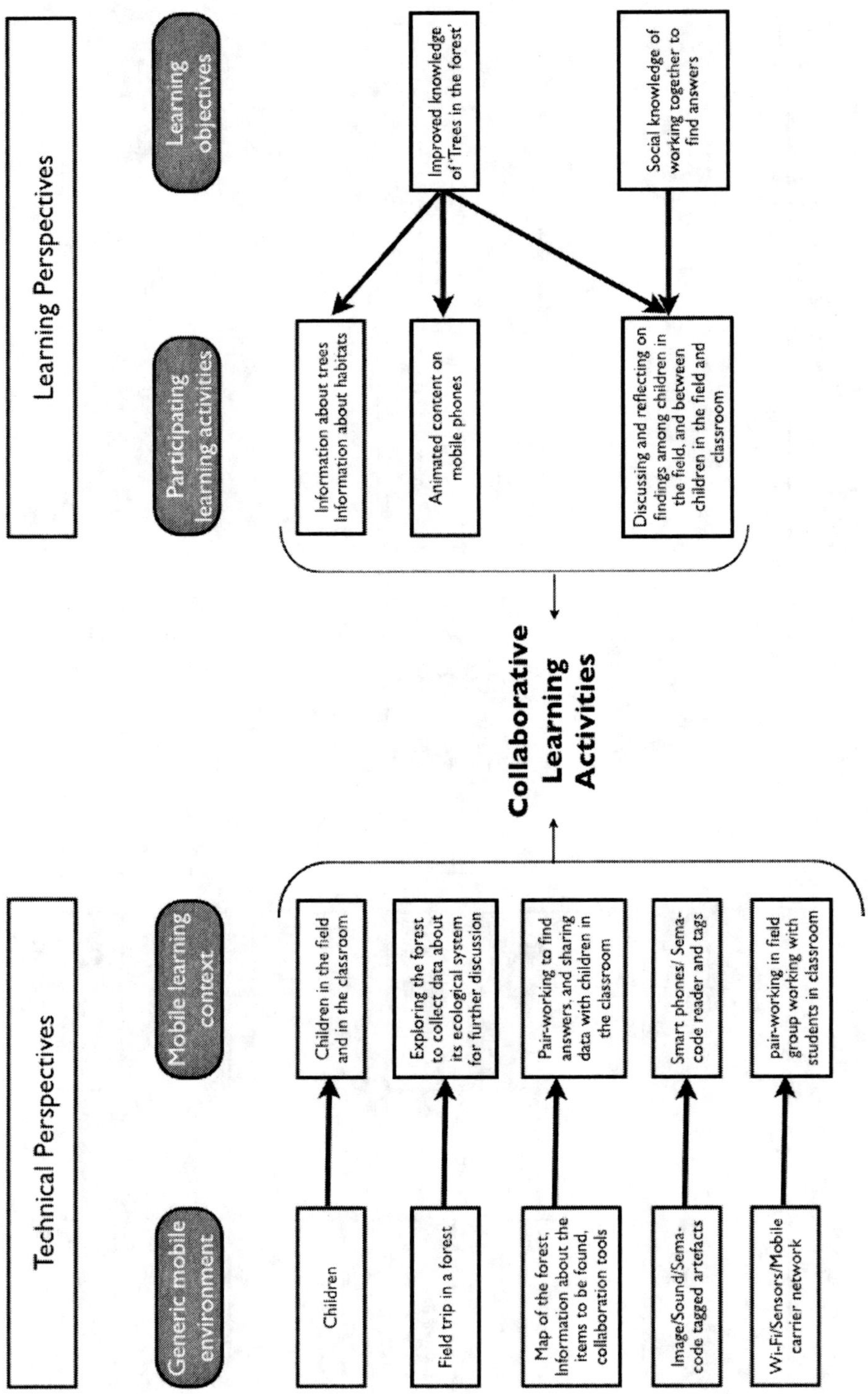

they solved the problems. This co-created content automatically encoded with metadata containing attributes such as GPS coordinates, time stamp, and the phone ID, provided rich contextual information for later use in the classroom. In particular, animated characters on the smartphones delivered content based on a specific location and tag triggered context to the smartphones. Once they identified the trees and received some additional information on the smartphone, they needed to scan the correct "semacode" tag placed on one of the trees. Upon selecting the correct tree, a video animation was sent to the smartphones in order to give the children the required information (i.e., the organized contents) to proceed to the next mission. At the end of this field trip, all the children were gathered together in the classroom. The post activities took place back in the school where all groups presented and discussed the content created during this field trip. The path from learning objectives through to mobile design issues is shown in Figure 4.

### Applying the Framework for Understanding Situated Expositive Learning Activities

In the previous two examples we took successful mobile learning projects and mapped their characteristics onto the framework. The purpose of these reverse engineering exercises was to attempt to validate the framework against previous work. However, the intention of the framework is that it should also be able to provide us with design guidelines for forward engineering new mobile learning activities. In Chapter XIII, the framework was used as a guide for requirements gathering in a system intended to provide location aware learning support for university students with specified organized contents. Figure 5 shows the results of this exercise in the context of the framework. Using the framework in requirements gathering helped to identify that junior and senior students had different needs in their mobile learning contexts. This insight enabled a system to be developed that could meet the disparate needs of different learner identities.

## CONCLUSION AND FUTURE RESEARCH DIRECTION

This chapter discussed what issues and contexts are important in designing learning activities with mobile technologies. It incorporated the best practices in mobile learning research into establishing a practical framework, but also provided a broader view to take account of the learning context, participating learning experiences and objectives beyond a single application domain. This framework was then applied as a post-hoc analytical tool and a guide for requirements gathering to three mobile learning systems for the purposes of validation.

It is of course noted that the framework has not yet been thoroughly assessed. Future work will involve the development of a generic design process to enable the application of the framework in a design context and the subsequent design and development of a complete mobile learning application. The following chapters in this book, in this respect, are used to represent the potential benefits of the framework.

In this book, our intention is to disseminate studies about how we can meet the challenges of mobile learning technologies, exploit practical experiences of the design of mobile learning environments, and further extend empirical research into contemporary mobile learning systems. Of course, the work selected in this book is still ongoing research but it shows a way forward for the design and development of future learning and education systems.

In effect, throughout this book, we would like to say that the challenges for the educators and technology developers for the future learning environment will be to find ways to ensure that this new learning is highly situated, personal, and

*Figure 5. Reviewing the learning space of the mobile helper project*

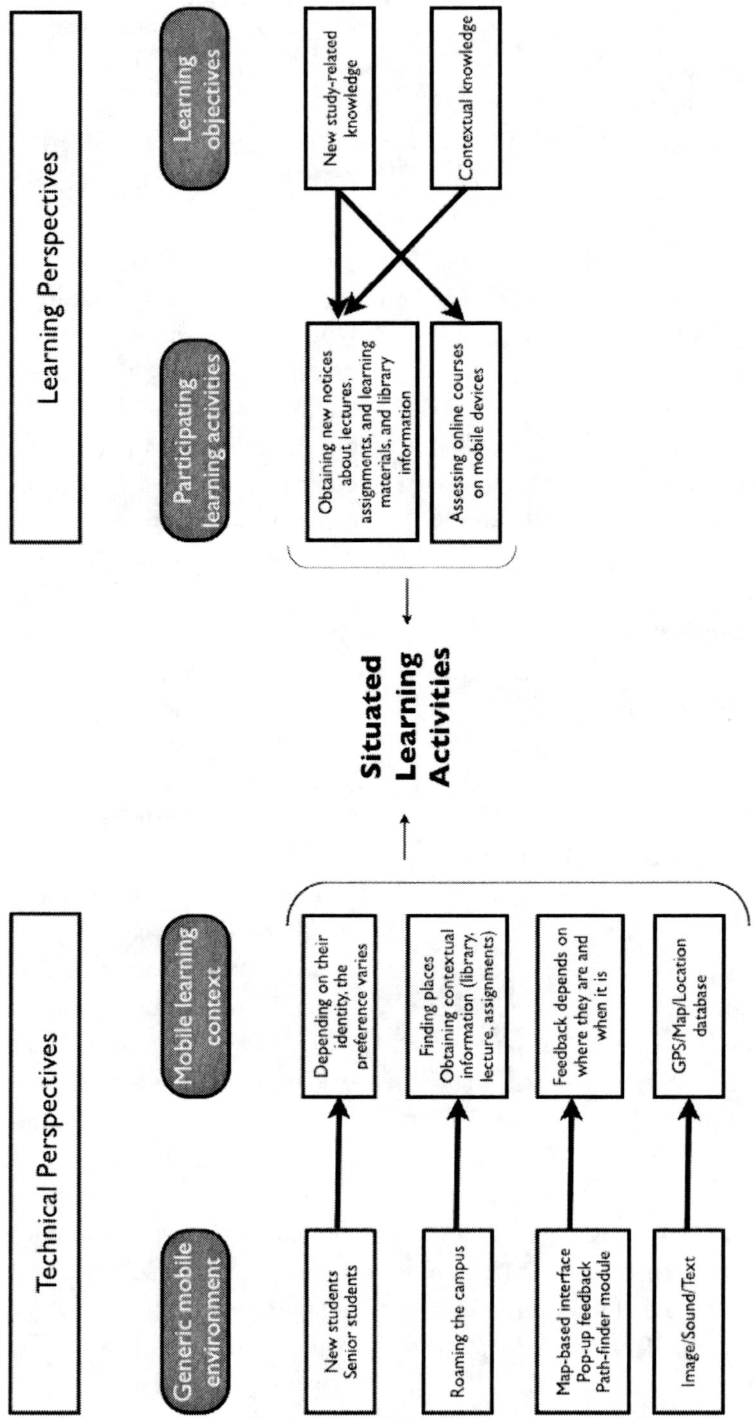

collaborative, in other words, truly learner-centered learning. As a consequence, the success of learning and teaching with mobile technologies will be measured by how seamlessly it weaves itself into our daily lives, with the greatest success paradoxically occurring at the point where we do not recognize it as learning at all.

## REFERENCES

Al-Dujaily, A. (2007). *Personality effect in the design of adaptive e-learning systems*. Unpublished thesis. Massey University, Auckland.

Beckman, M. (1990). Collaborative learning: Preparation for the workplace and democracy. *College Teaching, 38*(4), 128-133.

Benford, P., Magerkurth, C., & Ljungstrand, P. (2005). Bridging the physical and digital in pervasive gaming. *Communications of the ACM, 48*(3), 54-57.

Brown, J. S., Collins, A., & Duguid, P. (1989). Situated cognition and culture of learning. *Educational Technology, 18*(1), 32-42.

Brown, J. S., & Duguid, P. (2000). *The social life of information*. Cambridge, MA: Harvard Business School.

Bruce, B., & Hogan, M. (1998). The disappearance of technology: Toward an ecological model of literacy. In D. Reinking & M. McKenna & L. Labbo & R. Kieffer (Eds.), *Handbook of literacy and technology*. Mahwah, NJ: Lawrence Erlbaum Associates.

Bruner, J. (1966). *Toward a theory of instruction*. Cambridge, MA: Harvard University Press.

Bruner, J. (1986). *Actual minds, possible worlds*. Cambridge, MA: Harvard University Press.

Bruner, J. (1996). *The culture of education*. Cambridge, MA: Harvard University Press.

Brusilovsky, P. (2001). Adaptive hypermedia systems. *User Modeling and User-Adapted Interaction, 11*(1-2), 87-110.

Carroll, J., & Carrithers, C. (1984). Training wheels in a user interface. *Communications of the ACM, 27*(8), 800-806.

Colazzo, L., Molinari, A., Ronchetti, M., & Trifonova, A. (2003). Towards a Multi-Vendor Mobile Learning Management System. Paper presented at the World Conference on E-Learning.

Colella. (2000). Participatory simulations: Building collaborative understanding through immersive dynamic modeling. *Journal of the Learning Science, 9*(4), 471-500.

Collier, K. G. (1980). Peer-group learning in higher education: The development of higher-order skills. *Studies in Higher Education, 5*(1), 55-62.

De Jong, T. (2001). Knowledge construction and sharing with media based applications. *Journal of Educational Research, 21*(1).

Doswell, J. (2006). Context-aware mobile augmented reality architecture for lifelong learning. Paper presented at the sixth International Conference on Advanced Learning Technologies, Kerkrade, The Netherlands.

Hogan, K., & Pressley, M. (Eds.). (1997). *Scaffolding student learning: instructional approaches and issues*. Cambridge, MA: Brookline Books.

Laine, T. H., & Suhonen, J. (2008). Establishing a mobile blog system in a distance education environment. *International Journal of Mobile Learning and Organisation, 2*(2), 149-165.

Laurillard, D. (2002). *Rethinking university teaching: A conversational framework for the effective use of learning technologies* (2nd ed.). London, U.K.: Routledge Falmer.

Lave, J., & Wenger, E. (1991). *Situated learning: Legitimate peripheral participation*. Cambridge: University of Cambridge Press.

Luchini, K., Quintana, C., & Soloway, E. (2004, 24-29 April). *Design Guidelines for Learner-Centered Handheld Tools.* Paper presented at the Human Factors in Computing Systems, Vienna, Austria.

Moher, T., Hussain, S., Halter, T., & Kilb, D. (2005). *RoomQuake: Embedding dynamic phenomena within the physical space of an elementary school classroom.* Paper presented at the Conference on Human Factors in Computing Systems: ACM Press.

Patten, B., Arnedillo-Sanchez, I., & Tangney, B. (2006). Designing collaborative, constructionist and contextual applications for handheld devices: Virtual learning. *Computers and Education, 46*(3), 294-308.

Rogers, Y., Price, S., Fitzpatrick, G., Fleck, R., Harris, E., Smith, H. et al. (2004, June 1-3, 2004). *Ambient Wood: Designing New Forms of Digital Augmentation for Learning Outdoors.* Paper presented at the IDC 2004, College Park, Maryland, USA.

Rogers, Y., Price, S., Randell, C., Fraser, D. S., Weal, M., & Fizpatrick, G. (2005). Ubi-learning integrates indoor and outdoor experiences. *Communications of the ACM, 48*(1), 55-59.

Ryu, H., & Wong, L. (2008). Empirical studies on generic mobile environment issues that may affect mobile quiz implementation. *IEEE Wireless, Mobile, Ubiquitous Technologies in Education.* Beijing, China: IEEE Press.

Scornavacca, E., Huff, S., & Marshall, S. (2007). *Developing a SMS-based classroom interaction system, mobile learning technologies and applications.* Paper presented at the Conference on Mobile Learning Technologies and Applications.

Sharples, M., Taylor, J., & Vavoula, G. (2005). *Theory of learning for the Mobile age.* Paper presented at MLEARN.

Slavin, R. E. (1983a). Cooperative learning. *Review of Educational Research, 50*(2), 315-342.

Slavin, R. E. (1983b). When does cooperative learning increase student achievement? *Psychological Bulletin, 94*(3), 429-445.

Stone, A., Briggs, J., & Smith, C. (2002, August 29-30). *SMS and Interactivity - Some Results from the Field, and its Implications on Effective Uses of Mobile Technologies in Education.* Paper presented at the IEEE International Workshop on Wireless and Mobile Technologies in Education (WMTE'02), Vaxjo, Sweden.

Vavoula, G., & Sharples, M. (2002). *KLeOS: A personal, mobile, knowledge and learning organisation system.* Paper presented at the IEEE International Workshop on Mobile and Wireless Technologies in Education, Vaxjo, Sweden.

Vygotsky, L. S. (1926). *Pedagogical psychology.* Moscow: Educational Worker Publisher.

Wang, Y. (2004). *Context Awareness and Adaptation in Mobile Learning.* Paper presented at the 2nd International Workshop on Wireless and Mobile Technologies in Education.

# Chapter II
# Transforming the Practice of Mobile Learning:
## Promoting Pedagogical Innovation through Educational Principles and Strategies that Work

**Patrick Danaher**
*University of Southern Queensland, Australia*

**Raj Gururajan**
*University of Southern Queensland, Australia*

**Abdul Hafeez-Baig**
*University of Southern Queensland, Australia*

## ABSTRACT

*This chapter deploys Denning's (2004) powerful assertion that "an innovation is a transformation of practice in a community" (p. 1) through the elaboration of three key educational principles: engagement; presence; and flexibility. Each principle is accompanied by an elicitation of practical strategies that have proved effective in implementing the principles sustainably within particular courses and programs of study, as well as factors that inhibit that implementation. The authors use these principles and strategies that work as an evaluative lens for examining the pedagogical innovativeness of mobile learning and teaching environments. The application of that lens highlights a set of challenges and opportunities facing those technologies and their proponents, specifically in the authors' host institution and in higher education more broadly. Provided that those technologies can be used to engage with those challenges and opportunities, mobile learning can indeed contribute simultaneously to pedagogical innovation and to transformed practice in university learning and teaching.*

## INTRODUCTION

Mobile learning is increasingly associated with competing discourses about its relevance to and potential use in education. On the one hand, a celebratory discourse imagines mobile technologies revolutionizing how, when, where and by and with whom learning and teaching will take place (Goggin, 2005; Grohmann, Hofer & Martin, 2005). On the other hand, a resistant discourse positions many educators as skeptical of pedagogical claims made for new technologies and as seeking to preserve the centrality of the learner–educator relationship against perceived onslaughts from entrepreneurs and technophiles (Hodas, 1993).

It is therefore a complex and challenging terrain on which the promoters of mobile learning have to work to prosecute their arguments about the educational potential of such technologies and at the same time to compete with conflicting pressures that push educational policy-making simultaneously in different directions in response to diverse interests. While this no doubt helps to maximize the clarity and to strengthen the resolve of mobile learning advocates, it also presents a risk that the message of those advocates might be reduced in volume or even silenced by more vociferous voices endorsing alternative priorities.

One among several possible ways forward that would reduce that risk and avoid that outcome is to focus attention squarely on the pedagogically innovative character of mobile learning and teaching environments. If educators are convinced that such environments enhance learner engagement and increase learning outcomes, they are more likely to include those environments in their armories of pedagogical techniques and tools. They are also more likely to invest the energy and time needed to explore and enact the educational applications of those environments and to consider how such applications could potentially revolutionize pedagogical practice.

The objective of this chapter is to contribute to the project of conceptualizing mobile learning as pedagogical innovation. That is, whilst the previous chapter (chapter I) defined mobile learning with different learning activities, this chapter focuses more on learning outcomes from the different mobile learning activities. While many definitions of innovation abound, the authors deploy Denning's (2004) compelling notion of innovation as transformation of practice, which provides an opportunity to locate the interrogation of mobile learning and teaching environments foursquare within the situated learning of actual communities of practice in which those environments must function effectively, efficiently and equitably if they are to have any prospect of fulfilling their educational potential.

The authors approach the task of conceptualizing innovative mobile learning as transformation of pedagogical practice through their elaboration of three key educational principles: engagement; presence; and flexibility. These principles have been selected from several possible candidates as encompassing the key elements of the learner–educator relationship and as deriving from the authors' separate and shared experiential knowledge as university educators, as well as from their respective and collaborative research into mobile learning in the fields of education and health (see for example Gururajan, Kerr, Moloney & Houghton, 2005; Hafeez-Baig & Danaher, 2007; Howard, Hafeez-Baig, Howard & Gururajan, 2006). Each principle is accompanied by an elicitation of practical strategies that on the basis of that knowledge have proved effective in implementing the principles sustainably within particular courses and programs of study, as well as factors that potentially inhibit the implementation of those principles.

The chapter consists of three sections:

- The background to the study, highlighting its conceptual framework, selected literature about mobile learning and teaching environ-

ments, the three principles that constitute that framework and innovation, and the research method used to collect and analyze the data about those principles
- The main thrust of the chapter, focusing on the analysis of engagement, presence and flexibility and their potential relationship with mobile technologies, specifically wireless handheld devices
- Future research directions, directed at exploring further the innovative potential of, and the risks to, such technologies.

The authors argue that mobile learning and teaching environments are well worth pursuing and that their potential pedagogical applications can indeed be innovative, sustainable and transformative, but only if the effectiveness of the strategies is maximized and if the inhibiting factors are neutralized. If that outcome transpires, mobile learning can contribute simultaneously to pedagogical innovation and to transformed practice in university learning and teaching.

## BACKGROUND

This section of the chapter consists of three subsections:

- The study's conceptual framework
- A review of selected literature relating to mobile learning and teaching environments, to the three principles outlined above and to innovation
- The research method used to collect and analyze the data reported here about those three principles.

### The Conceptual Framework

The conceptual framework for evaluating mobile learning and teaching environments in university settings as innovation in and transformation of pedagogical practices is represented in Figure 1.

The authors conceived this conceptual framework in terms of a three legged figure in order to maximize the stability of the structure as a whole to which each leg contributes equally. As noted previously, the legs are portrayed as engagement, presence and flexibility. Thus the framework assigns equal priority and significance to these three selected educational principles as pre-requisites of effective university learning and teaching and as non-negotiable criteria for the design and enactment of mobile learning and teaching environments that are genuinely innovative and transformative. The authors recognize at the outset that this is no easy outcome, universities being prey

*Figure 1. Conceptual framework for evaluating mobile learning and teaching environments in university learning and teaching*

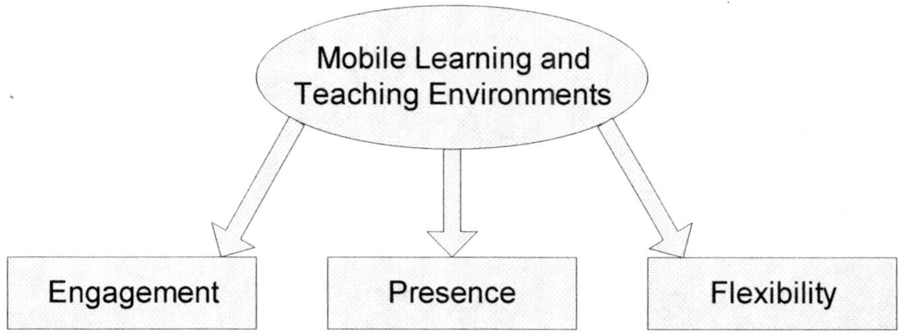

to a range of powerful and sometimes competing discourses where learning and teaching have to battle for attention and resources with pressures to enhance research profiles, boost external funding sources and so on. Equally significant is the fact that academics and students are far from autonomous in being able to select curriculum content, which is often mandated by external accreditation bodies. Nevertheless engagement, presence and flexibility are worth fighting for and insisting upon in the committees and teams that plan and implement university courses and programs.

## The Literature Review

As previously noted, this necessarily selected literature review has focused on five key ideas:

- Mobile learning and teaching environments
- Engagement
- Presence
- Flexibility
- Innovation

### Mobile Learning and Teaching Environments

Ubiquitous computing is an essential component of university learning and teaching environments if the concept of "any time, anywhere" is to be able to be put into practice (Weiser, 1998). In order for this benefit to be realized, these environments need to be interactive and to foster the active engagement of the learner. This can be achieved only through technology based infrastructure, technology supported classrooms such as eClass (Abowd, 1999) and computer integrated classrooms (Hoppe et al., 2000). The imperatives of an efficient learning and teaching environment are to capture and integrate lectures/tutorials, lecture notes, tutorial exercises, sounds and images in order to enhance the learning sequences. In the current climate, this is accomplished by Web-based learning management systems (LMSs) that are distributed and technology dependent (Abowd, 1999). While most of the current systems are predominantly connected using a wired network, the integration of information and communication technologies (ICTs) and wireless handheld devices can expand the network technologies. This in turn will enable the learner and instructor to use the computing power "any time anywhere," providing a truly ubiquitous feeling. The availability of the Internet and the affordability of wireless devices can interconnect with other computing resources seamlessly to provide a mobile learning and teaching environment for both learners and instructors (Liu et al., 2003; Tewissen et al., 2001).

Early studies have shown that the use of wireless, handheld devices in learning and teaching environments can enhance the availability and accessibility of information and engage students in learning related activities in diverse physical locations, promote group collaborative learning and reduce tedious work (Gay et al., 2001; Goldman & Kaufman, 2001; Liu et al., 2003). Therefore it is possible to conceive that, if used appropriately and effectively, such devices will facilitate mobility in student learning.

In a mobile learning and teaching environment, where students can interact and review resources "any time, anywhere," the learner and the instructor become clients for interactive learning and instruction centers. These centers can support teachers with the provision of controlling, monitoring and guiding activities for learning and teaching. In such a mobile environment the role of the teacher is that of mediator, supporter or facilitator, or to provide guidance for learning. To be successful in such a dynamic environment, a teacher must engage and monitor students' learning processes, assist them in their learning activities and help them to make meaningful connections with the real world and with prior and new knowledge (Copley, 1992). In this mobile learning and teaching environment a teacher's role can therefore be very important and challenging. For example, it is a teacher's responsibility to keep the learner focused and to intervene if a student deviates or develops misconceptions during the learning process (Applefield, Huber & Mahnaz, 2000; Brown & Campione, 1994). Such an environment needs to be supported by multimedia applications with Web pages, videos, sounds, online quizzes, presentation slides and case-based activities with two-way communication from students to teacher and from teacher to students.

Another characteristic of this mobile learning and teaching environment is the engagement of the learners to interact with their peers as well as with the learning materials (Greeno, 1997). Resources in this environment enable learners to enquire, discover and analyze problems critically. Furthermore, a learner is able to move swiftly among various sources of content materials and learning activities (Morrison & Lowther, 2005). The authors believe that ICT tools facilitate such learning activities and provide support for the learner and the instructor to engage in such activities in ways that facilitate effective mobile learning and teaching environments. The key components of those environments are the learner, the instructor, the technology and the processes involved in the creation of such environments. The intersections among these components are shown in Figure 2.

## Engagement

From the perspective of this chapter, engagement is defined as the active participation of the learner in the learning activities in mobile learning and teaching environments. In the context of this

*Figure 2. Components of mobile learning and teaching environments*

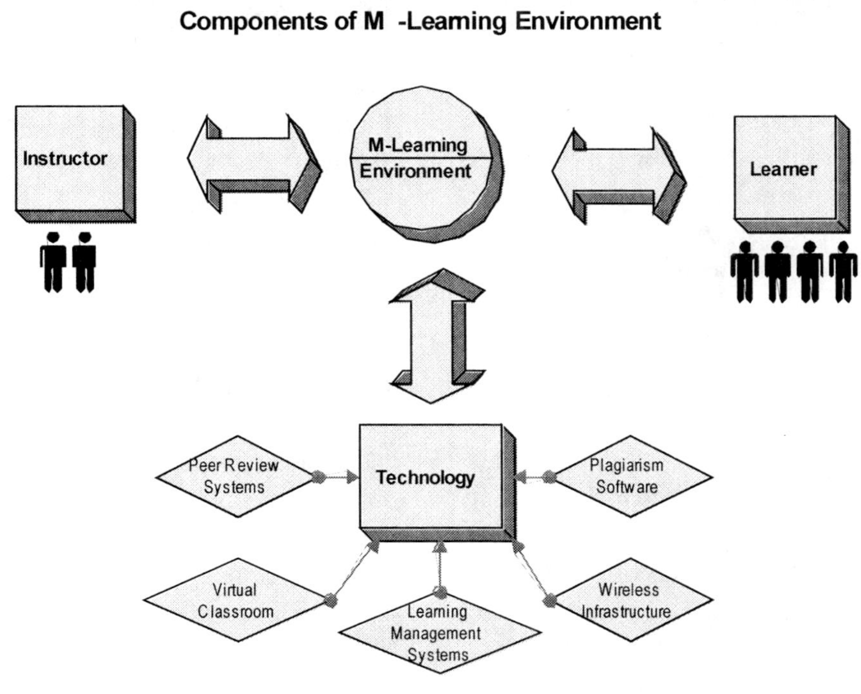

*Figure 3. Components of engagement in mobile learning and teaching environments*

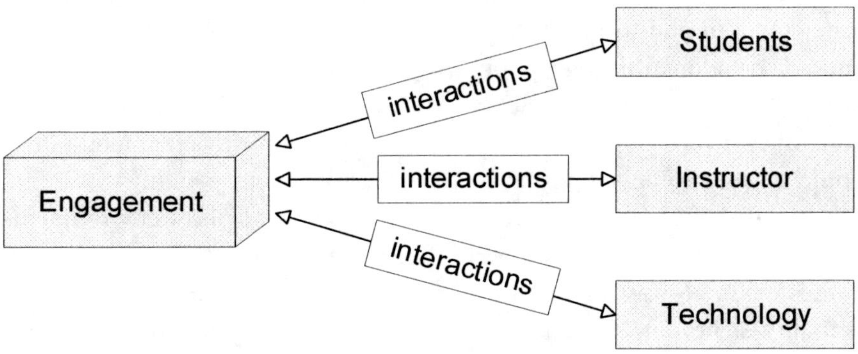


research, and following the evocation of such environments in Figure 2, the components of that engagement are students, instructor and technology. The relationships among these components are shown in Figure 3.

The growth of technology enables the gap between what occurs inside and outside the classroom in terms of learning and teaching experiences to be minimized. Recent developments in ICTs help to improve the activities associated with those experiences. Furthermore, wireless technologies and other LMSs improve such activities for students by actively engaging them in an interactive environment. In order for these objectives to be achieved and for the content to reach the learners, their active participation in the learning process is essential. In a context where provision can occur using ICT tools, it is critical for learners to engage actively in the learning process on their own initiative. This process of active engagement can be facilitated through the mobilization of peers and co-learners and instructors as well as the technology. For example, a learner can be engaged in the learning processes by actively participating in learning activities through the provision of feedback to her or his peers. This feedback provision can be built into an LMS and activated through a range of technologies. Such activities can then be shared with other learners and instructors. These ideas help to frame 'engagement,' depicted as the first leg of the conceptual framework in Figure 1.

## Presence

Presence refers to a simultaneous awareness and locatedness of self and others in a learning and teaching environment. It therefore "...evokes the engagement and interaction assumed to lie at the centre of the learner–educator relationship, whether face-to-face or mediated by space and/or time" (Danaher, Hickey, Brown & Conway, 2007, p. 211). Moreover, "Presence also betokens the empathy, encouragement, interest, and support and the emotional dimension of being human on which that relationship is presumed to be based" (p. 211). Presence is thus both central and crucial to the educational enterprise.

At the same time, it is appropriate to acknowledge some possible limitations of a focus on presence in mobile and online environments (see also Danaher et al., 2007, pp. 223-224). Firstly, the concept of social presence was developed in the mid 1970s and needs ongoing updating in the light of continuing technical developments in learning and teaching environments (Kehrwald, 2006). Secondly, while all media have the potential to be associated with presence, those media vary significantly in their specific characteristics (such as the degree of synchronous communication) and hence in their capacity to enhance particular kinds of presence (Hiltz, Coppola, Rotter & Turoff, 2000). Thirdly, as is elaborated later in the text, the positive valence of presence should not lead to the assumption that it is automatic or inherent in an educational environment or technology; on the contrary, it has to be designed and evaluated rigorously and fought for vigorously.

Those caveats notwithstanding, a particularly useful classification of different types of presence was developed by Anderson (2004), whereby cognitive, social and teaching forms of presence were linked respectively with student–content, student–student and student–teacher interactions. These interaction types thereby inform 'presence,' the second leg in the conceptual framework represented in Figure 1, and are illustrated in Figure 4.

## Flexibility

Flexibility encapsulates considerable theoretical potency in contemporary educational discourses. Flexible learning facilitates different modes of interaction between teacher and learner (Dimitrova et al., 2003; Payne, Ball & Snow, 2000). For example, e-learning allows learners with access to appropriate ICT tools to experience learning to suit their personal needs, thereby providing flexibility (Dimitrova et al., 2003; Goodison, 2001). In today's higher education context, it is critical to develop increased mobility and flexibility for students and teachers. Schellekens, Paas and van Merriënboer (2003) found that in Europe, educational programs are organized in rigid formats. They also concluded that operational flexibility among other aspects of educational programs may improve flexibility in education overall. Dimitrova et al (2003) showed that students deploy the learning and teaching environment in a multiplicity of ways and use various learning resources depending on their learning styles and strategies, thereby underscoring that mobility and flexibility are essential for learning. Similarly, Johnston (1999) argued that increased flexibility is an effective strategy in education to provide access to the wider community, to enhance equality of opportunity and to minimize the limitations of working, learning and leisure lifestyles. Van Meel (1992) believed that inflexibility in an educational environment can generate additional problems, whereas flexibility can have positive effects on that environment.

It can be inferred from this discussion that flexibility plays a critical role in mobile learning and teaching environments. These environments

*Figure 4. Components of presence in mobile learning and teaching environments*

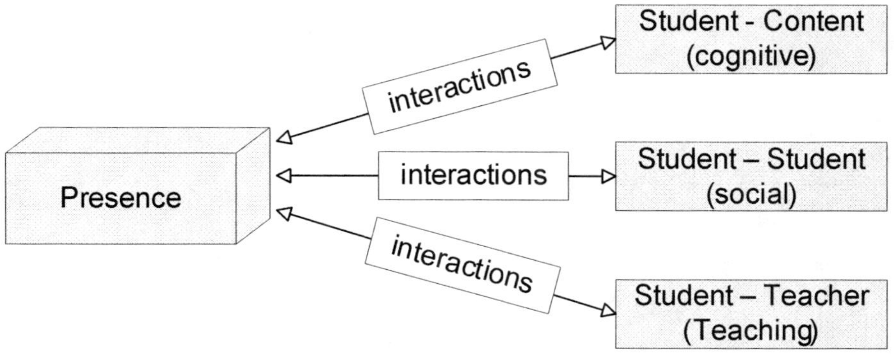

*Figure 5. Components of flexibility in mobile learning and teaching environments*

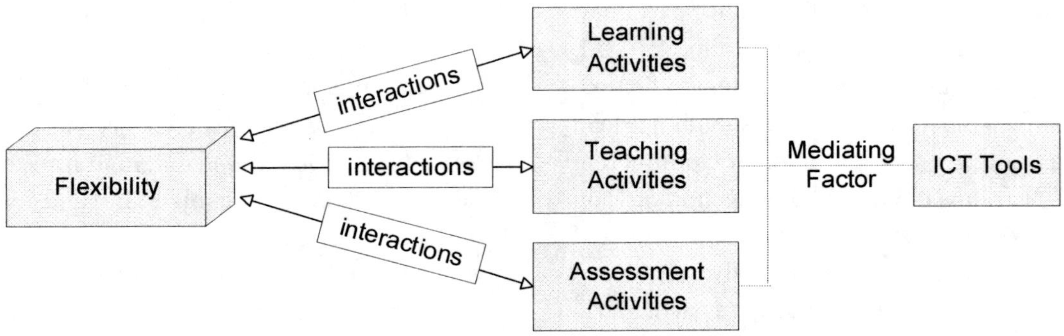


equally need to accommodate a diverse range of students and their varying needs. One way of achieving such an objective is to adopt flexibility in learning, teaching, and assessments activities through the use of ICT tools. The relationships among these factors are shown in Figure 5:

These ideas underscore 'flexibility,' the third leg in the conceptual framework represented in Figure 1.

### Innovation

Given the chapter's focus on the innovative potential of mobile learning and teaching environments, it is appropriate to consider briefly selected literature about innovation. While in scientific and technological fields an innovation is generally equated with invention and the creation of a new product or procedure, in education innovation is more likely to be associated with new or varied applications of curriculum, pedagogy and/or assessment. This recalls Denning's (2004) useful assertion that "an innovation is a transformation of practice in a community" that is not necessarily the same as "the invention of a new idea or object" (p. 1; see also Harreveld & Danaher, 2004, p. 10). Moreover, Denning claimed that "a transformation of practice in a community won't happen unless the new practice generates more value to the members than the old," and that that "value might not be economic; it may be pride, reputation, health, safety, freedom" (p. 2). Finally, he noted that both personal practices and organizational processes are necessary to establish a culture of innovation.

## The Research Method

The chapter explores the three selected elements of mobile learning and teaching environments outlined above by drawing on empirical evidence from three sites, using a hybrid research approach, as illustrated in Table 1.

*Table 1. Sources of empirical evidence for researching the three educational principles of mobile learning and teaching environments*

| Empirical Site | Research Method | Educational Principle |
|---|---|---|
| Masters business online course | Quantitative analysis | Engagement |
| Graduate entry teacher education online course | Critical reflection | Presence |
| Three focus groups about mobile learning and teaching environments | Qualitative analysis | Flexibility |

The purpose of this table is to illustrate the three selected elements of mobile learning and teaching environments. These, as explained earlier, are engagement, presence and flexibility. The concept of engagement is explored using a quantitative analysis of a Masters online course in the faculty of business at the university under review here. Presence is investigated via critical reflection on a graduate entry online course in the faculty of education there. Flexibility is studied using a qualitative analysis of three focus groups discussing mobile learning and teaching environments. The groups examined for the three concepts varied considerably, thus enhancing the overall understandings of such environments.

The authors recognize that this hybrid research approach is inevitably selective rather than comprehensive in its analysis of data from each site. Nevertheless it seems appropriate to weigh this potential limitation against the expected benefit of greater understanding and synthesis of knowledge about mobile learning and teaching environments to be gained from considering the three data sets in juxtaposition.

Exploratory research is suitable for the theory building stage of a project, and aims at formulating more precise questions that future research can answer (Neuman, 2003). Furthermore, Leedy and Ormrod (2004) indicated that case study is suitable for learning more about a little known or poorly understood situation. It provides understanding about complex phenomena within a given context. Riege (1996) incorporated ideas from five other definitions and defined the case study method as "A research method which focuses on a particular part of an organization or an industry within its context in order to rigorously explore and analyze contemporary real-life experiences in-depth using a variety of evidence" (p. 142). Moreover, in case study research, multiple cases can improve the quality of data collection and research findings (Bonoma, 1985; Merriam, 1988; Miles & Huberman, 1994; Patton, 1990; Robson, 1993; Yin, 2002). Therefore, a multiple case study approach is used here to explore the issues outlined in the introduction to this chapter.

In multiple case study research, each case must be chosen so that it either predicts similar results (a literal replication) or predicts contrasting results but for predictable reasons (a theoretical replication) (Yin, 2002). Therefore, to reflect the replication logic, the random selection of cases is neither necessary nor preferable (Madden, 2000). To illustrate the research method, a matrix (see Table 2) was built based on the literature (Amab, Mehmet & Schnitker, 2001; Leem, Suh & Kim, 2004).

This table presents the selection of cases. This is a customary business education style table where case selections are presented for better understanding. The left hand column details the type of selection. For example, Case A is constituted by a faculty of business student group. The right hand column portrays details of the type of study mode, for example, on-campus or online. When these two columns are read together, the reader is able to understand the types of groups chosen for the study.

The aim of this study was to contribute to elaborating the framework of m-learning (through a focus on the principles of engagement, presence and flexibility). Consequently the unit of analysis

*Table 2. Matrix of the multiple case study research method*

| | |
|---|---|
| m-learning organization | A regional university offering on-campus, off-campus, online and distance education with campuses in Australia and overseas |
| Case A – Business faculty (students)<br><br>Case B – Education faculty (students) | Case C – students and academic staff members |

was chosen as the stakeholder groups to whom those principles might be expected to apply most directly. In this project, therefore, the stakeholders were students and academic staff involved in the process of constructing mobile learning and teaching environments.

## The Faculty of Business Course (Case A)

The faculty of business Masters course that was interrogated for evidence of engagement was prepared in such a way that the entire course can be studied online. The instructional materials were weekly components about four specific emerging technologies: PDAs, voice recognition applications, RFID and plagiarism software. These technologies were discussed for the first four weeks based on their relevance, technical content and so on. The purpose of the materials in these weeks was to provide a good understanding of the four technologies to students. The next six weeks of materials included the management aspects of these technologies in an organization. These issues included cost aspects, licensing issues, implementation issues, change management issues, training issues, and so forth. The materials were derived from government white papers, journals and published sources. The materials were threaded in a PDF format based on the weekly schedule and students were given guidelines as to what was needed to be studied for a given week.

In terms of assessment, students were given two sets of assessments. The first set consisted of five weekly journal entries and a report on the enablers and inhibitors of these technologies. The second set consisted of five more weekly journal entries, a report on implementing these technologies in an organization and a consultant's report on using the technologies at enterprise level. Each assessment component was verified by plagiarism software prior to official submission. Students were asked to submit a plagiarism count report and were allowed to rectify the plagiarism issues by repeated submission to the plagiarism application. Each weekly journal was peer reviewed by two other students and awarded a score based on a marks template. If students failed to assess another student's work, they were awarded a penalty. The assessments were marked by a tutor as well as by other students. Weekly feedback was also provided to students to improve their presentation. Discussion board dialogues were encouraged for sharing ideas, seeking clarification and complying with course guidelines. Feedback about each journal entry was provided within 48 hours. A moderation process was executed when the difference between the two scores was over 10 percent.

## The Faculty of Education Course (Case B)

The faculty of education course that was examined for evidence of presence was a core component of a graduate entry teacher education program one year full time or equivalent part time in duration for prospective early childhood, primary, secondary and technical and further education teachers (see also Danaher, 2006; Danaher et al., 2007). It was available officially in either on-campus or online modes, but in practice the blended delivery was similar for all students, with the on-campus elements consisting of compulsory residential schools and regular seminars and the online element focusing on Web-CT asynchronous discussion boards and Macromedia Breeze presentations (PowerPoint slides accompanied by an audiotaped commentary).

The course was concerned with developing analytical skills in identifying and evaluating educational contexts and environments and their impact on students' learning. The four equally weighted modules examined relevant concepts and vocabulary, whole school approaches to inclusivity and social justice, the development of inclusive learning environments and the link between educators' work and their identities in relation to educational reform. The course had three items of summative assessment: the students' progress during their professional attachment in an educational site (50%), their proposal for a problem-based investigation during that attachment (10%) and their report of the outcomes of that investigation (40%). The first item was assessed by the student's supervising teacher in the educational site and the second and third items by the university staff members working in the course.

### The Focus Groups (Case C)

The three focus groups that were considered for evidence of flexibility were conducted in September and October 2006, with two groups of academics and one group of students (mostly from the faculties of business and education) at the authors' institution (see also Hafeez-Baig & Danaher, 2007, which highlights the third of these three focus groups). The goal was to contribute to evidence-based decision- and policy-making about the construction and expansion of mobile learning and teaching environments at the institution.

The focus group questions covered several issues, ranging from participants' understandings of mobile technologies and their current and potential educational applicability to concerns about access and digital divides to roles and responsibilities if such technologies are to be harnessed more comprehensively in the future. The discussion was audiotaped and transcribed, and then subjected to textual and thematic analysis, linking themes identified as significant by the participants with the research questions framing the broader study.

## CASE STUDIES

### 1. Issues, Controversies, Problems

This section of the chapter applies the research methods outlined to analyze selected data from the three nominated sites of investigation. The analysis is directed at three key tasks:

- Identifying strategies that work in implementing the respective educational principle in contemporary university mobile learning and teaching environments
- Eliciting factors that inhibit the deployment of such strategies
- Articulating implications for the innovative inclusion of mobile learning and teaching environments in contemporary university settings.

### Engagement

Engagement was evaluated in the faculty of business Masters online course represented in Table 1 and as Case A in Table 2 in terms of the three components depicted in Figure 3: students, instructor and technology. For example, analysis of the WebCT reports associated with the course demonstrated that 112 students accessed the course 12,164 times over a period of 12 weeks (90.27%). Engagement was evident also in the 1,130 discussion threads that emerged during the same period. The track tool indicated that there were 4,578 discussion messages during this time, with an average discussion time of 4.36 minutes per student. Over the life of the course, this became about 351 minutes of discussion per student. This was in addition to the time devoted to specific assignment queries (89 minutes per student) and reading various announcements posted by members of the teaching team (93 minutes per student). The course recorded a total of 46,366 message threads and a total of 2,270 hours for 112 students during the 12 week period (with an average of 1.7

hours per week per student engaging online with those message threads).

It must be acknowledged that, particularly initially, engagement was associated with a degree of confusion and anxiety among students in relation to specific aspects of the course, which differed from other courses in the way in which students were engaged. For example, a message thread entitled "A little confusion" associated with the new model of operating the course recorded 321 threads with an average discussion time of 1 minute 32 seconds. On the other hand, this thread was not found in later weeks of the course. Similarly, some initial confusion attached to marking standards as a result of varying referencing approaches was reflected in 1,802 discussion threads, again with an average discussion time per thread of 1 minute 32 seconds. An unanticipated technical problem in week 5 of the course (related to the system of automating submissions with date stamps unexpectedly not allowing students to submit or incorrectly levying an automatic penalty for timely submissions) also generated a number of threads discussing this problem and its potential effects on students, with 1,091 threads and an average discussion time of two minutes 34 seconds per thread.

Students' technology usage of the LMS also reflected a high degree of engagement in the course. Students spent an average of 89 minutes discussing the assignments on the discussion board, indicating their eagerness to do well in this part of the course's summative assessment. This was also indicated by students' engagement in e-mail communications with the lecturer, whereby they spent an average of 23 minutes accessing the email "send" button provided on the LMS. The LMS records for the course reported that various learning content was accessed 11,633 times, consuming a total time of 1,304 minutes to access and download the materials (reading and comprehension time was in addition to this figure). These LMS records demonstrate that students were actively and consistently engaged in the course in terms of materials provided, tasks completed and time spent interacting with other students.

In terms of the latter, as noted above the course assessment required students to review one another's submitted work, so that students acted as instructors in engaging with the weekly journals of other students in the course. The peer review system assigned these journals automatically to other students in the class, selected at random. These two students were entrusted with reading and commenting on the journals in terms of such criteria as quality, relevance and rigor. A simplified marking criteria sheet was used, whereby students ticked boxes against these criteria based on their reading of the journal. In addition to ticking the boxes, they were asked to provide written feedback so that a moderator (the course leader) could verify the level of marking. As noted above, if the level of discrepancy between the two markers' recommended marks exceeded 10 percent, a flag was sent to the course leader to review the journal independently and to overwrite the scores allocated by the two students. Moreover, students were able to use the LMS to comment on the overall quality of the journals. They were also encouraged to answer any queries in relation to the journals and other assessment specifications and to help other students in the course by sharing their resources. Students were thus able to analyze, evaluate critically and then judge one another's work in the weekly journals and in the associated LMS discussion pertaining to the journals' quality and contributions to knowledge construction by the students. These various strategies both facilitated and demonstrated high levels of engagement in the course.

This discussion of Case A has highlighted a number of potential difficulties that might also face mobile learning in seeking to enhance learner engagement. Prospective disengagement is often associated with feelings of confusion, uncertainty and lack of control or ownership, feelings that can as easily arise in mobile as in online learning environments. Equally this discussion has identified strategies related to course design and student assessment that proved effective in overcoming this prospective disengagement and that clearly worked well in relation to increasing student engagement. This augurs well for deploying such strategies in mobile learning environments, provided that equivalent principles of design and development are identified and implemented.

## Presence

This subsection is concerned with identifying evidence of presence in the graduate entry teacher education online course cited in Table 1 and identified as Case B in Table 2. Paradoxically, presence can sometimes be detected more readily through its absence, when the interaction types identified by Anderson (2004) and synthesized in Figure 4 are for various reasons missing from the pedagogical mix in a particular course or program. Nevertheless it is possible to discern specific strategies that underpin the successful enactment of presence and that also provide the foundation for the innovative application of mobile learning and teaching environments in university settings.

### Cognitive Presence (Student–Content Interactions)

On the basis of critical reflection on the teacher education course referred to previously, cognitive presence, which facilitates student–content interactions, is fostered by the following six actions. Firstly, it is important to bring the course content to life for students. Secondly, posing questions about the content can help students to align and integrate their experiential and theoretical knowledge. Thirdly, encouraging students to approach the content as open to contestation and critique facilitates their understanding of multiple perspectives and their intensive engagement with one or more of those perspectives. Fourthly, it is valuable to present the knowledge underpinning the content as constructed and situated rather than as timeless and universal. Fifthly, emphasizing higher order skills such as synthesis and evaluation assists students to engage with the content. Sixthly, designing summative assessment tasks that encourage and require the application of cognitive presence maximizes the two-way link between intended outcomes and student assessment.

Equally the following five factors have been found to inhibit the development of cognitive presence. Firstly, students and academic staff members vary considerably in the levels of time and energy that they can devote to engaging with the course content. Secondly, some students and academic staff members find the online learning and teaching unfriendly or even unnerving as an environment for engaging with the content. Thirdly, some students find difficult the process of acquiring new discipline-based concepts and vocabulary that frame the content. Fourthly, some students find difficult the process of making meaningful links between the theory as espoused in the content and the practice as experienced during their professional fieldwork. Fifthly, some students find uncomfortable the proposition of knowledge as constructed and situated and seek definitive answers from the content rather than creating those answers themselves.

### Social Presence (Student–Student Interactions)

Critical reflection on the teacher education course also elicits the following five strategies as having proved effective, if not for all students at all times, then certainly for some students some of the time, in helping to facilitate student–student interactions and hence to enhance the enactment of social presence. Firstly, asking students to introduce themselves at the beginning of the course seems an obvious technique but it is easily overlooked at such a busy time. Secondly, encouraging students to make links between the course's theoretical knowledge and their own experiential knowledge and hence valuing the sharing of personal experience, both prior to the course and during the professional attachment, heightens their involvement in the process. Thirdly, highlighting the value of learning through dialogue and discussion alerts students to the intended benefits of presence. Fourthly, decentering the position of academic staff members as founts of wisdom and instead promoting the practice of consulting multiple perspectives held by students on specific topics maximizes shared responsibility for learning. Fifthly, modeling and reinforcing online communications that are authentic, engaged and respectful helps to enact the educator's part of that shared responsibility.

By contrast, the development of effective and productive social presence was inhibited by the following four factors. Firstly, the course summative assessment program being based on individualized rather than group learning reduces the perceived importance of presence. Secondly, there is potential for at least some students to feel that they are in competition with one another and therefore to be disinclined to share ideas. Thirdly, the pressures of combining a busy course and program with a myriad of other demands leave little time to devote to interacting socially with one another. Fourthly, the immediate and public character of the online environment can occasionally generate posts that are not respectful of other course members.

## Teaching Presence (Student–Teacher Interactions)

Critical reflection on the course also yielded the following four strategies as effective facilitators of teaching presence or student–teacher interactions. Firstly, assigning explicit and public responsibility for facilitating each module of the course to a different staff member can maximize the sharing of multiple perspectives. Secondly, that staff member highlighting specific links between the respective module and her or his experiential knowledge and research and teaching interests can enhance her or his authenticity as well as interactions with students. Thirdly, staff members can increase both the volume and the value of those interactions by participating in the online discussion threads as module facilitator and/or as course learner, providing suggested responses to questions and posing further questions to elicit dialogue and engagement. Fourthly, encouraging students to pose questions via the discussion board rather than via individual email messages (unless they particularly need to do so) brings the student–teacher interactions more visibly into the public domain.

At the same time, three factors were identified as potentially restricting the development of teaching presence by means of these kinds of strategies. Firstly, contracted staff members generally contribute less visibly than full-time staff members to the discussion board (possibly because of their contracts containing insufficient time for such interaction and/or because of insufficient training in using the LMS). Secondly, a potential mismatch can develop between staff members seeking to maximize student engagement by posing higher order questions and students wishing to maximize their completion of summative assessment items by posing assessment-related questions of their own rather than responding to the staff members' questions. Thirdly, a related potential mismatch lies in the rhythms of students' and staff members' interactions with each other; particularly with the later modules in the course: while the respective staff member might seek to generate student enthusiasm about her or his allocated module, students are generally more concerned with the urgent demands of the professional attachment and/or of the summative assessment items.

While the focus in Case B was on fostering various forms of presence in an asynchronous online learning environment, the strategies identified as facilitating, and the factors identified as inhibiting, the enactment of such presence are equally applicable to mobile learning environments. Like online learning, mobile learning might be assumed to have to strive harder than face-to-face education to generate presence (although presence is no more inherent in face-to-face environments than it is in mobile or online environments). Also like online learning, mobile learning has at its disposal several technological attributes that with careful planning and implementation can be deployed in ways that maximize learners' interactions with the course content, instructors and fellow students.

## Flexibility

The three focus groups investigating students' and academic staff members' experiences and understandings of mobile learning and teaching environments identified in Table 1 and listed as Case C in Table 2 generated mixed evidence of flexibility, particularly in relation to the three components of such flexibility listed in Figure 5: learning activities; teaching activities; and assessment activities. Analysis of the focus group transcripts elicited 24 main themes, as portrayed in Figure 2.

These data indicate that the theme of "wireless" was the principal topic of discussion. This was followed by "access," "students" and "technology," and then by more focused topics. Figure 6 encapsulates a major discursive tension between an aspiration and expectation that mobile technologies such as wireless devices enable students to manage their time effectively and flexibly on the one hand and a concern that other factors such as institutional policies and resources (or lack thereof) might impede the implementation and operationalization of that flexibility.

Once the themes had been extracted from the transcripts of the three focus groups, a model was created using the Gaussian modeling process. This model, taking the form of a concept map, groups the themes based on the relative distance between each keyword as it appeared in the discussion. The model appears in Figure 7.

The concept map groups three major concepts that are closely interrelated and three minor concepts that are not related closely. The three closely related concepts of "wireless," "access' and "home" indicate that, according to this model of idealized mobile learning and teaching environments, students are able to access lecture and other learning materials from home using wireless

*Figure 6. Focus group themes*

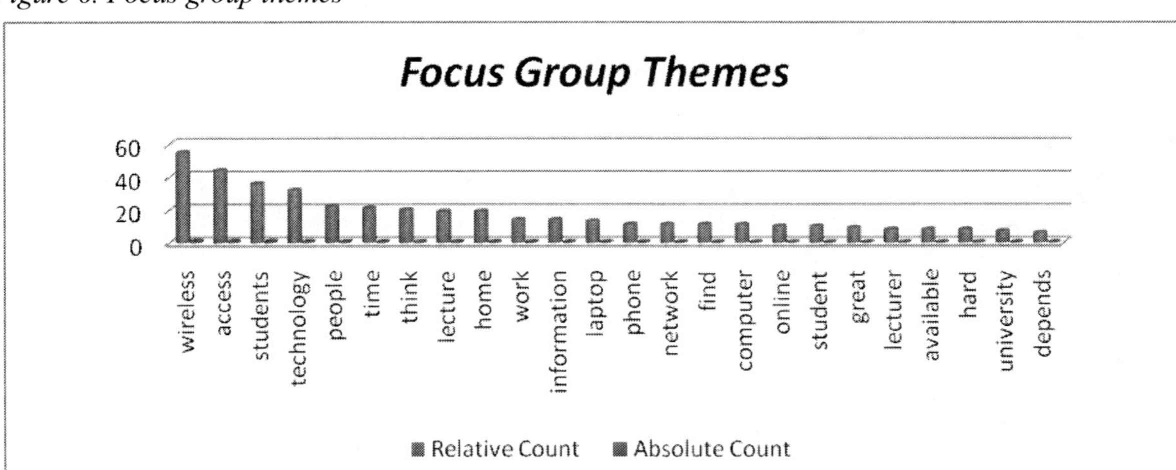

*Figure 7. Concept map of the relationships among the focus group themes*

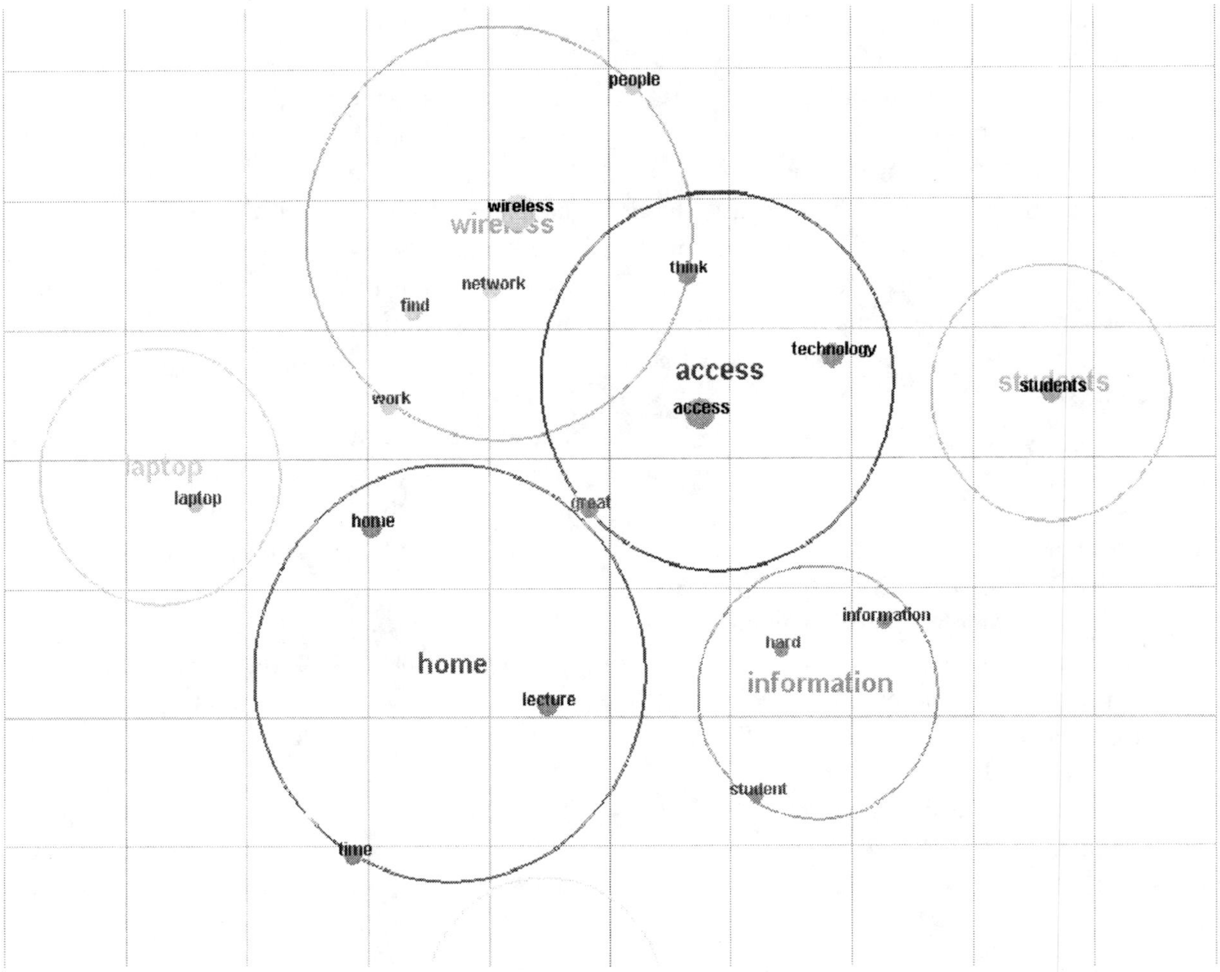


devices. This is clearly part of mobile learning, as students are able to access the materials when they are on the move (this is represented in the cluster related to the concept of "home"). The themes "work" and "find" also support the notion that students who have work commitments will be able to access learning materials using wireless devices; this is a crucial component of flexibility, given the diversity of the student populations of contemporary universities and the high proportion of such populations who travel frequently for work and other purposes.

The three minor concepts that are not related closely to one another are "students,' "information" and "laptop." This suggests the proposition of students being able to use mobile learning technologies to access materials within the university campus using wireless devices such as laptop computers. This proposition is noticeably predicated on the assumption of flexibility, although it was less explicitly discussed in the focus groups than the asserted relationship among "wireless," "access" and "home."

A further analysis of the focus group transcripts based on the themes listed in Figure 6 and depicted in Figure 7 enabled the identification of positive and negative sentiments associated with those themes. The theme of "access" was discussed positively because students and staff members believed that, in addition to enhancing access, wireless devices can facilitate discussion and other forums in real time, an outcome that could also enable off-campus students to participate and learn as flexibly as, and at the same time as, on-campus students. Similarly, a positive valance was attached to the theme of "lecture," because students were excited by the idea of being able to download broadcast lectures onto their computers and start their learning "any time anywhere" and thereby of being no longer constrained by physical location and time.

By contrast, students and staff members from the faculty of education tended to portray "technology" as a barrier that might leave particular groups of students behind. This was perceived as potentially operating in two ways: a disjuncture developing between the wireless devices and the educational resources needed for particular courses; and placing unsustainable demands for equipment and training that would threaten already straitened finances.

Of the three cases, Case C was the one located most directly in discussion of mobile learning technologies and environments. While allowing for disciplinary, paradigmatic and other differences across the three focus groups, a consensus emerged about the kinds of criteria that students and educators alike required from mobile learning if it is to be implemented on a wide scale in university learning and teaching. At one level this consensus was hardly surprising, focusing as it did on features that might be assumed of a dynamic, reciprocal and stimulating educational environment. From a different perspective, however, focus group participants were intrigued by the pedagogical possibilities of mobile learning technologies to create new and previously unimagined contexts less dependent on place and time than is currently the case.

## Solutions and Recommendations

It is clear that there are several strategies that work in fostering engagement, presence and flexibility in the mobile learning and teaching environments at the authors' institution. It is equally clear that such engagement, presence and flexibility are by no means automatic or assured. Fundamental to the enactment of successful strategies is a set of dispositions on the part of all participants and stakeholders, no easy task when those participants and stakeholders have multiple and sometimes competing aspirations and perspectives. It is these same dispositions that are likely to influence whether and how mobile learning and teaching environments are liable to be incorporated into the practice of current university settings, let alone to contribute to pedagogical innovation by helping to transform that practice.

As noted, the objective of this chapter is to contribute to the important project of conceptualizing mobile learning as pedagogical innovation. That conceptualization underpinned the analyses of the three cases presented here: the strategies that work in the two courses and as elaborated in the three focus groups do so partly because they constitute many of the key elements of transformed practice, whereby there are fundamental shifts in how, when, where and with whom and what learning and teaching take place. At the same time, the factors that inhibit the successful enactment of those strategies are equally likely to prevent or disrupt mobile learning becoming adopted and instantiated as pedagogical innovation.

Mobile learning applications are becoming much more prevalent. According to Education.au limited, PDAs are used increasingly to support learning and teaching in the educational domain. The United Kingdom Department for Education and Skills sponsored a handheld computing pilot with 150 teachers at 30 schools to explore issues associated with teaching aspects (Perry, 2003). Australia now has several similar activities in varied subject areas and levels of schooling in order to explore ideas and innovations. Applications of mobile learning can encourage disaffected young people (some of whom are homeless) to become involved in learning through games, quizzes and mini soap operas on mobile phones and PDAs. Mobile learning can facilitate context-aware, location-based services providing learning content anywhere, any time to anyone, as well as mobile collaboration and *ad hoc* networks to support organizational learning.

What then are the implications of the findings of this study for strategies that work towards and factors that inhibit the innovative inclusion of mobile learning and teaching environments in contemporary universities? The authors proffer three such implications for consideration here:

- Firstly, priority must be given to the educational character of the intended innovation; many of the inhibitors relate to concerns about and/or evidence of the privileging of the technical dimension of wireless devices that might cause them to become disconnected from the pedagogical environments in which they must become situated (see also Stone & Livingstone, 2004). Educational institutions should therefore focus on establishing fault tolerant technical infrastructure to support mobile learning as connectivity at the palm level is essential in this environment.
- Secondly, engagement, presence and flexibility have proven themselves worthy candidates for selection as criteria for evaluating whether and how particular wireless devices can become incorporated into sustainable and transformative mobile learning and teaching environments. This is no easy process because these educational principles are logically distinct as well as being philosophically interrelated – for example, a strategy for implementing wireless devices flexibly does not necessarily promote engagement or presence. From this perspective, instructional designers should develop innovative educational resources that can capture and engage students in mobile learning environments and articulate with the social dimension of mobile technologies (see also Carlen & Jobring, 2005; Looi, 2005). In those environments, it is easy to digress away from the main component of learning. It is still unknown how to arrest students within this context so that sufficient and suitable time can be devoted for learning. Future strategies should include group learning, games that relate to educational contexts and quizzes that can be conducted using the mobile infrastructure, among others.
- Thirdly, the data analysis presented has demonstrated the complexity of the contexts and networks in which mobile learning and teaching environments need to be established if they are to be successful. That complexity includes multiple and competing pressures and priorities across and among stakeholder groups, as well as a diversity of perceptions of the distinctive affordances of mobile technologies (Wiredu, 2006). Accordingly policy makers should decide the platform issues so that learning and teaching environments can be accessed. Assuming that these environments are applied to school as well as post-compulsory students who have limited cash power to procure expensive mobile devices and associated maintenance costs, policies should be developed to specify mobile platforms that can be made available to as many potential users as possible. Prior to the creation of any instructional resources, these policies should be developed and approved so that access to materials can be made available.

The authors also put forward three recommendations for dealing with the issues and problems outlined previously, each derived from one of the implications articulated in the previous paragraph:

- Firstly, any proposal for enacting wireless devices in university settings should be subjected to the fundamental question, "How will this proposal directly benefit and boost learning and teaching?" This question focuses attention on a range of policy issues, strategies that can take learning and teaching to new ways of thinking and ICT infrastructure that is essential to facilitate connectivity.
- Secondly, engagement, presence and flexibility should be among the criteria to be included in an evaluative checklist for activation to interrogate the innovative use of mobile learning technologies once they have moved beyond the proposal stage. This question deals directly with the innovative ways of developing curriculum and instructions as mobile devices come with limited processing power.
- Thirdly, decision and policy-makers must eschew a 'one size fits all' approach to planning, implementing and evaluating wireless devices and other mobile learning technologies in favor of a more situated method that acknowledges and even celebrates the complexity noted above. This notion links with the concept of providing a flexible learning and assessment environment using mobile technologies (Rushby, 1998). A key consideration here is the type of investment made in capital expenses versus infrastructural expenses.

## FUTURE RESEARCH DIRECTIONS

On the basis of the preceding discussion, the authors have distilled four possible future research directions pertaining to the innovative design and application of mobile learning and teaching environments in university learning and teaching. These posited research directions, the first three of which derive from and follow the implications and recommendations articulated at the end of the previous section, have been expressed in the form of an overarching research question and several proposed sub-questions:

- What are likely areas of convergence and divergence between educational imperatives and mobile learning technologies?
  What are the non-negotiable elements of successful learning and teaching practice? Which factors facilitate and inhibit the enactment of such practice? Which competing pressures are placed on learners and teachers? Which competing pressures are placed on the designers and implementers of mobile learning technologies? What other priorities, apart from educational imperatives, might motivate those designers and implementers? Which aspects of mobile learning technologies are most likely to contribute to meeting educational imperatives? Which aspects (if any) are likely to privilege technical capabilities at the cost of those imperatives?
- What other educational principles and strategies are automatic inclusions on lists of criteria for evaluating the educational effectiveness of mobile learning technologies?

  Do other researchers agree that engagement, presence and flexibility are pre-requisites of the effective inclusion of mobile learning technologies in learning and teaching? Which other educational principles might be considered pre-requisites of that outcome? Are these principles interdependent and mutually reinforcing or might they be positioned as potentially contradictory of one another? Which specific strategies are most likely to accompany the enactment of such principles? Are strategies tied to particular principles and/or sets of strategies liable to enhance the attainment of principles as a whole?

- What are effective techniques for mapping and engaging with the complexity of contexts and networks in which mobile learning technologies must become situated?

  What are the relative pros and cons of such potential techniques as scenario testing, forecasting, actor network analysis, policy needs analysis and social network analysis? To what extent might such techniques be able to elicit important nuances and subtleties while still allowing detailed and systematic planning? Under which circumstances (if any) might some contexts and networks be privileged or receive priority attention compared with others? Which evaluative tools are likely to be most successful in interrogating the equity and utility of these analytical and planning techniques?
- What are other possible approaches to understanding innovation and how might they contribute to understanding the potential effectiveness of mobile learning technologies?

  What are the relative pros and cons of such approaches to understanding innovation as creativity, diffusion, entrepreneurship, problem solving, strategic behavior and translation? In what ways do they converge with and/or diverge from innovation understood as transformation of practice? What similarities with and differences from the analysis of the three cases presented in this chapter might they offer compared with the authors' deployment of transformation of practice? What are the conceptual, methodological and empirical links between innovations and mobile learning technologies?

This chapter has explored three cases of university learning and teaching at the authors' institution as encapsulating evidence of strategies that work and factors that inhibit in enacting the three educational principles of engagement, presence and flexibility. These principles and strategies have been posited as pre-requisites of the successful implementation of mobile learning and teaching environments. They have also been linked directly with efforts to associate that implementation with pedagogical innovation, understood as the transformation of pedagogical practice.

The argument prosecuted in the chapter has been that both challenges and opportunities confront advocates of mobile learning technologies who wish to see a more widespread adoption of such technologies in the learning and teaching policies and experiences of contemporary universities. The authors contend that, on the basis of the analysis presented in the three cases explored here, mobile learning can indeed contribute simultaneously to pedagogical innovation and to transformed practice in university learning and teaching, but only if these and other educational principles and strategies that work are incorporated at the center of future decision and policy-making.

## ACKNOWLEDGMENT

The authors are grateful to their respective colleagues in the two courses reported here. They acknowledge thankfully the participation of the members of the three focus groups and the research assistance of Christine Knight and Robert White in the research project of which those groups formed a part. Hokyoung Ryu and David Parsons have been efficient and encouraging editors. The final version of the chapter benefited significantly from the constructive feedback of anonymous referees.

## REFERENCES


Abowd, G. D. (1999). Classroom 2000: An experiment with the instrumentation of a living educational environment. *IBM Systems Journal., 38*(4), 508-530.

Amab, D., Mehmet, P., & Schnitker, T. (2001). Could mobile banking go global? *The McKinsey Quarterly, 4*, 71-80.

Anderson, T. (2004). Teaching in an online learning context. In T. Anderson & F. Elloumi (Eds.), *Theory and practice of online learning* (pp. 273-294). Athabasca, AB: Athabasca University.

Applefield, J. M., Huber, R., & Mahnaz, M. (2000). Constructivism in theory and practice: Toward a better understanding. *High School Journal., 84*(2), 35-59.

Bonoma, T. (1985). Case research in marketing: Opportunities, problems and a process. *Journal of Marketing Research, 12*, 199-208.

Brown, A. L., & Campione, J. C. (1994). Guided discovery in a community of learners. In K. McGilly (Ed.), *Classroom lessons: Integrating cognitive theory and classroom practice* (pp. 229-272). Cambridge, MA: MIT Press/Bradford Books.

Carlen, U., & Jobring, O. (2005). The rationale of online learning communities. *International Journal of Web Based Communities, 1*(3), 272-295.

Copley, J. (1992). The integration of teacher education and technology: A constructivist model. In D. Carey, R. Carey, D. Willis & J. Willis (Eds.), *Technology and teacher education annual – 1992* (pp. 617-622). Charlottesville, VA: Association for the Advancement of Computing in Education.

Danaher, P. A. (2006, November). The social control–social capital debate about distance and online teacher education: Critical reflections on two courses at the University of Southern Queensland, Australia. *Journal of Open Learning and Teacher Education, 1*(3), 81-93.

Danaher, P. A., Hickey, A., Brown, A., & Conway, J. M. (2007). Exploring elements for creating an online community of learners within a distance education course at the University of Southern Queensland. In R. J. Luppicini (Ed.), *Online learning communities* (pp. 219-240). Charlotte, NC: Information Age Publishing.

Denning, P. (2004, April 21-27). Building a culture of innovation. *Ubiquity, 5*(8), 1-13. Retrieved May 31, 2004, from http://www.acm.org/ubiquity/issues5.html

Dimitrova, M., Sadler, C., Hatzipanagos, S., & Murphy, A. (2003). Addressing learner diversity by promoting flexibility in e-learning environments. In *Proceedings of the 14th international workshop on database and expert systems applications.*

Gay, G., Stefanone, M., Grace-Martin, M., & Hembrooke, H. (2001). The effects of wireless computing in collaborative learning environments. *International Journal of Human–Computer Interactions, 13*(2), 257-276.

Goggin, G. (2005, March). Mobile learning. *Professional Educator, 4*(1), 35-37.

Goldman, P., & Kaufman, B. J. (2001). How to push an elephant through a straw: Using wireless technology in a Web-enhanced skills program. *International Review of Law, Computers & Technology, 15*(3), 281-299.

Goodison, T. A. (2001). The implementation of e-learning in UK higher education. In *Proceedings of world conference on educational media, hypermedia and telecommunications 2001* (pp. 613-618). Chesapeake, VA: AACE.

Greeno, J. G. (1997). On claims that answer the wrong questions. *Educational Researcher, 26*(1), 5-17.

Grohmann, G., Hofer, A., & Martin, G. (2005, July 11). *ARIS MOBILE: Helping to define the future of mobile learning.* Paper presented at the 4th international conference on mobile business, University of Technology Sydney, Sydney, NSW.

Gururajan, R., Kerr, D., Moloney, C., & Houghton, L. (2005). An investigation into the factors of adoption of wireless applications for data management by nurses. In L. Eldon (Ed.), *Collaborative decision making in the internet era* (pp. 1-5). Asia-Pacific Decision Sciences Institute.

Hafeez-Baig, A., & Danaher, P. A. (2007, February). Future possibilities for mobile learning technologies and applications at the University of Southern Queensland, Australia: Lessons from an academic focus group. In D. Parsons & H. Ryu (Eds.), *Mobile learning technologies and applications (MoLTA) 2007: Proceedings of the conference on mobile learning technologies and applications (MoLTA) Massey University, Auckland, New Zealand February 19th 2007* (pp. 16-22). Auckland, New Zealand: Institute of Information and Mathematical Sciences, Massey University.

Harreveld, R. E., & Danaher, P. A. (2004, November). Innovation in open and distance learning and teacher education: The case of pre-service secondary vocational education and training at an Australian regional university. *Journal of Open Learning and Teacher Education, 1*(1), 1-13.

Hiltz, S. R., Coppola, N., Rotter, N., & Turoff, M. (2000, September). Measuring the importance of collaborative learning for the effectiveness of ALN: A multi-measure, multi-method approach. *Journal of Asynchronous Learning Networks, 4*(2). Retrieved June 25, 2006, from http://www.sloan-c.org/publications/jaln/v4n2/v4n2_hiltz.asp

Hodas, S. (1993, September 14). Technology refusal and the organizational culture of schools. *Education Policy Analysis Archives, 1*(10). Retrieved April 9, 2005, from http://epaa.asu.edu/epaav1n10.html

Hoppe, U., Lingnau, A., Machado, I., Paiva, A., Prada, R., & Tewissen, F. (2000). Supporting collaborative activities in computer integrated classrooms – the NIMIS approach. In P. Antunes & C. Salgado (Eds.), *Proceedings of CRIWG 2000* (pp. 94-101). Los Alamitos, CA: IEEE Computer Society Publication.

Howard, A., Hafeez-Baig, A., Howard, S., & Gururajan, R. (2006). A framework for the adoption of wireless technology in healthcare: An Indian study. In S. Spencer & A. Jenkins (Eds.), *Proceedings of the 17th Australasian conference on information systems (ACIS 2006).* Australasian Association for Information Systems.

Johnston, R. (1999). Beyond flexibility: Issues and implications for higher education. *Higher Education Review, 32*, 55-67.

Kehrwald, B. A. (2006, June). *Inhabiting online spaces: Reconsidering social presence in contemporary online learning environments.* Paper presented at the 4th international lifelong learning conference, Yeppoon, Qld.

Leedy, P. D., & Ormrod, J. E. (2004). *Practical research: Planning and design.* Upper Saddle River, NJ: Pearson Education.

Leem, C. S., Suh, H. S., & Kim, D. S. (2004). A classification of mobile business models and its applications. *Industrial Management & Data Systems, 104*, 78-87.

Liu, T. C., Wang, H. Y, Liang, J. K., Chan, T. W., Ko, H. W., & Yang, J. C. (2003). Wireless and mobile technologies to enhance teaching and learning. *Journal of Computer Assisted Learning, 19*, 371-382.

Looi, C.-K. (2005). Exploring the affordances of online chat for learning. *International Journal of Learning Technology, 1*(3), 322-338.

Madden, K. M. (2000). The role of communication in building customer relationships in the Australian financial services industry: An in-depth study. Unpublished Doctor of Philosophy thesis, Faculty of Business, University of Southern Queensland, Toowoomba, Qld.

Merriam, S. B. (1988). *Case study research in education: A qualitative approach.* San Francisco: Jossey-Bass Publications.

Miles, M. B., & Huberman, A. M. (1994). *Qualitative data analysis: An expanded sourcebook* (2nd ed.). Thousand Oaks, CA: Sage Publications.

Morrison, G. R., & Lowther, D. L. (2005). *Integrating computer technology into the classroom* (3rd ed.). Upper Saddle River, NJ: Prentice Hall.

Neuman, W. L. (2003). *Social research methods: Qualitative and quantitative approaches* (5th ed.). Boston, MA: Allyn & Bacon.

Patton, M. (1990). *Qualitative methods.* Newbury Park, CA: Sage Publications.

Payne, F., Ball, J., & Snow, R. (2000). A corporate approach to the introduction of flexible delivery education. In *Proceedings of the 30th ASEE/IEEE conference: Frontiers in education.* Kansas City, MO.

Perry, D. (2003). *Handheld computers (PDAs) in schools.* Coventry, UK: British Educational Communications and Technology Agency.

Riege, A. (1996). Marketing communication of international travel and tourism: A study of UK and German markets for Australia and New Zealand. Unpublished Doctor of Philosophy thesis, Queensland University of Technology, Brisbane, Qld.

Robson, S. (1993, Spring). The qualitative survey. *Survey,* 13-14.

Rushby, N. J. (1998). Learning without strings: Exploiting mobile technologies for flexible learning. *ACM-SIGCUE Outlook, 26*(2). Retrieved April 29, 2007, from http://www.eee.bham.ac.uk/handler/ePapers/rushby_1998.pdf

Schellekens, A., Paas, F., & van Merriënboer, J. J. G. (2003, April). Flexibility in higher professional education: A survey in business administration programmes in the Netherlands. *Higher Education, 45*(3), 281-305.

Stone, A., & Livingstone, D. (2004). Don't lose the learning from learning objects! A request for care in the adaptive context community. *International Journal of Continuing Engineering Education and Lifelong Learning, 14*(4/5), 318-330.

Tewissen, F., Lingnau, A., Hoppe, U., Mannhaupt, G., & Nischk, D. (2001). Collaborative writing in a computer-integrated classroom for early learning. In P. Dillenbourg, A., Eurelings & K. Hakkarainen (Eds.), *Proceedings of Euro-CSCL 2001* (pp. 593-600). Mahwah, NJ: Lawrence Erlbaum Associates.

Van Meel, R. M. (1992). *Perspectives on organization and management of higher education (OTIC research report 49).* Heerlen, the Netherlands: Open University of the Netherlands.

Weiser, M. (1998). The future of ubiquitous computing on campus. *Communications of the ACM, 41*(1), 42-43.

Wiredu, G. O. (2006). Perception, mobility and the affordances of portable computers: An historical epistemology of mobile computing. Retrieved July 29, 2007, from http://www.carstensorensen.com/download/Wiredu2006.pdf

Yin, R. K. (2002). *Case study research: Design and methods* (3rd ed.). Newbury Park, CA: Sage Publications.

## ADDITIONAL READING

Brown, K. M. (1996). The role of internal and external factors in the discontinuation of off campus students. *Distance Education, 17*(1), 44-71.

Bundy, A. (2000, February 13-14). *A window of opportunity: Libraries and learning in the 21st century.* Paper presented at the Australian Library and Information Association Distance Education Special Interest Group conference, Monash University, Gippsland, Vic.

Carlen, U., & Jobring, O. (2005). The rationale of online learning communities. *International Journal of Web Based Communities, 1*(3), 272-295.

Chen, Y. S., Kao, T. C., Yu, G. J., & Sheu, J. P. (2004). A mobile butterfly-watching learning system for supporting independent learning. In *Proceedings of the IEEE Computer Society 2nd international workshop on wireless and mobile technologies in education* (pp. 11-18). JungLi, Taiwan.

Clarke, B., & James, C. (1998). Flexibility in post-registration nurse education in England. *Innovations in Education and Training International., 35*(4), 292-301.

Collis, B., Moonen, J., & Vingerhoets, J. (1997, July). Flexibility as a key construct in European training: Experiences from the TeleScopia project. *British Journal of Educational Technology, 28*(3), 199–217.

Copley, J. (1992). The integration of teacher education and technology: A constructivist model. In D. Carey, R. Carey, D. Willis & J. Willis (Eds.), *Technology and teacher education annual – 1992* (pp. 617-622). Charlottesville, VA: Association for the Advancement of Computing in Education.

Elliott, G. (1999). *Lifelong learning: The politics of the new learning environment,* London: Jessica Kingsley Publishers.

Fozdar, B. I., & Kumar, L. S. (2007, June). Mobile learning and student retention. *International Review of Research in Open and Distance Learning, 8*(2). Retrieved June 20, 2007, from http://www.irrodl.org/index.php/irrodl/article/view/345/916

Gay, G., Stefanone, M., Grace-Martin, M., & Hembrooke, H. (2001). The effects of wireless computing in collaborative learning environments. *International Journal of Human–Computer Interactions, 13*(2), 257-276.

Gururajan, R. (2004). Emergence of the need of new legal & regulatory issues on mobile workforce. *International Journal of Applied Management & Technology, 2*(1), 71-87.

Gururajan, R., Kerr, D., Moloney, C., & Houghton, L. (2005). An investigation into the factors of adoption of wireless applications for data management by nurses. In L. Eldon (Ed.), *Collaborative decision making in the Internet era* (pp. 1-5). Asia-Pacific Decision Sciences Institute.

Hafeez-Baig, A., & Danaher, P. A. (2007, February). Future possibilities for mobile learning technologies and applications at the University of Southern Queensland, Australia: Lessons from an academic focus group. In D. P. Parsons & H. Ryu (Eds.), *Mobile learning technologies and applications (MoLTA) 2007: Proceedings of the conference on mobile learning technologies and applications (MoLTA) Massey University, Auckland, New Zealand February 19th 2007* (pp. 16-22). Auckland, New Zealand: Institute of Information and Mathematical Sciences, Massey University.

Haug, G., Kirstein, J., & Knudsen, I. (1999). Trends in learning structures in higher education [Project report]. Confederation of European Union Rectors' Conferences and the Association of European Universities. Retrieved October 31, 2000, from http://www.rks.dk/

Hiltz, S. R., Coppola, N., Rotter, N., & Turoff, M. (2000, September). Measuring the importance of collaborative learning for the effectiveness of ALN: A multi-measure, multi-method approach. *Journal of Asynchronous Learning Networks, 4*(2). Retrieved June 25, 2006, from http://www.sloan-c.org/publications/jaln/v4n2/v4n2_hiltz.asp

Howard, A., Hafeez-Baig, A., Howard, S., & Gururajan, R. (2006). A framework for the adoption of wireless technology in healthcare: An Indian study. In S. Spencer & A. Jenkins (Eds.), *Proceedings of the 17th Australasian conference on information systems (ACIS 2006).* Australasian Association for Information Systems.

Kehrwald, B. A. (2006, June). *Inhabiting online spaces: Reconsidering social presence in contemporary online learning environments.* Paper presented at the 4th international lifelong learning conference, Yeppoon, Qld.

Kirkpatrick, D., & McLaughlan, R. (2000). Flexible lifelong learning in professional education. *Education and Information Technologies, 3*(1), 24-31.

Ko, H. W., & Liu, T. C. (2001). When technology comes into classrooms. In G. Chen & J. Yang (Eds.), *Proceedings of GCCCE/ICCAI 2001* (pp. 17-20). Chung-Li, Taiwan: NCU Publications,

Kukulska-Hulme, A. (2007, June). Mobile usability in educational contexts: What have we learnt? *International Review of Research in Open and Distance Learning, 8*(2). Retrieved June 20, 2007, from http://www.irrodl.org/index.php/irrodl/article/view/356/879

Kukulska-Hulme, A., & Traxler, J. (2005), *Mobile learning: A handbook for educators and trainers.* London: Routledge.

Liu, T. C., Wang, H. Y., Liang, J. K., Chan, T. W., & Yang, J. C. (2002). Applying wireless technologies to build a highly interactive learning environment. In M. Milrad, U. Hoppe & Kinshuk (Eds.), *Proceedings of IEEE international workshop on mobile and wireless technologies in education (WMTE 2002)* (pp. 63-70). Los Alamitos, CA: IEEE Publications.

Looi, C.-K. (2005). Exploring the affordances of online chat for learning. *International Journal of Learning Technology, 1*(3), 322-338.

Martin, E. (1999). *Changing academic work: Developing the learning university.* Buckingham, UK: The Society for Research into Higher Education.

Meade P. H. (1997). *Challenges facing universities: Quality, leadership and the management of change.* Dunedin, New Zealand: University of Otago.

Rushby, N. J. (1998). Learning without strings: Exploiting mobile technologies for flexible learning. *ACM-SIGCUE Outlook, 26*(2). Retrieved April 29, 2007, from http://www.eee.bham.ac.uk/handler/ePapers/rushby_1998.pdf

Schellekens, A., Paas, F., & Van Merriënboer, J. J. G. (2003, April). Flexibility in higher professional education: A survey in business administration programmes in the Netherlands. *Higher Education, 45*(3), 281-305.

Stone, A., & Livingstone, D. (2004). Don't lose the learning from learning objects! A request for care in the adaptive context community. *International Journal of Continuing Engineering Education and Lifelong Learning, 14*(4/5), 318-330.

Taylor, P., & Joughin, G. (1997). *What is flexible learning? (Teaching through flexible learning resources series).* Brisbane, Qld: Griffith Institute of Higher Education, Griffith University.

Taylor, P., Peters, J., & Parra, D. (1998). *Supporting students for flexible learning (Teaching through flexible learning resources series).* Brisbane, Qld: Griffith Institute of Higher Education, Griffith University.

Traxler, J. (2007, June). Defining, discussing and evaluating mobile learning: The moving finger writes and having writ.... *International Review of Research in Open and Distance Learning, 8*(2). Retrieved June 20, 2007, from http://www.irrodl.org/index.php/irrodl/article/view/346/875

Van Meel, R. M. (1997). Management of flexible responses in higher education. Heerlen, The

Netherlands: Centre for Educational Technology and Innovation, Open University of the Netherlands.

Wiredu, G. O. (2006). Perception, mobility and the affordances of portable computers: An historical epistemology of mobile computing. Retrieved July 29, 2007, from http://www.carstensorensen.com/download/Wiredu2006.pdf

# Section II
# Enhancing Individual Learning Experiences

# Chapter III
# Understanding the Value of Interactive SMS for Large Classes

**Eusebio Scornavacca**
*Victoria University of Wellington, New Zealand*

**Sid Huff**
*Victoria University of Wellington, New Zealand*

**Stephen Marshall**
*Victoria University of Wellington, New Zealand*

## ABSTRACT

*The increasing sophistication and rapid uptake of mobile phones among university students is providing an unprecedented platform for the development of classroom interaction systems. This chapter describes the development of a SMS-based classroom interaction system and explores the impact that this application can have on students' learning experience. The findings indicate that instructors and students perceive a number of benefits from the additional channel of communication in the classroom. The chapter concludes with a summary and recommendations for future practice and research.*

## INTRODUCTION

Classroom interactivity has a number of significant benefits: it promotes an active learning environment, provides greater feedback for lecturers, increases student motivation, and enables a learning community (Angelo & Cross, 1993; Bishop, Dinkins, & Dominick, 2003; Hake, 1998; Mazur, 1998; McConnell et al., 2006). On the other hand, interactive activities for large classes (for instance, over 100 students) have proven to be quite difficult and, often, inefficient (Freeman & Blayney, 2005).

During the past six years, the rapid proliferation of mobile devices, particularly cellular phones, in the student demographic has changed the levels of student access to information and communications technology (ICT) in the classroom, presenting an extraordinary opportunity to develop interactive classroom systems and to enhance students' learning experience (Schwabe, Göth, & Frohberg, 2005; Scornavacca, Barnes, & Huff, 2006; Sheng, Nah, & Siau, 2005).

The present challenge for researchers is to go beyond anecdotal perceptions and obtain empirical evidence about the impact of these technologies in the classroom. This chapter aims to describe the development of a classroom interaction system as well as to understand the impact that mobile applications such as short-message-services (SMS) can have on students' learning experience.

The chapter is structured as follows: the next section illustrates the theoretical background used in this research. This is followed by a brief description of the system as well as the trial that took place. Subsequently a discussion of the survey results is presented. Finally, the chapter rounds off with a summary and conclusions.

## CLASSROOM INTERACTIVE FEEDBACK SYSTEMS

The traditional lecture theatre environment has provided universities with a cost effective and scaleable means of teaching students. Particularly, over the last century, as participation in university education has grown, the lecture theatre has coped with the growth in student numbers, but this growth has come at the price of making any form of interaction difficult and inefficient. The work of Laurillard (2002) and social constructivist theorists (Jonassen, Peck, & Wilson, 1999) suggest that a further price of the traditional lecture delivery is reduced engagement and student learning. This is not inevitable, however. The work of Mazur (1998) suggests that the use of interactive pedagogies that are able to engage students in peer discussion and activities, guided by the lecturer, can result in effective student learning outcomes. Using interactive classroom pedagogies it is possible to promote a more active learning environment, increase the motivation of students, inform the work of teachers and generally enable a genuine learning community in the classroom (Angelo & Cross, 1993; Bishop et al., 2003; Hake, 1998; Mazur, 1998; McConnell et al., 2006).

The decision to engage in a more interactive pedagogy is, however, not a trivial one, and it requires effort and application on the part of the teaching staff. Large classes of more than 100 students are clearly where the most benefits of a more interactive approach can be realized but the inherent inefficiency, inaccuracy and lack or anonymity present significant problems (Freeman & Blayney, 2005).

Classroom feedback systems (CFS) provide one possible technological affordance that can efficiently enable interaction in large classes. Known by a vast array of names and produced commercially by a range of vendors, CFS technologies have been used since the sixties (Judson & Sawada, 2002; Penuel, Roschelle, & Abrahamson, 2005) allowing students to respond to questions and have the results processed and displayed for use by the lecturer and the class as a whole. Modern systems provide the ability to answer a range of question types from simple yes/no through to detailed responses, free-form questions and role-playing (Bollen, Eimler, & Hoppe, 2004). Other media such as images are now also being used in particular contexts (Seppälä & Alamäki, 2003). The hardware used now ranges from small infra-red units like those used with televisions, through more reliable radio units, to the use of Web systems accessed by wireless personal digital assistants (PDA) or laptops.

These systems are generally well regarded by students when they are used, and many case studies describe the use of CFS technologies in disciplines ranging from the physical sciences (physics, chemistry, geology), through mathematics, accountancy and literature (Blodgett, 2006; Bollen et al., 2004; Carnaghan & Webb, 2005; Draper & Brown, 2004; Freeman & Blayney, 2005; Hall, Collier, Thomas, & Hilgers, 2005; Mazur, 1998; McConnell et al., 2006). There is also some evidence to suggest that teachers also benefit, not only from a better understanding of what their student's current knowledge is, but also by gaining the ability to vary their teaching in response to information from the students (Bishop et al., 2003; McConnell et al., 2006).

A variety of positive outcomes from the use of CFS technologies have been reported including improved understanding of important concepts (Mazur, 1998; Penuel et al., 2005), increased student engagement and participation (Freeman & Blayney, 2005; Penuel et al., 2005), improved quality of discussion in the classroom (Penuel et al., 2005) and a better teacher awareness of student difficulties (Penuel et al,, 2005).

However, not all studies of CFS have been positive. Using CFS technology without specific pedagogical changes appears to have no benefit on measured student performance and learning outcomes, despite the positive perception of the technology reported by the students (Blodgett, 2006; Carnaghan & Webb, 2005; Judson & Sawada, 2002; Penuel et al., 2005). It is clear that asking questions like "do you understand this" are not going to result in useful outcomes for either the student or the lecturer (Blodgett, 2006). Further, Slain et al. (2004) noted that when using CFS technologies, student outcomes in assessment tasks that operated at higher cognitive levels (Bloom, 1956) showed a much more significant improvement than the outcomes from assessment tasks that only test memorization. This suggests that CFS technologies have less value as a tool for supporting recall as opposed to understanding and thus will not be of great value when rote memorization is the only outcome expected from students. There is also the risk that the increased student engagement will expose the limitations of the lecturer, both in terms of their own depth of understanding but also in their ability to cope with a more dynamic classroom environment (Hall et al., 2005). Penuel et al. (2005) contends that the CFS technology does provide a vehicle for positive change, but that this will only occur if the actions of the teacher using the technology changes the culture of the classroom in a such a way as to support the student's engagement and peer interaction.

As well as the pedagogical challenges, the use of CFS hardware also introduces some difficulties for students and institutions. Historically, CFS systems depended on a significant wired infrastructure that was commonly built into lecture theatres (Judson & Sawada, 2002). The introduction of wireless units has helped reduce that need, but now means that the students either have to own or rent their own units or use a unit provided in the lecture. Either approach has problems, with students either losing or forgetting the units, or institutions needing to provide teaching assistants who hand out and collect units in class (Draper & Brown, 2004). Requiring students to purchase the hardware, even in environments where deals are made with textbook publishers to subsidize the equipment, can be problematic for student budgets (Greer & Heaney, 2004) and this is likely to be even more of a problem if wireless PDAs or Laptops are used.

Personal Web-enabled devices such as PDAs have a number of problems of their own which can reduce their utility as CFS devices. The larger display and more flexible programming environment means that PDAs can be used to provide a rich set of student feedback but this comes at the price of significant issues such as battery life and a complicated and unreliable wireless configuration (Bishop et al., 2003). There is also a problem, shared with laptops, that these devices are more capable and flexible which increase the temptation for students to engage in off-task activities (Ward, Reeves, & Heath, 2003). This latter issue may also be a symptom of a more general issue with some students preferring the anonymity of a traditional lecture that allows them to engage in activities other than participating in the class (Duncan, 2005; Mazur, 1998).

In the New Zealand context, despite significant growth in access to computers and wireless Internet access, students tend not to own portable devices such as PDAs and laptops (Barnes & Scornavacca, 2005). The preferred mobile communication device is undoubtedly the mobile phone (InfoSynch World, 2004). As with most of the world, mobile phone ownership and the use of short message service (SMS) communication has rocketed up over the past five years (Haghirian, Madlberger, & Tanuskova, 2005; Scornavacca & McKenzie, 2006). Mobile phones are regarded as essential devices by the student demographic and have the advantages of being familiar, permanently configured to work correctly, and battery lives generally measured in days rather than hours (Carroll, Barnes, & Scornavacca, 2005; Gopal & Tripathi, 2006; Scornavacca, Barnes, & Huff, 2006).

The potential advantages of mobile phones as educational devices enabling a form of mobile learning or m-learning outside of the physical classroom are well described in the literature (Quinn, 2000). Generally these have been aimed at the mobile professional, engaging in informal or on-demand learning but there are examples built around more formal e-learning settings such as the linking of Web discussions to mobile phones via filters and triggers. In one of the few examples of classroom use of mobile phones, Markett et al. (2006) describes the use of SMS to collect text in the form of a semi-structured discussion during class that can then be continued in a more traditional Web-based discussion afterwards. Sadly however, the problems caused by disruptive mobile phone use have more often led to the banning of mobile phones in classrooms and the use of jamming equipment. This can lead to unusual solutions such as that of Bollen (2004) who reported the development of an SMS-like system on PDAs in order to avoid restrictions on using mobile phones in class.

## THE TXT-2-LRN SYSTEM

The development of the TXT-2-LRN (text-to-learn) system is based on the assumption that nowadays most students have a SMS enabled mobile phone and that they bring it to the classroom. Therefore most of the necessary ICT infrastructure for a classroom interaction system is already in place. In order to take advantage of this opportunity it was necessary to enable the instructor to receive messages from students while lecturing. A very simple solution was adopted: it comprises of a mobile phone connected to the instructor's laptop and the installation of a SMS management tool, SMS Studio, which supports SMS-based voting,

*Figure 1. TXT-2-LRN infrastructure*

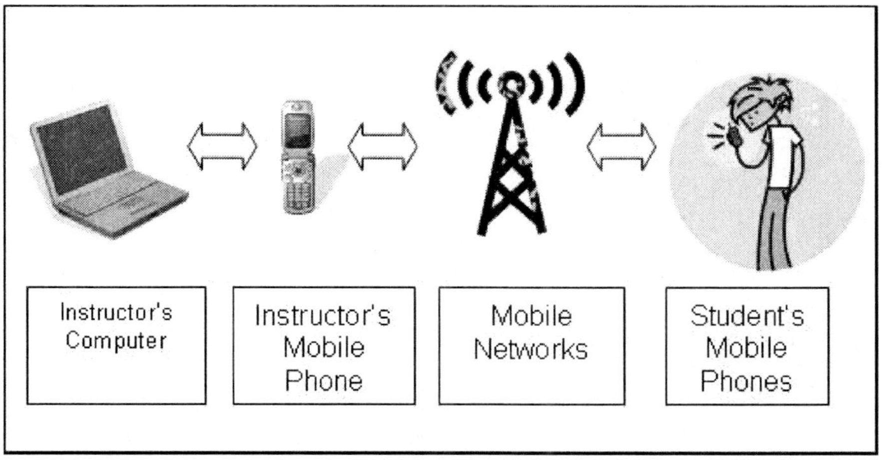

chat, information, quizzes and other services (see www.codesegment.com). An alternative to this set-up would be using an SMS-gateway instead of the phone. The SMS software enables the instructor to easily read incoming SMS messages on the computer screen as well as automatically analyze the results of polls. In addition, it also allows the instructor to send messages to any mobile phone. Figure 1 presents a diagram of the system's basic infrastructure:

Based on the literature review on mobile technologies, ICT in the classroom and pedagogy, two new classroom dynamics were designed:

Open Channel—This allows students to send questions or comments to the instructor's laptop via SMS without interrupting the class. The instructor was able to read the messages on the laptop screen and decide whether or when would be appropriate to comment on the message received. The instructor was able to easily manage the student messages by chunking them together and reviewing them at class "checkpoints," the points during a class when the instructor pauses, asks (verbally) whether there are any questions or comments, then proceeds to the next topic or segment. In order to make sure that no valid question or comment was overlooked, replies to questions that were not addressed during class could then be supplied by the instructor afterwards via SMS or a traditional discussion forum. Figure 2 presents how students' messages were displayed on the instructor's laptop.

M-quiz—during the class, the instructor presented a slide containing a question related to the topic and four options (A, B, C, D). Students discussed the question and were then able to use their mobile phones to cast votes and were able to see on the projector screen real-time graphics showing the results. This feature is used mainly for concept tests (Mazur, 1998) or to trigger class discussions. Figure 3 presents how the M-quiz results were displayed on the instructor's laptop,

*Figure 2. Screenshot - Open channel*

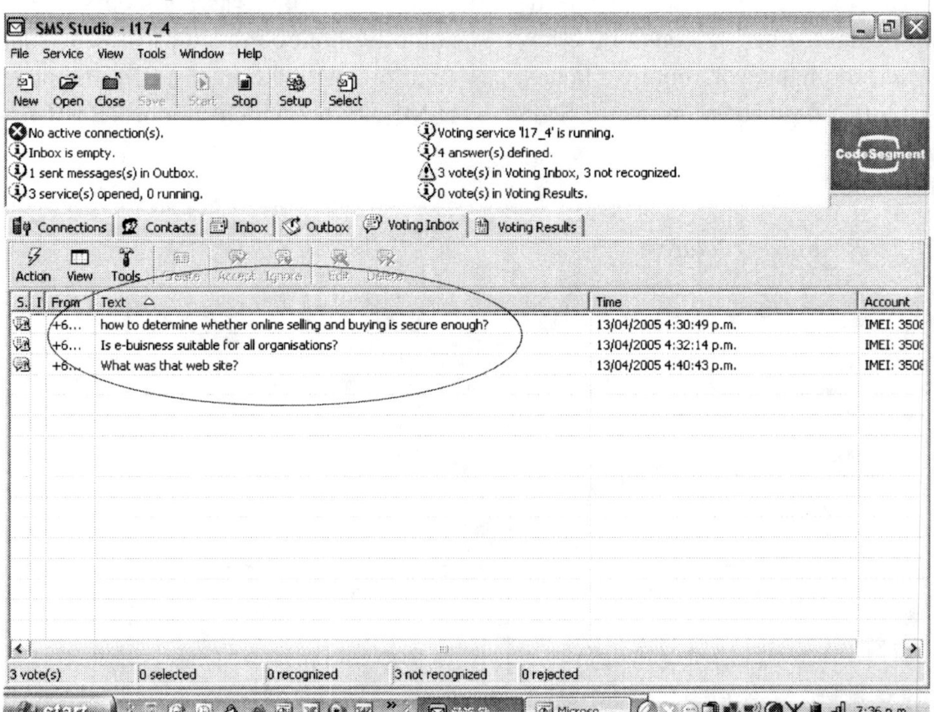

*Figure 3. Screenshot – M-Quiz*

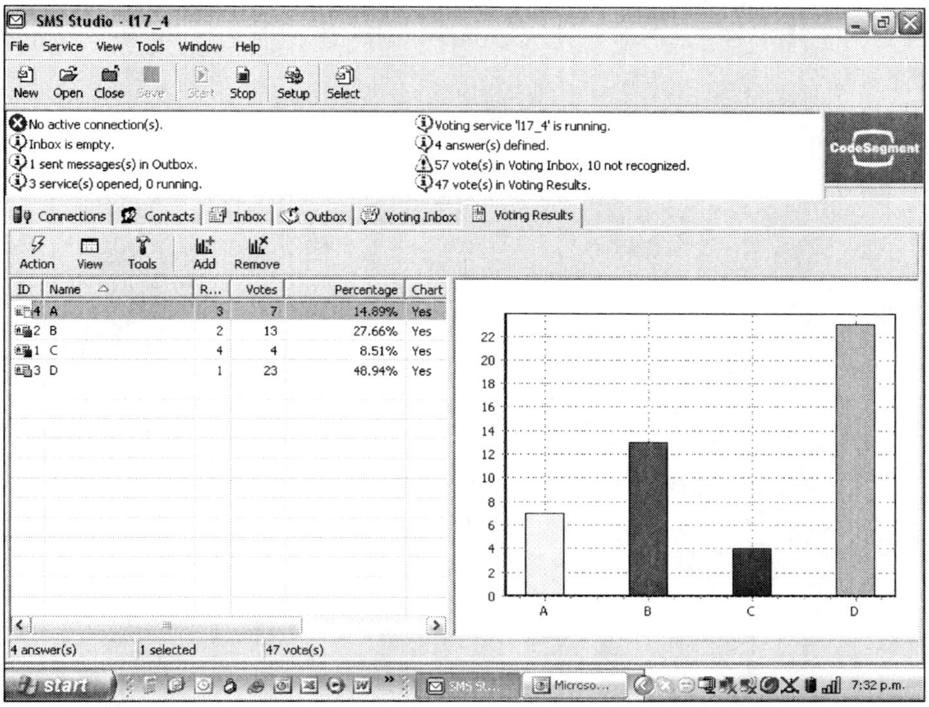


a separate screen containing only the graph was displayed to the students (using the projector).

Figure 4 presents an overview of the classroom activities using the system. Notice that in the beginning of each lecture a slide was presented reminding students that if they have a question or comment they may raise their hand or send an SMS. The instructor's mobile phone number also appeared in a corner of all remaining presentation slides. The interactive quiz (M-quiz) happened only once per lecture and students were invited to discuss their answers with the people next to them before casting their votes. In addition, students were informed that by sending a message their mobile phone number would be disclosed, however their identity would remain anonymous.

## USING TXT-2-LRN IN THE CLASSROOM

FCOM110 is a mandatory course for all bachelor of commerce and administration (BCA) students, which introduces them to the major areas of study of the faculty of commerce at Victoria University of Wellington, New Zealand. There are about 1,200 students enrolled in this course, making it one of the largest cohorts in the university. One-hour lectures were delivered three times a week to four groups of approximately 300 students. In April, 2005, during the five lectures on information technology, students were introduced to the TXT-2-LRN system and actively encouraged to use their mobile phones in class. During the trial,

*Figure 4. Overview of TXT-2-LRN classroom activities*

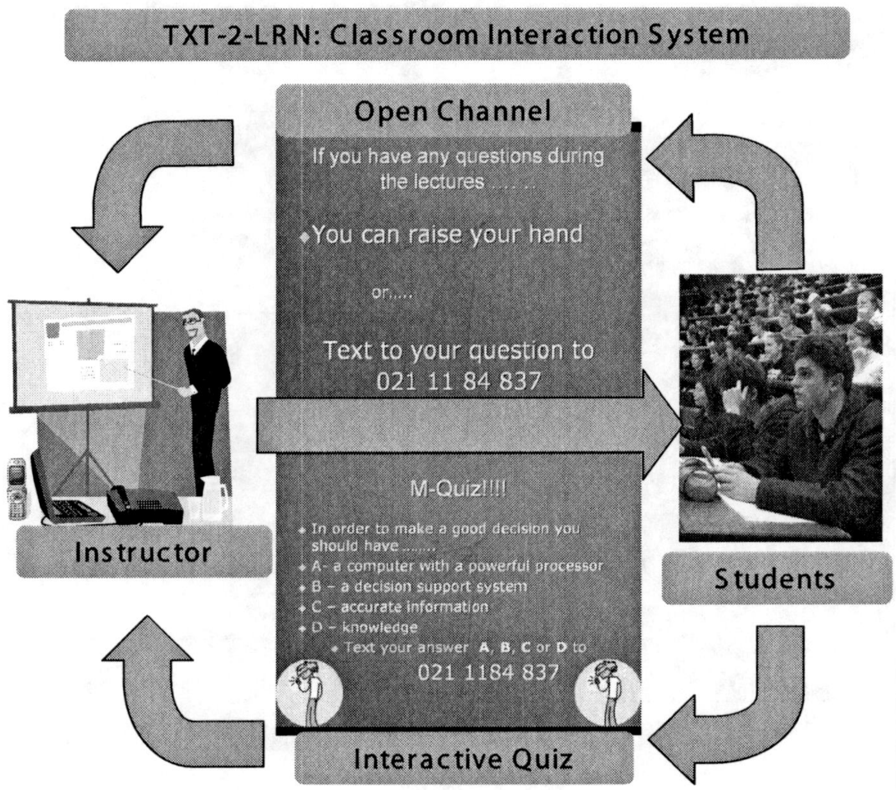


using the system was voluntary and students paid for their own messages. At the end of the fifth lecture, 600 students were randomly selected to answer a survey containing 22 questions measuring mobile phone usage, user acceptance of the system, and perceived impact on students' learning experience. A total of 569 valid questionnaires were returned. The questionnaire was developed in conjunction with Victoria University's Teaching Development Centre (UTDC) and it received face and content validation. The results are described in the following section.

## RESULTS

The sample presented an approximately equal gender distribution (52.7% female, 47.3% male), mostly aged between 18 and 22 years (72.9%), who were predominantly in their first year of studies (68%), and mainly native English speakers (64%). Most students (79.3%) attended more than four out of the five lectures involving the trial. These results are consistent with the overall student population enrolled in the course.

Almost all students (96.8%) owned a phone that they often carried in classes. Most students (64.9%) sent more than 14 messages per week and spent on average $35 per month with their phones. New Zealand has two mobile phone operators that offer quite different price structures on their SMS services. Vodafone NZ charges 20 cents per SMS, while Telecom NZ offers a flat rate of $10 per month for up to 500 messages. In our sample, despite the price difference favoring Telecom for heavy SMS users, Vodafone's market share was slightly greater (54.5%), and the majority of students (82.8%) used pre-paid cards instead of a monthly plan.

During the trial, approximately a quarter of the students used the open channel to send a question or comment in classes. However, over 90 percent of them perceived that the ability to send the instructor SMS's during class was in one way or another useful. The instructor involved in the trial had a very positive experience using the system and perceived an overall gain in quality and quantity of student feedback during class. In addition, the number of comments and questions received via the system (about 15 per class) was valuable without being disruptive. It is important to highlight that approximately 130 students at some point of time during the five lectures actively contributed with comments and questions. That is a substantial improvement for this large class environment where questions are not normally asked by the students.

Students also indicated their reasons for not using the open channel. Most (42.5%) responded that they had nothing to say, 33.6 percent gave cost as the main inhibiting factor, while only 4.6 percent believed it to be distracting. Analysis of results showed a highly significant dependence between cost as an inhibiting factor for using the open channel and the students' mobile phone operator. Approximately 40 percent of the Vodafone subscribers indicated cost was a factor compared to 26 percent of Telecom subscribers (Chi2= 13.96, d.f = 4, 1-p = 99.26).

Most of the students who used the open channel also participated in the M-Quizzes. The participation in M-Quizzes was higher than use of the open channel. More than half of the students entered at least one M-Quiz and 80 percent of them perceived that the M-Quizzes were somehow useful. On the other hand, the instructor noticed that this activity provided several benefits for the classroom environment (e.g., instantaneous feedback on concept tests or using results to stimulate class discussion). It was also noticed that students seemed to be very interested in the result of polls that reflected their collective opinion on a given issue. In the case of the M-quizzes, student participation was mainly inhibited by cost (55%) and lack of interest (20.7%). A small number of Telecom NZ users also reported having some issues regarding network availability.

So far, the survey questions have examined largely descriptive aspects of students' mobile phone usage as well as drivers and inhibitors for using the TXT-2-LRN. The remaining questions examine the impact of this application on respondents' learning experience. Table 1 summarizes the results. For these questions, respondents were asked to answer questions using a Likert scale from 1 (strongly disagree) to 5 (strongly agree). Reliability analysis was carried out to investigate the internal consistency of the scale used in this part of the questionnaire (Crombach's alpha = 0.92). A non-parametric sign test showed that the mean response to each of the first six questions was statistically significantly (at $p<0.001$) greater than the neutral point on the scale, in particular the increased engagement and interactivity, improved classroom discussions and the ability of the lecturer to react to the student's feedback effectively (Freeman & Blayney, 2005; Penuel et al., 2005). The perception that the system had less impact on their interest in the subject and the study of it is also consistent with many other studies (Blodgett, 2006; Carnaghan & Webb, 2005; Judson & Sawada, 2002; Penuel et al., 2005).

Response patterns were analyzed for co-dependence to determine the existence of socio-demographic effects. Gender, age, year of study and mobile phone usage did not present any significant dependence. However, students' native language not being English was a significant variable. The expectation was that English-as-a-second-language (ESL) students would find the system more useful than native English speaking students. This assumption was based on the idea that most of the ESL students would be more comfortable

*Table 1. Top ranked impacts of TXT-2-LRN in the classroom*

| Questions | Mean | s.d | Rank | Sig. |
|---|---|---|---|---|
| Using txt messages increased the levels of interaction in class. | 4.05 | 1.02 | 1 | p < 0.001 |
| Using txt messages during class made the classes more interesting. | 4.04 | 1.00 | 2 | p < 0.001 |
| Using txt messages in the classroom is a good idea. | 3.83 | 1.06 | 3 | p < 0.001 |
| I found this interaction method effective. | 3.76 | 1.03 | 4 | p < 0.001 |
| In general, I liked using txt facilities as part of this course. | 3.67 | 1.10 | 5 | p < 0.001 |
| I would like to see more use of txt facilities to assist my studies. | 3.67 | 1.09 | 5 | p < 0.001 |
| The use of txt messages during class increased my interest in the subject. | 3.58 | 1.06 | 6 | p < 0.001 |
| The use of txt messages during class enhanced my study. | 2.95 | 1.01 | 7 | --- |

interacting via the SMS channel as it would give them more opportunities to express themselves clearly (Elgort, Marshall, & Mitchell, 2003). Surprisingly, the result of the analysis found that native English speakers had a significantly more positive overall perception of the system than the ESL counterparts.

Students were also asked to indicate their preferred of method of communication if they would like to ask the lecturer a question (traditional vs. SMS). For this question, respondents were asked to answer on a Likert scale from 1 (raise my hand) to 5 (send an SMS). SMS appeared to be the strongly preferred method of communication for asking questions. Figure 5 summarizes the results.

Finally, students were asked for their opinion regarding the class size that this system would be most appropriate for. 74.8 percent believed that it should be used mostly in large classes.

## CONCLUSION

The rapid proliferation of mobile phones among the student population is generating a novel platform for the development of classroom interac-

*Figure 5. Preferred method of communication*

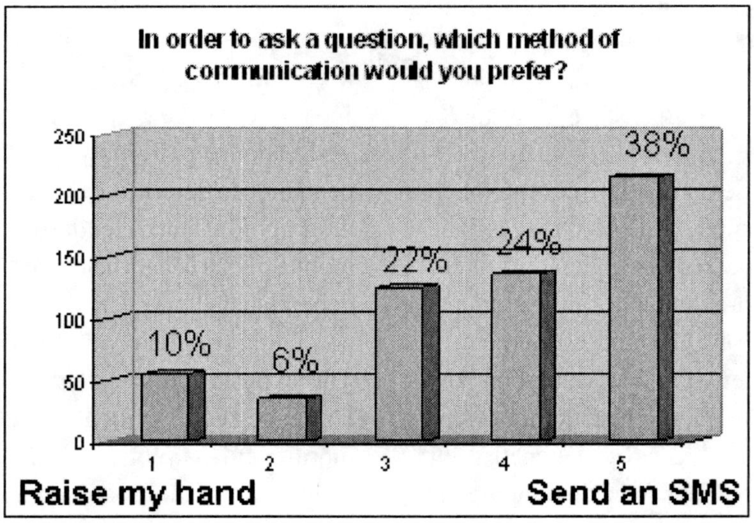

tion systems. This research described the use of a classroom interaction system and explored the impact that mobile applications such as SMS have had on students' learning experience.

The findings indicate that students and instructors can benefit from the additional channel of communication in the classroom, which implicitly points out individual learning experiences. The lecturer perceived a gain of quality and quantity of feedback from the students. Students indicated that the system was useful, making classes more interesting and interactive. The "open channel" was found to be an especially useful, efficient and preferred method of communication, in comparison to the traditional "raising hands" method of asking questions (see chapter I for the theoretical foundation of communication support in mobile learning). While students perceived only a moderately positive impact of the system in terms of increasing their interest in the subject and enhancing their study, they indicated that they would nevertheless like to see more use of this technology in the classroom. Overall, the main inhibitor for adoption of SMS in the classroom was the cost of text messages, rather than a perception of the systems value. Recent changes in the pricing structures offered for SMS services should see this barrier further reduced, increasing the usage of the system in the classroom.

## FURTHER RESEARCH DIRECTIONS

Further research should be undertaken regarding students' performance and learning outcomes versus use of the system, particularly longitudinal studies demonstrating a stable benefit and implementations of the system. Also further studies could investigate the use of the system in different socio-demographic contexts and disciplines. In addition, activities outside the classroom should be investigated in order to extend the boundaries of the classroom through mobile technologies.

## REFERENCES


Angelo, T. A., & Cross, K. P. (1993). *Classroom assessment techniques: A handbook for college teachers* (2nd ed.). San Francisco: Jossey-Bass Publishers.

Barnes, S. J., & Scornavacca, E. (2005). *Strategic Impact Of Wireless Applications In New Zealand Business.* Paper presented at the Hong Kong Mobility Roundtable, Hong Kong.

Bishop, A. L., Dinkins, R. K., & Dominick, J. L. (2003). Programming handheld devices to enhance learning. *EDUCAUSE Quarterly, 26*, 50-53.

Blodgett, D. L. (2006). *The Effects of Implementing an Interactive Student Response System in a College Algebra Classroom.* Unpublished Masters Thesis, the University of Maine, USA.

Bloom, B. S. (Ed.). (1956). *Taxonomy of educational objectives: The classification of educational goals. Handbook 1: Cognitive domain* (1st ed.). New York: Longmans Green.

Bollen, L., Eimler, S., & Hoppe, H. U. (2004). SMS-Based Discussions - Technology Enhanced Collaboration for a Literature Course. *Proceedings of the 2nd IEEE International Workshop on Wireless and Mobile Technologies in Education (WMTE?'04).*

Carnaghan, C., & Webb, A. (2005). Investigating the effects of group response systems on learning outcomes and satisfaction in accounting education. Retrieved 16 June, 2006, from http://www.learning.uwaterloo.ca/LIF/responsepad_june20051.pdf

Carroll, A., Barnes, S. J., & Scornavacca, E. (2005). Consumers Perceptions and Attitudes towards SMS Mobile Marketing in New Zealand. In *Proceedings of the Fourth International Conference on Mobile Business* (pp. 434-440). Sydney, Australia: IEEE Computer Society.

Draper, S. W., & Brown, M. I. (2004). Increasing interactivity in lectures using an electronic voting system. *Journal of Computer Assisted Learning 20*, 81-94.

Duncan, D. (2005). *Clickers in the classroom: How to enhance science teaching using classroom response systems*. San Francisco: Pearson.

Elgort, I., Marshall, S., & Mitchell, G. (2003). *NESB student perceptions and attitudes to a new online learning environment.*

Freeman, M., & Blayney, P. (2005). Promoting interactive in-class learning environments: A comparison of an electronic response system with a traditional alternative. *Proceedings of the 11th Australasian Teaching Economics Conference.*

Gopal, R. D., & Tripathi, A. K. (2006). Advertising via wireless networks. *International Journal of Mobile Communications, 4*(1), 1 - 16.

Greer, L., & Heaney, P. J. (2004). Real-time analysis of student comprehension: An assessment of electronic student response technology in an introductory earth science course. *Journal of Geoscience Education, 52*, 345-351.

Haghirian, P., Madlberger, M., & Tanuskova, A. (2005). *Increasing advertising value of mobile marketing - An empirical study of antecedents.* Paper presented at the 38th Hawaii International Conference on System Sciences, Big Island, Hawaii.

Hake, R. (1998). Interactive-engagement vs. traditional methods: A six-thousand-student survey of mechanics test data for introductory physics courses. *American Journal of Physics, 66*, 64-74.

Hall, R. H., Collier, H. L., Thomas, M. L., & Hilgers, M. G. (2005). A Student Response System for Increasing Engagement, Motivation, and Learning in High Enrolment Lectures. *Proceedings of the Eleventh Americas Conference on Information Systems, Omaha, NE, USA August 11-14, 2005.*

InfoSynch World. (2004). 1.87 billion mobile users by 2007. Retrieved 2006, from http://www.infosyncworld.com/news/n/5048.html

Jonassen, D. H., Peck, K. L., & Wilson, B. G. (1999). *Learning with technology: A constructivist perspective.* Upper Saddle River, NJ: Merrill.

Judson, E., & Sawada, D. (2002). Learning from past and present: Electronic response systems in college lecture halls. *Journal of Computers in Mathematics and Science Teaching 21*, 167-181.

Laurillard, D. (2002). *Rethinking university teaching: A conversational framework for the effective use of learning technologies* (2nd ed.). London: Routledge.

Markett, C., Sanchez, I. A., Weber, S., & Tangney, B. (2006). Using short message service to encourage interactivity in the classroom. *Computers and Education, 46*, 280-293.

Mazur, J. E. (1998). *Learning and behavior* (4th ed.). Upper Saddle River, NJ: Prentice Hall.

McConnell, D. A., Steer, D. N., Owens, K. D., Knott, J. R., Van Horn, S., Borowski, W., et al. (2006). Using conceptests to assess and improve student conceptual understanding in introductory geoscience courses. *Journal of Geoscience Education 54*, 61-68.

Penuel, W. R., Roschelle, J., & Abrahamson, L. (2005). Research on Classroom Networks for Whole-Class Activities. *IEEE International Workshop on Wireless and Mobile Teachnologies in Education (WMTE'05)*, (pp. 222-229).

Quinn, C. (2000). mLearning: mobile, wireless, in-your-pocket learning. *LiNE Zine* Retrieved 16 June, 2006, from http://www.linezine.com/2.1/features/cqmmwiyp.htm

Schwabe, G., Göth, C., & Frohberg, D. (2005). Does Team Size Matter in Mobile Learning? In *Proceedings of the Fourth International Conference on Mobile Business* (pp. 227-234). Sydney, Australia: IEEE Computer Society.

Scornavacca, E., Barnes, S. J., & Huff, S. (2006). Mobile business research published in 200-2004: Emergence, current status, and future opportunities. *Communications of the Association for Information Systems, 17*, 635 - 646.

Scornavacca, E., & McKenzie, J. (2006). *Perceived Critical Success Factors of SMS Based Campaigns: a managerial perspective.* Paper presented at the European Conference on Information Systems, Gotheburg, Schweden.

Seppälä, P., & Alamäki, H. (2003). Mobile learning in teacher training. *Journal of Computer Assisted Learning 19*, 330-335.

Sheng, H., Nah, F. F.-H., & Siau, K. (2005). *Using Wireless Technology to Facilitate Learning: A Grounded Theory Approach.* Paper presented at the 11th Americas Conference on Information Systems, Omaha.

Slain, D., Abate, M., Hodges, B. M., Stamatakis, M. K., & Wolak, S. (2004). An interactive response system to promote active learning in the doctor of pharmacy curriculum. *American Journal of Pharmaceutical Education, 68*, 1-5.

Ward, C. R., Reeves, J. H., & Heath, B. P. (2003). Encouraging active student participation in chemistry classes with a Web-based, instant feedback, student response system. *CONFCHEM: Non-Traditional Teaching Methods* Retrieved 16 June, 2006, from http://aa.uncw.edu/chemed/papers/srs/confchem/confchem_srs.htm

# Chapter IV
# Learning by Pervasive Gaming:
## An Empirical Study

**Christian Kittl**
*evolaris Privatstiftung, Austria & Karl-Franzens University, Austria*

**Francika Edegger**
*evolaris Privatstiftung, Austria*

**Otto Petrovic**
*evolaris Privatstiftung, Austria & Karl-Franzens University, Austria*

## ABSTRACT

*This chapter investigates how mobile games can be used for an efficient transfer of knowledge in learning processes that connect between the real world and the virtual world. In this chapter, the pervasive game concept is implemented on mobile phones as a means of enabling interaction and communication to support learning activities. The chapter presents the design of a new pervasive learning game, which was compared with a conventional pedagogical approach in terms of long-term learning results and learning efficiency. The empirical results revealed that the pervasive game led to higher energetic activation, more positive emotions and attitudes towards learning activities, and more efficient knowledge transfer than the conventional case-study approach.*

## INTRODUCTION

The current debate about effective and efficient learning can be characterized by the learning environments that support learners' autonomy and enable cooperative learning for authentic problems (Euler, 2005; Gräsel, 1997). This approach should allow the effective transfer of core competencies such as the capacity for teamwork, autonomy or readiness to take on responsibility, because in professional life these features are often even more important than simple professional competence. This claim originates from the view that knowledge imparted by conventional learning methods, such as pure "tell-test teaching," leads to accumulation of inert knowledge. This kind of knowledge is sufficient for passing exams in school, but fails when applied in real situations, especially in professional life (see also chapter XIV for this professional learning experience). In a more open and complex situation students (or learners) often fail to translate the things learned into real-life problem solving.

As a reaction to this situation, application-orientated learning environments, such as the traditional paper-based case study, which are conducted in teams, have been introduced. Empirical studies (e.g., Thomas, 2000) for learning by case studies have been in existence for many years. However, this approach neither enables a direct interaction with the real world nor includes technologies that are frequently used by students. In order to address this problem, one should look for Web, mobile, and interactive learning methods. We personally believe that this would be possible by means of a special type of digital game, namely pervasive games.

In general, digital games are considered to involve participants more actively in a learning process. However, commercial games often lack such pedagogically desired learning content. Digital game-based learning—the combination of digital games and high-quality learning content—would be in this context an invaluable pedagogical-didactic medium, engaging the student in the learning process.

The combination of digital games, in particular, with the real world, for example in the form of pervasive games, provides immense opportunities, but at the same time also represents a challenge. Mobile phones are the most important communication devices for young people, with penetration rates of over 100 percent in many European countries. It would therefore seem logical to use them for innovative learning models as well. However, up to now, mobile phones have mostly been used exclusively as devices for simply delivering certain learning modules, not as an interactive medium for learning. Consequently, we chose the "low tech–high involvement" approach in developing a system that does not require sophisticated devices and can be used in a large variety of situations.

The objective of this chapter is to employ currently available technological opportunities to develop an innovative learning platform that enables efficient, practice-oriented transfer of knowledge. In this context, learning games based on pervasive gaming would present a very promising approach. The concrete implementation of pervasive learning games would facilitate empirical research of this novel learning environment and an accompanying analysis of the games' potentials. The results could provide an innovative contribution to coping with current challenges in knowledge transfer.

## BACKGROUND

Traditional learning methods no longer suffice to fulfill the requirements of a modern teaching structure and do not sufficiently contribute to the acquisition of core competencies such as teamwork, independence and willingness to take on responsibility. It is against this background that current problems and constructivist solution approaches (see also chapter I of this book for more details) and their limitations are discussed. In addition, the potentials of mobile technologies, which present high-level, emotional information and communication channels, are discussed. Using these potentials in pervasive games leads to the development of a new type of game-based learning that can be employed in the creation of an innovative learning environment and for the expansion of constructivist solution approaches.

### Transferring Knowledge

Traditional teaching methods in schools and universities often build upon the principle of the "Nuremberg funnel" (Hirschfelder, 2006) in which all content can be learned as expertise through a methodology which is purely based on lectures. The teacher is assumed to know what knowledge needs to be acquired and how the learning objective can best be achieved. The learning material is thus split into smaller units, according to the needs of an average learner, and an attempt is made to "funnel" this knowledge into the learner using a suitable presentation format. The learner attempts to follow along with the lecturer and assimilates the learning material more or less passively (Thissen, 1997).

Although educational psychology has long since shown that this traditional teaching method exhibits inadequate effectiveness and suboptimal sustainability (Elbel & Laake, 2005; Edelmann, 2000; Edelstein & De Haan, 2004; Riedl & Schelten, 2000), this type of knowledge transfer remains the standard practice in almost all educational institutions. However, this leads to known problems such as the loss of motivation and lack of interest, knowledge gaps, diminished problem-solving capability and inadequate decision-making skills.

## Constructivism and its Limitations

The search for possible ways to address the shortcomings of traditional teaching methods has led to the development of constructivist learning approaches. The source of this observation is the assumption that knowledge is actively constructed by the learner. The development of knowledge is always closely associated with existing knowledge and can only proceed successfully if the new information can be inserted into known structures (Thissen, 1997). The low level of efficiency of the Nuremberg Funnel principle can also be derived from this constructivist approach, because the teacher cannot assume that the learned material will be perceived by the learner in the same manner as he himself perceives it. The teacher therefore cannot transfer the knowledge to the learner in an easy manner by means of lectures.

The constructivist notion of learning (Bruhn et al., 1998; Mandl & Gräsel, 1997; Notterman & Drewry, 1993; Thissen, 1997) is based on several basic assumptions, which are briefly summarized here. Firstly, it is assumed that learning is an active and constructive process. This implies that no direct transfer of knowledge is possible and that the teacher can only support the learner through the creation of suitable learning situations. Furthermore, learning is viewed as situation and context-related. The situation in which the knowledge transfer occurs is considered important for subsequent use of the learned material. Learning is also understood as a self-directed process in which the learner must be active in order to be able to associate existing knowledge with the learning material. Therefore, the greater the activity on the part of the learner the more successful the learning process. Finally, in constructivism, it is important that learning is a social process. That is, knowledge is only constructed in social exchange between the teacher and the learner and learning is strongly influenced by the social and cultural environment of the exchange.

In the scope of the constructivist approach, successful learning only occurs if the selected learning method corresponds to the above-mentioned basic assumptions and appropriately supports the knowledge transfer. Constructivist learning methods are available to fulfill these requirements, including self-directed, cooperative, problem-oriented and situated learning (Thissen, 1997).

Digital learning games are discussed in the following as a concrete example of constructivist learning (for both individual and collaborative learners). In addition to the significance of this approach in the constructivist learning theory, the limitations of this learning method are also discussed. Applying a learning method that transcends these limitations forms the central point of this chapter.

## A Popular Information and Communication Channel

One approach to designing a more efficient teaching structure is to search for new opportunities to create innovative learning environments. In this context, the use of mobile technologies, in particular, lends itself to surmounting the challenges of previous learning methods, for example by using digital learning games.

Today, mobile technologies present high-level, emotional information and communication channels that have changed the attitudes of primarily young people over the last ten years more than any other technology. For example, the penetration rate of SMS use among 29-year-olds is 98.4 percent in Germany (Talkline, 2006), significantly higher than the use of the wired Internet, which amounts to 81 percent (Barth & Cerny, 2006). Every fifth person in the same age group sends more than five SMS messages per day (Dimoco, 2004). Some 91 percent of under 24-year-olds have their own mobile phone turned on and carried 24 hours per day (Forum Mobilkommunikation, 2002).

No other medium has come close to attaining the same rate of uptake as the mobile phone in recent years. Currently, there are over 540 million GSM users in Europe, and the market penetration of GSM and 3G networks is over 85 percent, where 15 member states show a penetration rate of over 100 percent and only eight states a penetration of less than 60 percent (Research, 2007). Smartphones in particular, (i.e., mobile phones that typically use the operating system of a third-party provider and enable users to install programs themselves as required) are becoming more and more popular. According to the "Telephia European Subscriber and Device Report" (Q3 2006) (Telephia, 2006), 8.8 percent of recent mobile phone buyers selected Smartphones. In this respect, Europe is clearly the leader compared to the United States, which is often considered a benchmark in the digital economy (only 3.8 percent of United States mobile phone buyers opted for Smartphones).

The technical configuration of mobile phones at least potentially facilitates numerous usage possibilities. Most mobile phones owned by young people are Internet-capable and have a camera. About half the mobile phones are equipped with an infrared interface or Bluetooth for wireless data exchange. Almost 40 percent of users can utilize their phones as an MP3 player or radio. Mobile gaming is also a well-established phenomenon. For example, 29 percent of 16-24 year-olds use their mobile phone to play games (Glu Mobile, 2006). Playing mobile games is thus among the most popular activities on a mobile phone, in addition to SMS transmissions, voice telephony and taking pictures (Feierabend & Rathgeb, 2006).

## Advantages of Pervasive Games

The basic idea described in this chapter is to use the potential of the mobile phone communication channel and combine it with the possibilities of digital learning games in order to create a new and improved learning environment. The concept of pervasive gaming lends itself to this purpose, in that the real world fuses with the virtual world (Walther, 2005). In this context, one can speak of pervasive game learning, an expansion of mobile game-based learning.

Pervasive games take place in the real world, and the player communicates with his fellow players and the remote control system via wireless technologies. Various communication channels are available for this, among them, the mobile phone. Additional core technologies needed for pervasive gaming are portable displays, which render the digital content tangible in the real world independent of location, and sensor technologies, such as cameras, through which the status of the player can be ascertained (Benford et al., 2005). These technological prerequisites are fulfilled by modern mobile phones, which therefore offer the ideal medium for the execution of pervasive games.

Numerous characteristics distinguish pervasive games from conventional computer games. Important elements, which are consistently mentioned in the literature (Jegers, 2006; Jegers & Wiberg, 2006; Thomas, 2006; Walther, 2005), are the location-independent mobile playing field, the connection of the tangible real world with the intangible virtual world (i.e., augmented reality), and the increased social interaction between the players. Additional characteristics of pervasive games are the ubiquity of the game and the incorporation of the locality into the game. Pervasive games also offer the possibility of an uninterrupted, 24-hour-a-day game in which participation itself is the true motivation of the player.

Due to their special characteristics, pervasive games offer an optimal information system for the creation of a learning environment that is in complete agreement with constructivist learning theory. However, the resulting concept of pervasive learning games has not been widely researched to date. The only known implementations of pervasive game learning environments are the "Savannah" project (Benford et al., 2004), in which children were taught to understand the behavior of a lion, and "ProjectY," developed by Siobhan Thomas (2006b). The learning effects of these pervasive games were not empirically studied, however, which means that no empirical data are available regarding the effectiveness of pervasive learning games.

## DEVELOPING A PERVASIVE GAME FOR LEARNING PURPOSES

Due to the problems discussed in the previous section, the objective of the work described in this chapter is to create a mobile learning application that can be successfully employed in modern teaching. The focus is on supporting practical learning and the associated acquisition of the core competencies needed for everyday work activities. Furthermore, the playing mechanics of the developed pervasive game should be flexible in use and easily adaptable to new learning material. In order to meet these demands, a pervasive learning game is developed and directly evaluated in the learning context. A direct survey of students while and after they play the pervasive games provides indications of the advantages and disadvantages of the developed learning game. Additionally, a direct comparison with conventional case-study-based learning is made to provide further evaluation. With this systematic evaluation, this study should counteract the lack of empirical data for an efficiency analysis of pervasive games.

### Design Decisions

The basic idea of this chapter is to use the potential of the mobile phone communication channel and combine it with the possibilities of case-study-based learning and digital learning games in order to create a new and innovative learning environment. Unfortunately, although the use of pervasive games has numerous advantages, it is also often associated with high development and implementation costs and consequently has a limited mass appeal. Technologically complex equipment is often needed, which makes practicable and seamless application in a learning context nearly impossible. Because of this problem, our objective is to develop a pervasive game that does not require costly equipment, such as head-mounted displays. Above all we place value on a high degree of mass appeal and economic feasibility in the learning context because it is only under these basic conditions that comprehensive practical use is realistic.

Today's mobile phones are one of the few mediums that have high penetration rates for young people, the relevant target group, and simultaneously bring with them almost all the technological prerequisites for pervasive gaming. As previously discussed, most young people have high-tech phones that feature large color screens, allow digital picture-taking and transmission of MMS (multimedia messaging service) messages, and are Internet-capable. In combination with the capability for communication via voice telephony and SMS, today's mobile phones can thus provide the core technologies required for pervasive gaming. On this basis, the following section presents a game concept and its technical system implementation that only requires an MMS and camera-capable mobile phone from users, that is, the students. This approach guarantees a high degree of mass appeal in contrast to previously implemented pervasive games and thus has much potential to prove itself in practical use.

## A Sketch of the Game

The pervasive learning game 'Digital Economy' connects the real and virtual worlds, with the game system requesting students via SMS to find and analyze special everyday situations in the study area of the digital economy. The students work in teams and must document the specified situation and submit a proposal for improvement as quickly as possible by means of MMS. The contributions are then displayed on a Web site in a mobile blog and, based on their quality, earn the teams a varying number of points in the high score list. This game concept is presented in detail in the following section and can be easily adapted to other learning topics due to its versatile design.

The basis for the draft of the game idea was the creation of a practical learning environment in which the students can deal with the learning topic in the real world. In order to ensure simple and universal application of the game concept in various teaching materials, the game was developed based on the three core elements given in Table 1. These core elements can be easily adapted by teachers to various learning materials; they simply select a suitable problem with regard to the topic to be communicated.

In order to embed the core elements into the game, the students compete in teams to find the most innovative and fastest problem description and solution. The reference to a problem is communicated by the game leader in the form of an 'opportunity alarm SMS.' Then the students must hurry to the specified location and photograph the problem with their camera phone and send it to the game system via an 'opportunity MMS,' including the accompanying problem description. The students must then find innovative real world solution approaches. Each team is required to find a best-practice solution within a specified time frame and to send a description, including photo, via a 'solution MMS.' The opportunity and solution MMS are awarded 0-3 points by the game leader; these are immediately added to the team's high score. A single game cycle consists of several different opportunity alarm SMS, which are randomly sent out after the expiration of the preceding opportunity. This results in a

*Table 1. Core elements of the game*

| |
|---|
| **Indication of a problem:** The students are given a location where there is a problem connected with the learning topic. The students are asked to identify this problem. |
| **Description of the problem:** After identifying the problem, the students must describe the problem and send this description back to the game leader for assessment. The described problem represents a worst-practice situation. |
| **Problem resolution:** Finally, and most importantly, the students must find a best-practice situation for the resolution of the real world problem. |

continuously changing score, which contributes significantly to strengthening the gaming component. Additionally, the dynamics of the game are further increased through the continuous movement of the students in the real world as well as through notifications regarding the contributions by opposing teams.

The game idea proposed here is intended to lead to a unification of the advantages of case-study-based learning and learning games, and also to contribute to overcoming the limitations of both of these learning methods. For example, the learners come into contact with real problems through the pervasive game, which removes the limitation of a simplified representation of reality in conventional case studies. Furthermore, the learner can play the pervasive learning game at the same time as conducting other everyday activities (in the case discussed here, the game ran for approximately two weeks), which represents a time-saving compared to conventional case studies. The use of several different opportunity alarm SMS also counteracts a possible overemphasis of individual cases in case studies. The big advantage of the pervasive game developed here, compared to conventional digital learning games, is that it promotes the key qualifications for social competence and teamwork, because searching for and documenting solutions is coordinated and agreed upon within the group.

## Building a Connection to the Learning Context

The game idea previously discussed is not oriented to specific learning materials. However, in order to test the practical feasibility of the concept, it is necessary to focus on a specific learning topic and actively integrate the pervasive learning game into a teaching subject. To this end, we will focus on the topic of the 'Digital Economy.'

This novel learning approach was first tested during the winter semester 2006/07 at the Karl-Franzens University in Graz within the scope of both the 'Principles of Digital Economy' and 'Modeling of Business Models and Digital Economy' courses. A focal point of these courses was the communication of the methodological principles of networked thinking and acting in the digital economy. Using this practical learning, the objective is to show why it is important to concern oneself with business models and the digital economy, and how one can handle complex problems. An application-oriented presentation format should teach students networked thinking for the strategic management of a company. Students should also learn how to set objectives and circumvent problems in the digital economy. Moreover, the digital economy sector not only requires new knowledge but also offers completely new possibilities to constantly update this knowledge. Therefore, a central objective is the mediation of the skill of continuously undertaking a knowledge update in the digital economy area through independent work in a new field.

In addition to expositional teaching for the communication of basic facts, case studies and teamwork are suitable for achieving the course objectives, in order to apply the acquired knowledge to concrete problems in business practice. Because the developed pervasive learning game is considered as an optimization of case-study-based learning, courses on the topic of the digital economy are especially suited for use in this novel learning environment.

Therefore, the following procedure was employed in the courses. First, the principles of the digital economy were presented by means of a practice-oriented presentation format. Then the students were randomly divided into two groups and, in teams of five, dealt with the learning topic by means of either case studies or pervasive gaming. In order to evaluate the lecture, the results from the case studies and/or the pervasive learning game as well as from a class exercise were analyzed. Using this evaluation criterion, the students were supposed to present proofs about the skills of understanding, application, analysis, synthesis and evaluation of the learning material.

## Practical Implementation

Building upon the use cases, this section presents the system-oriented implementation of the pervasive learning game environment that was developed. Through the systematic stringing together of the use cases, the logical game cycle can be illustrated in an activation diagram. Based on this activation diagram and the use cases, the pervasive game is implemented via a software connection of the various user interfaces.

The use cases define the interactions between the actors and the observed system that occur during the course of the game. Each use case always describes exactly one action or one process. Illustration of the use cases therefore enables an elementary description of the developed game concept. Figure 1 shows an overview of the most important use cases that occur in the 'Digital Economy' pervasive learning game. The systematic cycle of the game that results from piecing together the use cases is illustrated in an activation diagram (Figure 2).

The backend system, which was developed in Java and deployed on the JBoss application server, is illustrated in Figure 3. The core of the system is the game engine, which controls the entire cycle. It is here that it is determined at runtime how incoming SMS or MMS with specific key words will be responded to, when certain time frames (for example, for the transmission of opportunity or solution proposals) are open, and which messages will be sent to the mobile terminals in the

*Figure 1. Use cases for the 'Digital Economy' pervasive game*

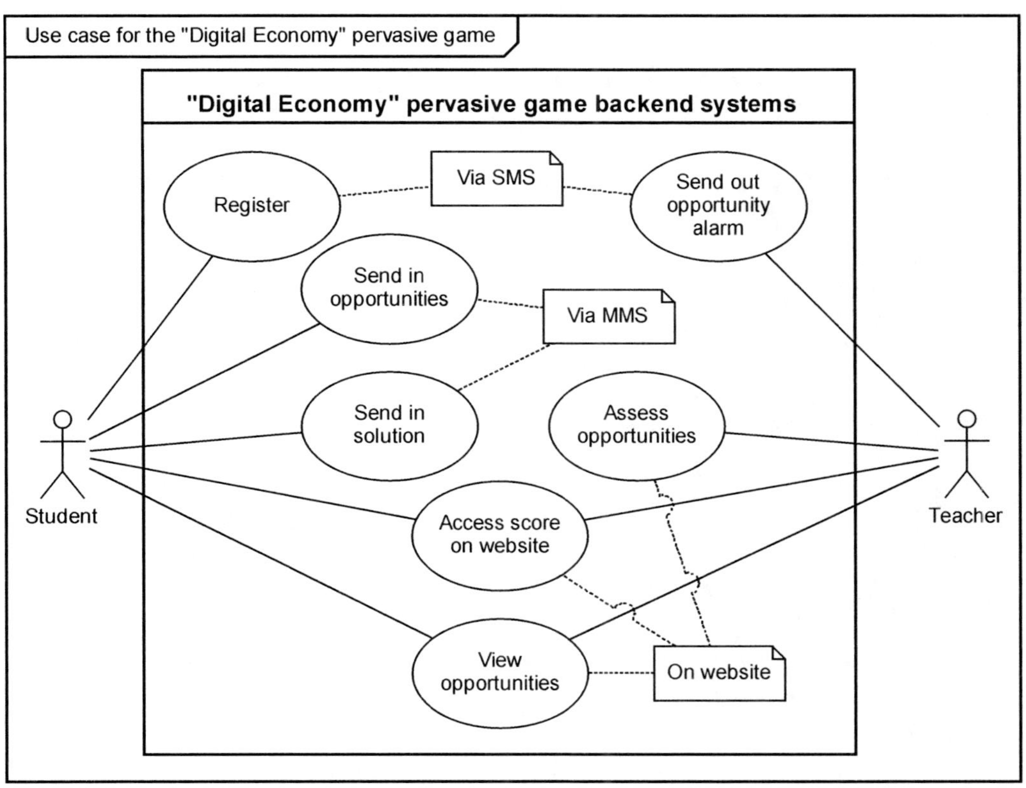

*Figure 2. Activation diagram of the 'Digital Economy' pervasive game*

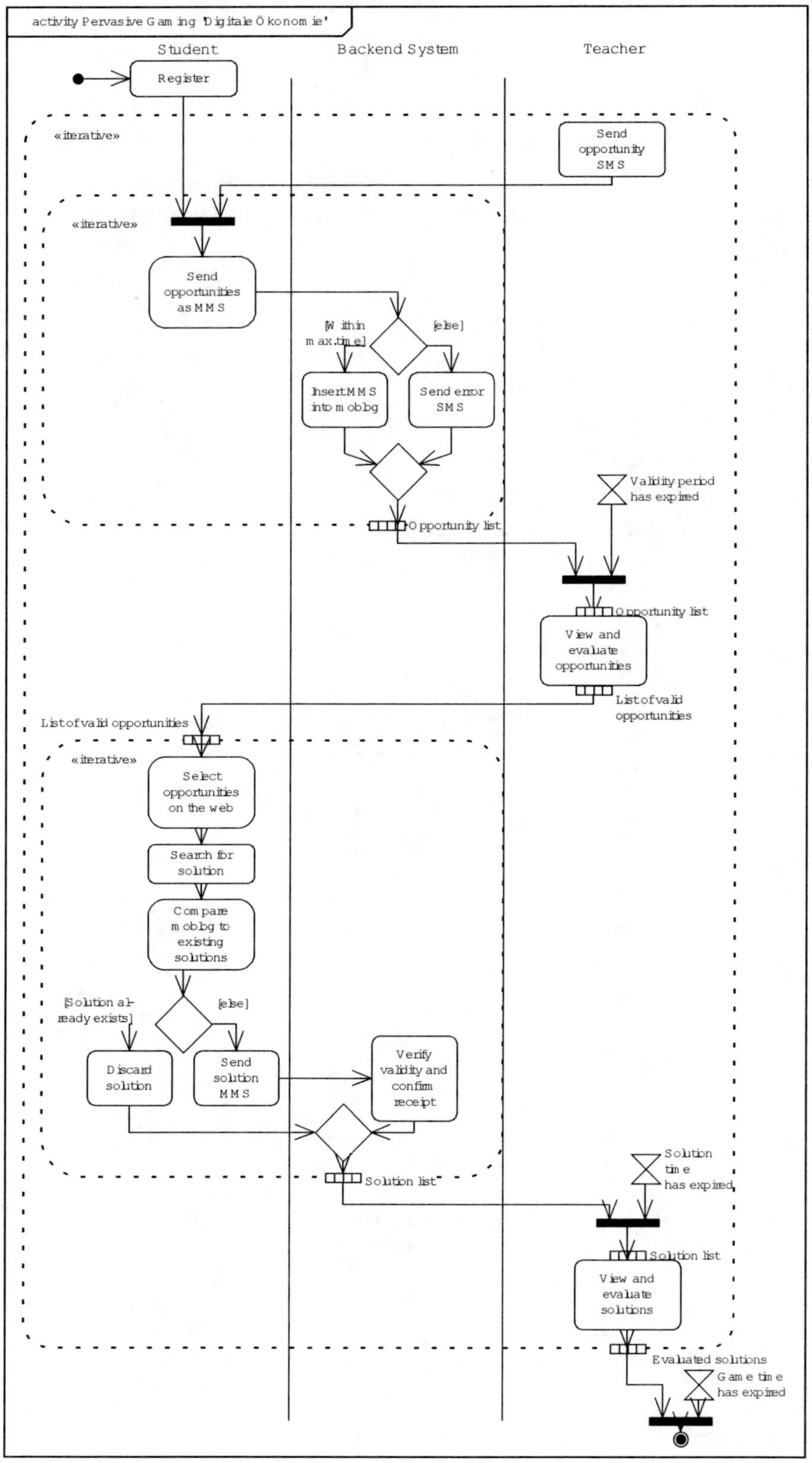

*Figure 3. System architecture of the "Digital Economy" pervasive game*

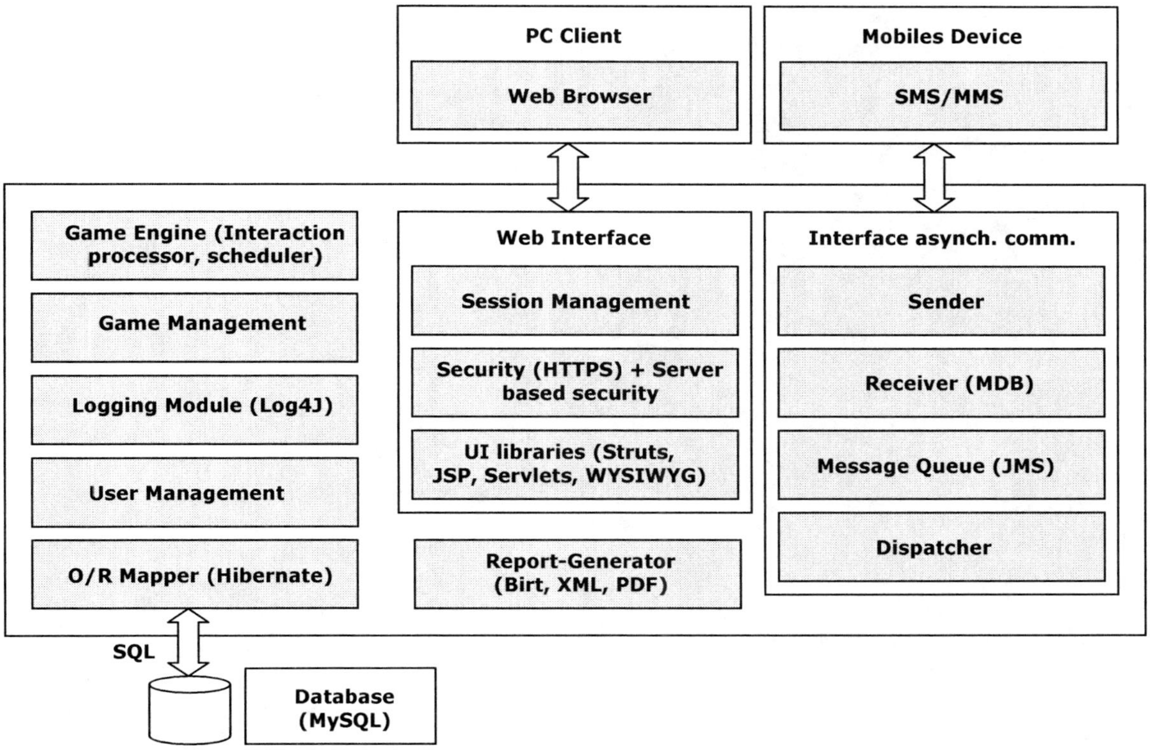


event of errors. The game management module enables easy management of these parameters via a Web interface. The overall playing mechanics as well as the transactions are stored in a relational database (MySQL), which is mapped to the Java layer using hibernate. Communication with the front end (PC via Web browser or mobile phone via SMS and MMS) is controlled by the respective interfaces. The backend also makes available the functionalities for the use of the moblog, via the UI libraries.

## An Example from the Game

In this subsection, an example from the game is given to illustrate the cycle from the posting of an 'opportunity alarm SMS' to the submission of an 'opportunity MMS' and a 'solution MMS.' In the present example, students can learn about the possibilities for user-friendly and economic guided tours through a museum by using the potential of the digital economy.

In this situation the text of the "opportunity alarm SMS" could read as follows:

Opportunity alarm 1! Location: Landeszeughaus, Herrengasse. Hint: What difficulties could you experience if you wander through a museum without a guide? Max. validity: 1/12 - 20:00. "Opportunity MMS" with "dec1" to 0676/800 812 00. Good luck!

A possible 'opportunity MMS' for this 'opportunity alarm' is shown in Figure 4.

A possible solution for this 'opportunity MMS' can be found in the art museum of Graz (Kunsthaus Graz). Students could formulate a suitable 'solution MMS' is shown in Figure 5.

*Figure 4. Example of an 'opportunity MMS'*

*dec1* If you go on your own, you have the advantage that you can determine how long you want to visit each exhibit. However, the disadvantage is that nobody provides you with information on the exhibits. That is a significant drawback of an individual tour.

## Testing the Effects of the Game

The pervasive learning game developed in the scope of this study is intended to serve efficient communication of long-term knowledge that can be successfully employed in subsequent professional life. Evaluation of the long-term effect and practicability is not immediately achievable, however, because as a rule the learners cannot be observed for the extremely long time frames required. Therefore, a general study methodology was selected that investigates the development of activating and cognitive processes in learners. Based on the effects of the learning method on the activating and cognitive processes, conclusions can then be drawn not only about the long-term effects and the practical feasibility of the knowledge, but also about the learning efficiency, for example. This approach may guarantee that evaluating the learning success is sensible, because activating and cognitive processes serve to describe and explain human behavior (Kroeber-Riel & Weinberg, 2003; Trommsdoff, 2004), and learning is defined as a change of behavior in specific situations (Edelmann, 2000).

## Measuring the Effect Constructs and the Flow Experience

The activating processes are critically important to the explanation of behavior, because they supply the individual with energy and place the individual in a state of motivation and productivity (Kroeber-Riel & Weinberg, 2003; Trommsdoff, 2004). Within the effect analysis we observe activating processes based on the "activation," "emotion" and "attitude" constructs. These constructs are split into several effect dimensions and can be

*Figure 5. Sample 'solution MMS'*

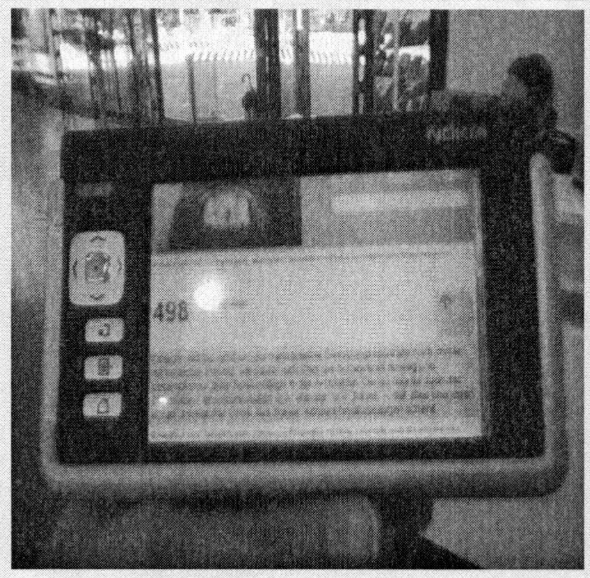

*dell* The drawback of moving independently without a "human" guide through the museum is removed by a digital guide! This example is given by the art museum Graz, where a Mobile Space Guide provides visitors with important information about all the exhibits. Every visitor is provided with this guide, implemented by a Nokia 770, when entering the museum. By pushing a single button, the visitor can request information on all objects in the exhibition.

determined by means of standard indicator scales. Details about the individual effect dimensions and the scales employed are illustrated in Table 2.

The cognitive processes, which serve primarily to deliberately control mental behavior, comprise the processes of information assimilation, information processing and information storage (Kroeber-Riel & Weinberg, 2003). Our study investigates the development of these processes by means of an individual knowledge test that focuses on the interrogation of practical knowledge (see Table 2). The students were also interviewed about their subjective learning successes.

In addition to the evaluation of the development of the effect constructs, we are interested in how much more pronounced the flow experience (Csikszentmihalyi, 2000) is for a pervasive learning game than for conventional learning methods. Therefore, the flow scale (short version) according to Rheinberg, Vollmeyer, and Engeser (2002) (see also Table 2) was used. The flow experience is described as a sense of the total submersion in an activity (Csikszentmihalyi, 2000), which occurs during an intrinsically motivating activity such as playing a game. Also, of interest is whether the flow experience is a decisive factor behind a possibly positive effect of the pervasive game in the learning process. This assumption stands to reason, as several studies (Drengner & Zanger, 2003; Schiefele & Roussakis, 2006) have reported a strong connection between the flow experience and mental activation as well as a connection with

*Table 2. Overview of the indicator scales used*

| Construct | Dimensions | Indicator scales |
|---|---|---|
| Activation | Energy activation<br>Tenseness | Activation-Deactivation Checklist (AD-ACL) (Thayer, 1989)<br>German translation: by Imhof (1998) |
| Emotion | Interest, joy, surprise, rage, disgust, disdain, shame, guilt, fear, sadness | Differential Emotions Scale (DES) (Izard, 1977), German translation: Merten/Krause (1993) |
| Attitude to learning material<br>Attitude to learning method | Activity, evaluation, strength | Semantic differential (Osgood et al., 1975) |
| Information assimilation, processing and storage | Individual knowledge test | Test of practical knowledge:<br>ten questions, assessment with grade |
| | Subjective learning success | Student interview |
| Flow experience | Flow value | Flow scale (short version) (Rheinberg et al., 2002) |

positive emotions. Such a connection would be reflected in correlations that transcend the effect of the learning method.

## Hypotheses

Based on the characteristics of pervasive games, it can be assumed that such pervasive games are more effective than conventional learning methods. The following hypotheses regarding the development of effect constructs can be derived from this in our empirical study:

*In relation to conventional case studies, the "Digital Economy" pervasive game produces:*

*(H1.1) An increased energy activation*

*(H1.2) More positive emotions*

*(H1.3) A more positive attitude vis-a-vis learning material*

*(H1.4) A more efficient transfer of knowledge*

Furthermore, the following hypotheses can be made in connection with the flow experience:

*(H2.1) The 'Digital Economy' pervasive game produces a stronger flow experience than the conventional case study.*

*(H2.2) A strong flow experience is associated with mental activation and positive emotions, which is reflected in positive correlations with these constructs that transcend the effect of the learning method.*

The correlations between the individual hypotheses are illustrated in Figure 6. It must be noted that the illustrated constructs comprise several dimensions in each case. This study does not include a detailed discussion or a verification of the measurement model for the individual effect dimensions (see Table 2), because the reliability of the employed standard scales has already been comprehensively tested in published literature.

*Figure 6. Correlations between the hypotheses*

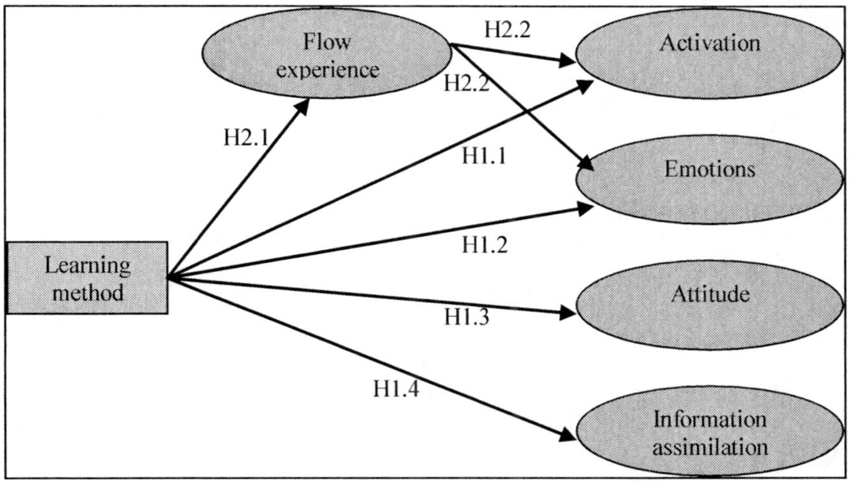


## Study Process

The hypotheses were tested by means of an empirical study. To this end, a total of 100 students were randomly assigned to two groups, the control and the experimental group, in two courses at the Karl-Franzens University in Graz. The control group had to deal with the topic of the digital economy within the scope of a conventional case study. In the experimental group, the conventional case study learning method was omitted; instead students were to be taught the same learning material by means of the 'Digital Economy' pervasive game. In order to enable all students of the second group to participate in the pervasive game, MMS-capable mobile phones were lent out to the students for the duration of the game. The pervasive game and the case study were both conducted at the same time over the course of two weeks in December 2006 and in January 2007.

## Results of the Empirical Study

### Multivariate Variance Analysis

The experiment conducted in the study design was conducted in two different courses for the same topic but at different times and with different students. Consequently it was not ensured *a priori* whether the data sets gathered from the two courses could be viewed in a uniform manner. A multivariate variance analysis (MANOVA) was therefore conducted at the start of the evaluations with the learning method (case study or pervasive game) and the course groups (first and second course) as fixed variables. The dependent variables employed were all effect dimensions illustrated in Table 2 (including flow values, knowledge test and subjective learning success). The MANOVA conducted clearly shows that the learning method affects the development of the knowledge constructs (probability of error less than 0.001). In contrast, the affiliation with a specific course group has no significant effects (probability of error

*Table 3. Development of the effect dimensions*

| Construct | Scale | Dimensions | Perv. game[1] | Case study[1] | Sig.[2] |
|---|---|---|---|---|---|
| Activation | 0 (not at all) - 3 (very strong) | En. activation | 2.282 (0.090) | 1.763 (0.091) | .000 |
| | | Tenseness | 1.153 (0.080) | 1.270 (0.081) | .302 |
| Emotion | 0 (not at all) - 4 (very strong) | Interest | 3.918 (0.107) | 3.370 (0.109) | .001 |
| | | Joy | 3.203 (0.119) | 2.499 (0.121) | .000 |
| | | Surprise | 2.374 (0.123) | 1.652 (0.125) | .000 |
| | | Sadness | 1.351 (0.084) | 1.708 (0.085) | .004 |
| Attitude to learning material[2] | 1 (very positive) - 7 (very negative) | Activity | 2.668 (0.176) | 2.991 (0.179) | .202 |
| | | Evaluation | 2.406 (0.156) | 2.838 (0.159) | .056 |
| | | Intensity | 2.556 (0.125) | 3.013 (0.127) | .012 |
| Attitude to learning method[3] | 1 (very strong) - 7 (not at all) | Activity | 2.531 (0.122) | 4.079 (0.141) | .000 |
| | | Evaluation | 2.705 (0.121) | 3.558 (0.110) | .000 |
| | | Intensity | 3.147 (0.101) | 3.626 (0.100) | .000 |
| Flow experience | 1 (low) - 7 (high) | Flow value | 5.033 (0.123) | 4.314 (0.125) | .000 |
| Information assimilation, processing and storage | 1 (very good) - 5 (insufficient) | Knowledge test | 3.401 (0.174) | 3.793 (0.177) | .117 |
| | 1 (very positive) - 5 (very negative) | Subjective learning success | 1.820 (0.068) | 2.180 (0.079) | .000 |

[1] Mean value (standard error)
[2] 2-sided
[3] Measurement after the end of the experiment (case study or pervasive game)

of 0.833) on the effect constructs. Likewise, the interaction effects between the learning method and the course group is negligible (probability of error of 0.995). Therefore, the data sets from both course groups can be combined and viewed in a uniform manner.

The following section discusses the effect constructs in detail, whereby the dependency of the learning method (case study or pervasive game) is analyzed. Differences in the individual constructs are considered significant when the MANOVA exhibits a probability of error of less than 5 percent for this dependent variable; a value of less than 1 percent is designated as highly significant. The mean values for the respective learning method, the employed scale range for illustrating the measurement, and the probabilities of error for significant group differences are illustrated for all effect dimensions (other than the less developed emotions of rage, disgust, disdain, fear, shame and guilt) in Table 3. The determined results confirm hypotheses (H1.1)-(H1.4) as well as (H2.1) and are discussed in more detail in the following section. Furthermore, Table 4 illustrates the correlations between the flow value and several effect dimensions. These results show significant positive correlations in both learning groups (pervasive game and case study) between the flow experience and the energy activation, interest and joy, which also corroborates hypothesis (H2.2).

*Table 4. Flow effect correlations*

| | | Energy activation | Tenseness | Interest | Joy | Surprise | Sadness |
|---|---|---|---|---|---|---|---|
| FLOW – Per. game | Correlation acc. to Pearson | .439** | -.133 | .369** | .413** | .135 | -.361* |
| | Significance[1] | .002 | .363 | .009 | .003 | .354 | .011 |
| FLOW - Case study | Correlation acc. to Pearson | .331* | -.356* | .534** | .325* | .043 | -.427** |
| | Significance[1] | .023 | .014 | .000 | .026 | .773 | .003 |

[1]Significance: ** Probability of error < 1 percent; * Probability of error < 5 percent.

## Activation Effects

The energy activation is significantly higher in the pervasive game than in the case study. This is important inasmuch as the energy activation dimension is viewed as a prerequisite for affective and cognitive processes. It significantly strengthens the processes of information assimilation, processing and storage. The energy activation is also considered a basis for a high degree of motivation and productivity (Kroeber-Riel & Weinberg, 2003). A state of high activation is also associated with a positive mood and a sense of well-being (Thayer, 1978). A high degree of energy activation while playing a pervasive game therefore has a positive effect on the learning capacity and learning efficiency. The high degree of energy activation also substantiates the assumptions of Thomas (2005, 2006a) regarding the activating effect of pervasive games.

Table 5 contains a comparison of the results of the activation measurement with the results gathered by Imhof (1998), who ascertained the activation dimension during a break and towards the end of a typical course lecture. Compared to conventional course lectures, the 'Digital Economy' pervasive learning game shows a clearly increased energy activation, which confirms the extraordinary activation potential of the game. Although the energy activation in the case study is clearly less than the value in the 'Digital Economy' pervasive game, it is still somewhat higher than it is in conventional lectures (see Table 5). In summary, it has been shown that case studies activate students more than conventional lectures, but not nearly to the level of the pervasive learning game.

The tenseness dimension is associated with anxiety, agitation, tension and stress (Thayer, 1978, 1989). Because this dimension is not very pronounced in either of the learning methods, it is assumed that the students did not find themselves in a direct stress situation in either of the learning methods. Despite the low tenseness values for the case study and the 'Digital Economy' pervasive game, these measurements are slightly higher than the values for conventional course instruction (see Table 5). This slight increase is comprehensible, because case-study-based learning and pervasive gaming requires more individual responsibility from students and thus also causes more tension and stress. Nevertheless, these tenseness values are still so low that they need not be associated with any negative effects.

## Affecting Emotions

The existence of very positive emotions among the participants of the pervasive game confirms hypothesis (H1.2) and is consistent with the experiences reported in previously conducted studies on learning games (Conati, 2002). Strongly pronounced emotions such as interest, joy and surprise also form an important prerequisite for

*Table 5. Comparison of various activation measurements*

| Permitted scale values between 0 (min.) and 3 (max.); measurements by Imhof (1998) were correspondingly scaled. | **Sample** | **Dim. A: Energy activation** | **Dim. B: Tenseness** |
|---|---|---|---|
| This paper; Topic: digital economy | **Pervasive game** | 2.28 | 1.15 |
| | **Case study** | 1.76 | 1.27 |
| Measurements during a break (A, at 12:00 p.m.) and towards the end (B, at 1:00 p.m.; C, at 3:00 p.m.) of the course lecture (Imhof, 1998) | **Group A** | 1.68 | 0.76 |
| | **Group B** | 1.65 | 1.00 |
| | **Group C** | 1.28 | 0.85 |

a positive attitude vis-a-vis the learning material and a longer memory retention regarding the experiences gained from the game.

## Affecting Attitudes

The 'Digital Economy' pervasive game changes people's attitude to learning material in a positive manner, which is especially and strikingly evident when comparing the measurements taken before and after the experiment. This quality exceeds all the previously known accomplishments of pervasive learning games and constitutes an important new finding of the present study. The positive attitude is particularly important for the penetration of the learned material into the long-term memory. If a student's attitude towards the learning material is low, then the student will only learn because of outside pressure, which negatively affects the subsequent retrievability of the acquired knowledge.

In addition to the attitude towards the learning material, the attitude towards the learning method was also studied. Measurements taken after the end of the experiment show that the attitude towards the pervasive game is clearly better than the attitude towards the case study. This indicates a high degree of acceptance and fun while playing, and substantiates previous experiences from pervasive learning games (Thomas, 2005). It has also been proven that learners generally respond more positively to learning games than they do to other learning methods. However an important factor in the present study is that the positive attitude toward the game can truly be converted into learning capacity.

## Effects on Information Assimilation, Processing and Storage

The knowledge test focused on interrogating practical knowledge, and yields better results for students in the pervasive game group. However, the 0.117 significance of the difference was above the 5 percent probability of error. As mentioned before, it must be noted that a knowledge test cannot evaluate all aspects of learning. For example, the penetration of the learned material into the long-term memory, the applicability in practical situations and the acquired experiences cannot be tested.

In this context, subjective learning success is an important indicator of learning capacity. In a student survey, the learning success achieved with the pervasive game was evaluated as being significantly better. This positive finding vis-a-vis the pervasive game is consistent with the other effect dimensions and indicates an efficient knowledge transfer.

## Effects of the Flow Experience

A great deal of attention was placed on the flow experience in this chapter due to its close association with intrinsic motivation and unforced action. In principle, the empirical assessment (see Table 3) shows that the pervasive game produces a significantly stronger flow experience than the case study simultaneously conducted. In addition to the associated positive effects, the following section compares the measured absolute values of the flow with other activities in order to gain a better idea of the motivating effect of the pervasive game.

The present data confirm existing papers regarding the flow experience in games (Rheinberg et al., 2003; Schiefele & Roussakis, 2006). The results show that the pervasive game produces a much stronger flow experience than the conventional case study. The strong flow experience points to a high degree of intrinsic motivation in the learner and shows that the game is being played for the game itself and not due to an external incentive, for example, a good grade. This free choice is an important basis for the transfer of practical knowledge and practical experience. The high flow value is also an indication for a sensible balance between the level of difficulty of the task and the skill level of the learner (Csikszentmihalyi, 2000). This equilibrium confirms that the 'Digital Economy' pervasive game neither overburdens nor insufficiently challenges the learner and thus creates good preconditions for motivated learning.

In order to better assess the observed flow values, Table 6 compares the results gathered within the scope of this paper with empirical data from other samples. It is evident that the flow value observed in the pervasive game is very high compared to other studies. A higher value (5.16) was observed only in graffiti sprayers. In contrast, the flow values for solving statistics problems are significantly lower, both during and after a lecture. Thus the flow value for the pervasive learning game clearly exceeds the flow values attained with conventional learning, which indicates a particularly motivating effect. In contrast to the pervasive game, the case study does not seem to differ from the flow values for a lecture or for the solving of statistics problems (see Table 6).

The correlations between the flow value and the effect dimension illustrated in Table 4 substantiate the assumption that the flow experience is an important factor for the development of the 'activation' and 'emotion' constructs. The present results are consistent with known research results that associate the flow experience with positive activation, a high degree of interest and a high degree of motivation (Schiefele & Roussakis, 2006).

*Table 6. Flow values for various samples (data for activities 3-6 from Rheinberg et al., 2003)*

| Activity (sample size) | Mean flow values |
|---|---|
| 1) "Digital Economy" pervasive game (N=49) | 5.03 |
| 2) "Digital Economy" case study (N=47) | 4.31 |
| 3) Graffiti spraying (N=292) | 5.16 |
| 4) Statistics problem (N=123) | 4.57 |
| 5) Lecture – middle (N=63) | 4.43 |
| 6) Lecture – end (N=63) | 4.21 |

## FUTURE RESEARCH DIRECTIONS

Based on the evaluation presented in this paper, the pervasive game represents a promising, novel learning environment that is capable of efficiently transferring knowledge. However, in order to ensure sustained diffusion of the pervasive learning game concept presented here, the following research questions should be investigated in detail.

### Configuration of the Pervasive Learning Game

A central question is what the institutional/organizational configuration of such systems should look like so that the learners and institutions in which they are involved actively use the possibilities of pervasive gaming. Together with the technical mass production capability of the system, this is a central success factor for sustained diffusion of the pervasive gaming technology. The objective of such a continuative study should be the formulation of configuration possibilities and recommendations as well as the justification thereof. It must also be determined in which teaching structure the potentials of the pervasive game can best be utilized. In this context it is the task of organizational research to determine the situational variables deemed as relevant and to subsequently draw the right conclusions for the organizational configuration.

### Exhausting the Technology

From a technological perspective, the main challenge is to implement expanded functions of pervasive gaming without needing dedicated front-end systems that restrict the mass production capability. Backend systems should be developed that support advanced pervasive games on existing terminals. A development of this nature includes the implementation of new game components that make the game even more realistic and dynamic. It is expected that such an expansion can lead to even higher flow and activation values.

In addition to improving the backend systems, the rapid technological progress in the mobile phone sector should not be disregarded. The continually changing technological basic conditions will increasingly offer new possibilities for pervasive games. Due to this factor, the concept of a pervasive learning game presented in this paper should be critically reviewed in two to three years with regard to new technology developments, and correspondingly updated.

### Only Good Because it's New?

An additional central research question is whether the positive results for the pervasive game can be attributed at least in part to the short-term effect of novelty or whether the pervasive game represents a sustainable new learning method. To answer this question, time series analyses must be conducted that analyze the change during repeated use of the pervasive game by the same test subjects.

Although it is likely that short-term effects play only a minor role due to the very clear effect of the pervasive game, a study of these effects is an important basis for the sustainable diffusion of pervasive learning games. Only if repeated use of the game yields the desired learning effect will it be appropriate for educational institutions to install the required backend system. Should the positive effects decrease sharply after only one use, the short-term improvement of learning capacities would not justify the costs of purchasing and/or supporting the required pervasive learning game infrastructure.

## CONCLUSION

Using the "Digital Economy" pervasive game, this study presents a modern, multifaceted learning environment that was developed to cope with current problems in knowledge transfer. The potentials of mobile phones were used in a targeted manner as an individual communication and information channel in a pervasive learning game. The results of current research were taken into account in the development of the game concept and the subsequent system implementation in order to gain the most benefit the positive characteristics of pervasive games.

Subsequent to the development of the system, the pervasive game was tested in the learning context using the topic of the digital economy, and compared to a conventional case study. The assessment of the accompanying empirical study showed a very positive effect for pervasive games. Compared to the conventional learning method, the pervasive game produced a higher degree of activation, more positive emotions, an improved attitude towards learning material and greater learning success.

Moreover, a significantly greater flow experience was observed in students in the pervasive game. Beyond the influence of the learning method, the degree of the flow experience correlated with the positive effects and could therefore be considered a decisive factor for the success of the pervasive game. This result re-confirmed the correlations between the flow experience, the activation and emotion constructs discussed in the literature (Drengner & Zanger, 2003; Schiefele & Roussakis, 2006).

The results show that the pervasive game offers a good alternative to the rigid structures of conventional learning methods. The pervasive game promotes active, cooperative and problem-oriented learning that first and foremost serves to generate experiences and consequently the transfer of practical knowledge. On the one hand, the pervasive game promotes networked thinking, and on the other hand it enables the intensified acquisition of key skills needed for everyday professional life.

An advantage of the game presented in this paper compared to previous pervasive games is the high degree of mass production capability of the system that occurs through the use of widely distributed mobile terminals. In contrast to technically complex pervasive games, the course instructor only needs an operational backend system with the corresponding software. In this manner, the basic economic preconditions for a sustainable diffusion of this technology are fulfilled.

## ACKNOWLEDGMENT

This research was partially supported by the European Community under the Information Society Technologies (IST) priority of the 6th Framework Programme for R&D. This document does not represent the opinion of the European Community, and the European Community is not responsible for any use that might be made of its content.

## REFERENCES

Aldrich, C. (2005). *Learning by doing: A comprehensive guide to simulations, computer games, and pedagogy in e-learning and other educational experiences.* San Francisco: John Wiley & Sons.

Amory, A., Naicker, K., Vincent, J., & Adams, C. (1999). The use of computer games as an educational tool: identification of appropriate game types and game elements. *British Journal of Educational Technology, 30*(4), 311–321.

Barth, B., & Cerny, P. (2006). *Kommunikation und IT in Österreich.* Wien: Integral Markt-und Meinungsforschungsgep.m.b.H.

Benford, P., Magerkurth, C., & Ljungstrand, P. (2005). Bridging the physical and digital in pervasive gaming. *Communications of the ACM, 48*(3), 54–57.

Benford, P., Rowland, D., Flintham, M., Hull, R., Reid, J., Morrison, J. et al. (2004). *Savannah, designing a location-based game simulating lion behaviour.* Paper presented at the Conference on Advances in Computer Entertainment.

Bruhn, J., Gräsel, C., Mandl, H., & Fischer, F. (1998). Befunde und Perspektiven des Lernens in Computernetzen. In F. Scheuermann, F. Schwab & H. Augenstein (Eds), *Studieren und weiterbildung mit mulitmedia* (p. 390). Nürnberg.

Conati, C. (2002). Probabilistic assessment of user's emotions in educational games. *Applied Artificial Intelligence, 16*(7-8), 555-575.

Csikszentmihalyi, M. (2000). *Das Flow-Erlebnis, jenseits von Angst und Langeweile - im Tun aufgehen.* Stuttgart: Klett-Cotta.

Dimoco. *Zukunft für Mobile-Marketing. SMS-Werbung auf dem Vormarsch.* Retrieved March 16, 2007, from http,//www.report.at/artikel.asp?kid=&mid=1&aid=6005.

Döring, S. (1997). *Lernen durch Spielen - Spielpädagogische Perspektiven institutionellen Lernens.* Weinheim:Beltz.

Drengner, J., & Zanger, C. (2003). Die Eignung des Flow-Ansatzes zur Wirkungsanalyse von Marketing-Events, *Marketing ZFP, 25*(1), 25-40.

Edelmann, W. (2000). *Lernpsychologie.* Weinheim: Psychologie Verlags Union.

Edelstein, W., & de Haan, G. (2004). Empfehlung 5, Lernkonzepte für eine zukunftsfähige Schule - von Schlüsselkompetenzen zum Curriculum. In Heinrich-Böll-Stiftung und Bildungskommission der Heinrich-Böll-Stiftung (Ed.), *Selbstständig lernen. Bildung stärkt Zivilgesellschaft* (pp. 130-188). Beltz: Weinheim.

Einsiedler, W. (1999). *Das Spiel der Kinder.* Bad Heilbrunn; Klinkhardt.

Euler, D. (2005). Didaktische Gestaltung von E-Learning-unterstützten Lernumgebungen. In Euler, D. & Seufert, C. (Eds): *E-learning in Hochschulen und Bildungszentren - Gestaltungshinweise für pädagogische Innovationen* (pp. 227–242). München: Oldenburg.

Feierabend, P., & Rathgeb, T. (2006). *JIM – Studie. Jugend, Information, (Multi-) Medi.* Stuttgart: Medienpädagogischer Forschungsverband Südwest.

Flitner, A.(2004). *Spielen - Lernen. Praxis und Deutung des Kinderspiels.* München: Beltz.

Forum Mobilkommunikation. (2007). *Klares Votum für das Handy.* Retrieved March 16, 2007, from http,//www.fmk.at/mobilkom/dl/FMK_4-6.pdf.

Garris, R., Ahlers, R., & Driskell, J. E. (2002). Games, motivation, and learning: A research and practice model. *Simulation & Gaming, 33,* 441 - 467.

Gee, J. P. (2003). *What video games have to teach us about learning and literacy.* New York: Palgrave.

Glu Mobile. (2007). *Challenges Mobile Myths With New Consumer Survey.* Retrieved March 13, 2007, from http,//www.glu.com/about/presp.php?press=100106b

Gräsel, C. (1997). Wir können auch anders: Problemorientiertes Lernen an der Hochschule. In Gruber, H. & Renkl, A. (Eds), *Wege zum Können: Determinanten des Kompentenzerwerbs.* Bern: Verlag Hans Huber.

Elbel, G., & Laake, A. (2002). *Handlungsorientierter Unterricht . Qualifikationen für den Arbeitsmarkt – Qualifikationen für das Leben.* Göttingen: Institut für berufliche Bildung und Weiterbildung.

Hirschfelder, D. (2006). Der „Nürnberger Trichter" – Ein Allheilmittel gegen die Dummheit?. In Großmann, U. (Ed), *KulturGUT - Aus der Forschung des Germanischen Nationalmuseums* (pp. 3-5)

Izard, C. E. (1977). *Human emotions.* New York: Plenum Press.

Jegers, K. (2006). *Pervasive GameFlow, Understanding Player enjoyment in Pervasive Gaming.* Paper presented at the 3rd International Workshop on Pervasive Gaming Applications.

Jegers, K., & Wiberg, M. (2006). Pervasive gaming in the everyday world. *Pervasive Computing, 5*(1), 78-85.

Kirriemuir, J., & McFarlane, A. (2006). *Literature review in games and learning.* London: Futurelab.

Kroeber-Riel, W., & Weinberg, P. (2003). *Konsumentenverhalten.* München: Vahlen.

Mandl, H., & Gräsel, C. (1997). Multimediales und problemorientiertes Lernen, Thyroidea – ein Lernprogramm für das Medizinstudium. In Hamm, I. & Müller-Bölling, D. (Eds), *Hochschulentwicklung durch neue Medien* (p.175). Gütersloh.

Merten, J., & Krause, R. (1993). *DAS (Differentielle Affekt Skala).* Arbeiten der Fachrichtung Psychologie, University of Saarland.

Notterman, J. M., & Drewry, H. N. (1993). *Psychology and education, parallel and interactive approaches.* New York and London: Plenum press.

Osgood, C.E., Suci, G. J., & Tannenbaum, P. H. (1975). *The measurement of meaning.* Urbana: University of Illinois Press.

Prensky, M. (2000). *Digital game-based learning.* New York: McGraw-Hill.

Randel, J.M., Morris, B.A., Wetzel, D.C., & Whitehill, B.V. (1992). The effectiveness of games for educational purposes: A review of recent research. *Simulation & Gaming, 23*, 261 - 276.

Research, Research and Consultancy Outsourcing Services (2007). *Europe Mobile Handset Market Analysis.* Retrieved March 16, 2007, from http://www.researchandmarkets.com/reportinfo.asp?report_id=337216

Rheinberg, F., & Vollmeyer, R. (2003). Flow-Erleben in einem Computerspiel unter experimentell variierten Bedingungen. *Zeitschrift für Psychologie, 211*(4), 161–170.

Rheinberg, F., Vollmeyer, R., & Engeser, P. H. (2002). Die Erfassung des Flow-Erlebens. In Stiensmeier-Pelster, J., & Rheinberg, F. (Eds), *Diagnostik von Selbstkonzept, Lernmotivation und Selbstregulation* (pp. 261-279). Göttingen: Hogrefe.

Riedl, A., & Schelten, A. (2000). Handlungsorientiertes Lernen in technischen Lernfeldern. In Bader, R. & Sloane, P. F. (Eds), *Lernen in Lernfeldern. Theoretische Analysen und Gestaltungsansätze zum Lernfeldkonzept* (pp.155 – 164). Eusl: Markt Schwaben.

Schiefele, U., & Roussakis, E. (2006). Die Bedingungen des Flow-Erlebens in einer experimentellen Spielsituation. *Zeitschrift für Psychologie, 214*(4), 207–219.

Talkline, *Talkline - 98 Prozent der jugendlichen Handybesitzer nutzen SMS.* Retrieved March 16, 2007, from http,//www.dsltarife.net/news/900.html

Telephia: *European Subscriber and Device Report, Q3 2006*, Retrieved March 12, 2007, from http://telephia.com/html/Smartphonepress_release_template.html

Thayer, R.E. (1978). Toward a psychological theory of multidimensional activation (arousal). *Motivation and Emotion, 2*(1), 1-34.

Thayer, R.E. (1989). *The biopsychology of mood and arousal.* New York: Oxford University Press.

Thissen, F. (1997). Das Lernen neu erfinden? Konstruktivistische Grundlagen einer Multimedia-Didaktik. In Beck, U. & Sommer, W. (Eds.), *LEARNTEC 97 Europäischer Kongreß für Bildungstechnologie und betriebliche Bildung* (pp. 69 – 79). Karlsruhe.

Thomas, P. (2005). *Pervasive, persuasive e-learning, modeling the pervasive learning space.* Paper presented at the 3rd International Conference on Pervasive Computing and Communications Workshops.

Thomas, P. (2006). Pervasive learning games, explorations of hybrid educational gamescapes. *Simulation & Gaming, 37*(1), 41-55.

Thomas, P. (2006). *From theory to practice, on designing a pervasive learning game.* Paper presented at the 4th Annual IEEE International Conference on Pervasive Computing and Communications Workshops.

Trommsdorff, V. (2004). *Konsumentenverhalten.* Stuttgart: Kohlhammer.

Walther, B.K. (2005). Atomic actions—molecular experience, theory of pervasive gaming. *Computers in Entertainment, 3*(3).

## ADDITIONAL READING

Attewell, J. (2005). *Mobile technologies and learning: A technology update and m-learning project summary.* Retrieved October 12, 2007, from http://www.m-learning.org/archive/docs/ The percent20m-learning percent20project percent20- percent20technology percent20update percent20and percent20project percent20summary.pdf, London: Learning and Skills Development Agency.

Csikszentmihalyi, M. (1996). *Creativity: Flow and the psychology of discovery and invention.* New York: Harper Collins.

Devinder, S., & Zaitun A. B. (2007). Wireless implementation of a mobile learning environment based on students' expectations. *International Journal of Mobile Learning and Organisation, 1*(2), 198-215.

Loveless, A. M. (2007). *Creativity, technology and learning—a review of recent literature.* Report 4 Update, Retrieved October 12, 2007, from http://www.futurelab.org.uk/resources/documents/lit_reviews/Creativity_Review_update.pdf, NESTA Futurelab. Bristol: NESTA Futurelab.

Maciej, K., & Scherm, E. (2005). Mobile Learning—Modetrend oder wesentlicher Bestandteil lebenslangen Lernens? Retrieved October 12, 2007, from http://www.fernuni-hagen.de/BWLOPLA/ME/Mobile-Education.de_Kuszpa_2005.09_Survey-Hagen.pdf

Naismith, L., Lonsdale, P., Vavoula, G., & Sharples, M. (2005). *Literature Review in Mobile Technologies and Learning.* Report 11, Retrieved October 10, 2007, from http://www.nestafuturelab.org/research/reviews/reviews_11_and12/11_01.htm NESTA Futurelab. Bristol: NESTA Futurelab.

Salen, K., & Zimmermann, E. (2003). *Rules of play. Game design fundamentals.* Cambridge, Mass: MIT Press.

Sharples, M. (Ed., 2007). *Big issues in mobile learning: Report of a workshop by the Kaleidoscope Network of Excellence Mobile Learning Initiative.* Nottingham, UK: University of Nottingham, Learning Sciences Research Institute.

# Chapter V
# iPods as Mobile Multimedia Learning Environments:
## Individual Differences and Instructional Design

**Peter E. Doolittle**
*Virginia Tech, USA*

**Danielle L. Lusk**
*Virginia Tech, USA*

**C. Noel Byrd**
*Virginia Tech, USA*

**Gina J. Mariano**
*Virginia Tech, USA*

## ABSTRACT

*In recent years, educators across the globe have begun to employ portable, digital media players, especially iPods, as educational platforms. Unfortunately, while the iPod grows in favor as a mobile multimedia learning environment, relatively little is empirically known about its educational impact. This chapter explores the use of the iPod as an educational platform and reports on a study designed to examine individual differences in iPod use as a mobile multimedia learning environment. This exploration into applied and basic research involving the iPod reveals that iPods are being used across a variety of content areas, educational levels and geographic locations, involving a variety of pedagogies. However, very little research has been conducted to establish the efficacy of the iPod for fostering learning. To address this need, the authors conducted a study that examined the effects of working memory capacity (WMC) on learning within an iPod-based mobile multimedia learning environment.*

## INTRODUCTION

Mobile learning, or m-learning, is typically defined as learning with mobile technologies (see Laouris & Eteokleous, 2005). This type of definition generally emphasizes the ability to move beyond place-bound teaching and learning environments (Goh & Kinshuk, 2006; Seppala & Alamaki, 2003) based on the application of wireless educational technologies (e.g., mobile phones, personal digital assistants, laptop computers, portable digital media players). Educational research into the efficacy of mobile learning and mobile technologies tends to focus on "their use embedded in classroom practice, or as part of a learning experience outside the classroom" (Naismith, Lonsdale, Vavoula, & Sharples, 2006, p. 11). One arena in which this is especially the case is the use of portable digital media players (e.g., iPods, Zunes, MP3 players). In recent years, educators across the globe have begun to employ portable digital media players, especially iPods, as educational platforms (see Belanger, 2005; Cebeci & Tekdal, 2006; Trelease, 2006).

The use of the iPod for educational purposes has included lecture capture at Duke University (USA), podcasting at Auckland University of Technology (New Zealand), foreign language instruction at Astley Community High School (U.K), math instruction at Apollo Parkways Primary School (Australia), and an entire degree at Sligo Institute of Technology (Ireland), to name only a few. Unfortunately, while the iPod grows in favor as a mobile multimedia learning environment, relatively little is known about its educational impact. How well do students learn from podcasts? How are students using iPods to view or re-view lectures? Does listening to native speakers on the iPod affect learners' foreign language listening, writing, or speaking skills? This chapter explores the use of the iPod as an educational platform and reports on a study designed to examine individual differences in iPod use as a mobile multimedia learning environment.

## IPOD RESEARCH

Over the past several years, research addressing the use of the iPod as a mobile multimedia learning environment has included both applied research, which is designed to solve problems, produce products, or fulfill a specific need; and, basic research, which is designed to expand the current knowledge base regarding learning in iPod-based mobile multimedia learning environments. There is, however, a disparity between the depth of applied and basic research, with there being much more applied research than basic research.

### Applied Research: iPods in the Classroom

Mobile multimedia learning environments can take on many forms as technological advancements abound and are being used to supplement and even replace some forms of formal classroom education. The use of iPods and podcasts for educational purposes is a growing trend in the realm of education from primary school through college. Even the military has incorporated these educational tools for learning.

The Navy College Program for Afloat College Education (NCPACE) teamed with Dallas Tele-College to institute educational programs using the iPod for deployed sailors, thereby reducing the number of computers needed on board ships while still allowing the sailors the opportunity to learn (Jay, 2007). In addition, the National Defense University's Information Resources Management (IRM) College not only uses video iPods to deliver education but also to allow students to create assignments for their courses, such as recording interviews with officers.

New Mexico State University (NMSU) offers educational programs to the airmen on Holloman Air Force Base so that they can continue to pursue their education while on deployment (Venegas, 2007). Their iPod program begins with "sociology in a sack" in which the iPod is loaded with sociology lectures and includes support materials such as a syllabus, sample quizzes, and instructions for the program. Upon the end of their offsite operation, the airmen return the iPods to the school and take their exams for course completion. Considering distance education courses often utilize the Internet for most of their materials and lectures, this program can meet the needs of those airmen who are deployed in places where Internet connections are scarce and therefore limit the opportunities to continue their education.

In addition to military use, Apple has reported that there are a plethora of elementary schools, secondary schools, and universities that have reported success with the use of their iPod devices for educational delivery and production. For example, Bob Sprankle, a teacher at Wells Elementary School in Maine, has been using iPods and podcasts with his third and fourth-grade students to teach research, writing, and presentation skills as well as the embedded technology expertise that inevitably accompanies projects such as these (Apple Education, 2007a).

Vanderbilt University (2007) has partnered with Sheridan School District in rural Arkansas to transform students' use of a lengthy bus ride into educational time. This project, the Aspirnaut Initiative, was formed as a way to elevate the mathematics and science achievement of the rural community base and one aspect of this project includes the use of iPod podcasts for the students (Clark, 2007; Vanderbilt University, 2007). This three-year project includes iPods programmed with lessons that focus on science and math and as a bonus the students get to keep the iPods if they complete the full study. The school district's superintendent states that beyond the focus of the current project, the students may actually receive course credit for the time they spend learning while commuting to and from school (Paul, 2007).

High school students preparing for college admissions are also a targeted audience for iPod use. Kaplan, a college-preparation company, recently began to offer customizable iPod programs to students via Apple's iTunes store (Man, 2007). This innovative SAT preparatory program allows students to purchase materials for viewing and practice with timed and untimed tests and the student can even vary the level of difficulty of the programs they choose. Kristen Campbell, the national director of SAT and ACT programs for Kaplan Test Prep, thinks that students will embrace these types of programs because they allow students to study in a medium in which they are familiar and comfortable. Currently, Kaplan is the only test-prep company offering these types of study programs via the iPod iTunes store.

The University of Minnesota's School of Public Health (SPH) features podcasts for many different uses for students and the general public (University of Minnesota School of Public Health, 2007b). They feature weekly health segments (Public Health Moment) and public lectures sponsored by SPH (Public Lectures). Their "Public Health Matters" section contains student and faculty-created podcasts that cover a wide variety of topics related to current research within the department and other public health issues (University of Minnesota School of Public Health, 2007a).

The University of Utah (UU) uses podcasts for a wide variety of information-sharing endeavors. They present podcasts of lectures, interviews, speeches, performances, and have developed several series of lectures for download (University of Utah, 2007). UU features five major categories for podcast feeds including humanities, politics and society, health and wellness, science and technology, and arts and culture. They also have several other podcast program offerings to include the Genetic Science Learning Center, Hinckley Institute of Politics, Humanities Happy Hour, and Sci-Fi Fridays.

The University of Wisconsin-Madison began their podcasting integration with iPods in 2005 to assist students with language acquisition in a German language class (Apple Education, 2007b). Their goals included incorporating class content (language instruction) into the daily lives of their students outside of the traditional classroom models and making homework more fun for the student. Instructors report that their project actually helps students learn the German language more easily and increases student motivation because they format some of their podcasts in a fun radio show style with fictional callers; thus decreasing straight lecture format and time for the students.

iPod initiatives at Georgia College and State University (GCSU) began in 2002 and have evolved into more than lectures and music to enhance the classroom education of its students (Georgia College and State University, 2006). By the 2006-2007 academic year, there were nearly 50 iPod-based projects designed to reduce in-class lecture time and increase time for students and faculty members to collaborate on experiences and discuss the most important course content areas. These innovative iPod projects have spawned other technological advancements for the university as well, such as their iColony project, a virtual learning community for students with similar interests; and iDreamers, a group of faculty and staff members interested in further developing plans for iPod usage in the future.

The University of Virginia's (UVA) Web site boasts an enormous amount of iPod-based information available for their students as well as the general public. They list 20 categories of information to choose from ranging from academic lectures to science and research (University of Virginia, 2007). The available podcasts also include recurring series, such as Engaging the Mind, Technology in World History, and others. In addition, each school at UVA is represented with links for individual podcasts related to topics of interest for that particular area of expertise and interest.

While many higher education institutions are employing iPod technology, there are few who are formally evaluating their iPod and/or podcasting programs. Duke University was one of the first higher education institutions in the United States to implement an iPod program. In August of 2004, more than 1,600 entering first-year students were given iPods (Duke University, 2005). Approximately 48 courses implemented iPod content during the first year of the program. These courses included music, foreign languages social science, and humanities (Duke University, 2005). Additionally, first-year engineering students used their iPods in the engineering programs' required computational methods course.

As part of the program evaluation conducted by Duke's Center for Instructional Technology, data were gathered via several methods: student and faculty focus groups; a survey of first-year students and faculty; course feedback; and conversations with staff, administrators, and campus stakeholder groups. As part of the evaluation, the Center for Instructional Technology wanted to discover how iPods were instructionally used. They found five categories of academic use for the iPod: disseminating course content, recording classroom activities, recording field notes, supporting study of the content, and storing and transferring files (Duke University, 2005). The evaluation also revealed several benefits in utilizing the iPods. The overall benefits reported after the first year of the program included the convenience of portability, easy digital recording, flexible location, increased student interest and engagement, and enhanced support for individual learning differences.

Despite the benefits of using the iPod in the classroom, Duke students and faculty reported several problems (Duke University, 2005). One problem concerned the challenges of integrating multiple systems for storing, accessing, sharing, and distributing content. Another issue was the limited pre-existing documentation and training resources regarding academic uses. Furthermore, integrating the iPods with Duke's existing technology infrastructure was difficult. PC users were also affected by the limited documentation and training. The inadequacy of documentation and training may have contributed to another problem: a lack of awareness and accurate knowledge among students and faculty regarding how iPods function and their academic applications. A final set of difficulties with the iPods was bulk purchasing and licensing issues.

At the institutional level, Duke University reported several benefits from incorporating iPods into the classroom. The evaluators found four major institutional impacts (Duke University, 2005). The first was increased collaboration and communication between campus technology groups, which resulted in wider planning and improvement of Duke's infrastructure and technology services. The second impact was the widespread attention resulting from the program. Duke gained new contacts, partnerships, and collaborations with other institutions of higher education, publishers, and technology vendors. Third, the program sparked conversations between faculty, administrators, students, and staff about how to best implement technology in the classroom. Consequently, the 2005-2006 Duke Digital Initiative was created with continued iPod use and further incorporation of other technologies. Finally, the iPod program increased overall awareness of Duke's commitment to technology.

However, there were some problems with the evaluation, especially regarding the survey response rate. Only 28 percent of first-year students involved in the program responded to the survey, while only 13 percent of first-year student faculty responded (Duke University, 2005). The low numbers suggest that perhaps not all benefits or problems were accounted for. The program also began with no formal objectives, a limitation remedied in the 2005-2006 Duke Digital Initiative. Finally, cost was not addressed in the evaluation. While the implementation program cost $500,000, there were no data included regarding how much money the university would have to spend to continue such a program.

Cost was addressed, however, in the 2005-2006 program evaluation report (Duke University, 2006). One of the main challenges mentioned in the report was the difficulty of acquiring multimedia course content in a cost-effective and timely manner. Additionally, cost was alluded to in other challenges. Ongoing planning, training, and support is an expensive challenge due to how rapidly technology changes. Furthermore, the university lacked the classroom capabilities to meet the demand for video recording of classroom presentations. Upgrading classroom infrastructure to meet this demand would be an expensive endeavor, although as part of the 2007-2008 Initiative infrastructure development is planned.

Cost was not the only challenge. The evaluators again found that faculty and students need more training on how to use the iPod and to transition classroom pedagogy into the multimedia environment (Duke University, 2006). Improvements slated for the 2007-2008 year include more development and utilization of support services and teaching strategies in addition to further research of technology in learning and instruction.

Besides Duke University, many other American colleges and universities have implemented iPod programs and podcasting as a means to advance instruction. The University of Michigan's School of Dentistry started a podcasting program in response to first-year dental students' anxiety regarding the quantity of information presented in their courses (Brittain, Glowacki, Van Ittersum, & Johnson, 2006). Originally, the students asked for video recordings, so the School of Dentistry asked an advisory group to study the situation and recommend solutions. In the first pilot study, the focus was on what type of media format was best for lecture review. The results revealed that students preferred an audio-only format over video or audio-synced PowerPoint. The preference was due largely to the mobility of the audio files.

Two more pilot studies were conducted to fully examine the issue (Brittain et al., 2006). The second pilot study focused on the costs involved with recording lectures. Recording via the computer, although more expensive than recording with the iPod, was deemed more suitable by students and faculty due to its superiority in audio recording quality. The School of Dentistry decided to use a computer and automated recording, processing, and posting tools to reduce staff time, thus decreasing the overall costs involved with the project.

Finally, the third pilot study looked at the dissemination of recordings, in particular the file format students preferred most (Brittain et al., 2006). MP3 and AAC files were both found to be equally well-liked, so the school decided to post both file types to a centrally located Web site for downloads. Due to convenient access, students' usage of the recordings rose from 28 percent to 57 percent .

In addition to the many American higher education institutions using iPods and podcasting, international institutions are also bringing iPod technology into their classrooms. One such institution is Charles Stuart University in Australia. Researchers investigated and evaluated the use of podcasting in an undergraduate and postgraduate class by studying the listeners, producers, and educators involved in podcasting the course (Chan, Lee, & McLoughlin, 2006). Some of the students were part of the production team, which created podcasts targeted towards other students in the class, who served as the listeners.

To evaluate the "listening" students, an online survey was employed to gather data concerning students' podcast use, the educational value of the podcasts, and the knowledge and skills gained through listening to the podcasts (Chan et al., 2006). The researchers had a 42 percent response rate to their survey. The majority of students used the available podcasts frequently, felt they were useful, and helped with assignments.

To evaluate the "producers," the researchers used a focus group interview (Chan et al., 2006). The focus group questioned participants about motivation, benefits related to involvement, skills developed, lessons learned, and suggestions for improvement. After coding the transcripts, the researchers found that the main area the producing students elaborated on was the skills they developed from participating in the production of the podcasts, especially with regard to teamwork, communication, and critiquing others.

A study at another international institution looked at an undergraduate course in communication (Edirisingha, Rizzi, Nie, & Rothwell, 2007). At Kingston University in London, students enrolled in the Introduction to Intercultural Communication course served as participants. Students were surveyed and interviewed as part of the study. The survey revealed that 50 percent of the students had not listened to any of the podcasts available to them via the virtual learning environment (VLE). The main reason given for not listening to the podcasts was a lack of time. Other reasons included; they did not know the podcasts were available, they had technical difficulties accessing the materials, and they did not view the podcasts as relevant for learning. Additionally, of those who did listen to the podcasts, many did not download the podcasts because they could access the material any time they wanted. Furthermore, 31 percent noted that they preferred their MP3 player to be dedicated to music listening only, indicating they did not want the podcasts residing on their portable players.

Students were also surveyed regarding their activities while listening to podcasts (Edirisingha et al., 2007). The majority of respondents (47%) reported they only listened and did not participate in other activities, while 33 percent noted that they took notes while listening. Only seven percent reported doing other activities while listening to the podcasts. These findings suggest that the students, for the most part, engaged with the podcasts.

In addition to the previous findings, the researchers asked students about the key learning objectives related to the podcasts (Edirisingha et al., 2007). Students rated the following statements favorably: (a) Podcasts were useful for me to know more about the assessed work, (b) Podcasts were useful in preparation for workshops/seminars, and (c) Podcasts provided a good supplement to the other learning material for this module. Other statements regarding motivational aspects were rated highly as well. The students, however, did not believe that the podcasts were helpful in organizing weekly activities or made good use of their time. A study by Edirisingha, Salmon, and Fothergill (2007), looking at learners' attitudes, found similar results.

Although students may remark that the podcasts are useful and aid their learning, there is little empirical evidence to support these claims. While many institutions are now using podcasts and iPod technology, more information about how humans learn from mobile technology is needed before such programs are implemented in the curriculum.

## Basic Research: iPods as Mobile Multimedia Instructional Environments

Part of the allure of iPod-based mobile technology is the mobility (being able to use the technology in multiple and mobile environments). Unfortunately, while there is substantial applied research involving the use of iPods in education, there is very little basic research examining how students learn from iPods. Due to the paucity of educationally related basic empirical literature on the use of the iPod, the following studies are taken from a wide array of disciplines.

Salvucci, Markley, Zuber, and Brumby (2007) examined participants' ability to drive under the distraction of iPod usage. Participants were asked to search for and play music, podcasts, and videos while using a driving simulator. Salvucci et al. (2007) found that searching for and the selection of all three media types had a significant detrimental effect on driver performance, measured by lateral deviation (drivers ability to maintain a central lane position) and speed change (drivers ability to maintain a constant speed). Overall, Salvucci et al. (2007) discovered that simply listening to songs or podcasts had no significant effect on driving performance, but that searching for and selecting media led to an increase in driver distraction and a decrease in driver performance.

O'Hara, Mitchell, and Vorbau (2007) also found distraction during mobile technology use in a study addressing individuals' daily use of mobile technologies, including iPods. O'Hara et al. (2007) studied the behaviors of established users of mobile technologies and found that the technology was being consumed in a variety of places and times such as at home, work, cafes, gyms, and airport lounges, as well as while driving to work, eating lunch, and going to bed. Reasons for using these mobile technologies included avoiding others, blocking out the surrounding environment, sharing videos, using time more effectively, and managing solitude. During the study, O'Hara et al. also found that the use of mobile technologies could lead to divided attention; that is, users often needed to divide their attention between watching a video using the mobile technology and monitoring others around them or being aware of events such as the arrival of a desired bus (2007).

While Salvucci et al. (2007) and O'Hara et al. (2007) found that iPod-based mobile technology is used in a variety of locations, for a variety of reasons, and that iPod use may be distracting, Hürst, Lauer, and Nold (2007) focused on screen size or the presentation modality effect on learning. Hürst et al. (2007) had participants view an animation on algorithm application using either an iPod, with a 2.5" screen, or a laptop computer. In addition Hurst el al. (2007) had participants view either an animation with on-screen text explanations or animation with audio narration. Hürst et al. (2007) deduced that there was no difference in student learning based on screen size (i.e., iPod versus laptop), but that students who viewed the animation with audio narration presentation learned more than students who viewed the animation with on-screen text presentation. Hurst et al.'s (2007) findings that learning was improved through the use of animation with audio narration, as compared to animation with on-screen text, is in agreement with previous research (Ginns, 2005; Mayer & Moreno, 1998). Also, when asked which of the devices participants preferred for learning, half expressed no preference and the other half preferred the device they had used to engage the material with audio narration presentation. Ultimately, Hürst et al. (2007) concluded, "maybe the most important observation from our experiment is that people actually were able to learn something by using mobile devices" (p. 164).

The findings of Hürst et al. (2007) have demonstrated that iPods may be used effectively to foster learning and the findings of Salvucci et al. (2007) and O'Hara et al. (2007) lead credence to the question of whether or not attention, or distractibility, is a variable of interest in the use of iPods in education. With the issues of learning and attention in iPod-based mobile multimedia instruction environments in mind, a study was conducted.

## IPODS AND INDIVIDUAL DIFFERENCES: A STUDY OF WORKING MEMORY

iPods are currently being used in many primary and secondary schools, as well as colleges and universities, as an educational platform. As the previous discussion of basic research into iPod use demonstrates, however, little is known regarding how the iPod-based educational environment interacts with the learner. One aspect of the learner that may bear on the use of the iPod as an educational platform is the learner's attention. Attention has previously been demonstrated to be an essential component of learning in desktop multimedia instructional environments (Mayer, 2005b; Mayer & Moreno, 1998; Moreno & Mayer, 2000). This previous research has demonstrated that when attention is split between two sets of stimuli, such as an animation with on-screen narrative text (Mayer & Moreno, 1998); or when attention is diverted from instructional content to non-instructional content, such as when an animation is accompanied by non-essential sounds, music, or videos (Moreno & Mayer, 2000); learning suffers. In addition, Mautone and Mayer (2001) have indicated that when attention is guided *toward* instructional content, learning improves.

Given the iPods' mobility and its 2.5" screen, it is reasonable to question whether attention plays a special and important role in learning complex mental tasks in an iPod-based multimedia learning environment. In general, learning complex mental tasks has been sensitive to individual differences in attention; specifically, in individual differences in the ability to control attention (Daneman & Carpenter, 1980; Oberauer, Süb, Schulze et al., 2007; Wilhelm, & Wittmann, 2000). The ability to control attention is a central component of *working memory capacity* (WMC; Kane, Bleckley, Conway, & Engle, 2001; Kane & Engle, 2003), where WMC is defined as the ability to (a) maintain the current cognitive task in working memory, (b) maintain task relevant information in working memory, and (c) retrieve task relevant information from long-term memory (Feldman Barrett, Tugade, & Engle, 2004). A study was completed to assess the importance of attentional control, low WMC versus high WMC, in stationary and mobile multimedia learning environments using the iPod.

### Working Memory Capacity

Working memory capacity (WMC) represents the ability of an individual to control his or her attentional resources relative to working memory, especially in the presence of distraction (Feldman Barrett et al., 2004; Unsworth & Engle, 2007). That is, WMC moves beyond working memory storage capacity (see Miller, 1956) to include both storage *and* processing capacity (see Daneman & Carpenter, 1980; Kane & Engle, 2003). Thus, ultimately, WMC is a measure of control: the ability to control the maintenance of information in working memory (storage) and the retrieval from long-term memory of information relevant to a current problem or situation (processing). This control is most evident when there are distractions, internal (e.g., thoughts, drives, feelings) or external (e.g., talking, music, motion), challenging the attentional system (Unsworth & Engle, 2007).

Is this type of attentional control important? Previous research has indicated that individuals with high WMC outperform individuals with low WMC on measures of general fluid intelligence (Conway, Cowan, Bunting et al., 2002; Kane et al., 2004), long-term memory activation (Cantor & Engle, 1993), attentional control (Kane et al., 2001; Rosen & Engle, 1997), resistance to proactive interference (Kane & Engle, 2000; Lustig, May, & Hasher, 2001), primary memory maintenance and secondary memory search (Unsworth & Engle, 2007), and resistance to goal neglect (Kane & Engle, 2003; Roberts, Hager, & Heron, 1994). While these studies have indicated a positive effect for WMC on memory attributes, there is also extensive literature indicating that high WMC positively affects more academic pursuits, such as reading comprehension (Daneman & Carpenter, 1980), language comprehension (Just & Carpenter, 1992), vocabulary learning (Daneman & Green, 1986), reasoning (Conway et al., 2002; Kyllonen & Christal, 1990; cf. Buehner, Krumm, & Pick, 2005), computer language learning (Shute, 1991), lecture note taking (Kiewra & Benton, 1988), Scholastic Aptitude Test performance (Turner & Engle, 1989), mnemonic strategy effectiveness (Gaultney, Kipp, & Kirk, 2005), and story telling (Pratt, Boyes, Robins, & Manchester, 1989). This research has demonstrated a strong, positive relationship between variations in WMC and variations in complex cognitive task performance.

## A Study of iPods and Individual Differences

To investigate the effects of high and low WMC students' learning in stationary and mobile iPod-based multimedia learning environments, 76 undergraduate students (52 males and 24 females) were taken from a larger population of 294 students who were administered the OSPAN working memory span test. The OSPAN test measures working memory capacity and requires students to solve math sentences aloud (e.g., is (2 + 6) – 2 = 4?) while also maintaining a list of unrelated words in working memory (La Point & Engle, 1990; Turner & Engle, 1989). Using the OSPAN test scores, the upper and lower quartiles of the original 294 students were designated as high (n = 38) and low (n = 38) WMC, respectively. These 76 students were then randomly assigned to either the stationary instruction group (n = 38) or the mobile instruction group (n = 38).

All students, regardless of WMC (high or low) or instructional group (stationary or mobile) engaged in a multimedia tutorial designed to address the workings of a car brake. The multimedia tutorial consisted of a narrated Flash® animation explaining how a car brake works (see Mayer & Anderson, 1992). This explanation included drawings of a foot pressing a brake pedal, a piston moving inside a master cylinder, brake fluid moving out of the master cylinder, brake fluid expanding smaller pistons in the wheel cylinder, and these smaller pistons pushing the brake shoes against the brake drum (see Figure 1). The animation from the foot stepping on the brake pedal to the brake shoes pressing against the brake drum lasted 30 seconds; however, this 30 second animation was played three times in order to accommodate the narration. Thus, each multimedia instructional episode lasted 90 seconds. Finally, this car brake multimedia tutorial was presented on a 5$^{th}$ generation iPod (i.e., video iPod) with a 2.5" screen and Altec Lansing® headphones.

*Figure 1. Sample animation screens and accompanying verbal content from the brake tutorial (animations adapted from and text quoted from Mayer & Anderson, 1992, p. 446)*

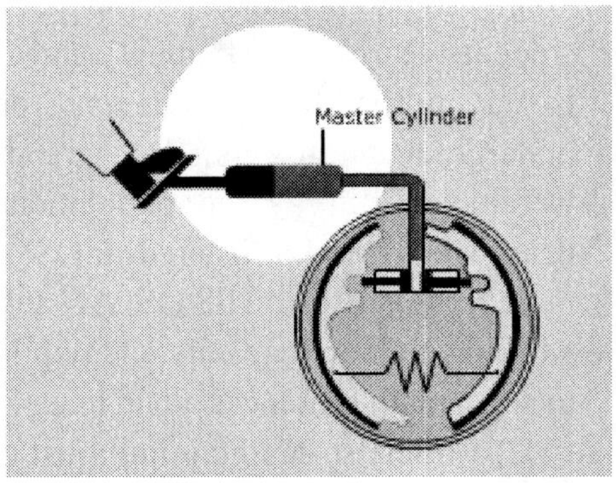


1. "When the driver steps on the car's brake pedal, a piston moves forward inside the master cylinder."

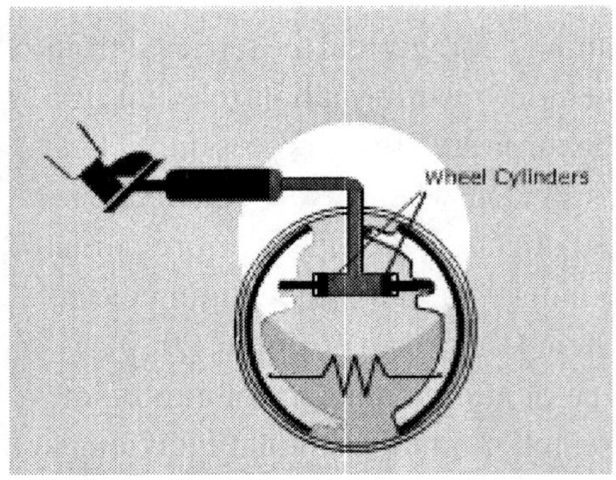


2. "The piston forces brake fluid out of the master cylinder and through the tubes to the wheel cylinders."

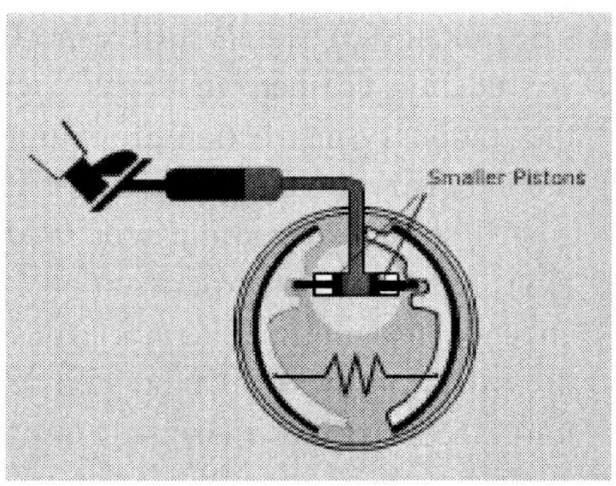


3. "In the wheel cylinders, the increase in fluid pressure makes a set of smaller pistons move."

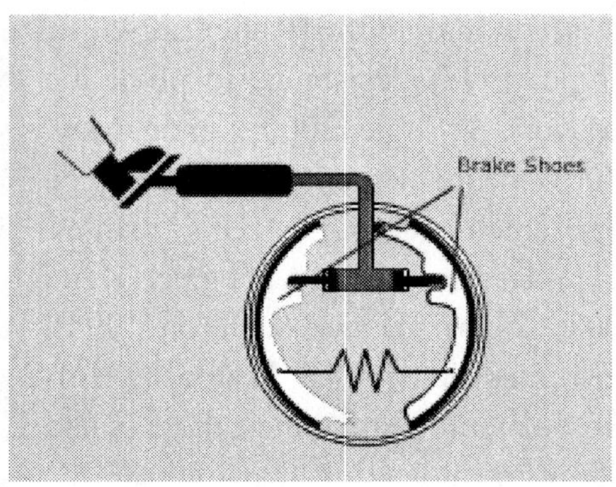


4. "These smaller pistons activate the brake shoes."

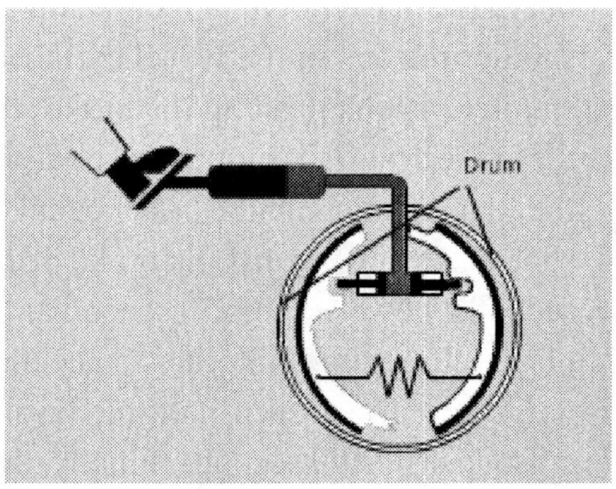


5. "When the brake shoes press against the drum, both the drum and the wheel stop or slow down."

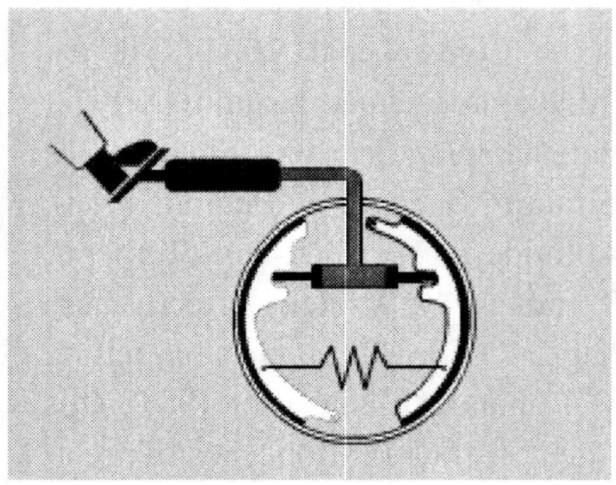

Students in the stationary instruction group viewed the multimedia tutorial on an iPod while sitting in a chair at a desk in a computer lab. Students in the mobile instructional group were first provided with a random number from 1 to 3 and then asked to walk 25 yards down a hallway, and back, repeatedly, until the multimedia tutorial came to an end. Every five yards along this walk was a two-sided sign on the floor that included the numbers 1, 2 and 3, and above each number an arrow pointing left or right (see Figure 2). Participants were instructed to walk to the side of the sign indicated by the arrow above the number to which they were assigned. This walking and navigating, while engaging in the multimedia tutorial, simulated the type of mobile environment one might encounter while trying to walk across a college campus.

After students finished engaging in the multimedia tutorial, whether in the stationary or mobile instructional group, they were seated and asked to answer both recall questions assessing students' ability to remember the essential knowledge to which they were exposed and their ability to express this knowledge, and transfer questions assessing students' ability to apply the knowledge recalled to new, yet similar situations (i.e., near transfer). Students' recall of the car brake tutorial was assessed using a single open-ended question: "Please provide an explanation of how a car brake works." Two trained raters evaluated each student's recall response (r = .90) and provided one point for the inclusion of each of the following idea units, regardless of wording: "(a) driver steps on brake pedal, (b) piston moves forward inside master cylinder, (c) piston forces brake fluid out to the wheel cylinders, (d) fluid pressure increase in wheel cylinders, (e) small pistons move, (f) small pistons activate brake shoes, (g) brake shoes press against drum, and (h) drum and wheel stop or slow down" (Mayer & Anderson, 1992, p. 450).

Students' ability to transfer knowledge from the car brake tutorial was assessed using four questions (see Mayer & Anderson, 1992, p. 449): (a) "Why do brakes get hot?"; (b) "What could be done to make brakes more reliable, that is, to make sure they would not fail?"; (c) "What could be done to make brakes more effective, that is, to reduce the distance needed to bring a car to a stop?"; and (d) "Suppose you press on the brake pedal in your car but the brakes do not work. What could have gone wrong?" Two trained raters evaluated each student's responses (r = .83) based on acceptable answers established by Mayer and Anderson (1992). Acceptable answers to the first transfer question, "Why do brakes get hot?" included friction causes brakes to get hot; acceptable answers to the second transfer question, "What could be done to make brakes more reliable, that is, to make sure they would not fail?" included maintain a back up system or use a system to cool the brakes; acceptable answers to the third transfer question, "What could be done to make brakes more effective, that is, to reduce the distance needed to bring a car to a stop?" included using a brake shoe that is more sensitive to friction or providing a smaller gap between the brake shoe

*Figure 2. Directional signs used in the mobile learning environment*

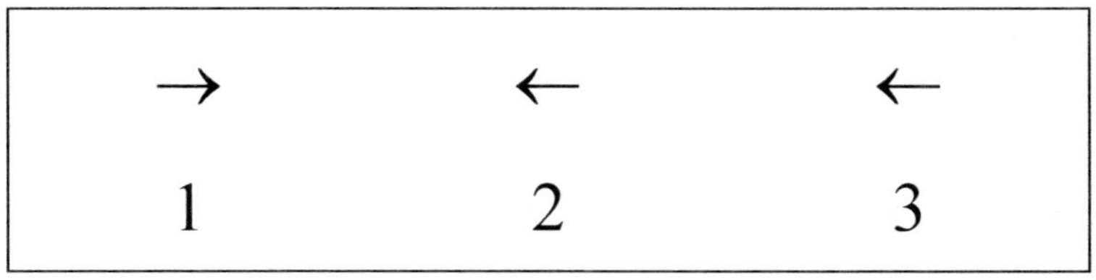

and brake drum; and finally, acceptable answers to the fourth transfer question, "Suppose you press on the brake pedal in your car but the brakes do not work. What could have gone wrong?" included that there may be a leak in the brake fluid line or that the brake pads are worn.

## Results and Discussion

The results of the study addressed high and low WMC students' cognitive performance in stationary and mobile multimedia learning environments. Overall, there was a significant difference in cognitive performance between participants who learned in a stationary learning environment as compared to participants who learned in a mobile learning environment. Specifically, participants in the stationary learning environment recalled more idea units (M = 5.74, SD = 1.75) from the car brake multimedia tutorial than participants in the mobile learning environment (M = 4.92, SD = 2.96), F(1,72) = 4.72, MSE = 25.94, Cohen's *d* = 0.33, p = .03. In addition, participants in the stationary learning environment transferred more car brake knowledge (M = 5.58, SD = 0.59) than participants in the mobile learning environment (M = 4.97, SD = 1.21), F(1,72) = 19.02, MSE = 12.36, Cohen's *d* = 0.69, p = .00. These results indicated that participants recalled and transferred more information when learning in a stationary versus a mobile learning environment. Specifically, it was found that students who learned about car brakes using a portable digital media player (i.e., iPod), while navigating a walking course that required attention to the path taken, performed significantly more poorly on measures of recall and transfer than students who learned while simply sitting at a desk.

In addition to the stationary/mobile learning results, an analysis of the WMC data indicated that high WMC students recalled more idea units (M = 5.92, SD = 2.67) that low WMC students (M = 4.74, SD = 2.07), F(1,72) = 7.27, MSE = 39.94, Cohen's *d* = 0.49, p = .00. Similarly, for transfer, high WMC students generated more valid transfer responses (M = 5.61, SD = 0.67) than low WMC students (M = 4.95, SD = 1.16), F(1,72) = 29.96, MSE = 13.63, Cohen's *d* = .69, p = .00. These results are consistent with previous findings regarding a general WMC effect (Doolittle, 2007; Unsworth & Engle, 2007), that high WMC students outperform low WMC students on recall and transfer after engaging in a multimedia tutorial.

The preceding results demonstrate that students learn more from a short cause-and-effect multimedia tutorial when stationary, as opposed to when mobile, and that high WMC students learn more than low WMC students. However, a significant question yet to be answered is whether or not there are individual differences in stationary/mobile learning when WMC is taken into account. The answer is mixed. Specifically,

*Table 1. Means and standard deviations of recall and transfer from engaging in a car brake multimedia tutorial for instructional groups (stationary, mobile) and WMC groups (low, high)*

| | Instructional Environment | | | | | | | |
|---|---|---|---|---|---|---|---|---|
| | Stationary | | | | Mobile | | | |
| | Recall | | Transfer | | Recall | | Transfer | |
| | M | SD | M | SD | M | SD | M | SD |
| Low WMC | 5.33 | 1.68 | 5.50 | 0.51 | 3.71 | 2.33 | 4.00 | 1.35 |
| High WMC | 6.43 | 1.69 | 5.71 | .072 | 5.63 | 3.10 | 5.54 | 0.65 |

*Note: Maximum score for recall was 8. Maximum score for application was 7.*

there was no interaction between instructional group (stationary, mobile) and WMC (high, low) for recall of the car brake tutorial. There was, however, a significant interaction for transfer (see Table 1), $F(1,72) = 11.98$, $MSE = 7.79$, $p = .00$. An examination of the means indicates that mobile learning environments are less advantageous to low WMC students than high WMC students. This examination was confirmed using a contrast analysis comparing the mobile-low WMC group to the remaining three groups (i.e., mobile-high WMC, stationary-low WMC, stationary-high WMC), $F(1,72) = 27.53$, $MSE = 0.65$, Cohen's $d = 1.33$, $p = .00$. These interaction results indicate that if deep learning, as measured by transfer, is the goal of instruction, mobile multimedia learning environments significantly disadvantage students with low WMC. Namely, students with the least attentional control and the most susceptibility to distraction (low WMC students) performed the poorest under conditions that required significant attentional control: learning while mobile.

## IPODS AS MOBILE MULTIMEDIA LEARNING ENVIRONMENTS

The preceding discussion and research has indicated that (a) iPods are currently being used as educational platforms; (b) podcasting is the typical, although not the only, pedagogy of choice; and (c) students can and do learn from engaging with the iPod. However, previous research also indicates that (d) little basic empirical research is being conducted on the efficacy of the iPod as a learning platform; but (e) recent empirical research has demonstrated that students with low working memory capacity perform less well in mobile learning environments than students with high working memory capacity. These findings begin to bring into focus the landscape of iPod-based instruction, although a crisp and clear picture is still distant.

At the most basic level, it is important to note that Hürst et al. (2007) and the study reported within this chapter both found that students learn as the result of engaging in an iPod-based instructional episode. In the case of Hürst et al., students learned strategic knowledge, a search algorithm, while students in the current study learned procedural knowledge, how a car brake works. These foundational findings provide support for the use of the iPod as an instructional platform.

The research of Salvucci et al. (2007) and the current study, however, also demonstrate that the iPod must be used with caution and that all students may not benefit equally from the instructional use of the iPod. Salvucci et al. found that, generally, using the iPod can require significant attentional resources in conditions of distraction, for example, using the iPod while driving. Further, the present study revealed that students with considerable attentional resources—high working memory capacity—were unaffected by distractions when using the iPod; however, students with fewer attentional resources—low working memory capacity—were negatively affected by distractions while using the iPod. These findings demonstrate that students' attentional control has a significant impact on the efficacy of the iPod as an instructional platform.

As mobile learning opportunities have become ubiquitous with the advent of portable, lightweight devices (e.g., mp3 players, mobile phones, PDAs) it is imperative that educators, instructional designers and instructional technologists maintain a 360° view of the educational playing field. At this nexus of pedagogy, design and technology "there is a tension… that comes from the fact that most mobile devices in current use are not designed specifically for education or training but rather for personal information management or personal communication" (Kukulska-Hume & Taxler, 2005, p. 3). Kukulska-Hume and Taxler goes on to state that "future projects ought to address learning gains more directly, to gather evidence of what can be learnt using these [mobile] devices" (p. 189). These comments emphasize that for mobile learning to be educationally successful there is a need to focus on the development of educationally appropriate technology, individually appropriate pedagogy, and the interaction between the technology and the individual.

## FUTURE RESEARCH DIRECTIONS

Research and explorations involving mobile multimedia learning environments utilizing portable digital media players is just beginning. While the initial findings are promising, much is still unknown. What specific types of pedagogies work best with specific types of mobile technologies based on specific areas of instructional content? What individual difference variables (e.g., spatial ability, reading ability, prior knowledge, working memory capacity) affect students' success when learning in mobile multimedia learning environments? How can mobile multimedia learning environments be made more accessible to students with special needs? How can, or should, the use of mobile multimedia learning environments be integrated into the daily educational environments of students?

Currently, educators are forced to create mobile multimedia learning environments within mobile devices that were not originally designed to foster learning; cell phones, MP3 players, and PDAs were designed more as communication tools than educational tools. Given that these communication tools are being adapted for educational purposes, it is essential that educators share their adaptations so that others may benefit from and expand these initial efforts. In addition, however, these case studies of adaptation must include rigorous evaluation and assessment so that there is evidence from which to make educated decisions. Without rigorous evaluation and assessment the field of mobile learning runs the risk of devolving into a faddish application of technological flash without substance.

## REFERENCES

Apple Education. (2007a). Elementary school students become podcasting pros. Retrieved August 7, 2007, from www.apple.com/education/profiles/wells

Apple Education. (2007b). Podcasts: Earmarked for success. Retrieved August 7, 2007, from www.apple.com/education/profiles/wisconsin-madison

Belanger, Y. (2005, June). *Duke University iPod fist year experience final evaluation report.* Retrieved July 1, 2007, from http://cit.duke.edu/pdf/ipod_initiative_04_05.pdf

Brittain, S., Glowacki, P., Van Ittersum, J., & Johnson, L. (2006). Formative evaluation strategies helped identify a solution to a learning dilemma. *EDUCAUSE Quarterly, 29*(3). Retrieved August 6, 2007, from http://www.educause.edu/apps/eq/eqm06/ eqm0634.asp?bhcp=1

Buehner, M., Krumm, S., & Pick, M. (2005). Reasoning = working memory ≠ attention. *Intelligence, 33*, 251-272.

Cantor, J., & Engle, R. W. (1993). Working-memory capacity as long-term memory activation: An individual-differences approach. *Journal of Experimental Psychology: Learning, Memory, and Cognition, 19*(5), 1101-1114.

Cebeci, Z., & Tekdal, M. (2006). Using podcasts as audio learning objects. *Interdisciplinary Journal of Knowledge and Learning Objects, 2*, 47-57.

Chan, A., Lee, M. J. W., & McLoughin, C. (2006). Everyone's learning with podcasting: A Charles Sturt University experience. *Proceedings of the 23rd Annual Ascilite Conference: Who's Learning? Whose technology?* Retrieved August 1, 2007, from http://www.ascilite.org.au/conferences/sydney06/proceeding/pdf_papers/p171.pdf

Clark, J. (April 10, 2007). School bus becomes mobile classroom thanks to iPods. Retrieved

August 7, 2007, from http://www.todaysthv.com/news/news.aspx?storyid=44254

Conway, A. R., Cowan, N., Bunting, M. F., Therriault, D. J., & Minkoff, S. R. (2002). A latent variable analysis of working memory capacity, short-term memory capacity, processing speed, and general fluid intelligence. *Intelligence, 30*, 163-183.

Daneman, M., & Carpenter, P. A. (1980). Individual differences in working memory and reading. *Journal of Verbal Learning and Verbal Behavior, 19*, 450-466.

Daneman M., & Green, I. (1986). Individual differences in comprehending and producing words in context. *Journal of Verbal Learning and Verbal Behavior, 19*, 450–466.

Doolittle, P. (2007). *The Effect of Working Memory Capacity on Multimedia Learning: Does Attentional Control Result in Improved Performance?* Manuscript submitted for publication.

Duke University. (2005). iPod first year experience final evaluation report. Retrieved August 1, 2007, from http://cit.duke.edu/pdf/reports/ipod_initiative_04_05.pdf

Duke University. (2006). Duke Digital Initiative end of year report. Retrieved August 10, 2007, from http://www.duke.edu/ddi/pdf/ddi_exec_report_05_06.pdf

Edirisingha, P., Rizzi, C., Nie, M., & Rothwell, L. (2007). Podcasting to provide teaching and learning support for an undergraduate module on English language and communication. *Turkish Online Journal of Distance Education, 8*(3). Retrieved July 22, 2007, from http://tojde.anadolu.edu.tr/tojde27/articles/article_6.htm

Edirisingha, P., Salmon, G., & Fothergill, J. (2007). Profcasting – A pilot study and guidelines for integrating podcasts in a blended learning environment. In U. Bernath & A. Sangrà (Eds.) *Research on competence development in online distance education and e-learning.* Oldenburg: BIS-Verlag. Retrieved July 20, 2007, from http://hdl.handle.net/2381/404

Feldman Barrett, L., Tugade, M. M., & Engle, R. A. (2004). Individual differences n working memory capacity and dual-process theories of the mind. *Psychological Bulletin, 130*(4), 553-573.

Gaultney, J. F., Kipp, K., & Kirk, G. (2005). Utilization deficiency and working memory capacity in adult memory performance: Not just for children anymore. *Cognitive Development, 20,* 205-213.

Georgia College and State University. (2006). The GCSU iPod story. Retrieved August 7, 2007, from http://ipod.gcsu.edu/GCSU%20iPod%20Story/index.html

Ginns, P. (2005). Meta-analysis of the modality effect. *Learning and Instruction, 15*, 313-331.

Goh, T., & Kinshuk. (2006). Getting ready for mobile learning – adaptation perspective. *Journal of Educational Multimedia and Hypermedia, 15*(2), 175-198.

Hürst, W., Lauer, T., & Nold, E. (2007). A study of algorithm animations on mobile devices. *Proceedings of the ACM Technical Symposium on Computer Science Education* (SIGCSE), *39*(1), 160-164.

Jay, E. F. (2007). iPods viable for military education. *Military Advanced Education Online Edition.* Retrieved August 7, 2007, from http://www.military-advanced-education.com/print_article.cfm?DocID=2043

Just, M., & Carpenter, P. A. (1992). A capacity theory of comprehension: Individual differences in working memory. *Psychological Review, 99*, 122-149.

Kane, M. J., & Engle, R. W. (2000). Working-memory capacity, proactive interference, and divided attention: Limits on long-term memory retrieval. *Journal of Experimental Psychology: Learning, Memory, and Cognition, 26*(2), 336-358.

Kane, M. J., & Engle, R. W., (2003). Working-memory capacity and the control of attention: The contributions of goal neglect, response competition, and task set to Stroop interference. *Journal of Experimental Psychology: General, 132*(1), 47-70.

Kane, M. J., Bleckley, M. K., Conway, A. R., & Engle, R. W., (2001). A controlled-attention view of working memory capacity. *Journal of Experimental Psychology: General, 130*(2), 169-183.

Kane, M. J., Hambrick, D. Z., Tuhoski, S. W., Wilhelm, O., Payne, T. W., & Engle, R. W. (2004). The generality of working memory capacity: A latent-variable approach to verbal and visuospatial memory span and reasoning, *Journal of Experimental Psychology: General, 133*(2), 189-217.

Kiewra, K. A., & Benton, S. L. (1988). The relationship between information-processing ability and notetaking. *Contemporary Educational Psychology, 13,* 33-44.

Kukulska-Hulme, A., & Traxler, J. (Eds.). (2005). *Mobile learning: A handbook for educators and trainers.* London: Routledge.

Kyllonen, P. C., & Christal, R. E. (1990). Reasoning ability is (little more than) working-memory capacity?! *Intelligence, 14*, 389-433.

La Pointe, L. B., & Engle, R. W. (1990). Simple and complex word spans as measures of working memory capacity. *Journal of Experimental Psychology: Learning, Memory and Cognition, 16,* 1118-1133.

Laouris, Y., & Eteokleous, N. (2005). *We need an educationally relevant definition of mobile learning.* Paper presented at the mLearn2005: 4th World Conference on mLearning, Cape Town, South Africa.

Lustig, C., May, C. P., & Hasher, L. (2001). Working memory span and the role of proactive interference. *Journal of Experimental Psychology: General, 130*(2), 199-207.

Man, L. (2007). Software lets students use iPods to cram for SATs. *The Kansas City Star.* Retrieved August 14, 2007, from http://www.technewsworld.com/ rsstory/58464.html

Mautone, P. D., & Mayer, R. E. (2001). Signaling as a cognitive guide in multimedia learning. *Journal of Educational Psychology, 93*(2), 377-389.

Mayer, R. E. (Ed.) (2005a). *The cambridge handbook of multimedia learning.* Cambridge: Cambridge Press.

Mayer, R. E. (2005b). Principles of multimedia learning based on social cues: Personalization, voice, and image principles. In R. E. Mayer (Ed.), *The cambridge handbook of multimedia learning* (pp. 201-212). New York: Cambridge.

Mayer, R. E., & Anderson, R. B. (1992). The instructive animation: Helping students build connections between words and pictures in multimedia learning. *Journal of Educational Psychology, 84*(4), 444-452.

Mayer, R. E., & Moreno, R. (1998). A split-attention effect in multimedia learning: Evidence for dual processing systems in working memory. *Journal of Educational Psychology, 90*(2), 312-320.

Miller, G. (1956). The magical number seven, plus or minus two: Some limits on our capacity for processing information. *Psychological Review, 63*, 81-97.

Moreno, R., & Mayer, R. E. (2000). A coherence effect in multimedia learning: The case for minimizing irrelevant sounds in the design of multimedia instructional messages. *Journal of Educational Psychology, 92*(1), 117-125.

Naismith, L., Lonsdale, P., Vavoula, G., & Sharples, M. (2006). Literature review in mobile technologies and learning (Report 11). Bristol, UK: Futurelab.

Oberauer, K., Süß, H. M., Schulze, R., Wilhelm, O., & Wittmann, W. W. (2000) Working memory capacity—facets of a cognitive ability construct. *Personality and Individual Differences, 29*, 1017-1045.

O'Hara, K., Mitchell, A. S., & Vorbau, A. (2007). Consuming video on mobile devices. *Proceedings of the SIGCHI Conference on Human Factors in Computer Systems, 1*, (pp. 858-866).

Paul, R. (April 17, 2007). Rural Arkansas school turns school buses into classrooms. *Ars Technica.* Retrieved August 7, 2007, from http://arstechnica.com/news.ars/post/ 20070417-rural-arkansas-school-uses-ipods-to-turn-schoolbuses-into-classrooms.html

Pratt, M.W., Boyes, C., Robins, S., & Manchester, J. (1989). Telling tales: Aging, working memory, and the narrative cohesion of story retellings. *Developmental Psychology, 25*, 628-663.

Roberts, R. J., Hager, L. D., & Heron, C. (1994). Prefrontal cognitive processes: Working memory and inhibition in the antisaccade task. *Journal of Experimental Psychology: General, 123*, 374-393.

Rosen, V. M., & Engle, R. W. (1997). The role of working memory capacity in retrieval. *Journal of Experimental Psychology: General, 126*(3), 211-227.

Salvucci, D. D., Markley, D., Zuber, M., & Brumby, D. P. (2007). iPod distraction:

Effects of portable music-player use on driver performance. *Proceedings of the SIGCHI Conference on Human Factors in Computer Systems, 1*, 243-250.

Seppala, P., & Alamaki, H. (2003). Mobile learning in teacher education. *Journal of Computer Assisted Learning, 19*, 330-335.

Shute, V. J. (1991). Who is likely to acquire programming skills? *Journal of Educational Computing Research, 7*, 1-24.

Trelease, R. (2006). Diffusion of innovations: Anatomical informatics and iPods. *The Anatomical Record (Part B: New Anatomy), 289B*, 160-168.

Turner, M. L., & Engle, R. W. (1989). Is working memory capacity task dependent? *Journal of Memory and Language, 28*, 127-154.

University of Minnesota School of Public Health. (2007a). Public health matters. Retrieved August 7, 2007 from http://blog.lib.umn.edu/sphpod/sphprograms/

University of Minnesota School of Public Health. (2007b). Tune in to our podcasts and blogs. Retrieved August 7, 2007 from http://www.sph.umn.edu/podcast/home.html

University of Utah. (2007). Podcasting from the U. Retrieved August 7, 2007 from http://www2.utah.edu/podcast/info.php

University of Virginia. (2007). University of Virginia podcasts and Webcasts. Retrieved August 7, 2007, from http://www.virginia.edu/uvapodcast/

Unsworth, N., & Engle, R. W. (2007). The nature of individual difference n working memory capacity: Active maintenance in primary memory and controlled search from secondary memory. *Psychological Review, 114*(1), 104-132.

Vanderbilt University. (2007). Aspirnaut Initiative: A pilot educational initiative to elevate the mathematics and science achievement of rural community K-12 students who commute on long bus rides to school. Retrieved August 7, 2007, from http://www.aspirnaut.org/

Venegas, V. (January 27, 2007). Teaching the troops: NMSU offers distance education through "Sociology in a Sack". Retrieved August 7, 2007, from http://www.nmsu.edu/ ~ucomm/Releases/2007/january/sociologysack.htm

## ADDITIONAL READING


Ackerman, P. L., Beier, M. E., & Boyle, M. O. (2002). Individual differences in working memory within a nomological network of cognitive and perceptual speed abilities. *Journal of Experimental Psychology: General, 131,* 567–589.

Attewell, J. (2005). *Mobile technologies and learning: A technology update and m-leraning project summary.* London, UK: Learning and Skills Development Agency.

Baddeley, A.D., & Hitch, G.J. (1974). Working Memory, In G.A. Bower (Ed.), *The psychology of learning and motivation: advances in research and theory* (Vol. 8, pp. 47-89), New York: Academic Press.

Chen, J., & Kinshuk (2005). Mobile technology in educational services. *Journal of Educational Multimedia and Hypermedia, 14*(1), 91-109.

Chinnery, G. M. (2006). Going to the MALL: Mobile assisted language learning. *Language Learning & Technology, 10*(1), 9-16.

Conway, A. R. A., Cowan, N., & Bunting, M. F. (2001). The cocktail party phenomenon revisited: The importance of working memory capacity. *Psychonomic Bulletin & Review, 8,* 331–335.

Daneman, M., & Carpenter, P. A. (1983). Individual differences in integrating information between and within sentences. *Journal of Experimental Psychology: Learning, Memory, and Cognition, 9*, 561-584.

Daneman M., & Green, I. (1986). Individual differences in comprehending and producing words in context. *Journal of Verbal Learning and Verbal Behavior.* 19, 450–466.

Daneman, M., & Tardif, T. (1987). Working memory and reading skill re-examined. In M. E. Coltheart (Ed.), *Attention and performance XII: The psychology of reading* (pp. 491-508). Hove, UK: Lawrence Erlbaum.

Friedman, N. P., & Miyake, A. (2000). Differential roles for visuospatial and verbal working memory in situation model construction. *Journal of Experimental Psychology: General, 129*, 61-83.

Hambrick, D., & Oswald, F. L. (2005). Does domain knowledge moderate involvement of working memory capacity in higher –level cognition? A test of three models. *Journal of Memory and Language, 52*, 377-397.

Kalyuga, S., Chandler, P., & Sweller, J. (1999). Managing split-attention and redundancy in multimedia instruction. *Applied Cognitive Psychology, 13*, 351-371.

Kane, M. J., Poole, B. J., Tuholski, S. W., & Engle, R. W. (2006). Working memory capacity and the top-down control of visual search: Exploring the boundaries of "executive attention." *Journal of Experimental Psychology: Learning, Memory, and Cognition, 32*, 749-777.

Maag, M. (2006). iPod, uPod? An emerging mobile learning tool in nursing education and students' satisfaction. Proceedings of the 23rd annual conference of the Australasian Society for Computers in Learning in Tertiary Education: Who's learning? Whose technology? Retrieved from http://www.ascilite.org.au/conferences/sydney06/proceeding/pdf_papers/p92.pdf

Massey, A. (2006). Design, development, and assessment of mobile applications: The case of problem-based leanring. IEEE Transactions on Education, 49(2), 183-192.

Mayer, R. E. (2001). *Multimedia learning.* Cambridge, UK: Cambridge University Press.

Mayer, R. E., & Anderson, R. B. (1991). Animations need narrations: An experimental test of a dual-coding hypothesis. *Journal of Educational Psychology, 83*(4), 484-490.

Mayer, R. E., Heiser, J., & Lonn, S. (2001). Cognitive constraints on multimedia learning: When presenting more material results in less understanding. *Journal of Educational Psychology, 93*(1), 187-198.

Mayer, R. E., & Moreno, R. (1998). A split-attention effect in multimedia learning: Evidence for dual processing systems in working memory. *Journal of Educational Psychology, 90*(2), 312-320.

Mayer, R. E., & Sims, V. (1994). For whom is a picture worth a thousand words? Extensions of a dual-coding theory of multimedia learning. *Journal of Educational Psychology, 86*, 389-401.

McQuillan, J. (2006). *iPod in education: The potential for language acquisition.* Cupertino, CA: Apple, Inc.

Metcalf, D. S. (2006). *mLearning: Mobile learning and performance in the palm of your hand.* Amherst, MA: HRD Press.

Moreno, R., & Mayer, R. E. (1999). Cognitive principles of multimedia learning: The role of modality and contiguity. *Journal of Educational Psychology, 91*(2), 358-368.

Oberauer, K., Süß, H.-M., Wilhelm, O., & Wittmann, W. W. (2003). The multiple faces of working memory: Storage, processing, supervision, and coordination. *Intelligence, 31,* 167–193.

Pasnik, S. (2007). *iPod in education: The potential for teaching and learning.* Cupertino, CA: Apple, Inc.

Pedersen, J. E., & Marek, E. A. (2007). Technology integration: PDAs as an instructional and reflective tool in the science classroom. *Contemporary Issues in Technology and Teacher Education, 7*(1), 521-528.

Roschelle, J. (2003). Keynote paper: Unlocking the learning value of wireless mobile devices. *Journal of Computer Assisted Learning, 19*, 260-272.

Rouet, J., Levonen, J. J., & Biardeau, A., (2001). *Multimedia learning: Cognitive and instruction issues.* London: Pergamon.

Shah, P., & Miyake, A. (1996). The separability of working memory resources for spatial thinking and language processing: An individual differences approach. *Journal of Experimental Psychology: General, 125*, 4-27.

Sharples, M. (2006). *Big issues in mobile learning.* Nottingham, UK: Learning Sciences Research Institute.

Swan, K., Kratcoski, A., & van 't Hooft, M. (2007). Highly mobile devices, pedagogical possibilities, and hoe teaching needs to be reconceptualized to realize them. *Education Technology, 3*, 10-12.

Unsworth, N., & Schrock, J. C., & Engle, R. W. (2004). Working memory capacity and the antisaccade task: Individual differences in voluntary saccade control. *Journal of Experimental Psychology: Learning, Memory, and Cognition, 30*(6), 1302-1321.

Wagner, E. D. (2005). Enabling mobile learning. *EDUCAUSE Review, 40*(3), 41-52.

# Chapter VI
# From Individual Learning to Collaborative Learning—Location, Fun, and Games:
## Place, Context, and Identity in Mobile Learning

**Martin Owen**
*Medrus Learning, UK*


## ABSTRACT

*This chapter explores mobile learning where location and game like or playful activity adds to the context of learning. The relationship between space, play, and the development of context and learner identity is explored through an examination of the issues concerning the context of space, narratives, and engagement. There is a discussion of the meta-knowledge and specific learning attributes we would want to encounter in mobile game like learning. These issues are further explored in three case studies of learning activities which have been designed such that the context of location and game like or playful learning is significant. The examples include simple games based on multiple choice questions, a complex multi-role simulation and an environmental tagging and hypermedia project. The case is made for the potential of the context of location, and game-like learning in mobile learning.*


## INTRODUCTION

This chapter looks at two specific aspects of mobile learning. First, learning where location plays an important part in the context of learning, and subsequently the chapter will refer to locative learning. Second, the chapter will also explore (though not in every case describe) elements of fun and playfulness that are explicitly designed into the learning experience. The case examples, described in the chapter, are all drawn from the work of Futurelab, an independent research and development organization based in Bristol, UK during its first five years. Futurelab was set up to catalyze and network education, technology and media organizations into innovative activity that had the potential to transform learning and teaching (see also Owen 2005). It was funded mainly from government sources, but there has also been funding from commercial sources (detailed on Futurelab's Web site.)

Specifically, this chapter investigates some general research questions about technology and learning and then looks at some particular research questions in each of the case studies. The major questions we have are:

- Can we make playful, game-like activities that deliver educational experiences we value?
- Can we use technology that adds to the context of space, which increases the learning potential of that space?

And,

- Can we use technology to add new playfulness to space—technologies that add learning context to space—that significantly change that space for learning purposes?

These questions broaden into examining issues like:

- What relationships are important between space, play, and the development of context and learner identity?
- What meta-knowledge and specific learning attributes can we foster in mobile game like learning?
- What is the importance of narrative, the temporal rhythms, the context, projected identities, collaboration, and creativity in the design of mobile learning experiences?

Three cases will be described. Each case contributes in its own way to increasing our understanding of the ways that technology and location can work together to provide new learning experiences. Each case study in itself generated its own research questions and the approach has been bottom-up pragmatic understanding rather than any attempt at grand theory building. However, from these concrete examples we can begin to look for a coherent picture of some of the issues and opportunities that locative learning presents. However, it is also fair to say that there is a fairly explicit commitment to constructivist learner centric approaches and to a strong element of user-involved design in all the cases.

The three case studies are:

- Light Aircraft Pilot: where we use a particular approach to multiple choice quizzes to encourage game like learning based in a real world context;
- Savannah: where we use GPS enabled, wirelessly networked HP iPAQs to create a multi-player role playing game where the school soccer pitch is turned into an African Savannah and the students become a pride of lions; and
- Mudlarking in Deptford: The use of GPS enabled handheld devices to add information to the spaces that the learner is in.

Light aircraft pilot is part of the COLLAGE (collaborative learning platform using game-like enhancements) project, where students visit historic sites or sites of ecological interest and play out roles mediated by hand-held computers and mobile phones. *COLLAGE* was funded by the European Union (EU) e-learning program. The chapter also draws on two other EU funded programs, which had Futurelab as a partner. These ran over a longer timescale than most Futurelab projects, and all three had a specific intent of outlining research themes in learning with technology in location. The other two projects were *La Piazza* (part of the Kaleidoscope Network of Excellence) which was a pre-research exercise in intergenerational learning in technology enhanced public spaces; and *Puente* (funded by the EU Minerva program) which has been a continuation of *La Piazza* with an opportunity to implement some ideas. The research questions that arise from the EU projects are focused on how to design experiences. Specifically these experiences were for locative learning with fun and games as additional dimensions and designing for intergenerational learning experiences.

This chapter will not address technological issues other than is necessary to describe the application. Typically any project suffers the issues of working at the limits of what technology can do or in using technology in new ways. There is a confident belief that it will get better and what we need for locative learning can and will be delivered at a price we can afford.

## BACKGROUND

Located learning is a specific subset of what has been called mobile or m-learning. M-learning has been defined by a list of potential applications—quizzes, games and information services provided on handheld devices (see mlearning.org). A great advantage of learning with these devices is that they permit learning on the move and learning anywhere. The fact that the learning is anywhere is what distances such mobile learning from what is described in this chapter. This chapter describes learning *somewhere*, where the location and the context of the learner in that location matter to the learning process.

There are both similarities and differences between *learning somewhere* and *learning anywhere*. There are many interesting problems to be solved by the *anywhere learning* scenario. Access to such learning is easy in the developed world with mass ownership of phones and the importance of phones in the culture of young people (see, for instance, Prensky (2004)). The mass ownership of powerful interactive pocket devices can have a significant impact on learning. However, there is also something powerful about learning from where you are, either because of the place itself, what is in the place, or the people who are sharing the place with you. Alternatively, it could be the journey you are on, and that you are leaving a trace or trail of your presence. There are very significant overlaps between learning anywhere and located learning. In *La Piazza* we describe Learning Nomads (see Ackermann et al., 2006) who are characterized by the emergence of new habits of learning because learning and communication technology is always with them. The learner will be able to make a choice when they feel the context of the place is a necessary part of that which they want to learn or just choose to learn anything (or nothing).

Technology-supported, located learning is not new. There have been audio tours of historic attractions and museums for some time. The addition of wireless networked learning is the innovative feature. In 1996, Jim Spohrer envisioned a portable educational device, the Worldboard, which would overlay the virtual and physical worlds (Spohrer, 1999). The Worldboard would give its user access to the history of its surroundings, make visible hidden features such as buried pipelines, and allow the user to add virtual annotations to that place. Such devices are now almost realized as exemplified by the research of the Ordnance Survey in the UK, and the work of Chris Parker, Chris Phillips and Andrew Radburn in their Augmented Reality and Zapper experiments (Bryant, 2003; Revell, 2004).

Located learning, using a mobile phone or similar devices can allow fieldwork to take on completely new dimensions. When the whole world becomes a classroom, it seems irrelevant to sit in school to learn history, geology, botany, or environmental studies. Our research questions however demand we take more account of learner motivation and the context of space in a way that it becomes meaningful to the students socio-cultural context; space becomes *place*.

*Figure 1. Ordnance survey experiment with information overlays from mobile devices from direction sensitive GPS systems (Zapper)*

*Figure 2. Ordnance survey experimental system for augmented reality urban information overlays*

## Games and Play

Another significant thread in the mobile activities we have developed is playfulness, fun, and games in the learning activities. Mobile learning provides interesting contexts and new affordances for the design and implementation of games. The location, mobility and the inter-player communication capacity of the devices all contribute to innovative possibilities for game and play designs. There has been an increased interest in the educational use of computer games. A lot of this has to do with motivation (see Prensky, 2001), which is a significant part of a successful learning intervention, however there are other reasons as well; play allows the learner to try out roles and project themselves into new identities (Gee, 2003; Shaffer, 2007). An additional factor is that play with *mobile* technologies also allows for a more embodied experience. In other words, you learn with your whole body, you do not just squat in front of a screen (Schusterman, 2005)

There are some characteristics identified in games research that are useful to designing mobile games. Some of the earliest work on computer games (Malone, 1981) focused on how they might sustain intrinsic motivation. Malone describes three factors: challenge, fantasy and curiosity. Challenge implies that the outcome of an activity should be at the edge of the potential performance of the learner as undertaking tasks well within their capability removes the game element. Fantasy implies that we can project our current identity into some new role through a narrative supplied by the game. Curiosity implies that the outcomes are varied and uncertain as it is less entertaining to watch and participate in a game where the result is already clear. Malone and many recent writers have drawn on ideas presented by Csikszentmihalyi (1992) who describes entry into *flow*, a state where the mental (and embodied) self is absorbed in an activity to produce a virtuous circle of mental and physiological engagement. This is common to the games. The video associated with the Savannah project has students claiming as much in their own language; "it was good being a lion" etc. In turn Csikszentmihalyi describes the influence of the anthropologist Roger Callois (1958/2001) in describing four broad classes of games that are endemic in human society:

- *Agon*, which have competition as their main element;
- *Alea*, which are games of chance;
- *Ilnix*, or vertigo, the fun to be had on a roller-coaster; and
- *Mimicry,* in which alternative realities such theatre and dance are performed in groups.

Clearly all forms are possible with mobile devices. I shall describe *Savannah* in a case study below that demonstrates *agon, alea, mimicry* and, if the students are particularly active, *ilnix* as well. At the time of writing, the Wii product from Nintendo is popular especially because of new game play arising from physical movement and *ilnix*. The games developed in COLLAGE tend to involve mimicry, however they are organized as teams of students and are therefore open to competition.

Shaffer (2007) also discusses the value of games in allowing students to project identities and to adopt the behavior of others. He describes *epistemic games*, in which school students take on and practice adult roles like *investigative science journalist* or *town planner* and use the computer tools that are appropriate to that role. In the COLLAGE project we have specific examples of *environmental journalist* and *light aircraft pilot* that are mobile enhanced, location based learning games that also demonstrate Shaffer's epistemic concept and are examples of Callois' mimicry. The intention of gaming in education is clearly more than motivation, and in the cases we describe there is an element of role projection and the development of identity. Vygostsky (1933) and Singer et al. (2006) point to the importance of play in coming to understand the world both cognitively and in social-emotion growth, and we can infer from Callois's *ilnix* that we also come to understand the world's kinaesthetics through play as well. Vygotsky reinforces the importance of play in understanding the world at an early age because play allows us to disassociate thought from current perceptions:

"*Since a situation is always communicated, psychologically through perception, and perception is not separated from affective and motor activity, it is understandable that with his consciousness so structured, the child cannot act otherwise than as constrained by the situation—or the field—in which he finds himself. In play, things lose their motivating force. The child sees one thing but acts differently in relation to what he sees. Thus, a situation is reached in which the child begins to act independently of what he sees*" Vygotsky (1933).

However the importance of this continues in later life. In the same paper Vygotsky (1933) notes:

"***Henceforth play is such that the explanation for it must always be that it is the imaginary, illusory realization of unrealizable desires.*** *Imagination is a new formation that is not present in the consciousness of the very young child, is totally absent in animals, and represents a specifically human form of conscious activity. Like all functions of consciousness, it originally arises from action. The old adage that children's play is imagination in action can be reversed: we can say that imagination in adolescents and schoolchildren is play without action.*" Vygotsky (1933) (emphasis added)

Thus, by structuring learning in play, we are allowing engagement in fantasy and actions that would otherwise be impossible. The concept of *Savannah* came to me when considering how I might have improved playing "cowboys" when I was a child had I access to the technology we used in Savannah. By using mobile technology and game-attributes there is a major reconnection with the games and play that have featured as human learning over the eons to the concepts and methods of computer game play of today.

However, as well as traditional values in play, there are all the reported advantages of the computer game generations. There are meta-learning effects in learning from computer games. Gee (2003) provides a list of 36 things we learn from computer games. These are Prensky's useful translations of the 36 things and an attempt to classify them by attributes that may model experiential learning.

1. Doing and reflecting
2. Appreciating good design
3. Seeing interrelationships
4. Mastering game language
5. Relating the game world to other worlds
6. Taking risks with reduced consequences
7. Putting out effort because they care
8. Combining multiple identities
9. Watching their own behavior
10. Getting more out than what they put in
11. Being rewarded for achievement
12. Being encouraged to practice
13. Having to master new skills at each level
14. Tasks being neither too easy nor too hard.
15. Doing, thinking and strategizing
16. Getting to do things their own way
17. Discovering meaning
18. Reading in context
19. Relating information
20. Meshing information from multiple media
21. Understanding how knowledge is stored
22. Thinking intuitively
23. Practicing in a simplified setting
24. Being led from easy problems to harder ones

25. Mastering upfront things needed later
26. Repeating basic skills in many games
27. Receiving information just when it is needed
28. Trying rather than following instructions
29. Applying learning from problems to later ones
30. Thinking about the game and the real world
31. Thinking about the game and how they learn
32. Thinking about the games and their culture
33. Finding meaning in all parts of the game
34. Sharing with other players
35. Being part of the gaming world
36. Helping others and modifying games, in addition to just playing.

In designing our games such as Savannah and those in COLLAGE this list provide a cross check that the game experience is richer than the supposed learning content alone. These attributes are not intrinsic to games, however. In our experience, it seems that what is a good game design allows these attributes to be addressed explicitly. One of the goals of the COLLAGE project is to provide some support for teachers to develop game like learning themselves and checklists like this provide explicit support for their design processes.

Almost inseparable from the notion of play is the role of narrative in playful learning scenarios. Part of the toolkit developed in the COLLAGE program is material support for workshops to derive narrative from place and context and then in turn map the narrative to learning goals and a game design. Game and location significantly come together. Stories have always been important for education, and *place* (as opposed to plain "space") has a central role in any story. Geography has always been an implicit component of education as explored by Casey (1997) who stresses the importance of place in human activity. Learning in place is critically part of human learning and mobile technology clearly has a lot to offer. In developing this link between location and playfulness we have used location in two ways. In Savannah we take an ordinary space—in this case a school sports field—and use technology to convert it into a different space; we add *unreality*. In most other cases, particularly in the Mudlarking example, we use technology to add to what is already in the location. The key integrating word between learner, location and the mediating narratives is *context*, and getting the technology to support context. This is one of the major non-trivial problems for ubiquitous mobile learning.

The critical issue for "learning in place" activities in mobile learning is fundamentally linked to the concept of context. Dourish (2004) develops two distinct elements to explore the issue of context in ubiquitous computing which is a fundamental of located learning. He says:

*"First, I have argued that context is an emergent property of occasions of interaction, rather than being a stable, objective set of features that externally characterize activity. Context remains critically important for understanding, contextualizing and disambiguating forms of activity and information, but it is in the nature of context to be continually negotiated and redefined. Second, I have argued that these contextual properties take on their meaning or relevance through their relationship to forms of practice; that is, it is engaged action around artifacts and information that make those artifacts meaningful and relevant to people. The meaning of a technology, then, cannot be divorced from the ways that people have of using it"* Dourish (2004).

The tension Dourish presents between the two aspects of context is reflected in the experiences described here. Even in a game, where we have strongly mediating narratives (suggesting and scaffolding practice) there is a constant to and fro from being in and out of a game character or role - being participant or pupil - and personal re-configuring of the space that the learner is in. This tension shows up in the interaction between imagination as described by Vygostsky on play, and the immediacy of location. Young (2002) points in particular to the perception-action behavior in computer game play, simultaneity of what we might term reflection and action.

Some of our tools developed in COLLAGE and Create-A-Scape (developed from the Mudlarking experience) provide environments that can also be authored by learners. There is symmetry between user and creator. This is a particular feature of modern Internet software (sometimes referred to as Web 2.0). This seems particularly true of locative applications, and in particular, Google's opening of their maps and Google Earth applications has lead to a flowering of locative-based creative collaborative activity (see for instance http://googlemapsmania.blogspot.com/#top). The overall creativity emerges from a network effect; a small action by the many creating a resource that is valued by all. Thus by adding personal annotations to some ubiquitous information system attached to some location, repeatedly by many visitors, new knowledge and patterns emerge. Further, more traditional views of creativity are possible with symmetrical tools and systems. The toolsets allow students to create their own location based games and other media to share with others.

In conclusion to this overview of the issues, when we make a holistic combination of the issues in location, games and learning we note the overlaps between place, the narrative of place,

*Table 1. The co-design method sequence*

| Stage | Method |
|---|---|
| Deciding what to research | Consulting with stakeholders: (Government, media and technology industry, educationalists) |
| Understanding what needs to be done | Commissioning leading experts in the field to undertake a literature survey written to be understood by stakeholders |
| Calling for Ideas | Publishing an open call for ideas- providing context information from lit survey |
| Selecting ideas | Filtering over 100 ideas down to 3, Using representatives of stake-holders to judge what ideas to take forward based on originality, potential educational benefit, potential contribution to knowledge, technical feasibility, and the nature of the potential development and research partnership |
| Discovery | A deeper literature review in relation to the specific idea rather than the generic area. A commercial review of existing need/provision<br>Produce a background paper. |
| Design workshops round 1 | Bringing in multidisciplinary team (subject experts, designers, technology experts, education researchers, teachers) to contribute knowledge to paper-prototype some concepts to expand the idea.<br>Taking the idea to children (or whoever end-users will be) for additional paper-prototyping.<br>Consolidating this knowledge as design brief. |
| Wire-framing | Developing an elaborated model of what we expect. |
| Design Workshop 2 | A repeat of previous workshops with more concrete examples including children's work. |
| Build | Initial working prototype |
| Trial 1 to trial n | Iteratively test and review prototype with end users. Gather data from trials: video, interview, questionnaires – whatever is appropriate |
| Project review | Original stakeholders, potential new end-users. |
| Post-project publication | In video, print and web form. Material presented both for academic publication and for a wider audience. Conference and more public presentation. Press releases and interviews. |

identity and the narrative of identity. Narrative seems to be a powerful linkage, and attention to the learners' engagement with, or creation of, narratives, seems to provide a thematic undercurrent to the design and development of activities.

## METHODOLOGIES

Futurelab works to different rhythms and outputs when compared to university research laboratories. Everything Futurelab does is done in collaboration with others. Partners can be individuals, freelance workers, large or small media companies, large or small technology companies, universities and of course schools and other educational institutions. All projects, including those described, are relatively short in the first instance, usually six months, and if they demonstrate value they are extended. The case studies do not present a model of implementation of theories; neither do the theories above have explicit empirical evidence derived from the implementation of the cases. Rather, it is a case of a dynamic between developing practice and reflecting upon practice. It is an iterative process that is grounded in practice and the role of reflection. Building theory and confirming experimental hypotheses are part of the iteration, however the main aim is to improve practice (teaching and educational media design) as born out of experience.

The process is of co-design. Design with inventors, technologists, artists, academic researchers, and importantly end users, typically teachers and learners. The sequence of the method is shown in Table 1.

This complex sequence is subject to rigorous project management methodologies. The frequent reference back to stakeholders ensures a high level of triangulation and questioning of evidence. It is better suited for the development of activity, tools and practice than producing scholarly reports. However, what is lacking in formal types of experimental rigor such as the use of control groups and pre-and post testing is compensated for by stake-holder validation. In the case examples that follow, the first is taken from a part of the process, whilst Savannah and Mudlarking have been through the whole of the cycle.

## THE CASE EXAMPLES

### Light Aircraft Pilot

This is a simple illustration of being *game like*; choosing to implement as narrative and focusing on identity. The example arises in an early stage in the methodology process, using the software with teachers and students to discover issues that might arise. Some of the issues were about the limitations of mobile learning solutions such as legibility of screens in bright sunlight situations, finding situations where having both hands free was important, and so on. This helps discover very practical points to address in implementation such as how and when do you expect the student to consult the mobile device. However going trough the iterative process in co-design also sensitizes the co-design partner to possibilities, as in this case.

The scenario involves the use of a simple multiple choice question (MCQ) creation tool implemented on a mobile device. High school physics students used the application on a field visit to a private airfield. The aim of the experience is to provide contextualization of physics knowledge gained in the classroom and to stimulate further interest in science by investigating the science of flight. The students would gather information using the hand-held device and test their knowledge *in situ*. The teacher had devised a series of questions that could be answered in the field based on information that would be supplied by knowledgeable people in the location either as part of their demonstration or asking practitioners directly while on location. The students would have a guided tour around the airfield and an extended period examining a light airplane. The original intention of the teacher was to use the MCQ authoring system of the mobile device to produce a list of questions that would focus and direct the student learning to key points. However, the teacher's questions mainly asked for factual recording or recall of what the students had, or will have, seen and heard. However it is possible, using the technology available to the teacher, to reconfigure the activity in such a way that the narrative of the location can create a role-play where the solutions to the problems arise in the role.

A specific piece of information that is part of the experience is the fact that when airspeed falls below a level that flight can be sustained (i.e., the aircraft will stall), a horn sounds in the wings that is solely powered by the movement of air across the device. This horn scream is a highly audible failsafe device.

In the original formulation of the system, a question on this topic looked like the following:

When the airspeed falls below 50 Knots

a) The altimeter will show...
b) A horn will sound on the wings...
c) Wrong answer (c)
d) Wrong answer (d)

However it is possible to create questions that fit a more authentic role playing game based on the narrative of the location and context. We have to imagine we have been constructing the questions in the way a pilot (or engineer or whatever) would be given a task. The student acting on knowledge gained on the field trip is expected to act in ways determined by context and role. With a role/play narrative approach the questions would be formulated like this:

(As a pilot) you hear a loud scream of a horn coming from the wings. You realize you have to

a) *Incorrect answer*
b) Pull back the throttle to increase airspeed
c) *Incorrect answer*
d) *Incorrect answer*

This in itself now becomes part of a problem solving narrative that the student needs to engage in. The student is placed in a situation of applying their knowledge in the way that a pilot has to use knowledge.

At this stage in our understanding we are acting on a hypothesis that learning through a problem solving narrative is *better* than learning that just requires recall. We do not have hard evidence to reveal any such improvement (the number of subjects answering this question, or most of our mobile learning experience, hardly allows the construction of scientific, validated, replicable studies). All we can say is that by reformulating questions that require the same level of factual knowledge it is possible to engage the student into different patterns of activity and approach to knowledge. This example demonstrates that with a little thought a simple MCQ program on a mobile device can be turned in to a locative narrative role playing game.

## Savannah

The simplicity of the previous example contrasts starkly with Savannah, which took advantage of new affordances of connectivity of handheld games devices, and shifted the focus away from the individual player to the connected player able to find fellow gamers on any city street (Flintham et al., 2003). The game was a product of the collaboration of Futurelab, HP labs (Bristol), the BBC Natural History Unit and the Mixed Reality Laboratory of Nottingham University. In this case we had two iterations of trials with 35 children from six different schools.

The game consists of two related areas of activity. In the first, children are able to play at 'being a pride of lions' outside in a playing field (100m by 50m), interacting with a virtual Savannah and exploring the opportunities and risks to lions in that space. Children are given global positioning system (GPS) devices linked to personal digital assistants (PDA) through which they 'see,' 'hear,' and 'smell' the world of the Savannah as they navigate the real space outdoors. The second location, the 'den,' is an indoor space where children can reflect on how well they have succeeded in the game, access other resources to support their understanding, and develop strategies for surviving as lions in the virtual Savannah.

In order to 'sense' the Savannah, children use their handheld PDAs (with headphones), moving around the playing field outdoors acting as lions (Figure 3). They hear the sounds of the Savannah relating to the specific zones or wildlife there, they see still images of the environment and animals to be found in the zones and they 'smell' the scents to be found in those zones, through still pictures of animal paw prints. On these PDAs, the children can also 'mark' specific information and send it back to the Den for later analysis; in later levels they can also 'attack' specific features of the Savannah.

They also have an energy bar that lets them know their specific energy levels at any time (Figure 4). The PDA also receives messages sent by facilitators in the den, such as 'you are too hot,' 'you are hungry,' or on occasion 'You are dead – return to the den.' In the den, there is an interactive whiteboard and flip chart. The whiteboard comprises a map of the Savannah, and a series of 'energy bars,' each of which relate to the relative energy of each individual lion. On returning to the den, children can pull up on the whiteboard the trails that they have made in the Savannah, and recall the sounds and images that they encountered at specific points, 'marked' and sent back to the den.

The virtual Savannah map comprises a number of zones including long grass, short grass, gully, kopje, spring, and trees. These areas are also populated by the wildlife that one would find in those different parts of the Savannah. All these aspects of the gaming environment are based on research carried out as part of the BBC Natural History Unit's work in this area, and a specialist BBC 'lion expert' with over 10 years of experience working in African Savannahs was involved in designing the play space. The main challenge for the children in the game is to understand and

*Figure 3. Students playing Savannah*

*Figure 4. Savannah PDA showing energy bar*

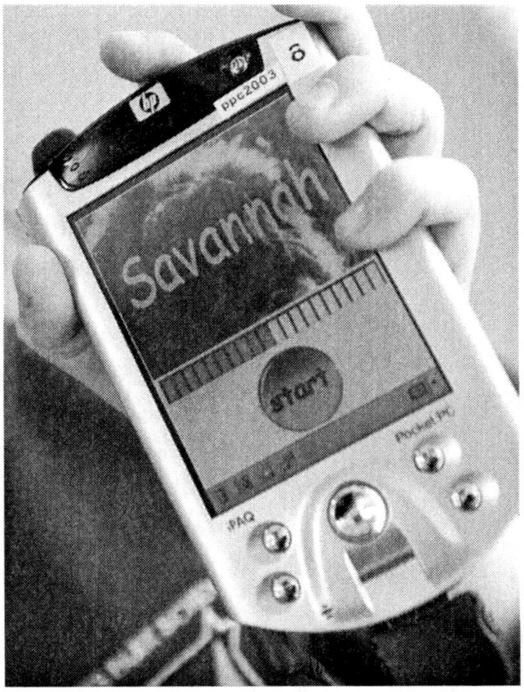


survive in this territory. The environment contains a number of threats: for example, an angry elephant, an old water buffalo, bush fire and, on the fringes of the territory, the Masai.

All of these have to be avoided. Attacking them would lead to certain death. There is also another pride of lions and they have to kill any of the pride's cubs. A number of animals can be found in the Savannah that they could successfully kill and eat, essential for their survival. Some of these require one lion, and others require the whole pack to successfully kill them. There are also sources of water and shade. The children have to balance the costs and benefits of different types of activity, whether attacking, drinking, sleeping or running, in order to maintain their energy levels.

They also have to negotiate with each other, in order to decide whether or how to collaborate in achieving the game's objectives.

Savannah was designed to explore the overarching question: "Is it possible to create an integrated mobile gaming and research environment which encourages children to develop self-motivated and collaborative approaches to developing conceptual understanding of animal behavior?" The specific learning objectives informing the Savannah design were developed iteratively with the technical development of the project.

The aim of the Savannah experience is:

1. To encourage understanding of animal behavior and longer-term survival. By the end of playing Savannah, children should have:
   a. Developed an understanding of the key features of the Savannah environment,

   b. Developed a conceptual understanding of the opportunities and threats faced by lions in the Savannah environment,
   c. Developed a conceptual understanding of the strategies and choices faced by lions in interaction with the Savannah environment, and
   d. Developed a conceptual understanding of the relationship between individual lion behavior and longer-term survival of the species.
2. To encourage strategies for working with problems as a team. By the end of playing Savannah, children should be able to use a range of different strategies for problem solving, including:
   a. Research
   b. Discussion
   c. Hypothesis generation and testing
   d. Revision of ideas
   e. Requesting expert advice
   f. Listening to each other and building on each others' ideas.

In addition to this, there was the overarching objective of exploring the extent to which features of computer games design and mobile computing could contribute to creating an engaging and enjoyable experience for children, and encourage self-motivated learning in the process of playing the game. These issues map onto the generic learning from games list from Gee (2003).

Savannah is a clear attempt to develop a learning resource that makes extensive use of locative technology, game play and fun with a strong element of collaboration, role-play and identification. It provided for multiple rhythms: in the moment in the field and reflective learning and research back in the Den. It required identification, imagination, and reutilization of learned knowledge. There is a strong narrative of place, role and time. The research has been described in a full report (Facer et al., 2004) and in a number of journal and conference articles.

On a very basic level, both of the first two objectives (1(a) & (b)) were met in the trial: the evidence of children's creation of maps of the Savannah, their discussions and planning suggest that by the end of playing Savannah children had developed an understanding that the Savannah consists of a number of different types of environment, including trees, grassy areas, marsh/lake areas, and areas of threat, such as rivers, human villages and other lions' territories. The children's discussion of these features while placing cubs and planning hunting, and while out on the field, suggests that this mixed landscape, by the end of the play, had become an underlying assumption in children's planning and play activities. That there were threats and dangers, as well as opportunities for food and water in the environment, and that these threats took a number of different forms was also an underlying assumption by the time children reached the later levels of the game. The children's engagement with the game—the running away from elephants or fire or humans, the attacks on animals—embody in practice children's understanding of these features of the environment and their potential implications for lions.

The children's planning activity also evidenced an awareness of different strategies for survival (1(c)), primarily based around different hunting techniques. The discussions amongst some of the groups about group success, once they entered the den, suggested awareness that success was not simply individual (1(d)).

However, it is clearly possible, on observing children's play in the Savannah to take a more critical approach to the actual achievements of the experience in supporting children's understanding of the complexity of animal behavior and its relationship with the environment, in particular in respect of objectives 1(b)-1(d). This may be a function of the trial nature of the activity. Most gaming activity allows for repeating actions over again at increasing levels of complexity and this was not possible in our trials. We suspect that as a result of this limitation, children's awareness of opportunities and threats in the environment was almost wholly structured around the notion of what might attack them and what they might attack; there was no opportunity to discuss, for example, the different threats to the animals posed by changes in the season, competition for water, and impact of human occupation of the Savannah. Similarly, the strategies explored for survival were comprised predominantly of hunting. The focus on conserving energy, for example, was subordinated to the demand to gain food. The key strategy of expending very little energy as a lion was not explored in the game, the negotiation of demands of cubs and hunting was sidelined, and the only choices open to the players were related to hunting and drinking tactics. Indeed, if we consider objectives 1(c) and 1(d), it could be said that the notion of lions as strategists and lions as part of a pride with long term survival as a goal, were almost wholly subordinated to 'game goals,' with the children's focus almost exclusively on game play rather than linking that game play in with a wider conception of animal behavior, evolution and habitat.

The game very successfully offered the children the opportunity to play at being lions, to experience first-hand some of the challenges and difficulties that lions might experience, but it also encouraged a focus on the mechanics of the game play (getting together in one place at one time), on simple game rules and game objectives (to some extent, rules that were significantly more simplistic than the children's pre-existing expectations of animal behavior).

In essence, it offered children the opportunity to learn about the Savannah environment in the same way that lions themselves might learn about it. It might not matter, for example, to a lion why particular animals can be found in particular places, how their individual survival is linked with species survival, in what ways human presence can operate as a threat. Ironically, this sort of approach is advocated as an important aspect of game-based learning, drawing for its justification the ways in which animals might learn:

*"Games are thus the most ancient and time-honored vehicle for education. They are the original educational technology, the natural one, having received the seal of approval of natural selection. We don't see mother lions lecturing cubs at the chalkboard; we don't see senior lions writing their memoirs for posterity. In light of this, the question, 'Can games have educational value?' becomes absurd. It is not games but schools that are the newfangled notion, the untested fad, the violator of tradition. Game-playing is a vital educational function for any creature capable of learning"* (Crawford 1982, Chapter 2).

The den environment successfully offered children the resources and the time to reflect on their success and failure out on the field, but again, this environment was prestructured by the objectives of the game itself. Reflection, then, was organized around the narrative and achieving game objectives to such an extent that these other questions, the why, the how, and so on, were effectively irrelevant. What is clear is that if we wish children to develop conceptual understanding of these relationships there are two options: (1) model this into the game's rules or (2) provide these opportunities for thinking before or after the game in a different environment altogether. Without either of these two, the level of understanding is likely to make for an effective lion, but it won't make for a particularly rich human understanding of these phenomena.

The second set of objectives was related to providing a resource that would enable children to develop strategies for working with problems as a team. By the end of playing Savannah, it was intended that children should be able to use a range of different strategies for problem solving. From the evaluation, at a relatively superficial level, we can confidently say that Savannah provided an opportunity for children to practice these different skills. It provided sufficient challenge to encourage children to research, to talk about different strategies, to ask for help, to listen to each other (with some prompting), to test out ideas in the field and to refine them. We clearly saw all of these activities taking place. However, what was clear was that the experience as currently structured provided little scaffolding to encourage children to develop these skills further and to reflect upon them. There is a need, therefore, to revisit the design. The injunction to 'research' at the outset and throughout, for example, was not accompanied by support in reflecting on what 'research' actually means. The design offered insufficient help in formulating specific questions or hypotheses and defining how these could be explored. Similarly, although the children were encouraged to consider listening to each other and to work as a team, we provided few structuring elements in the environment to facilitate this amongst children who had little experience of this way of working before they started. If we are to encourage these skills, it is clear that some degree of reflection on the processes of working together and of problem solving needs to be made explicit. The emphasis in our original design of a scoring system based on lion survival and the insufficiently designed 'team work scores' need to be revised, and opportunities for reflection on these processes offered before or after participation in the experience.

Savannah has provided an interesting testbed for exploring ideas in the three areas of mobility, collaborative learning and games. It was an early experiment in mobile locative technology with strong narrative game elements and has provided an interesting platform on which we can build for further learning. Unfortunately, it has not been possible to continue to use Savannah itself as a testbed for refining our understanding of these mobile games worlds. Nevertheless, Savannah has highlighted the real strengths and weaknesses of games-based learning, in particular, it has emphasized the real engagement and enthusiasm that children demonstrate in these environments, but also the relative limitations of rule-based worlds for teaching complex issues where children act as participants within, rather than manipulators of, the narrative of the game's world (i.e., where they cannot control or reflect upon the rules of the world).

A video of children playing Savannah can be viewed on the Futurelab Web site.

## Mudlarking in Deptford

Field trips have long been a part of education, but the full potential of this activity is not always realized since it is often a static, one-way experience. The Mudlarking in Deptford project was designed to enable young people aged 11-14 to actively engage with an area of historic and educational interest, Deptford Creek, by using geo-positioning (GPS) mobile technology to design and produce their own guided tour. Students helped to create multimedia 'enhanced landscapes' which were combined to create a wider historical tour that combines text, drawings, images (still and video) and audio, and which future visitors to the creek can access and perhaps add their own reflections to (Figure 5).

The outcome was a rich and fluid journey companion for all visitors to the creek. The context of the project is the multiple layers of history that exist in any rich urban and metropolitan environment like London. Mudlarks were young children (from the seventeenth to the twentieth century) who lived alongside the Thames who would work in the tidal mud to discover things of potential value to sustain themselves and their family. Mudlarks would sift through the many layers of silt in a manner we replicate with the layers of cultural history of a location. Juliet Sprake from Goldsmith's College London, who brought the project to Futurelab said:

*"The Mudlarking in Deptford project encourages students to actively discover their environment by traveling through the area, noting the Creek's various smells, sights, sounds, objects and buildings. Students are then encouraged to make wider associations with issues such as regeneration, industrial heritage and natural habitats"*

The key phrase is *actively discover.*

We used the functionality of the technology provided by HP Labs that we used in the Savannah experiment to add another dimension to field work experience: mobile geo-positioning systems (GPS) on handheld computers allowed students to record their observations in real-time, and to see immediately the connections between the layers of history before them. They added their own texture to the layers of history and experience around the site. Thus the concept was born of using technology to aid the modern mudlarker to not just access the past but extend it, enabling today's students to leave virtual objects for other visitors about their environment, as well as access possible future scenarios put there by colleagues or strangers (artist's impressions of future ideas such as floating allotments). This activity led to responses like:

*Figure 5. A screenshot of the map of the creek as it appears on the PDA and the authoring screen. Areas that trigger media events are indicated by the ovals.*

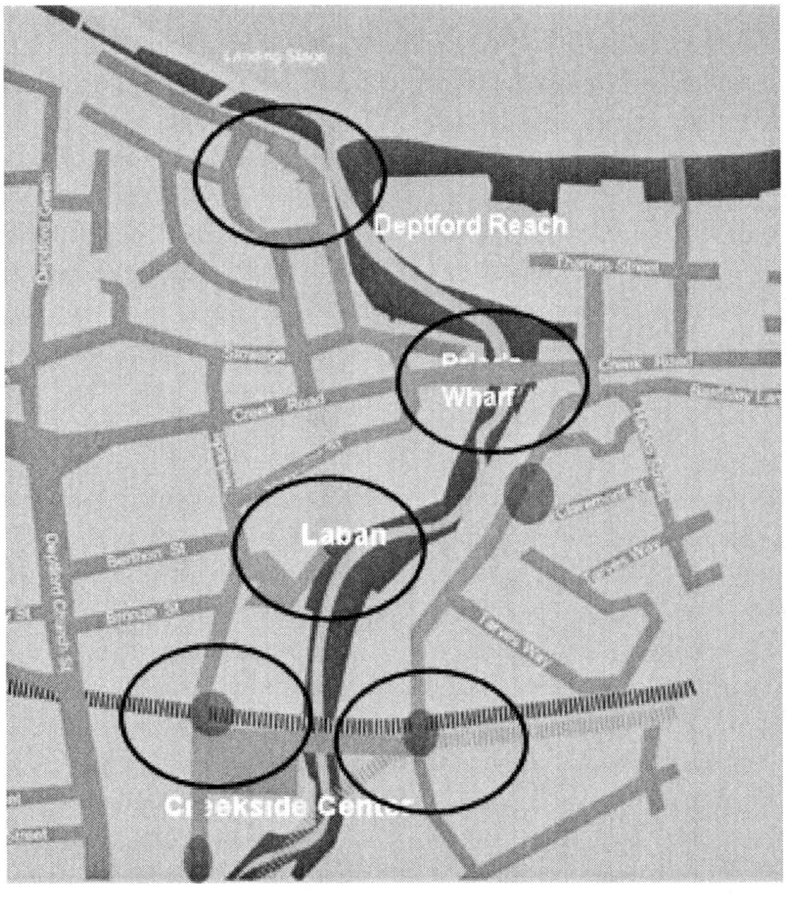

*"At first I thought that doing a tour for other people would be boring but it has been fun and I have learned more about the environment. I also learned about the history of the creek"*

However at the heart of the project was a desire to give students an opportunity to author their interpretation of their location, relating back to the key points of stories, identities, traces and trails. Inasmuch as we already had working technology the purpose of the experiment was to discover what else do we needed to do. How do we support learners in this new design/expressive experience in new media?

Key research questions project were:

- How can mobile technologies enable young learners to interact with an everyday urban site differently?
- How can sharing experiences through location-sensitive technologies create a common understanding between users?
- How far can the application enable users to represent their knowledge in an innovative way?

Additional aims of the project included:

- Encouraging local people to re-imagine an otherwise familiar environment
- Working with local students at the design stage.

Much of the focus was on the process of design, considering the intellectual and software tools that would scaffold the expression of the students. Generating media was more technically complex than the locative tools. We brought in professional designers to work alongside the student to teach techniques.

The experts collaborated with the student design group to develop an interactive landscape, centered around four major destination points (or nodes) on the journey; navigating from their present position using GPS, which was one of the features of the trial that most excited the students, against a suggested map. Dynamic activities were suggested around each of the nodes. One of the most popular was the 'wobbleometer,' which takes place on a bridge that vibrates in response to local traffic in a spectacular way. Another was the 'tarantella,' named after the famous Spanish dance, that referred to the mesh structure of "The man-made hills near a dance centre which students thought looked like webs made by a giant spider."

The trial of the students work was split into three parts. 23 students from the local area were involved, around half of which were from the co-design team who had originally designed the tour. A second group was selected from a different school, which had not been involved in the project prior to the trial. Students set off on suggested routes for the tour. During the tour, a teacher and an observer, who ensured the students' safety and watched their behavior during the tour, accompanied all students. Finally, the whole group met back at the Creekside Centre for a post-tour debriefing with interviews and questionnaires.

Overall, responses to the Mudlarking in Deptford project were extremely positive. Using four nodes to structure the tour allowed students to take a break from the technology to converse with fellow students, and to complete other activities while traveling between nodes. The students were interested in the technology and being able to identify their position on a map gave them a sense of being truly involved in the tour. It also acted as a 'safety net' to reaffirm that the tour was working even when no particular content was being delivered to the handsets. The students enjoyed taking part not only in activities they had worked on themselves, but also those designed by other students. In addition, because the tasks are interactive and fun, they are likely to serve as more effective memory prompts in the classroom. There was a lot of interaction and collaboration at the nodes along the route. Students shared 'hidden stories' from the environment, both during and after the tour. In this way the flexibility of the experience is demonstrated. The ability to record their own experiences for a 'real' audience was an attractive feature for the students, and the variety of methods available to record stories meant that all students were able to capture at least some experiences to share with others.

The emphasis on co-design introduced in Mudlarking is also strongly represented in the La Piazza and COLLAGE programs. From the La Piazza program we developed a series of workshop prompts (see Ackermann et al., 2006) and in COLLAGE we have developed a series of workshop tools to elicit roles, narratives and game mechanics.

These were tentative attempts to devise the toolkit that will allow learners to draw from and contribute to pervasive layers of electronic information in their environment. We found that thinking through the issues, that is the design of that electronic information, is the most significant issue that links with the issues set out in the introduction. It has been the focus of Futurelab's contribution to COLLAGE. It is the key issue of La Piazza with the additional dimension of building stories between generations. A practice is emerging to enable learners to focus on narrative and role as enablers to move from space to place.

As a result of the Mudlarking experience, Futurelab has gone on to produce the Create-A-Scape website which provides the authoring tool and contextualizes it with workshop materials and case studies of use. COLLAGE will produce tools to develop location based game-like learning and it will come with materials to support workshops on game design that embody some of the concepts outlined in this chapter.

## FUTURE RESEARCH DIRECTIONS

Although this chapter focused on the foundations of articulated philosophical positions such as Casey's understanding of place, and the basis of well developed learning theories (e.g., Vygotsky on play) our experimental work, although principled and well grounded in the reality of real children and real teachers, is experimental. There has been no widespread trial of Savannah. Although Create-A-Scape (Mudlarking software) is now widely available there has been no scientific affirmation of the value of "location," "game" or "play." There is no uncontestable evidence about role, identity, and collaboration. It would be unreasonable to expect them. We have been at a stage in the cycle when we are only beginning to try to find if there are new possible practices, let alone being able to prove their efficacy. What we have is a dynamic between experiment and potential practice. This must continue without apology for lack of rigorous scientific validation. Waiting for confirmation requires large-scale implementation that is unlikely to be funded, and by the time it has been achieved the world will have moved on.

Having said that, there is a clear need for metrics that can point to success and failure in things that are not normally or easily assessed. We need to know more about assessing the emergence of identity rather than just the acquisition of content knowledge. We need metrics for knowing that someone is a better or weaker collaborator, a better or weaker strategist, better or weaker place sensitivity and so on. This may emerge from work on serious games as Gee (2003) and others make claims for games developing some of these attributes. However, as we are interested in identity, more longitudinal studies of learning and other habits in mobile lifestyles are needed.

Setting up experiments like Savannah is difficult and therefore their replication is difficult. This does not make for a good experimental method. We may need a research infrastructure (indoor facilities, outdoor facilities, good wireless networking, reliable equipment, on-the-spot technical help) that provides a research arena before adequate research can be done. However, on the other hand, as free open access Wi-Fi technology and reliable low cost mobile hardware proliferates, the requirement for a research facility may not be so essential. We should be able to set up and repeat experiments more easily in future.

Notwithstanding the need for better assessment of what we do, the market and technology sustain a rapid cycle of development. New mobile technologies appear; new forms of visual display will be significant for learning, processors are getting smaller and cheaper and can be embedded in any everyday object. There will be a continued need, therefore, to experiment and use the affordances of new technology to create new learning experiences. This will continue the dynamic between technology, experimentation and pedagogic practice. I expect more interpersonal connectedness, more augmentation of reality, smarter access to information about place in place and so on. The growth of interest in social software adds new layers of user habits and possibilities. We live in interesting times.

## ACKNOWLEDGMENT

I am grateful to Keri Facer for making available her research findings on the Savannah project and Dan Sutch for his research findings on Mudlarking. Futurelab funded both projects. The European Union Directorate General for Education and Culture funded COLLAGE and Puente. La Piazza was a project within the Kaleidoscope Network of Excellence funded by the European Union IST Programme.

Figures 1 and 2 are images created by Julie Buck-Rogers, Ordnance Survey. Figures 3, 4 and 5 are courtesy of Futurelab.

## REFERENCES

Ackermann, E., Chabert, A., Decortis, F., Magli, R., Napoletano, L., & Owen, M. (2006). "La piazza": Convivial spaces for intergenerational learning: What places for the digital age? In A. Mendez-Vilas, A. Solano Martin, J. Mesa-Gonzalez, & J.A. Mesa Gonzalez (Eds.) *Current developments in computer-assisted education,* (pp. 1359-1363), Badajoz, Spain: FORMATEX.

Bhakhtin, M. (1981). *The dialogical imagination: four essays.* trans. Emerson C. and Holquist M. Austin: University of Texas Press.

Bruner, J. (1990). *Acts of meaning.* Cambridge, MA: Harvard University Press,

Brynat, J. (2003) *Map data on the move.* Retrieved November 20th 2007 http://www.ordnancesurvey.co.uk/oswebsite/business/sectors/wireless/news/articles/pdf/Map%20data%20on%20the%20move.pdf

Caillois, R. (2001). *Man, play and games.* Urbana: The University of Illinois Press.

Casey, E. (1979). *The fate of place: a philosophical history.* Berkeley, The University of California Press.

Crawford, C. (1981). *The art of computer game design.* New York, McGraw Hill

Csikszentmihalyi, M. (1992). *Flow: The psychology of happiness.* London, Rider.

Dourish, P. (2004). What we talk about when we talk about context. *Journal of Personal & Ubiquitous Computing, 8*(1), 19-30.

Erikson, E.H. (1963). *Childhood and society* (2nd ed). New York: Norton.

Facer, K., Joiner, R., Stanton, D., Reid, J., Hull, R., & Kirk, D. (2004). Savannah: mobile gaming and learning, *Journal of Computer Assisted Learning, 20*, 399-409.

Faux, F., McFarlane, A., Roche J., & Facer, K. (2006). *Learning with handheld technologies.* Bristol: Futurelab.

Flintham, M., Anastasi, R., Benford, S.D., Hemmings, T., Crabtree, A., Greenhalgh, C.M. et al. (2003). Where on-line meets on-the-streets: Experiences with mobile mixed reality games. *Paper presented at the Conference on Human Factors in Computing Systems CHI 2003,* Fort. Lauderdale, FL.

Gee, J. P. (2003). *What video games have to teach us about learning and literacy.* New York: Palgrave/Macmillan.

Gibson, J.J. (1977). The theory of affordances. In R. Shaw & J. Bransford (Eds.), *Perceiving, acting and knowing.* Hillsdale, NJ: Erlbaum.

Malone, T.W. (1981). What makes computer games fun? *Byte*, 6, 258-277

Naismith, L., Lonsdale, P., Vavoula, G., Sharples, M. (2004). *Mobile technologies and learning,* Bristol: Futurelab.

Nonaka, I., & Takeuchi, H. (1995). *The knowledge crating company.* New York: Oxford University Press,

O'Malley, C., & Stanton Fraser, D. (2004). *Learning with tangible technologies.* Futurelab Bristol.

Owen, M. (2005). Driving demand for broadband. *ITSE, 3*(1)

Owen, M. (2006). *Social software and learning.* Futurelab, Bristol

Prensky, M. (2001). *Digital game-based learning.* New York, McGraw-Hill Education.

Prensky, M. (2004) *What can you learn from a cell phone?* Retrieved November 20th 2007. http://www.marcprensky.com/writing/Prensky What_Can_You_Learn_From_a_Cell_Phone-FINAL.pdf

Revell, S. (2004). From Hide and Seek to AR-Quake: Considering the Challenges of Location-Based Gaming. *Proceedings of GI Days 2004,* Muenster, Germany.

Shaffer, D. W. (2007). *How computer games help children learn.* New York: Palgrave Macmillan.

Shusterman, R. (2005) Somaesthetics and social theory. *Human Affairs, 2,* 105-115,

Singer, D., Golinkoff, R. M., & Hirsh-Pasek, K. (2006). *Play=learning: How play motivates and enhances children's cognitive and social-emotional growth.* New York, NY: Oxford University Press.

Spohrer, J. (1999). Information in places. *IBM Systems Journal, 38*(4), 602-628.

Vygostsky, L. (1933). *Play and its role in the mental development of the child.* First Published: 1933, Source: Voprosy psikhologii, 1966, No. 6; Translated: Catherine Mulholland; Transcription/ Markup: Nate Schmolze; Retrieved November 20th 2007, http://www.marxists.org/archive/vygotsky/works/1933/play.htm

Young, M. F., De Palma, A., & Garrett, S. (2002). An ecological psychology perspective on situations, interactions, process and affordances. *Instructional Science, 30,* 47-63.

# Section III
# Enhancing Collaborative Learning Experiences

# Chapter VII
# Collaborative Technology Impacts in Distributed Learning Environments

**Martha Grabowski**
*Le Moyne College, USA & Rensselaer Polytechnic Institute, USA*

**Greg Lepak**
*Le Moyne College, USA*

**George Kulick**
*Le Moyne College, USA*

## ABSTRACT

*Many studies have examined the impact of collaborative technology in distributed learning environments. Few of those studies involved new collaborative technologies such as mobile computing, and few were empirically tested. This chapter addresses the need to empirically examine the impacts of new collaborative technologies including mobile, wearable, embedded, and ubiquitous technologies, on distributed learners. The chapter introduces a technology-independent framework for considering collaborative technologies, including mobile technology; it relates expected technology impacts to user preferences using a generalizable research framework rooted in the social science, communication and technology acceptance literature. The framework is updated to include the lens of contextualization richness, and the results of an empirical test of the framework are presented. The results show user preferences for technologies with a high range of design features to support cognitive learning, while showing preference for technologies with a low range of design features to support perceived learning. Next steps and implications for future work conclude the chapter.*

## INTRODUCTION

Academic and business professionals are increasingly tethered to their mobile, wearable, and collaborative computing devices, yet we have few empirical studies examining the impact of such technologies on their users (Althaus, 1997; Benbunan-Fich & Hiltz, 1999; Cisco, 2003; Johanson & Torlind, 2004; Mark, Fisher & Poltrock, 2002). Increasingly, these technologies are found in distributed learning environments, where users are offered the possibility of distributed information exchange, knowledge management and enhanced learning (Hiltz & Wellmann, 1997; Kim & Bonk, 2002; Roschelle, 2003).

These technologies have been traditionally described as differing in media richness (Daft & Lengel, 1986; DeSanctis & Gallupe, 1987), as well as in synchronicity—the degree to which information is exchanged instantaneously—and the extent to which the technologies enable distributed collaboration. Of late, these technologies have also been described as differing in degrees of contextualization richness; the degree to which the technologies provide information about a situation, intentions, and feelings about information, as represented by members using the technologies (Majchrzak, Malholtra, & John, 2005).

No matter how these technologies are distinguished from one another, however, our understanding of the impacts of newer collaborative technologies such as mobile devices on distributed users is incomplete. Thus, although some studies evaluating the impact of traditional collaborative technologies on distributed learners have been undertaken (e.g., Alavi, Marakis & Yoo, 2002), little empirical research has been undertaken assessing the impact of newer forms of collaborative technology on distributed international learners.

Our research addresses this need by presenting the results of an empirical analysis evaluating the impacts of newer and traditional collaborative technologies on users in a distributed learning setting. We begin with an overview of previous work, and then outline our research framework. The results of an assessment of different technologies in the setting using the framework are presented. We conclude by discussing implications of the research.

## BACKGROUND

Collaborative technologies are traditionally defined as tools that enable individuals to jointly engage in the active production of shared knowledge. New collaborative technologies are rooted in this heritage, and are defined as those technologies such as wearable, ubiquitous, and mobile computing that offer their users the benefits of any-time, any-space, any-distance communication and collaboration. When first introduced, new collaborative technologies were described as differing from existing audio and video systems in terms of media richness, or the degree to which a technology offers multiple cues, immediate feedback, natural language interfaces, and message personalization (Daft & Lengel, 1986; Galegher & Kraut, 1992). *Media richness theory* proposes that communication effectiveness improves if the technology used by participants matches the information processing requirements of the tasks to be performed, and suggests that rich media is appropriate for equivocal communication activities, such as negotiation, belief monitoring, analysis, decision-making and reflective interaction. Similarly, media richness theory suggests that leaner media is appropriate for unequivocal activities, such as message passing, identifying information, or storing text, data, or messages. When technology capabilities are matched appropriately to task and environmental requirements, media richness theory suggests that user performance and processes will be enhanced.

Many studies have been undertaken to study the impact of conventional technologies on users. Unfortunately, few of these studies have examined the impact of newer collaborative technologies on users who use them frequently; users engaged in distributed learning. Distributed learning is the process by which individuals acquire knowledge and understanding, primarily through social interactions across time and/or geographic distance, using information and communication technologies (Alavi et al., 2002). Following Johnson et al. (1991), distributed learning involves the social creation of knowledge through instructional strategies that emphasize small-group learning among students:

*Learning arises from the opportunity for the group members to monitor each other's thinking, opinions, and beliefs, while also obtaining and providing feedback for clarification and enhancement of comprehension. An individual's exposure to the group members' points of view may challenge his/her understanding, and can motivate further learning (Glaser & Bassok, 1989). Thus, in collaborative distributed environments, learning occurs through communication and collaborative interactions, often those that are technology-mediated* (Alavi et al., 2002).

New collaborative technologies such as mobile computing are increasingly being used in distributed learning environments to enhance learning and to facilitate knowledge exchange, communication, participation, and community building. However, the impacts of technology introduction in distributed learning settings have been portrayed in contradictory ways: on one hand, technologies have been found to be integrative, connecting disparate others and mitigating location effects of distance or geography (Constant, Kiesler, & Sproull, 1996), increasing the power of marginal group members (Sproull & Kiesler, 1986), providing rich communities (Wellmann & Guila, 1999), providing support not found in traditional learning environments (Bonk & Kim, 1998), and nurturing the development of online community between participants, increasing the flow of information, as well as enhancing group support, commitment, cooperation and satisfaction with the group effort (Brufee, 1993; Dede, 1996; Harasim et al., 1995; Haythornthwaite, Kazmer & Robins, 2000; McGrath, 1984; Wellmann, 1999).

On the other hand, collaborative technologies have been criticized for providing a reduced cues environment ill-suited to emotional, expressive, or complex communications, and for providing an environment with longer decision times, anti-social flaming behaviors, and decreased social involvement (Daft & Lengel, 1986; Kiesler & Sproull, 1992). The critical literature acknowledges that collaborative technologies foster interactions among participants, but there are questions about whether the increased or enhanced interactions promote knowledge exchange and/or learning because of the absence of non-verbal cues, which limits the modes of communication among participants (Curtis & Lawson, 2001).

Early work in assessing technology impacts in distributed learning environments measured student performance outcomes, such as grades on tests or grade point averages (GPAs) (Alavi, 1994). In other studies, a comparison was made between learning outcomes (students' grades and/or perceived learning) in a distributed environment and learning outcomes obtained in a traditional face-to-face environment (Hiltz & Wellman, 1997; Storck & Sproull, 1995). More recently, collaborative technology has been seen to impact both cognitive and perceived learning outcomes in distributed settings (Alavi et al., 2002). *Cognitive learning* involves changes in an individual's mental models; that is, internal representations of knowledge elements comprising a domain as well as interrelationships among those knowledge elements. *Perceived learning* involves changes in a learner's perceptions of skill and knowledge levels before and after the learning experience. Approaches to the measurement of these two variables differ: cognitive learning measures are often outcome or performance-related, while perceived learning measures are often process-related (Alavi et al., 2002).

Decision researchers have long been interested in the impact of technology introduction on users and on the systems where the technology is introduced, considering variables such as an individual's decision-making performance (Biros, Daly, & Gunsch, 2004; Chen & Lee, 2003; Gosh & Ray, 1997; Smith, Johnston, & Paris, 2004; Todd & Benbasat, 1999); confidence (Adidam & Bingi, 2000; Cowley, 2004; Morton, Ackermann, & Belton, 2003) and satisfaction (Chang, Chang & Paper, 2003; Muylle, Moenaert & Despontin, 2004; Wang, 2003;Wixom & Todd, 2005; Woodroof & Burg, 2003;), as well as user workload (Bunce & Sisa, 2002; Kristensen et al., 2004; Sauer et al, 2002; Sebok, 2000; Sohn & Jo, 2003; Xie & Salvendy, 2000); vigilance (Wright & McGown, 2001); effort (Kuo et al., 2004); fatigue and stress (Chenoweth, Dowling, & St. Louis, 2004; De Croock, Merrienboer & Paas, 1998; Goode, 2003).

To summarize, then, many studies have examined the impact of collaborative technology in distributed learning environments, but few of those studies involved new collaborative technologies such as mobile computing, and few were empirically tested. Thus, this research addresses the need to empirically examine the impacts of new collaborative technologies including mobile, wearable, embedded, and ubiquitous technologies, on distributed learners. In the next section, we outline the research framework for this study, which compares the impacts of newer collaborative technology with the impacts of more traditional collaborative technology on distributed learners.

## RESEARCH FRAMEWORK

Following the literature just surveyed, we initially distinguished collaborative technology design features along two dimensions; media richness and synchronicity. Media richness is defined as the degree to which the technology provides multiple cues, immediate feedback, natural language interfaces, and message personalization (Daft & Lengel, 1986; Galegher & Kraut, 1992). Synchronicity is defined as the degree to which communication is exchanged instantaneously and in real-time.

We hypothesized that the design features of collaborative technologies would have non-random impacts on two measures important in distributed learning settings; cognitive learning outcomes and perceived learning processes (Alavi et al., 2002). For instance, the richer and more synchronous the technology, the greater we hypothesized the positive impacts on learner outcomes and processes in the distributed learning environment would be. Positive impacts on learning processes are defined as improved perceived learning; improved communication, collaboration and participation; as well as improved information gathering, analysis and decision-making. Positive impacts on learning outcomes are defined as greater technology use; more user satisfaction and confidence; and reduced effort, workload and frustration/stress when using the technology. These constructs are illustrated in Figure 1.

In this study, following Alavi et al., (2002), both input-output and process measures were used to measure learning impacts. Table 1 lists the propositions and Table 2 the operationalizations, dependent variables, and data collection methods for each proposition. As seen in Figure 1, we hypothesized that technology design features, operationalized in a range from low to high, and covering the degrees of media richness and synchronicity of the technologies, would impact the learning outcomes and processes of distributed learners in non-random ways. We acknowledge the impact of learning inputs, such as knowledge, skills and abilities, motivation and effort in our research design, but those variables were not measured in this study.

*Figure 1. Research framework*

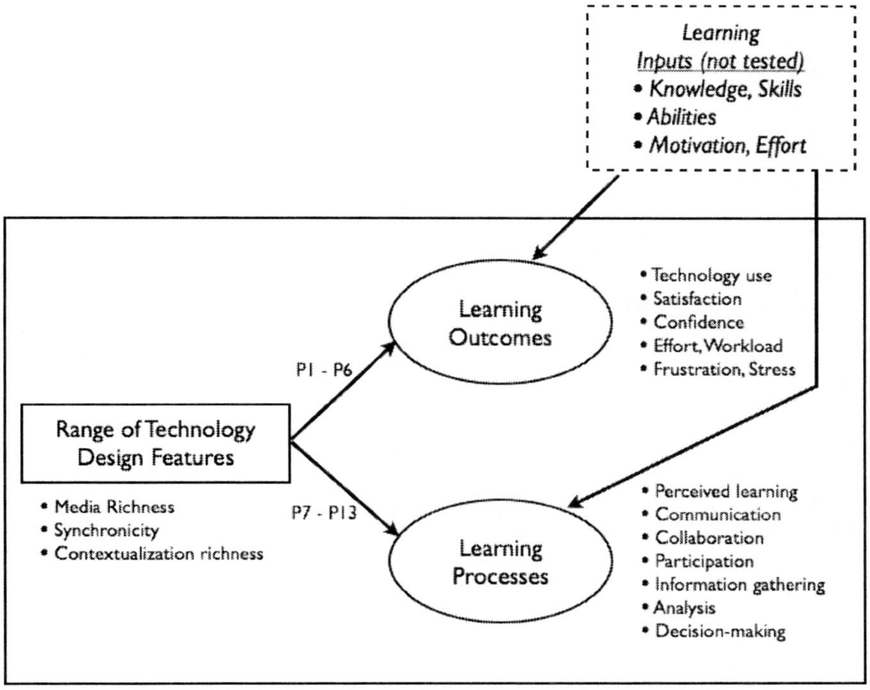


## PROCEDURE

An empirical assessment of the research framework was undertaken, using the propositions outlined in Tables 1 and 2. Data were collected in a distributed learning environment, a senior undergraduate/fifth year course in information technology strategy and structure that was team taught at two institutions in the spring 2003 semester. One institution was in the eastern United States; the other institution was in Warsaw, Poland.

The course was designed to promote learning and understanding of the processes involved in organizations seeking to use information technology as a strategic competitive asset. The 16-week course employed a mix of pedagogical techniques: classroom lectures and discussion, case analyses, group working exercises, and guest speakers. Eight of the 16 weeks were distributed across the two learning settings; the remaining eight weeks were face-to-face sessions involving only the resident learners in each location. At the beginning of the semester, participants were organized in small groups of three to four members in order to collaboratively analyze cases of organizations faced with various information technology (IT) strategy dilemmas.

Groups were asked to collaborate in their case analysis assignments using three technologies—audio technology, high bandwidth videoconferencing, and mobile technology—and were required to jointly analyze and present case analyses to students in both settings simultaneously, using real-time high bandwidth videoconferencing. We were interested in examining the impact of mobile technology on distributed learners. Following our research framework and hypotheses, the other two technologies were chosen to represent a range of technology design features, from low to high. The case analysis groups included

*Table 1. Propositions*

**Input-Output Propositions**

1. Participants using technology with a higher range of design features will show more technology use than participants using technology with a lower range of design features.
2. Participants using technology with a higher range of design features will report more satisfaction in the technology than participants using technology with a lower range of design features.
3. Participants using technology with a higher range of design features will report more confidence in the technology than participants using technology with a lower range of design features.
4. Participants using technology with a higher range of design features will report less effort required to use the technology than participants using technology with a lower range of design features.
5. Participants using technology with a higher range of design features will report less workload required to use the technology than participants using technology with a lower range of design features.
6. Participants using technology with a higher range of design features will report less frustration/stress in using the technology than participants using technology with a lower range of design features.

**Process Propositions**

7. Participants using technology with a higher range of design features will report improved perceived learning when using the technology than participants using technology with a lower range of design features.
8. Participants using technology with a higher range of design features will report improved communication when using the technology than participants using technology with a lower range of design features.
9. Participants using technology with a higher range of design features will report improved collaboration when using the technology than participants using technology with a lower range of design features.
10. Participants using technology with a higher range of design features will report improved participation when using the technology than participants using technology with a lower range of design features.
11. Participants using technology with a higher range of design features will show improved information gathering when using the technology than participants using technology with a lower range of design features.
12. Participants using technology with a higher range of design features will show improved analysis when using the technology than participants using technology with a lower range of design features.
13. Participants using technology with a higher range of design features will show improved decision-making when using the technology than participants using technology with a lower range of design features.

members from both the United States and Polish institutions. Participants in these groups were also required to prepare a written summary of their joint case analysis, and were tested twice during the semester on their knowledge of IT strategy concepts and their ability to apply those concepts in a case analysis. Completion of the case analyses required application of the concepts, ideas, techniques and skills taught during the semester, and constituted 80 percent of the student's grade for the course.

*Table 2. Operationalizations and measurement of dependent variables by proposition*

**1. Input-Output Assessment Measures**

| Proposition | Dependent Variable | Variable Operationalization |
|---|---|---|
| **Proposition 1** | **Technology use** | • Self-reported technology use |
| **Proposition 2** | **Satisfaction** | • Self-reported user satisfaction |
| **Proposition 3** | **Confidence** | • Self-reported user confidence |
| **Proposition 4** | **User Effort** | • Self-reported physical effort |
| **Proposition 5** | **User Workload** | • Self-reported perceived workload |
| **Proposition 6** | **User Frustration/ Stress** | • Self-reported user frustration/stress with media |

**2. Process Assessment Measures**

| Proposition | Dependent Variable | Variable Operationalization |
|---|---|---|
| **Proposition 7** | **Perceived learning** | • User's perceived learning |
| **Proposition 8** | **Communication** | • User's self-reported communication |
| **Proposition 9** | **Collaboration** | • Self-reported user collaboration |
| **Proposition 10** | **Participation** | • Self-reported user participation in group activities |
| **Proposition 11** | **Information Gathering** | • Self-reported information gathering during group activities |
| **Proposition 12** | **A nalysis** | • Self-reported user analysis during group activities |
| **Proposition 13** | **Decision-making** | • Self-reported user decision-making during group activities |

Participants were provided with access to three different collaborative technologies as tools for their team case analyses. *Audio technology* involved voice telephone interactions between team members on the collaborative projects. The *handheld technology* was a Dell Axim X5 personal digital assistant (PDA) with voice, wireless Internet and e-mail, text, simple message system (SMS) texting, graphics and animation capabilities. Participants were asked to use only handheld technology to communicate by e-mail so as to clearly distinguish technology use and communication channels by participant. *High bandwidth videoconferencing* provided real-time voice, video and audio communication between the United States and Polish participants over a dedicated T1 line for periods of one hour at a time. No costs were incurred by team members to use this technology, although scheduling videoconference time in advance was required. All participants were provided with access to the high bandwidth videoconferencing media. Because of technology deployment difficulties in Poland, however, only the United States-based participants were provided with access to the audio and handheld technologies. This was disappointing in terms of the research design, as the research could only examine mobile learning impacts for the U.S. subjects.

The technologies were characterized by their range of design features, which included media richness and synchronicity (Table 3). Overall, the high bandwidth videoconferencing presented the highest range of system design features, as well as the highest degree of media richness, with real-time audio and video cues for its users, natural language interfaces (voice), message personalization (calling participants by name, and participant recognition over time), and almost immediate feedback due to the T1 link. High bandwidth videoconferencing was highly synchronous, but, because of the slight transmission delays of real-time audio and visual data across the ocean, was not characterized as the most highly synchronized technology. However, overall, the high bandwidth videoconferencing technology was characterized as the technology with the highest range of design features.

At the other end of the spectrum, the handheld technology presented the lowest range of design features: it provided a medium degree of media richness, as its text, graphics, SMS, voice and electronic mail capabilities provided multiple cues to users, almost immediate feedback, and a medium degree of message personalization and natural language interfaces. Of the three technologies, however, the handheld technology provided the least synchronicity, as participants were forced to endure time lags between transmission of most messages (text attachments, SMS, graphics, e-mail or animation). Overall, therefore, the range of design features, as operationalized by media richness and synchronicity, was least for the handheld technology. In the middle, the audio technology provided a medium range of technology features. The audio technology was the leanest of the technologies, providing only audio cues to its users. However, the audio technology was highly synchronous. Taken together, the audio technology therefore presented a medium range of design features, as summarized in Table 3.

## Participants

A total of 49 individuals participated in the study: 24 were students from the United States-based college and 25 students from the Polish university. Ninety percent of the U.S. college students were United States nationals; 50 percent of the Polish university students were Polish nationals; the remaining students at the Polish institution were of Greek, Singapore, Dutch and German nationality. The average age of the United States college students was 20.5 years; of the Polish university students, 24.5 years. All had completed at least six courses in business and information technology, and all participants reported more than average fa-

*Table 3. Technology design feature characterization (*following Hancock, Thom-Santelli & Ritchie, 2004; 1 = highest, 3 = lowest)*

| Media Features* | Handheld (Text, graphics, SMS, e-mail, voice, animation) | Audio (voice phone) | High Bandwidth Videoconferencing (Audio, video, real-time T1 link, distributed graphics) |
|---|---|---|---|
| Media Richness | 2 | 3 | 1 |
| Synchronicity | 3 | 1 | 2 |
| TOTAL RANGE OF DESIGN FEATURES | Low (5) | Medium (4) | High (3) |

miliarity with technology. None had been involved in previous examinations of technology impact in a distributed learning environment. Sixteen percent of the United States college students and 80 percent of the Polish university students had previous experience with handheld technology prior to the semester when the evaluation was conducted. One of the challenges associated with this study was the small sample size, which limits the strength of the study's results.

## Method

All participants were provided with e-mail accounts, Internet access, and Blackboard accounts on the same server at the U.S. institution. Internet access was provided as an information source and as a publishing medium for posting project and background information. Blackboard accounts provided each participant with a publishing medium, calendar, announcement capability, a common course reference and a common document and file repository. All participants had previous experience using the Internet and e-mail; all United States subjects had previous experience with Blackboard. None of the Poland participants had previous experience using Blackboard.

Participants in both locations were required to analyze identical case studies of organizations utilizing information technology as part of the organization's business strategy. Groups analyzing each case were comprised of students from both distributed settings. Groups were required to collaborate on case analyses and jointly present the results of their case analysis in one-hour real-time sessions using high bandwidth videoconferencing between the two distributed learning sites.

Unfortunately, there were deployment difficulties with some of the technologies in the Polish university. As a result, although the United States college participants had access to all three technologies, handheld, audio and high bandwidth videoconferencing, students in the Polish university only had access to the high bandwidth videoconferencing technology. Thus, the impacts of different collaborative technologies were only studied using data from the United States college participants. The data from the Polish university participants is used for comparison of the high bandwidth videoconferencing results.

At the end of the semester, all participants completed the same seven-point Likert-scale survey that queried participants about their learning processes within and between groups. The survey solicited participant opinions about the effects of the technologies utilized on the dependent variables listed in Table 2. If participants did not use a particular technology, they were asked to disregard that section of the survey. For each question, the answers ranged along a 7-point scale from "strongly disagree" to "strongly agree."

Statistical analyses of the data were performed and the results are presented in the next section. First, contingency table analysis was used on the responses by United States participants to individual survey questions to test whether support existed for the propositions in Table 1. The analysis takes into consideration the repeated measures nature of the data whereby the same participant was exposed to different technologies and provided responses that were classified into one of several categories. Second, categorical data analysis was then used to test whether differences exist between the perceptions of participants at the two locations.

# RESULTS

## Technology Analysis

To assess the different technology contributions to learning outputs and processes, we first examined technology use by participants at the United States college who utilized three technologies in a variety of learning contexts, and responded to survey questions about their use and perceptions of the technologies. The American participants'

*Table 4. Significant findings and support for propositions*

1. Input-Output Measures

| **Dependent Variable** | **Technology Preferred** | **Strongly Agree (%)** | **Strongly Disagree (%)** | **Technology** | **Strongly Agree (%)** | **Strongly Disagree (%)** | **Statistics** | **Proposition** |
|---|---|---|---|---|---|---|---|---|
| Greater control over tasks | Handheld | 64% | 14% | Over High Bandwidth Video | 29% | 14% | $\chi^2 = 2.8$* | -- |
| | Handheld | 62% | 8% | Over Audio | 15% | 0% | $\chi^2 = 3.6$* | -- |
| More improvement in task performance | Handheld | 62% | 8% | Over Audio | 15% | 8% | $\chi^2 = 6.0$** | -- |
| Increased productivity | Handheld | 57% | 14% | High Bandwidth Video | 29% | 21% | $\chi^2 = 3.6$* | -- |
| More satisfied with info provided by technology | High Bandwidth Video | 50% | 0% | Audio | 29% | 0% | $\chi^2 = 3.0$* | Supports P3 |
| | Handheld | 69% | 0% | Audio | 31% | 0% | $\chi^2 = 5.0$** | -- |
| More confident in performance of technology | Handheld | 62% | 0% | Audio | 23% | 0% | $\chi^2 = 3.6$* | -- |
| More reduction in physical effort required to complete tasks | Handheld | 29% | 7% | High Bandwidth Video | 21% | 36% | $\chi^2 = 2.8$* | -- |
| More reduction in mental effort required to complete tasks | Handheld | 29% | 21% | High Bandwidth Video | 14% | 43% | $\chi^2 = 3.6$* | -- |
| | Audio | 15% | 15% | High Bandwidth Video | 15% | 38% | $\chi^2 = 3.0$* | -- |
| More stress | High Bandwidth Video | 0% | 54% | Handheld | 0% | 92% | $\chi^2 = 3.6$* | -- |
| | Audio | 8% | 69% | Handheld | 0% | 100% | $\chi^2 = 3.6$* | -- |
| More frustration | Audio | 8% | 62% | Handheld | 0% | 92% | $\chi^2 = 2.8$* | -- |

*continued on following page*

2. Process Measures

| Dependent Variable | Technology Preferred | Strongly Agree (%) | Strongly Disagree (%) | Technology | Strongly Agree (%) | Strongly Disagree (%) | Statistics | Proposition |
|---|---|---|---|---|---|---|---|---|
| Improved learning over distances among team members | High Bandwidth Video | 33% | 8% | Over Handheld | 8% | 25% | $\chi^2 = 3.6$* | Supports P7 |
| Improved collaboration among team members | High Bandwidth Video | 33% | 17% | Handheld | 8% | 25% | $\chi^2 = 4.0$** | Supports P9 |
| Improved collaboration between teams | High Bandwidth Video | 42% | 8% | Handheld | 25% | 25% | $\chi^2 = 4.0$** | Supports P9 |
| Improved analysis across time among team members | High Bandwidth Video | 27% | 0% | Handheld | 18% | 36% | $\chi^2 = 5.0$** | Supports P12 |
| Improved analysis across time among team members | High Bandwidth Video | 31% | 0% | Audio | 15% | 23% | $\chi^2 = 2.8$* | Supports P12 |
| Improved analysis over distances between teams | High Bandwidth Video | 36% | 9% | Handheld | 18% | 18% | $\chi^2 = 3.0$* | Supports P12 |
| Improved decision-making across time among team members | High Bandwidth Video | 18% | 18% | Handheld | 9% | 45% | $\chi^2 = 4.0$** | Supports P13 |
| Improved decision-making over distances among team members | High Bandwidth Video | 36% | 18% | Handheld | 27% | 36% | $\chi^2 = 3.0$** | Supports P13 |
| Improved decision-making across time between teams | High Bandwidth Video | 36% | 18% | Handheld | 18% | 36% | $\chi^2 = 4.0$** | Supports P13 |

** = $p < 0.10$; ** $p < 0.05$ ^ Chi-square statistic based on matched pairs design (Fleiss et al., 2003)* responses were interpreted as follows: "strongly agree" was interpreted to correspond to a score greater than or equal to 6 on the 7-point scale; "strongly disagree" was interpreted to correspond to a score of less than 3 on the 7-point scale; and "partially agree" was interpreted to correspond to a score of 3 or greater, but less than 6, on the 7-point scale. Two technology treatments for each survey question were compared.

Data for each survey question were analyzed by using a chi-squared test for matched pairs (Fleiss, Levin & Myunghee, 2003); in the context presented here, each participant served as his/her own control. As such, the null hypothesis of principal interest is the homogeneity of technology effects at the two ends of the ordered agreement scale using responses for each survey question. Thus, support for each alternative proposition in Table 1 exists when more respondents strongly agree (or less strongly disagree) on a particular survey question that the technology with a higher feature range resulted in an improved learning process or outcome. For this type of analysis, the value of the test statistic is approximately distributed as chi-squared with one degree of freedom if the null hypothesis is true. Table 4 presents a summary of the statistically significant results.

Table 4 shows very interesting results regarding their support for the propositions in Table 1. First, Table 4 shows that there were stark contrasts between the input-output and process results. For instance, with the exception of the proposition concerning satisfaction with the technology, there were no statistically significant input-output results with the Unite States participants using the three technologies. At the same time, participants expressed almost uniform preference for the handheld technology over high bandwidth videoconferencing, relating that the technology with the lowest range of design features provided greater task control, greater productivity, and required less physical and mental effort to use. This result directly contradicts the input-output propositions, which suggested that the technologies with the higher range of design features would be associated with enhanced performance and greater satisfaction.

This trend was not isolated to the preference of the participants for the handheld technology over high bandwidth videoconferencing, however. Table 4 also shows that participants similarly showed significant preference for handheld technology over the technology with a medium range of design features, the audio technology, relating that, again, the handheld technology provided greater task control, improved task performance, greater satisfaction with the information provided, greater confidence in the technology, and less frustration or stress with the technology being used. Thus, participants almost uniformly reported improved performance with the handheld technology, contrary to the stated input-output propositions.

These findings might be explained as an overwhelming endorsement of the novel handheld technology by learners, if it were not for the process results reported in Table 4. In contrast to the input-output results that reported participants' preference for the technology with the lowest range of design features, the same participants expressed significant preference for the technology with the highest range of design features in the process results; learning over distances, collaboration between and among teams, analysis across time and distance, and decision-making across time and distance.

In almost all of the significant process findings, the technology with the highest range of design features (high bandwidth videoconferencing) was preferred over the technology with the lowest range of design features (handheld technology). In addition, the technology with the highest range of design features was preferred over the technology with a medium range of design features (audio technology). Thus, participants expressed significant preferences for different types of technology in different learning states, expressing preferences for technology with a low range of design features (handheld technology) to improve input-output performance, or cognitive learning, while expressing preference for technology with a high range of design features (high bandwidth videoconferencing) to improve their learning processes, or perceived learning. Both results have significant implications for teaching, learning, and future research, and will be discussed in the following section.

Interestingly, Table 4 reports no significant findings for process measures that are critically important in learning: communication (P8), participation (P10), and information gathering (P11). From this study, we cannot make any inferences about the impact of a range of a technology's design features on the participants' communication, participation and information gathering.

## Location Analysis

The results just reported focus on the impacts reported by United States participants utilizing three technologies in a distributed learning setting. In this section, we discuss the results reported by both United States and Polish university participants utilizing technology with a high range of design features (high bandwidth videoconferencing) in a distributed learning setting. Although not related to our interest in mobile technology impacts, we include the results here to broaden our understanding of collaborative technology impacts for technologies with a high range of design features, a range not shown with the handheld technology examined in this study, but a range that is seen in other mobile technology.

Analyses were conducted using data from the two independent samples (United States and Polish groups), with the responses by participants from each sample distributed over the categories of agreement used in the previous section. Traditional chi-square tests for independence between location and strength of agreement were conducted for separate survey questions. Results of the data analysis are phrased in terms of rejecting the null hypothesis of independence. Only results that are significant at the five percent level are summarized in Table 5.

Note that the only significant input/output findings in Table 5 were Polish students who reported improved satisfaction with and confidence in the high bandwidth videoconferencing technology, when compared to the U.S. students. These results run counter to the U.S. students' input-output results reported in Table 4, as the U.S. students reported statistically significant improved satisfaction and confidence with the handheld technology, rather than the audio technology, and no statistically significant results for high bandwidth videoconferencing. These results may reflect novelty effects for the Polish students, as the handheld results may reflect for the U.S. students; it may also be the case that the Polish students, who used leaner technologies in their daily work, may have expressed a preference for technology with a higher range of design features in this study.

On the process side of Table 5, the results are similar, with Polish students reporting significant improved learning, information gathering and decision-making among team members using the high bandwidth videoconferencing, compared to the results reported by the U.S. students. In this case, these results echo the findings in Table 4, where U.S. students reported statistically significantly improved learning and decision-making among team members when using the high bandwidth videoconferencing. Thus, with the process measures, both U.S. and Polish participants agree that the high bandwidth videoconferencing technology, with a high range of design features, supports improved learning and decision-making, with the Polish participants' reports significantly stronger than the U.S. participants'. This may suggest that participant reports of improved learning and decision-making with high range design feature technologies may transcend locational or geographic differences, a finding examined in the next section. It should again be noted that both sets of participants preferred technology with a high

*Table 5. Significant findings for U.S. vs. Polish participants for high bandwidth videoconferencing*

1. Input-Output Measures

| Dependent Variable | Country | Strongly Agree (%) | Strongly Disagree (%) | Vs. Country | Strongly, Partially Agree (%) | Strongly Disagree (%) | Statistics* | Finding |
|---|---|---|---|---|---|---|---|---|
| Satisfied that high bandwidth video helped increase productivity on projects | Poland | 73%, 13% | 13% | U.S. | 25%, 56% | 19% | $\chi^2 = 7.9$** | Stronger agreement from Polish students than U.S. students |
| Confident in the reliability of the information provided by high bandwidth video | Poland | 64%,29% | 7% | U.S. | 25%, 75% | 0% | $\chi^2 = 6.8$** | Stronger agreement from Polish students than U.S. students |

2. Process Measures

| Dependent Variable | Country | Strongly, Partially Agree (%) | Strongly Disagree (%) | Vs. Country | Strongly, Partially Agree (%) | Strongly Disagree (%) | Statistic^* | Finding |
|---|---|---|---|---|---|---|---|---|
| High bandwidth bideo improved learning over distances among team members | Poland | 58%,33% | 8% | U.S. | 13%,67% | 20% | $\chi^2 = 6.1$** | Stronger agreement from Polish students than U.S. students |
| High bandwidth bideo improved information gathering over distances among team members | Poland | 64%,18% | 18% | U.S. | 13%,63% | 25% | $\chi^2 = 8.1$** | Stronger agreement from Polish students than U.S. students |
| High bandwidth bideo improved decision-making among team members | Poland | 75%, 17% | 8% | U.S. | 20%,60% | 20% | $\chi^2 = 8.2$** | Stronger agreement from Polish students than U.S. students |

*** $p < 0.05$ ^ Chi-square statistic based on conventional test for independence*

*Table 6. Revised technology design feature characterization (*following Hancock et al.,2004; 1 = highest, 3 = lowest)*

| Media Features* | Handheld (Text, graphics, SMS, e-mail, voice, animation) | Audio (voice phone) | High Bandwidth Videoconferencing (Audio, video, real-time T1 link, distributed graphics) |
|---|---|---|---|
| Media Richness | 2 | 3 | 1 |
| Synchronicity | 3 | 1 | 2 |
| Contextualization Richness | 2 | 3 | 1 |
| TOTAL RANGE OF DESIGN FEATURES | Low (7) | Low (7) | High (4) |

range of design features for perceived learning or process measures, an outcome first identified with the U.S. participants in Table 4. This, too, may represent a location-independent finding that transcends geographical differences.

It is interesting that Table 5 reports no significant finding with respect to whether U.S. or Polish participants' effort, workload or frustration (in the input-output measure analysis) or communication, collaboration, participation or analysis (in the process measure analysis) were improved. These results are similar to the results in Table 4, where no significant results for U.S. participants' communication and participation were found. No inferences can be made in this study, therefore, about the impact of collaborative technologies with varying design features on users' communication or participation, a surprising result.

## Contextualization Richness Analysis

Recently, calls in the literature have questioned the value of media richness models in general (Carlson & Zmud, 1999; Markus, 1994) and specifically for explaining how individuals perform in distributed groups (Saunders, 2000). Instead, it has been suggested that the concept of media richness should be replaced with the concept of *contextualization richness,* particularly for individuals in distributed learning groups (Majchrzak et al., 2005). Contextualization richness can be defined as the degree to which context information about messages is structured for easy absorption, with context defined as information about a situation, intentions, and feelings about the message as owned, evolved and represented by members in the communication process (Majchrzak et al., 2005).

As reported in earlier sections, the technologies under study were characterized by their range of design features, which included media richness and synchronicity, as described in Table 3. Following the recent call in the literature to incorporate contextualization richness in technology design feature analyses, we described the technologies using the lens of contextual richness, as seen in Table 6: the handheld and audio technologies were both characterized as having a low range of design features, and the high bandwidth videoconferencing was characterized as before, as having a high range of design features. Audio technology, providing single (audio) channel message formatting, is clearly the leanest technology with respect to contextual information. Handheld technology, which provides a mix of message formats for contextual information from various participants, text, graphics, animation, e-mail, texting, voice (over IP), as well as video playback capabilities, provides a medium degree of contextual richness. High bandwidth videoconferencing, which provides a wide array of message formats from multiple participants in real time, voice, audio, video, real-time graphics, animation and videoconferencing, provides the highest degree of contextual richness of the three technologies.

Interestingly, review of Table 6 shows that the addition of a contextual richness construct would not have changed the characterizations or the results reported in Tables 4 and 5. This suggests that the use of the contextual richness construct may not provide any additional explanatory power to the conclusions or results of this research; it also suggests that the concept of using a range of design features from low to high may be a more robust measurement construct than simply evaluating technologies in terms of their individual features (media richness, degree of distribution, synchronicity, or contextual richness), a notion the literature supports (Carlson & Zmud, 1999; Majchrzak et al., 2005; Markus, 1994).

## DISCUSSION

Table 7 summarizes the significant results of this study. Participants expressed significant preferences for technology with a low range of design

*Table 7. Significant preferences summary*

| **U.S. Participants** | **Handheld (Text, graphics, SMS, e-mail, voice, animation)**<br>*Low Range* | **Audio (voice phone)**<br>*Medium Range* | **High Bandwidth Videoconferencing (Audio, video, real-time T1 link, distributed graphics)**<br>*High Range* |
|---|---|---|---|
| **Cognitive Learning** | • Task Control<br>• Task Performance<br>• Productivity<br>• Confidence<br>• Satisfaction<br>• Effort<br>• Frustration<br>• Stress | | |
| **Perceived Learning** | | | • Learning*<br>• Collaboration<br>• Analysis<br>• Decision-making* |
| **Polish Participants** | | | |
| **Cognitive Learning** | | | • Productivity<br>• Confidence |
| **Perceived Learning** | | | • Learning*<br>• Information-gathering<br>• Decision-making* |

= *Location-independent significant findings*

features (i.e., handheld technology) to improve cognitive learning, as measured by metrics such as task performance, confidence and satisfaction; they expressed significant preferences for technology with a high range of design features to improve their learning processes, such as perceived learning, decision-making and analysis. All participants reported improved learning and decision-making when using technology with a high range of design features. However, no support could be found for the propositions that suggested that technology with a high range of design features improved communication, collaboration or participation, which suggests that further investigation is warranted.

These results suggest three conclusions: first, different ranges of technology design features can support different learning outcomes and processes; second, a mix of technology design features may be important to enhance learning; and third, more investigation is required to understand the impact of technology design features on improved communication, participation and collaboration among distributed international learners. These conclusions are broadly supported by the existing literature, as noted earlier.

There are alternative explanations for the results that were observed. We were not able to measure or assess cultural differences between the Polish and U.S. students, which is clearly an area for future research. In addition, it was not possible to measure the novelty effects of the technologies, particularly for the Polish students; novelty effects may clearly have played a role in determining the impacts of the technologies on their users, both Polish and American.

Considering the results more broadly, this study offers a number of research contributions: it introduces a framework for considering collaborative technologies, including mobile technology, which transcends any particular technology; it also relates expected technology impacts to user preferences using a generalizable research framework rooted in the social science, communication and technology acceptance literature. The framework is updated to include the lens of contextualization richness, and the results of an empirical test of the framework are presented; the results show user preferences for technologies with a high range of design features to support cognitive learning, while showing preference for technologies with a low range of design features to support perceived learning.

## FUTURE RESEARCH DIRECTIONS

Overall, this study suffers from a number of limitations. First, as mentioned earlier, the number of participants in the study is small, limiting the strength of the conclusions that can be drawn from the results. In addition, the technology infrastructure in Poland was significantly different than that available in the United States, which caused a modification in our original experimental design: instead of evaluating three new collaborative technologies in a distributed international learning setting, we were able to assess three technologies in one setting, and one technology in the other setting. The two groups of participants shared one technology, but the technology infrastructure limitations in the Polish setting clearly impacted our experimental design.

Our conclusions would have been strengthened if were able to explore more fully the complex relationships between technology preferences and learning. For instance, users do not always prefer rich or high bandwidth technology, as evidenced by the lack of consumer adoption of videophones, the preference of young people for text messaging, and the results of this study. Future work should explore the complexities of the impact of user control on communications, the convenience of synchronous vs. asynchronous communication, and the interaction between tasks and technology preferences in distributed learning environments. We suspect, although we were not able to demonstrate in this study, that these variables play a significant role in determining technology preference and use in choices of collaborative technology in distributed learning settings.

In short, even though the results were significantly impacted by the available technology infrastructure and small sample size, they still have important implications for researchers and those interested in the impacts of newer collaborative technologies on distributed learners. In this study, participants expressed clear preference for the handheld technology in order to complete the task at hand. Participants using the handheld technology indicated that their task performance, task control, confidence, satisfaction, effort, stress, and frustration were all significantly improved with the use of the new technology. However, at the same time, the same participants indicated a preference for more traditional, but richer, technology, in order to improve their learning processes, particularly their analysis, decision-making, collaboration and learning. These results suggest that the impacts of new collaborative technologies, including mobile technology, on distributed learners can be significant, particularly in terms of accomplishing tasks, but that learners express a preference for technologies with a high range of design features in order to support their learning processes, and they express a preference for a mix of technologies in accomplishing learning tasks and improving their learning processes. Researchers and managers interested in improving distributed learning performance and processes might therefore consider a mix of low and high-range design features in the technologies they deploy. These results also have significant import for researchers interested in the role and impact of mobile technologies on distributed learners.

This study also provides empirical support for the strategy of coupling the introduction of handheld technologies that promote successful task accomplishment with more traditional technologies with a high range of design features that enhance the learning processes. Finally, the technology and location analyses found no support for participants' improved communication, collaboration and participation when using traditional or new collaborative technologies. This is surprising, given the previous findings in many studies (Alavi, et al., 2002; Hiltz & Wellman, 1997;Sproull & Keisler, 1986 ), and suggests that further analysis of these constructs with new collaborative technologies is warranted.

## ACKNOWLEDGMENT

The authors would like to acknowledge the assistance of Bob Clapp and Shaun Black, from the Le Moyne College Information Technology organization, as well as Michael Fronmueller and Bernard Arogyaswamy, our colleagues who facilitated the experiment at Warsaw University. The assistance of the Warsaw University faculty and staff is also gratefully acknowledged.

## REFERENCES

Adidam, P., & Bingi, R. (2000). The importance of decision confidence to strategy outcomes. *The Journal of Applied Business Research, 16*(2), 35-50.

Alavi, M. (1994). Computer-mediated collaborative learning: An empirical evaluation. *MIS Quarterly, 18*, 159-174.

Alavi, M., Marakas, G.M., & Yoo, Y. (2002). A comparative study of distributed learning environments on learning outcomes. *Information Systems Research, 13*(4), 404-415.

Althaus, S.L. (1997). Computer-mediated communication in a university classroom: An experiment with on-line discussion. *Communication Education, July,* 158-174.

Benbunan-Fich, R., & Hiltz, S.R. (1999). Educational applications of CMCS: Solving case studies through asynchronous learning networks.

*Journal of Computer-Mediated Communication, 4*(3), March, http://www.ascusc.org/jcmc/vol4/issue3/benbunan-fich.html

Biros, D., Daly, M., & Gunsch, G. (2004). The influence of task load and automation trust on deception detection. *Group Decision and Negotiation, 13,* 173-189.

Bonk, C.J., & Kim, K.A. (1998). Extending sociocultural processes to adult learning. In M.C. Smith & T. Poiurchot (Eds.), *Adult learning and development: Perspectives from educational psychology* (pp. 67-88). Mahwah, NJ: Lawrence Erlbaum Press,

Brufee, K.A. (1993). *Collaborative learning: Higher education, interdependence, and the authority of knowledge.* Baltimore: Johns Hopkin University Press.

Bunce, D., & Sisa, L. (2002). Age differences in perceived workload across a short vigil. *Ergonomics,* 45(13), 949-960.

Carlson, J.R., & Zmud, R.W. (1999). Channel expansion theory and the experiential nature of media richness perceptions. *Joural of the Academy of Management, 42*(2), 153-170.

Chang, R., Chang, Y., & Paper, D. (2003). The effect of task uncertainty, decentralization and AIS characteristics on performance of AIS: An empirical case in Taiwan. *Information & Management,40*, 691-703.

Chen, J., & Lee, S. (2003). An exploratory cognitive DSS for strategic decision making. *Decision Support Systems, 36*, 147-160.

Chenoweth, T., Dowling, K., & St. Louis, R. (2004). Convincing DSS users that complex models are worth the effort. *Decision Support Systems, 37,* 71-82.

Cisco. E-Learning. (2003). Retrieved 5 April 2003 from http://business.cisco.com/prod/tree.taf%3Fasset_id-44748&public_view=true&kbns=1.html

Constant, D., Kiesler, S.B., & Sproull, L.S. (1996). The kindness of strangers: The usefulness of weak electronic ties for technical advice. *Organization Science, 7*(2), 119-135.

Cowley, E. (2004). Recognition confidence, recognition accuracy and choice. *Journal of Business Research, 57,* 641-646.

Cramton, C. (2001). The mutual knowledge problem and its consequences for dispersed collaboration. *Organization Science, 12,* 346-371.

Curtis, D.D., & Lawson, M.J. (2001). Exploring collaborative on line learning. *Journal of Asynchronous Learning Networks*, 5(1), 21-34.

Daft, R., & Lengel, R. (1986). Organizational information requirements, media richness and structural design. *Management Science, 32*(5), 554-571.

Dede, C.J. (1996). The evolution of distance education: Emerging technologies and distributed learning. *American Journal of Distance Education, 10*(2), 4-36.

De Croock, M., Van Merrienboer, J., & Paas, F. (1998). High versus low contextual interference in simulation-based training of troubleshooting skills: Effects on transfer performance and invested mental effort. *Computers in Human Behavior, 14*(2), 249-267.

DeSanctis, G., & Gallupe, R.B. (1987). A foundation for the study of group decision support systems. *Management Science, 33*(5), 589-609.

Fleiss, L.F., Levin, B., & Myunghee, C.P. (2003). *Statistical methods for rates and proportions* (3rd ed.). New Jersey: John Wiley & Sons, Inc.

Galegher, J., & Kraut, R.E. (1992). Computer-mediated communication and collaborative writing: Media influence and adaptation to communication constraints. *Proceedings of the 1992 ACM*

*Conference on Computer-Supported Cooperative Work.* Toronto, Ontario, Canada. Pp. 155-162. Retrieved 16 November 2004 from http://delivery.acm.org/10.1145/150000/143474/p155-galegher.pdf?key1=143474&key2= 6761560011&coll=GUIDE&dl=ACM&CFID=31360932&CFTOKEN=55657576,.

Glaser, R., & Bassok, M. (1989). Learning theory and the study of instruction. *Annals of Psychology, 40*, 631-666.

Goode, J. (2003). Are pilots at risk of accidents due to fatigue? *Journal of Safety Research, 34*, 309-313.

Gosh, D., & Ray, M. (1997). Risk, ambiguity, and decision choice: Some additional evidence. *Decision Sciences, 28*(1), 81-104.

Hancock, J.T., Thom-Santelli, J., & Ritchie, T. (2004). Deception and design: The impact of communication technology on lying behavior. *Proceedings of CHI2004.* Vienna, Austria: The Association for Computing Machinery, 24-29 April. Retrieved 16 November 2004 from http://delivery.acm.org/10.1145/990000/985709/p129-hancock.pdf?key1=985709&key2= 11205600&coll=GUIDE&dl=ACM&CFID=31360932&CFTOKEN=55657576,

Harasim, L., Hiltz, S.R., Teles, L., & Turoff, M. (1995). *Learning networks: A field guide to teaching and learning on line.* Cambridge, MA: The MIT Press.

Haythornthwaite, C., Kazmer, M.M., & Robins, J. (2000). Community development among distance learners: Temporal and technological dimensions. *Journal of Computer-Mediated Communication,* 6(1), September. Retrieved 10 April 2003 from http://www.ascusc.org/jcmc/vol6/issue1/hathornthwaite.html

Hiltz, S.R., & Wellman, B. (1997). Asynchronous learning networks as a virtual classroom. *Communications of the ACM, 40*(9), 44-49.

Johanson, M., & Torlind, P. (2004). Mobility support for distributed collaborative teamwork. *Proceedings of the 2004 Information Technology Conference (ITCON),* August. Retrieved 23 January 2006 from http://www.itcon.org/data/works/att/2004_25.content.00610.pdf

Johnson, D.W., Johnson, R.T., & Smith, K. (1991). Creative learning: Increasing college faculty productivity. Washington, DC: ASHE-ERIC Higher Education Report (4), Clearinghouse on Higher Education. The George Washington University.

Kiesler, S., & Sproull, L. (1992). Group decision making and communication technology. *Organization Behavior and Human Decision Processes, 52*, 96-123.

Kim, K.J., & Bonk, C.J. (2002). Cross-cultural comparisons in on-line collaboration. *Journal of Computer-Mediated Communication,* 8(1) http://www.ascusc.org/jcmc/vol8/issue1/ kim.html

Kristensen, B., Laursen, B., Pilegaard, M., & Jensen, B. (2004). Physical workload during use of speech recognition and traditional computer input devices. *Ergonomics, 47*(2), 119-133.

Kuo, F., Chu, T., Hsu, M., & Hsieh, H. (2004). An investigation of effort-accuracy trade-off and the impact of self-efficacy on web searching behaviors. *Decision Support Systems, 37*, 331-342.

Majchrzak, A., Malholtra, A., & John, R. (2005). Perceived individual collaboration know-how development through information technology-enabled contextualization: Evidence from distributed teams. *Information Systems Research, 16*(1), 9-27.

Mark, G. (2002). Conventions for coordinating electronic distributed work: A longitudinal study of groupware use. In P.J. Hinds and S. Keisler (Eds.), *Distributed work.* Cambridge, MA: MIT Press.

Markus, M.L. (1994). Electronic mail as the medium of managerial choice. *Organization Science, 5*(4), 502-527.

McGrath, J.E. (1984). *Groups, interaction and performance.* Englewood Cliffs, NJ: Prentice-Hall.

Morton, A., Ackerman, F., & Belton, V. (2003). Technology-driven and model-driven approaches to group decision support: Focus, research philosophy, and key concepts. *European Journal of Information Systems*, 12.

Muylle, S., Moenaert, R., & Despontin, M. (2004). The conceptualization and empirical validation of web site user satisfaction. *Information & Management, 41*, 543-560.

Roschelle, J. (2003). Unlocking the learning value of wireless mobile devices. *Journal of Computer-Assisted Learning, 19*, 260-272.

Sauer, J., Wastell, D., Hockey, G., Crawshaw, C., Ishak, M., & Downing, J. (2002). Effects of display design on performance in a simulated ship navigation environment. *Ergonomics, 45*(5), 329-347.

Saunders, C. (2000). Virtual teams: Piecing together the puzzle. In R. Zmud (Ed.), *Framing the domains of IT management.* Cincinnati, OH: Pinnaflex Publishing.

Sebok, A. (2000). Team performance in process control: Influences of interface design and staffing levels. *Ergonomics, 43*(8), 1210-1236.

Smith, C., Johnston, J., & Paris, C. (2004). Decision support for air warfare: Detection of deceptive threats. *Group Decision and Negotiation, 13,* 129-148.

Sohn, S., & Jo, Y. (2003). A study on the student pilot's mental workload due to personality types of both instructor and student. *Ergonomics, 46*(15), 1566-1577.

Sproull, L., & Kiesler, S. (1986). Reducing social context cues: Electronic mail in organizational computing. *Management Science, 32*(11), 1492-1512.

Storck, J., & Sproull, L. (1995). Through a glass darkedly: What do people learn in videoconferencing? *Human Communication Research, 22*, 197-219.

Taylor, S., & Todd, P. (1995). Assessing IT usage: The role of prior experience. *MIS Quarterly, 19*(4), 561-570.

Todd, P., & Benbasat, I. (1999). Evaluating the impact of DSS, cognitive effort, and incentives on strategy selection. *Information Systems Research, 10*(4), 357-374.

Wang, Y. (2003). Assessment of learner satisfaction with asynchronous electronic learning systems. *Information & Management, 41*, 75-86.

Wellmann, B. (1999). The network community: An introduction to networks in the global village. In B. Wellmann (Ed.). *Networks in the global village* (pp. 1-48). Boulder, CO: Westview Press

Wellmann, G., & Guila, M. (1999). The network basis of social support: A network is more than the sum of its ties. In B. Wellmann (Ed.). *Networks in the global village* (pp. 83-118). Boulder, CO: Westview Press

Wixom, B.H., & Todd, P.A. (2005). A theoretical integration of user satisfaction and technology acceptance. *Information Systems Research, 16*(1), 85-102.

Woodroof, J., & Burg, W. (2003). Satisfaction/dissatisfaction: Are users predisposed? *Information & Management, 40*, 317-324.

Wright, N., & McGown, A. (2001). Vigilance on the civil flight deck: Incidence of sleepiness and sleep during long-haul flights and associated changes in physiological parameters. *Ergonomics, 44*(1), 82-106.

Xie, B., & Salvendy, G. (2000). Review and reappraisal of modeling and predicting mental workload in single-and multi-task environments. *Work & Stress, 14*(1), 74-99.

## ADDITIONAL READING

Alavi, M., Marakas, G.M., & Yoo, Y. (2002). A comparative study of distributed learning environments on learning outcomes. *Information Systems Researc, 13*(4), 404-415.

Majchrzak, A., Malholtra, A., & John, R. (2005). Perceived individual collaboration know-how development through information technology-enabled contextualization: Evidence from distributed teams. *Information Systems Research, 16*(1), 9-27.

Roschelle, J. (2003). Unlocking the learning value of wireless mobile devices. *Journal of Computer-Assisted Learning, 19*(3), 260-272.

# Chapter VIII
# Constructing Mobile Technology-Enabled Environments for an Integrated Learning Approach

**María Felisa Verdejo**
*Dep. LSI. LTCS Group, UNED, Spain*

**Carlos Celorrio**
*Dep. LSI. LTCS Group, UNED, Spain*

**Emilio Julio Lorenzo**
*Dep. LSI. LTCS Group, UNED, Spain*

**Marta Millán**
*IES Diego Velázquez, Spain*

**Sergio Prades**
*Dep. LSI. LTCS Group, UNED, Spain*

**Javier Vélez**
*Dep. LSI. LTCS Group, UNED, Spain*

## ABSTRACT

*This chapter presents a review of the approach, design and implementation of the networked educational infrastructure of the ENLACE project that supports ubiquitous learning processes. The design phase in ENLACE was conceived based on a Learning Object Repository and the learning tools are presented as well as the approach for integrating these tools into the networking infrastructure. Considerations about the evaluation and the evolution of the system are also discussed.*

## INTRODUCTION

Mobile technologies have demonstrated the potential to enrich learning experiences outside the classroom (Faux, McFarlane, Roche, & Facer, 2006; Soloway et al., 1999), blurring the boundary between formal and informal learning (Scanlon, Jones & Waycott, 2005) and shifting the best learning opportunities from teacher-centered practices. Most of the work so far has focused on instruments and handhelds used for data collection and multimedia information access, while visiting museums and field settings (Soloway et al., 2001). Moreover, the trend of handhelds as student personal devices has opened new possibilities of supporting synchronization and interactivity for a variety of collaborative classroom activities (Tatar, Roschelle, Vahey, & Penuel, 2003). However, the provision of technology to support learning in different spaces, still remains as a range of unintegrated options, and typically there is not computational support for a seamless workflow of activities where data, content, and artifacts could be shared and reused by long-term learning processes, either by the students, teachers, or other contributing peers in a learning community.

In the ENLACE Project[a] we are exploring the design and implementation of a technological infrastructure as an educational networking platform offering services that will provide ubiquitous Web applications for learning activities both inside and outside the classroom in order to improve student learning and teachers' work (Verdejo, Celorrio, Lorenzo & Sastre, 2006a). A guiding principle in our approach is to imbue the technology design with a set of rich scenarios grounded on current learning sciences research on socio-constructivist learning, that is, learning is considered both a social and a cognitive activity (Koschmann, 1996). We are designing, with teachers and experts, technology-enabled scenarios providing facilities to foster experiential and constructive learning activities, meaningfully related to different subjects in the curriculum, where the study of a problem is considered through different facets and methods.

In our vision students are engaged both in individual and collaborative activities that promote their curiosity by inquiring about open questions (Branch, & Oberg, 2004), directly exploring and observing their environment, compiling information, analyzing and representing data, building models to explain and predict behavior, exchanging and discussing their findings, linking their views with real problems faced by professionals, and opening and contributing to create a shared memory with personal traces in a learning community. For example, a middle school teacher proposes a question to their students: "Why does this bird live in this place/habitat in the spring and summer time?" This question acts as the thread for a series of activities in differently localized scenarios, including a field trip to a nature park to experience and identify that particular habitat, and raises topics not only related to the natural science subjects (flora, fauna, habitat, and climate), but also to music (motivated by the need to identify the sounds of the birds), to geography (topographic and relief maps) and to mathematics (to calculate and represent distances using different measures). This also motivates teachers to coordinate themselves and integrate their respective agenda into a larger perspective.

The learning workflow includes activities for a long term period (to be carried out either in sequence, parallel, or overlapping in time) in different scenarios (classroom, home, computer lab, field trips, and exploratorium). The activities involve teachers and learners belonging to a school and wildlife monitors from an association. The educational network enables the collection of data in a site of interest, supporting its recording and reuse later on. The purpose is to facilitate the articulation of physical exploration of a site of special interest, with analytical reflection in the school. In order to permit a smooth flow throughout the scenarios of activities, the technological infrastructure should grant a ubiquitous context addressing the integration issue in a number of dimensions. These include: the school curriculum, location and time, social organization levels, as well as the heterogeneity of devices, taking into account their interoperability and connectivity conditions.

An essential component in the ENLACE educational networking infrastructure is the "learning object repository" (LOR) (Verdejo, Celorrio, & Lorenzo, 2006b). Technically, the LOR is an interoperable persistence mechanism, integrating data and artifacts created with heterogeneous tools. These user-created artifacts are meant to be reused by students in diverse contexts, using other tools or different devices (Hoppe et al., 2005). Conceptually, the LOR provides facilities to create customized workspaces either individual or social (a kind of structured memory) and the mechanisms for artifacts to be automatically metadescribed with shared conventions. For example, using the LOR, students can exchange different annotated and conceptualized pictures to cover views of a tree in different seasons, mine (their own or peers) information to complete sample measurements; search, query and aggregate collected data. The LOR functionality allows students to create and enrich incrementally the models of the targeted ecosystems, so that they can reflect on different observations taken at various sites, at different times, all throughout the year.

In this chapter, we present the evolving state of the project, highlighting the main issues. First, we point out how our work relates to recent developments reported in the literature. Then, the design and implementation approach is described. The current technical framework is presented afterwards. Some issues related to evaluation and technical requirements for the sustainability of the approach beyond the research project are outlined in our future research directions. A final consideration about reusability is discussed in the conclusion.

## BACKGROUND

In the last decade mobile learning initiatives, from small-scale case studies to large projects, have flourished almost everywhere in the world. On the one hand many of them are presented as opportunities for designing novel learning experiences that move out of the classroom to situate learning in a context (Chen, Kao, & Sheu, 2003; De Crom, & De Jager, 2005). On the other hand, there is an emerging community of practice, where mobile learning is often described as informal learning, in contrast to other conventional technological options with an emphasis on "personal" (portable, tailored to the needs of the individual) "ubiquitous" (accessible and embedded in everyday context of use) and "non-intentional" (to be used as the occasion arise) attributes.

Furthermore, at the research level, this relatively new area is increasingly an object of analysis and reflection. A number of papers have raised the issue of what is the special nature of mobility (Roschelle, 2003) as well as attempts to define the still controversial concept of mobile learning (Sharples, Taylor, & Vavoula, 2005). The current literature also offers reviews of the pedagogical theory in the light of m-learning (O'Malley et al., 2005), and discusses the methodological approaches to design (Stringer, Fitzpatrick, Halloran, & Hornecker, 2005) and evaluation of these kinds of environments (Taylor, 2003).

Some authors, including many from this book, consider the setting in which learning occurs as a continuum from formal settings to social structures (Sefton-Greene, 2004). As Faux et al. (2006) points out, one of the challenges is to create coherent links across different sites, and for this purpose, the need is to provide a technological infrastructure that supports the development of learning services and their deployment for networked collaboration. This implies support for mobile personal devices but also enriching the switch from individual to collaborative activities, from personal to shared information and reflection in small groups as well as participation in larger ones such as the class or a community. Interestingly, in many reported experiences with handheld devices, the main reason for discontinuing their use is the problem of software availability. Designing the technological support as a flexible and contextualized composition of services for different learning contexts is the driving force of our approach.

## THE DESIGN AND IMPLEMENTATION PROCESS

The design phase in ENLACE was conceived as a set of participatory cycles involving students and teachers in a secondary school, monitors from SEOBirdlife[b], as well as project researchers. First of all, a feasibility study, testing the mobile infrastructure in a variety of settings was carried out. Small activities were deployed with different technical configurations to evaluate the quality of the wireless connection, the period of autonomy of the handhelds and portable devices in stand alone and connected mode, the robustness and the scope and portability of the whole infrastructure on a field trip. For example we tested (a) students with handhelds, a teacher with a laptop computer and a smart blackboard in a classroom with a wireless internet connection; (b) students with handhelds in a stand alone mode working in the school courtyard and in the lab; and (c) a Wi-Fi ad-hoc network with a laptop and handhelds in a field setting. Evaluation results indicated that the use of the PDAs enhanced learners' experience and their motivation, but the initial idea of creating a permanent Wi-Fi network with a laptop working as a server for the PDAs in the field setting was unsuitable with the current state of the technology. PDAs ran out of battery in a short period and the laptop was a burden for the monitors.

For the first design cycle, a collaborative project involving music, mathematics, science and geography as part of the curriculum was devised. Adopting an inquiry-based approach, a main question to investigate was given to the students aged 11-12 years, launching the learning processes to be developed in a range of school settings: the classroom, the computer room and the science lab; at home and on the field trips, throughout an academic year. Three teachers and 27 students were involved as well as four monitors. The selected settings were provided with an educational technological network, supporting both the individual and the collaborative learning workflow with different facilities including a variety of Web resources and cognitive tools. These were deployed on different devices operating with and without permanent connectivity. The next section describes the scenarios (A to E illustrated by Figure 1) implemented in the first formative design cycle (2005/2006). The formative evaluation and evolution of these scenarios has become essential to refine and develop a user-centered system.

### The First Cycle

This first cycle was a pilot study including the core pedagogical ideas: inquiry and collaborative long-term activities, involving both a holistic view across the curriculum, and a continuum between formal and informal settings. It included five scenarios (Figure 1), three of them with mobile devices.

From the technological side, the learning object repository (LOR) provided the essential functionality for interoperability, as well as mechanisms to automatically integrate the outcome of different tools. The idea was also to use existing cognitive tools, as is the case of ModellingSpace (Avouris, Margaritis, Komis et al., 2003), that was chosen by the teachers as an adequate modeling tool for the age of the targeted students. The added-value here was to integrate this "external" tool in the flow of the networked support. Other software allowing authoring facilities for flexible data col-

*Figure 1. ENLACE Scenarios in the first cycle*

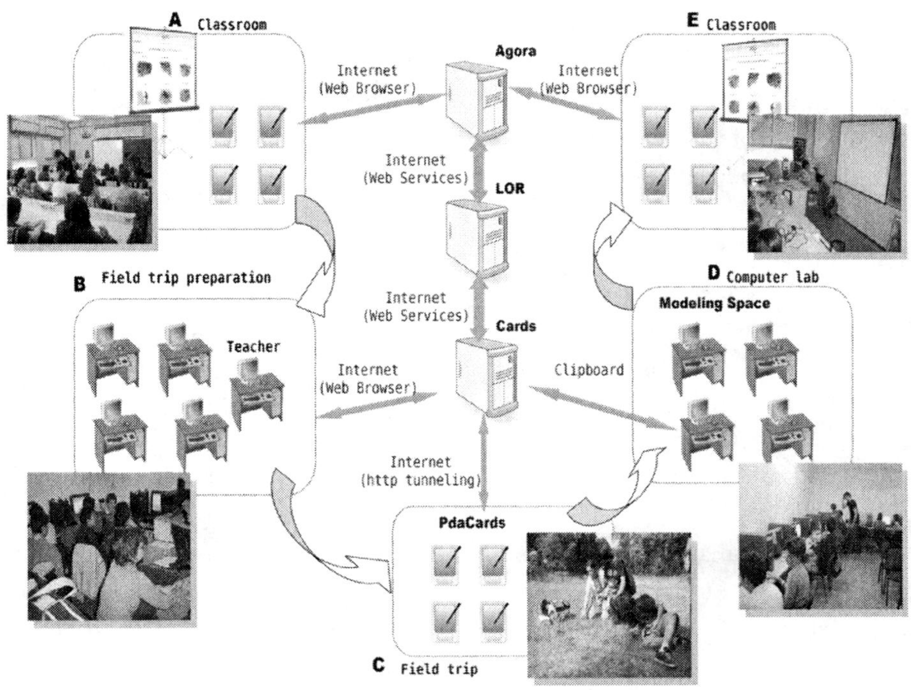


lection and for the management of basic activity workflow was designed and implemented.

## Scenario A: First Contact and Warm-Up Activity in the Classroom

The activity started with an initial session, which took place in the classroom with the music teacher. In this session students learnt to recognize the bird-songs that will later be heard on the outdoor trip. For this test we used a multimedia content created by the SEOBirdlife organization. It features a data sheet for every bird detailing its habitat, customs, special features and identification tips. It also includes pictures and sounds. While working with this material (students had the opportunity to work with it at home also), several interactive tests were performed with the Agora voting software tool. Agora is a Web tool developed in the framework of the ENLACE project. The purpose of Agora is to give technological support to a wide range of possible discussion and voting scenarios that can come up in an educational environment.

In each test, the teacher showed pictures of a set of birds, then played a sound and asked the students to guess the corresponding bird. Each student voted for a bird using his own handheld device. Agora then automatically summarized and projected the voting results and the teacher led a discussion about the results. These tests helped to measure the students' bird identification skills. Eventually, the learning session was repeated until the students reached a certain level of understanding.

The inclusion of a technological background in this scenario is important as it provides the students with an introduction to both the handhelds and the Agora tool, which will be used in many other activities (see scenario E in Figure 1). In a voting context, the technology grants a formal and fair framework, while students feel handhelds are both familiar and exciting. Agora has been designed to consider two main factors that change between different scenarios: the nature of the candidates (for instance names of students for a delegate election, pictures of observations, artifacts such as an activity plan, a concept map or any other model) and the student participation in the candidate proposition and selection. These factors can be modified through the session manager interface. Furthermore, Agora is interoperable with the learning object repository so that candidates could be searched and uploaded from the LOR.

Once the teacher has created and customized the voting session through the session manager it is available to be used in the classroom through the Web player. The teacher role in Agora can switch between two different views: the teacher view and the projector view. The projector view is the default and is intended to be used with a projector showing the whole class the workflow of the activity, the current stage, the candidates, the voting results, and so forth. The teacher view can be used occasionally and provides access to more detailed information (for example, finding out which students have not yet voted so that the teacher can give them a verbal reminder).

This simplicity allows the teacher to place workflow control on a student, who can act as an assistant. The workflow changes dynamically if, for example, a draw occurs, allowing a new voting stage to begin to resolve the draw. The student view allows the student to follow the workflow and participate when the time comes. For example, when the teacher switches to the

*Figure 2. Agora projector and student's view on PDA*

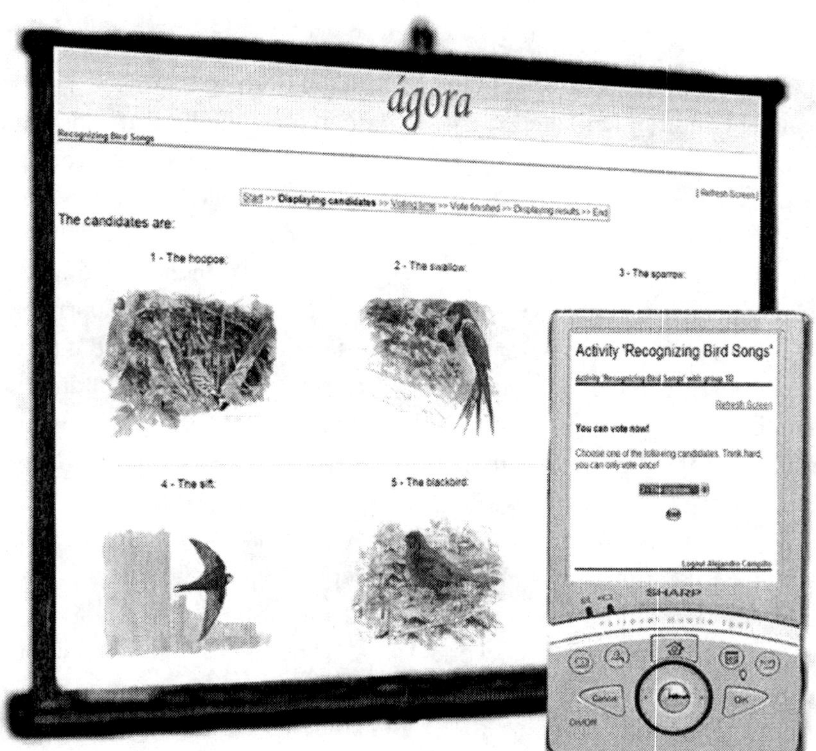

"voting time" stage the student view is refreshed, asking the student to check the candidates on the projector and then vote through a combo box. The projected images serve here as an anchor for the activity: a shared reference around which action and attention can be coordinated (Crook, 1994). Figure 2 shows a picture of both the projector and the student (handheld) view.

## Scenario B: Preparing the Field Trip in the School Computer Room and at Home

The next focus was a field trip to a nature park called "El Monte de El Pardo" with a variety of ecosystems, in order for the students to collect different data *in situ*. Observations and experiments to be carried out were discussed and planned collaboratively in small groups (see scenario B in Figure 1) in the computer room at school and occasionally at home. In this planning phase students actively explored the topic to establish the information they would need, in particular the concept of a habitat, and the observations needed to identify and describe it. This took into account flora and fauna behavior, meteorological conditions, as well as physical measures or evidence for the models students had to build as part of their solution. They used a variety of sources, including the SEOBirdlife Web site and a professional botanical database. As a result of this step, students created the set of card templates to be used during the field trip to collect data. For this purpose they used the tool CARDS and with this tool they were able to define the different fields each card template comprises. Each field can have a different answer type (free text, an element of vocabulary, number, a range, a picture, etc.). For the birds' example the observation card included a field for the name, the GPS location of the spot where the birds were heard and the altitude of that location. Once a card template had been defined it was saved by CARDS in the LOR via Web services and it was available to be downloaded to the PDAs later on for the field trip. Eventually the teacher could run an Agora session with a group of students to revise and come to an agreement about the fields a card template should have. Furthermore, an activity plan for the field trip had to be created. In this case it included a set of potential tasks for each one, plus its description and a set of card templates to be used for data collection. These activity templates were also created with the application CARDS and stored in the LOR for further use.

## Scenario C: The Field Trip

PDACARDS is an application allowing the same operation of the Web interface of CARDS to fill a card using a PDA (Figure 3). PDACARDS needs to establish communication with CARDS before the field trip in order to download the required background images and forms for the chosen activities. After this initialization, students can collect their data with the PDA in a standalone mode. A connection is needed only for a short period of time at the beginning of the field trip and at the end for uploading filled cards, thus a permanent connection is not necessary. This is the best solution with the current state of the technology (a very high rate of battery consumption for a permanent Wi-Fi connection, high cost and unreliable telecommunications network coverage in unpopulated areas).

Once the first synchronization was done, and with the students in the forest, they were organized into several groups with a SEOBirdlife instructor. Groups were dispersed and coordinated to cover a variety of ecosystems. After searching and choosing a suitable place they stopped to listen to the birds in absolute silence. At this point they filled the card with the name of the birds that they thought they were listening to, and with their GPS position. They carried out other activities, collecting a variety of data, depending on the

*Figure 3. Filling a card on the field trip*

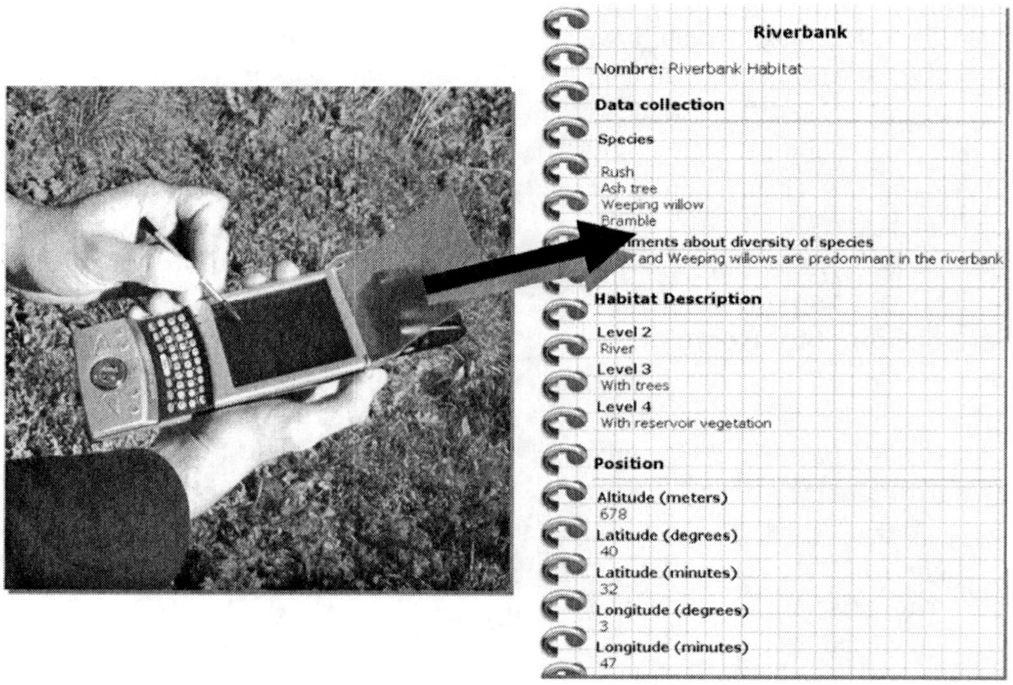


itinerary and the questions raised in their activity plan. For example, selecting a representative tree for an ecosystem, deciding which method to use to determine its height, taking the measurements needed to calculate the solution later on, its location, taking pictures of animal tracks or reminders, and counting exemplars in a selected area to establish the level of biodiversity.

The filled cards are stored in the PDA. The last synchronization occurs at the end of the activity and consists of uploading the filled cards from the PDAs to the LOR. The data is submitted, via http, to the CARDS application which makes a learning object with the card as content and adding as metadata information from the context session, like author, activity or date. After that the LO is sent to the LOR, via Web services, and is available for further use.

## Scenario D: Analyzing Data and Creating Models

Once students had finished the field trip activity where they had collected data about the birds and the habitat, they proceeded to analyze and elaborate the data, building different models, in order to present and justify the answers to the main research questions. For example, in the pilot one of the models was a topographic profile of the visited area showing the location of the observations. This learning activity involved building a graphical representation of an itinerary using the experimental data students had collected themselves. First, they had to select and aggregate the appropriate data (from objects in the repository), then they had to construct the graphical representation, and finally they placed their observations on the representation. For this purpose, three modeling tools interoperating with the LOR were available:

CARDS, ModellingSpace and OXOtool. With the ModellingSpace tool, students created an itinerary entity, which included the two variables (position and altitude), establishing a "table relationship" to link these two variables. Once the table was filled with the collected values, they could visualize the path profile. Students, using CARDS, had to search the LOR for their observations, analyze their cards, select and aggregate the relevant data to generate a position-altitude table representing each one of the reference points in their field trip path where they have gathered the birdsong information. Then they exported this tabulated data to the clipboard to be used as input for the ModellingSpace. The generated profile can be stored as an image file, which can be sent back to CARDS, and transformed into a LO with contextual generated metadata to be stored in the LOR for further use. In this case, students used the profile they have created as a background for the final habitat map where they placed "active" birds icons, linked to their respective cards, allowing them to show the LO complete information. OXOtool is another tool developed in ENLACE which provides functionality for aggregating "active" icons over a background to create a model and/or a new composition object.

## Scenario E: Discussing and Choosing a Representative Profile in the Classroom

The last phase of the learning project took place in the classroom again. At this point, the different objects representing the habitat profiles built by the groups of students were stored in the LOR. In these sessions the whole class, using the Agora voting tool, proceeded to discuss and evaluate which habitat profile was more accurate and eventually selected one as a class representation: this profile became a new shared learning object and was stored in the LOR. Comparing and reflecting is not a straightforward step: the technology facilitates knowledge sharing, management, and organization of the discussion but for the process to be focused and fruitful, it needs to be carefully managed by the teachers.

The following sketchy script summarizes the first cycle:

- The first phase of problem refinement and background information collection; students worked individually and in small groups, CARDS offered them facilities for authoring cards where they could gather and organize the information. In parallel, students joined for "discussion and consolidation" sessions in the networked classroom (Agora, connecting a shared board, a laptop and the PDAs) working together in a large group. These activities were carried out for different but related subject areas, fostering a cross-curriculum perspective. The products created are auto-documented digital objects, open to inspection and available to be consulted, shared and elaborated incrementally with different tools.
- Planning the field trip, collecting data *in situ* and building models to elaborate, explain, and justify answers to the questions investigated. Students were supported by the following tools: CARDS, ModellingSpace, and OXOtool networked with the LOR, (running on desktop PCs) and PDACards for the PDAs with GPS cards, working individually and in collaboration, both in standalone mode, and synchronized with the network. The networked tools support situated data collection, its aggregation across individuals and groups over time and space, and input/output links to modeling spaces. It is the first step to assemble a collaborative activity across a variety of technologies.
- Revising and reflecting in a large group, using Agora with the LOR, deployed in the networked classroom. Students had the opportunity to present their viewpoint to the entire class, and carry out an interactive peer evaluation.

### The Second Cycle

On the basis of findings from the first cycle, some modifications were made to the design and deployment of the second one, involving new scenarios and applications.

From the learning point of view the need emerged to conceptually organize and relate the knowledge and the artifacts created along the inquiry process. To this end a version of a collaborative concept map tool named COMET, interoperable with the LOR, was designed and built. Furthermore, this tool enables us to reify the mappings between representations (data and models). The graphical representation provides opportunities for students to present, compare, negotiate and share their views on the data and models fostering the co-construction of knowledge. As well this, a new version of the OXOtool, renamed COMPO, which supports remote collaborative modeling and improved functionality for discussion and sharing was released. This tool allows students to aggregate information from individual contributions connecting geospatial representations with data or more abstract entities. COMPO has been designed to run clients on different devices (including PDAs).

In order to support the management of learning activities, from the social organization and technical point of view, a Web based e-learning platform, PELICAN, was developed. This platform (Vélez, Barros, & Verdejo, 2006) provides support for the definition (and reuse) of projects, activities, and groups (through a portal user interface) facilitating the deployment of the tools mediating the learning workflow.

Other mobility scenarios with PDAs were included to enhance the cross-disciplinary approach: from a visit to an interactive science museum to record the data from the in site simulations students could experiment with, to a geomining museum where they could examine and identify samples of minerals relating to the ecosystems studied. A visit to the Museo Del Prado was also planned, to look at paintings from the sixteenth century where students could find the same landscape and think respectively about the historical evolution of the area and the factors involved. For each scenario, students have designed and authored the cards for data collection to be used to record the observations and information captured in the different visits with the PDAs and used them later on. These facilities have supported a more student driven approach for explorations, giving them direct control and leading to an appropriation of the technology. The current system is technically described in the next section.

## THE SYSTEM AT WORK

The ENLACE system has been designed to facilitate the flow and transformation of data and artifacts across tools and scenarios for all the actors involved in the learning activities. The educational network of ENLACE includes several learning tools (Figure 4, ❶). The artifacts and outcomes generated by those tools are encapsulated as LOs with a set of metadata according to different LO types (Figure 4, ❷).

A LO type defines the description for a class of LOs as an aggregation of a general metadata schema plus other metadata schemas purposefully defined to characterize these objects. The LO type specification also includes the format of content (for example, multimedia material, XML files and models following tool specific formats) described with a MIME type. The first step for integrating a tool with the infrastructure consists of defining the new LO type in the LOR that will be used for storing the outcome generated by the tool. This LO type should contain specific metadata that will make it possible to customize searching and handling of the data according to the functionality requirements of the tool, and will enable artifact interoperability among other tools.

For instance, CARDS generates 'observation' objects that represent direct observations made

*Figure 4. Global vision of ENLACE system*

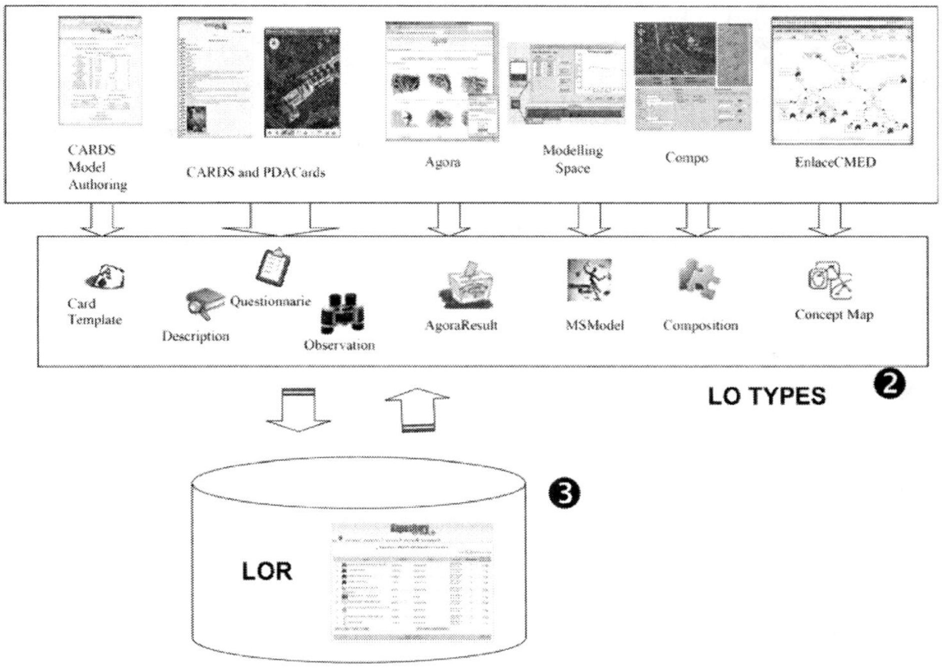


by students during a field trip. This type of LO contains metadata about the observation localization coordinates and ecosystem. Another tool like COMPO, can search only for the LO made in a concrete ecosystem or place and reuse them to generate a composition.

Thus, the LOR (Figure 4, ❸) provides the support for interoperability, together with external tools and other tools developed in ENLACE, based on artifacts and its metadata. The next sections will take a closer look at the LOR and its interface between the ENLACE tools.

## The LOR Service

The learning object repository service (LOR) was developed as the result of the need for a persistence mechanism for collaborative learning in experimental domains and evolved during research undertaken in two European projects. It was important for students to reuse partial results created in their learning activities in further experimentation and reporting, and also it was advantageous for the teachers to reuse their learning designs for other related activities instead of starting a new definition from scratch. The LOR has been designed in order to provide a generalized, configurable and interoperable storage framework based on metadata for external learning tools. From the functionality point of view, interface facilities and an API are provided to guide and simplify the creation of LO types and the tool's specific metadata schemas in a declarative way. Also, the creation of repository views is possible in the LOR in order to build separate workspaces for students and teachers. For instance, a group repository view can be created to store the artifacts generated in a learning activity by a group, a class repository view can be defined for the students to share results with the whole class, or a private teacher's repository view can be created where students can send their assignments but they are only viewable by the teacher.

*Figure 5. Web portal interface of the LOR*

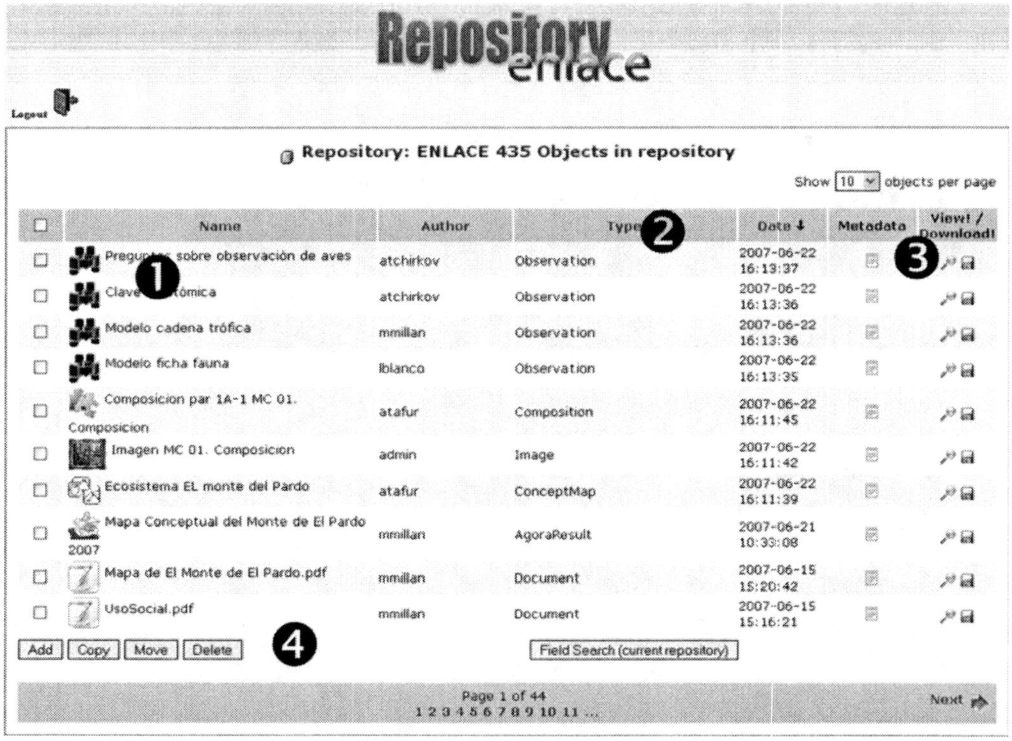


The LOR service provides functionality directly accessible from a Web portal interface, and remotely available via Web services. The Web portal interface of the LOR, shown in Figure 5, offers the users access to the list of LOs accessible in a repository view ❶ and information about their corresponding types ❷. The LOR service offers facilities for managing LOs, inspecting their metadata and viewing and downloading their content from the Web (Figure 5, ❸). Other operations include: adding, modifying, deleting or searching for objects in the repository and moving and copying objects from one repository view to another (Figure 5, ❹).

Searching can be done by querying by example (i.e., selecting an object, and using its metadata as a starting template), filters (predefined combination of templates that provide useful and recurrent queries) and query patterns (parameterized combinations of templates, that the user can modify and instantiate) (Mayorga, Barros, Celorrio, & Verdejo, 2006).

The main reason for searching is for reuse. Thus the role of metadata is to assist users and software tools to query and retrieve potentially useful objects for a specific purpose in the learning process. The goal is to obtain rich descriptions for different kinds of artifacts created by the learners, minimizing the manual labor of filling in metadata, and, thanks to the definition of LO types and specific metadata schemas, granting a generalized framework for the external tools to metadocument their working objects based on their needs.

In summary, the LOR service, in its current state, offers:

- A persistence mechanism for community sharing and reuse of learning resources

- Concern for standards: LOM metadata compatibility and content packaging
- Interoperability via Web services with external tools
- Mechanisms for handling automatic generation of metadata for objects created with external tools via application profiles for tools (Verdejo et al., 2006b).
- A service for synchronizing distributed repositories in a network (Celorrio, Verdejo, & Barros, 2006).
- Enhanced searching facilities based on metadata, including query by example, relevance feedback, filters and query patterns.
- Heterogeneity, that is, the possibility of storing objects belonging to a variety of Learning Resources: LO types/metadata schema defined in a declarative way.

## The Approach for the Tools

One of the main characteristics of the LOR, and what makes it the backbone of the ENLACE networking infrastructure, is that it provides mechanisms for external tools to take advantage of its storage and retrieval facilities. Although the LOR incorporates a Web interface for accessing the repository objects, its most important feature is the tool-adapted persistence mechanism which it exposes through Web services for uploading, retrieving and searching for the stored objects from any computer tool. Many diverse tools can be potentially integrated in the networking infrastructure via the LOR's Web services. Furthermore a "wrapper" approach is also provided for those tools not able to interoperate directly with the LOR's external interface (Verdejo et al., 2006b).

All the main functionalities of the LOR are provided to any external application in this way. By using these Web services the LOR offers a remote interface for storing, searching and retrieving any LO in an adaptive way. These operations use the LO types and metadata previously defined for the tool's objects, which can also be requested remotely.

The approach for the tools is to provide added-value functionalities to the infrastructure in a modular way enriching the collection of services of the educational network. These tools can be individually set up in a learning scenario, using the LOR to integrate their generated outcomes through the learning workflow. Next we will describe the tools developed in ENLACE to illustrate this approach.

## CARDS and PDACards Tools

CARDS is a software tool for authoring models of forms and creating instances of them, such as observations, description cards or questionnaires. It also provides operations for the user to fill these models of forms interactively.

In addition, a set of operations on collections of instances are defined in CARDS, such as selecting, filtering, aggregating fields and mapping them to generate new instances of other forms (for example, a table of position/altitude for the different points of the observations cards related to an itinerary to be exported to the Modelling-Space tool). Furthermore, CARDS can be seen as a metamodelling system for wrapping other tools products because it allows importing and exporting data in different formats, acting as a bridge to the LOR functionalities for those standalone applications not able to generate descriptions for their input/output.

CARDS includes a Web-based authoring system (Figure 6 ❶) for defining models for data collection and the metadata associated with these models. These models are named card templates and they are basically a set of fields grouped under sections. Each of the fields of a card template is parameterized with a type that determines the type of content for the answer slot (i.e., a text, a number, a string from a vocabulary, an image, etc.) and attributes that describe whether the field is mandatory and if multiple answer values

*Figure 6. CARDS tool authoring system*

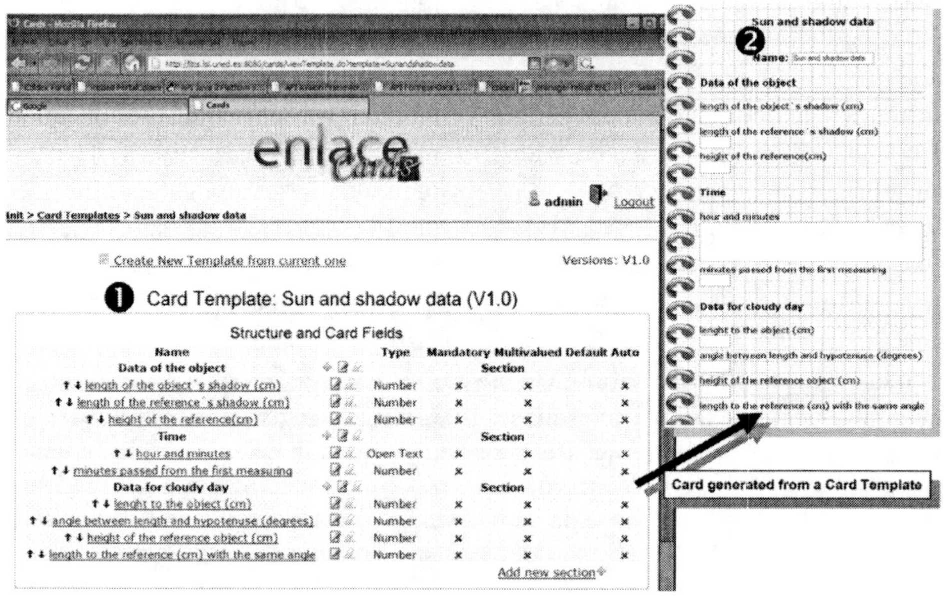


or default values are permitted. Card templates can also contain association information between fields and metadata items of a specified LO type of the LOR, enhancing automatic metadata generation when filling cards. Teachers, students and instructors are able to create new card templates in the authoring system from scratch or just by instantiating and stating parameters from previously created ones.

Once defined, card templates are stored in the LOR via Web services and they are ready to be used in a data collecting activity. A card is a filled instance of a card template. CARDS Web interface (Figure 6 ❷) allows users to fill cards and generates as a result an LO with an associated type and some automatically generated metadata according to the information of the card template.

PDACards is a stand-alone application designed for PDAs that allows the same operations of the Web interface of CARDS for filling cards. It resolves the problem of filling observation cards in scenarios where no internet connection is available and therefore there is no access to the CARDS Web application In this way, a permanent connection is not necessary, except for a short period of time at the beginning and end of the field trip for synchronization.

When PDAs are given to the students they need to connect to the CARDS application in order to load the required card templates for the chosen activities. After this initialization students can collect their data with the PDA in a standalone mode. When the field trip has ended another connection is established and all observations are sent back to CARDS, which stores them in the LOR.

## COMPO

COMPO is a collaborative tool, implemented as a Web application, that allows several users to create visual associations between LOs online, which are stored in a LOR repository, and a background image that can be a LO too.

To establish this visual relationship, COMPO associates with each LO:

- A representative icon (chosen from a large collection related with ENLACE)
- A size for the icon (five sizes, from XS to XL)
- A descriptive text
- A localization within the background image

In this way the relationship between the concepts and the improved meaning gained by the LO visualization comes from the shape of the icon, its size (representing the importance of this LO in comparison with others LOs) and its localization in the background image. It has, therefore, a different meaning if the icon is (Figure 7, ❺) near the riverbank or in a forest clearing, or if it is greater or smaller. These associations are conceptually very easy to understand, in fact, with a simple look the user can see the relationship between the LO and the background image and the relationship among the LOs. In addition, with a double click on the icon the LO content associated in the repository is displayed in another browser tab.

The COMPO interface is divided into two large areas: one private and personal (Figure 7, ❶, ❷, ❸ and ❼) and one shared and collaborative (Figure 7, ❹ and ❺). In the private area the user, selecting filters for metadata (Figure 7, ❶), can query the LOR (through Web services) for LOs that comply with the search conditions given in

*Figure 7. An example of the Manzanares river composition*

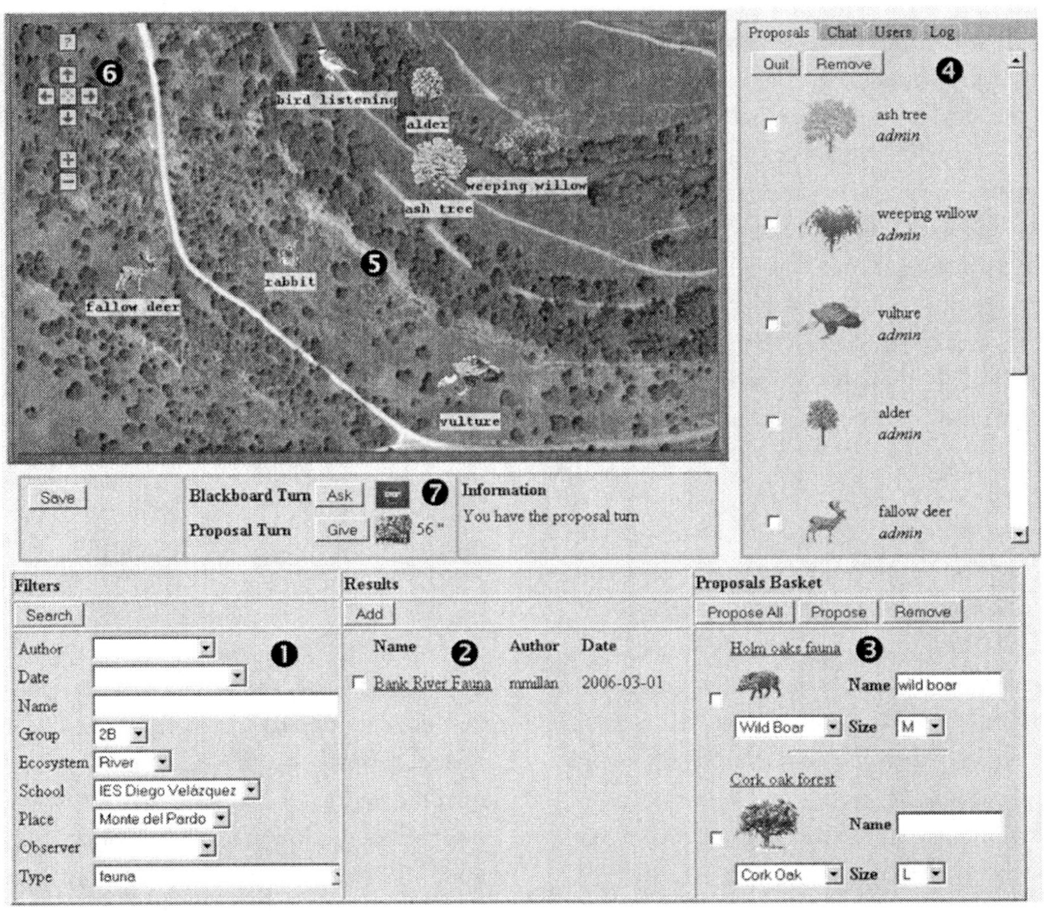

the filters, and then choose one or several of them from the results (Figure 7, ❷). To speed up the proposal of multiple LOs, COMPO provides a proposals basket (Figure 7, ❸) where it stores the selected LOs of different searches. In the basket the user selects the icon, the size and the text for each LO, and then adds them to the proposal zone (Figure 7, ❹).

Every change made in the collaborative area is refreshed in quasi real time in every browser connected to the composition, so that when a user adds a proposal to the proposal zone (Figure 7, ❹) it appears in every browser at the same time. To place a proposal in the image a user can drag and drop the icon from the proposal zone (Figure 7, ❹) to the background image (Figure 7, ❺). He or she can change the localization of the icon or remove it from the image altogether. These changes are refreshed in the other browsers as well. In order to obtain precision when placing or moving icons, the map allows zoom in/out and provides arrows to move in four directions (Figure 7, ❻).

For the correct management of concurrent changes, COMPO implements a writer-reader system. Only one user can propose new proposals and only one user can move icons to the image at any time. The users can see who currently has these turns (Figure 7, ❹, tab users) and the state of their own turns (Figure 7, ❼).

To fulfill all these functionalities COMPO implements the following series of modules:

- A persistent central server contained in a servlet (Tomcat Web server using Java, Struts and Velocity) that centralizes the composition data, the users, logs, chats, events, LOR communication, and so forth.
- A refreshing module that determines which part of a browser composition data needs to be refreshed because of the changes made in the composition.
- An asynchronous data flow system between the browsers and the server to refresh in quasi real time every change made by a user, sending in both directions only the new data (using AJAX for sending and receiving data and DHTML for modifying the object changed in the browser).
- A turn based interface to guarantee correct concurrency (one user modifies the composition, the rest watch the changes) allowing the users to ask for the turn and give it to the next user.
- A chat module integrated into the Web application (also based on AJAX) to provide the users with a direct communication platform (Figure 7, ❹, tab chat).
- A log module to register the actions performed by any user (Figure 7, ❹, tab log).
- A connection module with the LOR using Web services to read the metadata filter, make queries, see LO content and save the compositions.

In order to offer flexibility for using different schemas of collaboration, COMPO allows a series of parameters to be configured such as the maximum time of the turns, the maximum number of proposals for a single user, the background image, the repository in the LOR where the composition will be saved, different roles for the users (manager, normal user, watcher) and different rules to allow flexible policies like "everyone can modify the proposal of everyone else" or "everyone can only modify his or her own proposal" prompting the users to negotiate and come to an agreement using the chat tool.

## COMET Tool

COMET (concept maps for enlace tools) is a concept map tool with extended features for using artifacts generated from other learning tools. A concept map (Novak, 1977) is a graphical way of representing and organizing knowledge. It is comprised of textual-labeled nodes (concepts) and links (relationships) organized with the most general concepts at the top of the map and the more specific, less general concepts arranged hierarchically below.

The main goal of this cognitive tool is to provide a knowledge representation for a higher level of abstraction: a metalevel to link the models that learners and teachers have constructed through the learning activities. In this way they can explicitly interrelate and describe the outcomes generated previously with other tools. In order to facilitate this reflective process three core functionalities are provided, and to illustrate them we will use the concept map in Figure 8. It is one of the maps generated by the students when building their description of "El Monte de El Pardo," as a Mediterranean forest ecosystem.

The first functionality allows students to associate one of the learning objects stored in the LOR with a node on the concept map. When the user clicks a node on the map and selects the "add LO" option from the menu, a new window, called "search LOR," appears (S at the bottom of Figure 8). It is possible to search for objects according to different metadata criteria as well as seeing the content of a concrete object before linking it to the selected node on the map. When the object is associated, an icon, representing the type of the learning object, appears on the right of the node. For example, observations are represented by a binoculars icon (Figure 8, ❶), descriptions with a book icon (Figure 8, ❷), ConceptMaps with a map icon (Figure 8, ❸), and compositions created with COMPO, by a jigsaw icon (Figure 8, ❹).

A snapshot of the searching interface is shown by S in the Figure 8. The node selected ❶ is a "riverbank transect" node, the query is for all the learning objects with the name "Transecto en ribera" (Spanish translation of riverbank transect) with "LO Type" observation from the activity "Excursión al Monte de El Pardo" carried out by "2°F" Group. All the learning objects that match these constraints appear in a list with further information. Once one of the objects is selected by clicking the "add" button in the search LOR window, the selected learning object is linked to the concept node. Afterwards, the user can see the content of the associated learning object by clicking on that node.

In this way it is possible to organize the models generated by tools as CARDS or COMPO as well as to associate observations to concept descriptions. In the example the "riverbank transect" observation was data collected on the field trip and is linked to nodes representing the descriptions of the kinds of flora located in this transect. A description card of the ash tree is displayed in Figure 8 ❷. The different pieces of information are now in a broader context with explicit links, created by the students themselves, to represent their global vision of the knowledge construction process. Furthermore, the concept map could be saved and loaded as another learning object in the LOR, and then be associated with another concept on a different map. In this case the link is represented by a map icon, as shown in ❸. This is a powerful mechanism to encapsulate and organize more complex structures, allowing different levels of abstraction.

The second functionality is the possibility of associating a set of learning objects with a node according to several criteria (in the functionality described before the association was only with one object). An example is shown in Figure 8 ❺ where the "observations of el Monte de El Pardo" concept has a box icon with several other icons inside. In this example all learning objects with type 'observation' and with activity metadata equal to "Excursión al Monte de El Pardo" are associated with that concept. By clicking the node, a window with a list containing all the retrieved objects is displayed and each object can be inspected. Thus, all the data collected by students can be grouped here, defining a kind of "view" of the LOR. COMET in this way allows the building of a kind of personalized conceptual indexing to objects in the LOR. For example, a student can build a concept map with all the activities he or she has performed, each one in a node, and with all the objects created in each one respectively associated.

*Figure 8. COMET tool*

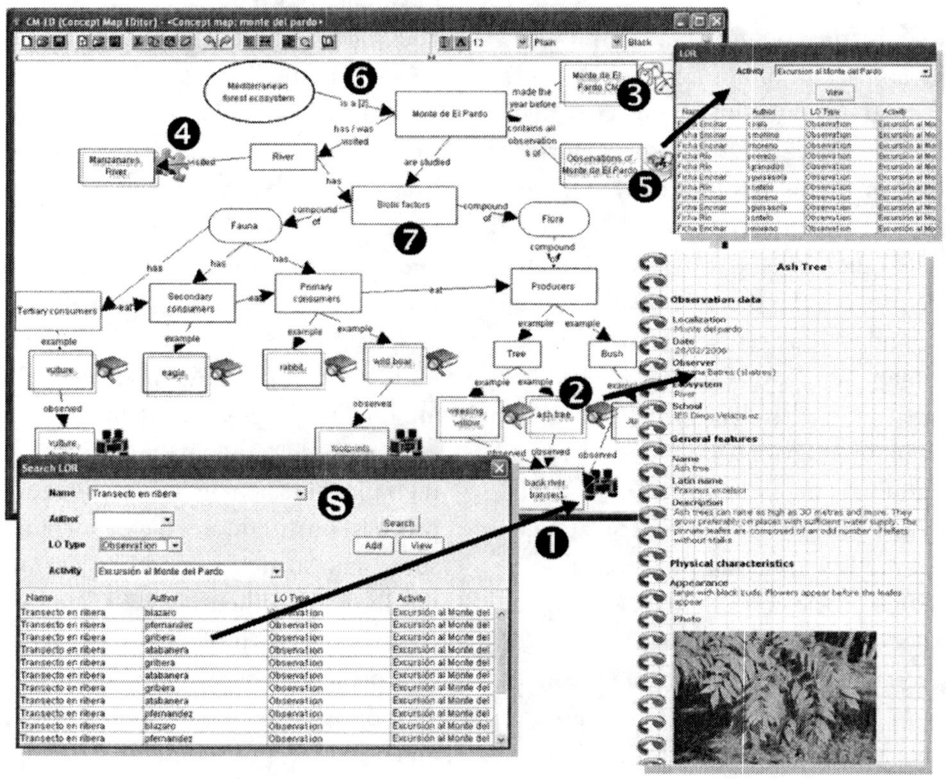


The third functionality provides automatic support for merging maps to facilitate the task of collaborative concept mapping construction. Depending on the collaboration schema, it is the case that students have to join, intersect, complement or select alternatives from different partial maps. The algorithm inputs two maps and generates a proposal for a merged one. The merging algorithm is based on Liu (2002). Roughly speaking it is an iterative process where the basic step is to compare groups of two concepts joined with an edge on a map with another map and change the first one taking into account the following cases:

- If the edge and the concepts linked are the same (equal concept names and labeled edge) on both maps, then it adds a number two (in brackets) on the edge. This means that the two maps share the exact link.
- If they have the same concepts but different labels on the edge, then it writes both labels on the edge and the number two in brackets.
- If only one concept is the same, then it adds the different concept (and its edge)
- If both concepts in the second concept map are not found in the first map, then an unconnected new group for these concepts and their edge is added.

Figure 8 graphically illustrates the result of a merging process. In this example two groups of students made a different map: one about the flora and the other about the fauna of el Monte de El Pardo. Both concepts appear in the figure connected by a common concept "Biotic factors" (Figure 8, ❼). In the top of the map (Figure 8, ❻) there is an edge labeled "is a [2]," This means that the two compared maps had the same two concepts linked with the word "is a." Below this link there is another with the label "has/was visited," this is interpreted as this edge appears in both maps linking the same concepts but with two different labels: "has" and "was visited." This merged map helps students to detect disparities and it is a starting point to work together to refine it and come up with a consistent and negotiated outcome. In the figure, the relationship "eat" (below the "biotic factors" node) between the "primary consumers" and "producers" concepts is a result of that refinement process performed by all the students in the classroom.

The COMET tool has been developed on the CM-ED concept map editor engine (Larrañaga, Rueda, Elorriaga, & Arruarte, 2002). This editor is a Java stand alone application, which allows the drawing of different kinds of concepts as nodes, and relationships as edges, and both of them can be displayed in several shapes. This editor was chosen for its intuitive and friendly interface and the potential for extending the code to create new functionalities. A number of modules have been added: the LORManager is used to interoperate with the LOR services and ModelManager to render the content of the learning objects. The MergingManager is used for the merging process and Java WebStart allows the opening of the application like a Web page. Furthermore, a SocialManager allowing integration with the social model in PELICAN is described in the next section.

## The Pelican Platform

Pelican is an e-learning platform that supports several processes related to collaborative learning experiences. The goal is to provide a central Web site where users involved in a learning process can keep in contact forming a virtual community where they jointly undertake a flow of collaborative activities.

One of the core capabilities of Pelican is to allow instructional designers and teachers to describe learning activities and arrange them within a learning project according to certain instructional workflow settings. When a new activity is created, a set of resources (that is, Web references), role allocation information, and accessible tool-provided services can be specified. A learning design is usually created to fit the pedagogical requirements of a learning situation. Nevertheless, each one of its components constitutes an artifact that could be reused in another different social and cognitive context. To demonstrate, we will show how the scenarios previously described in Figure 1 can be specified in Pelican. Firstly, a new learning activity has to be created in the platform for each scenario, indicating the references, roles and services that are available for the participants. For instance, as can be seen in snapshot A of Figure 9 (Pelican manager interface), for the learning scenario A, an activity named "recognizing bird songs" has been created. Within it, (enlarged on the right hand side for better visualization) are two references to a relevant Web database of birds ➊, and two services, one for the Agora voting tool ➌ and the other one for the CARDS tool have been added. In addition, two roles must be defined to allow the Agora tool to map between the different Pelican actors and those that play the role of voter and manager in the voting system, typically the teacher but occasionally the students ➋. Once all activities have been created they can be aggregated to define a larger learning project. One of the added values of the platform for the users is that services are launched automatically with the corresponding parameters once an activity starts.

Another important aspect that instructional designers must pay special attention to is the social context where the collaborative experience is to be deployed. It includes considering whether participants are to be arranged into several working groups, and if this is the case, their properties, such as, size, heterogeneity level among members, lifespan, and so forth. These factors depend strongly on the settings of the learning scenario, that is, the characteristics of the student collective and the type of learning tasks to be undertaken. Pelican provides a framework to design a "society model" specifying the structure of the virtual community and then, populate it. The former task requires defining the potential different actors and group templates. An actor is a stereotype of a user such as a teacher, student or monitor, and represents a set of permissions and social responsibilities within the community. Group templates are stereotypes of groups containing structural information such as maximum and minimum number of members, types of allowed actors, inner group templates, and so on. The latter task consists of registering new users, creating fresh groups and joining users within groups embodying one or more actors, considering group factors such as size, heterogeneity level among members, lifespan, and so forth. For instance, the social model defined in Pelican to support ENLACE scenarios was created to match the actual arrangement of learners

*Figure 9. PELICAN tool*

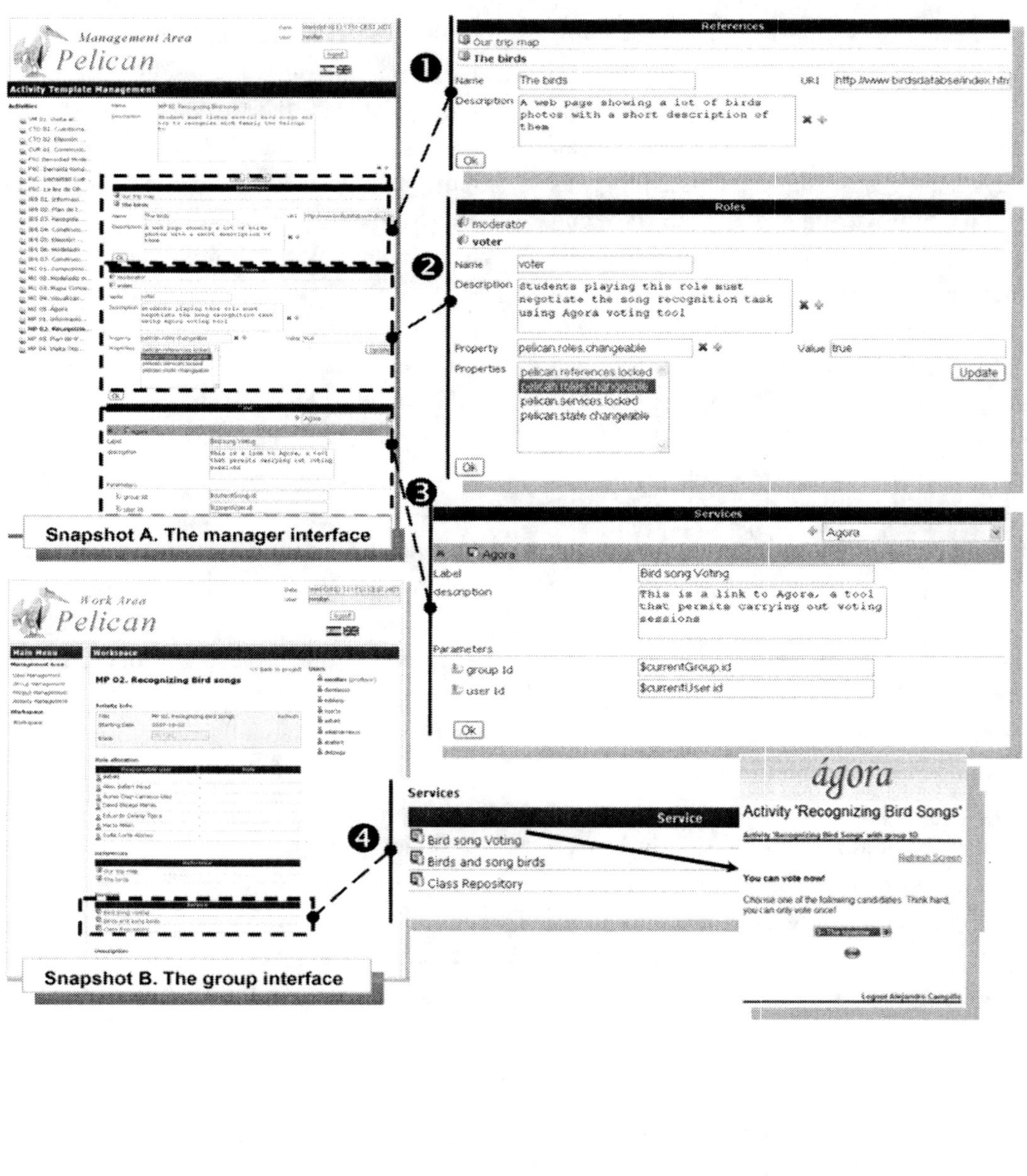

in the secondary school. In that model the root group template is the school. It is built of several classroom groups; each classroom group has a set of associated teachers, up to 30 students and, optionally, a set of working groups. Once a virtual community has been arranged a learning project can be deployed across different groups. When this is finished a fresh set of learning activities are created to furnish the Pelican workspaces of target groups. A learning workspace in Pelican is a virtual environment (snapshot 9B) where collaboration takes place. As can be seen in Figure 9 ❹, users interact within learning activities through the embedded services provided by external tools (Agora, LOR, CARDS, etc.).

Pelican provides means to integrate different supporting tools within workspaces so collaboration can take place across a transversal use of them. Nevertheless, in order to offer continuity in their use, a proper level of interoperability must be supplied. To achieve this goal, other proposals (IMS Tools Interoperability Guidelines, 2006) support some integration standards (Bote et al., 2004). Pelican uses a scripting language to describe how external tools must interact with each other and a rule based mechanism to schedule their execution along the scenario lifetime. Advanced administrators can define a set of rules that capture all the dynamic aspects of such a scenario. These aspects are coded as scripts that use Web services technology to communicate in the background with integrated external tools in order to carry out some coordination tasks such as transferring and transforming generated products among the tools or preparing them for the next phase of the scenario. In the ENLACE experience, for instance, all described scenarios present an intrinsic continuity between them. Scenario A is training for bird recognition, scenario B prepares the material for scenario C, and so on. The rule based mechanism supplied by Pelican can be used to guarantee continuity among those scenarios. For example, models stored in the group LOR within scenario D must be transferred to the class LOR in scenario E and a new Agora session must be dynamically set up with the generated products.

## FUTURE RESEARCH DIRECTIONS

### Evaluation Design

The final design of the system functionality has been achieved through a formative evaluation process in each of the two development cycles, keeping users involved in the design process. Usability testing has been performed on the software and the network support in the different scenarios. The use of multiple interfaces and a combination of devices was perceived as natural. In parallel, a pedagogic analysis of the learning activities and its deployment has been carried out. There was evidence of a positive impact on the students' engagement, enhancing their ability to control their tasks and their collaborative skills. The benefits of *in situ* data collection combined with subsequent modeling, review and reflection, were also felt by the teachers to be a very effective approach for knowledge co-construction, and to have great potential for an integrated use of the technology across the curriculum. Outdoor activities with monitors succeeded in both bridging the move from informal to formal learning, and raising awareness on professional work. Technical issues such as short battery life, and unreliable connections were identified as problematic. Infrastructure management was also an issue, though some attempts to involve students in the technical set up seemed to be a feasible approach; they become familiar with the technology very quickly. In some cases the timing of the activities was in conflict with the tight school schedule, overwhelming teachers with timetable negotiations with other teachers and groups of students. Teachers pointed out the need for software tools facilitating the monitoring of the tasks and the annotation of students outcomes.

A final evaluation within the project involving a larger number of students and teachers, and an external evaluation specialist will be carried out in the 2007/08 academic year. We will address three major targets: teachers, students, school and community. Accordingly, the main questions to be investigated are (1) What is the impact on teaching behaviors and instructional practices? (2) Is the system used in such a way that it will lead to enhanced student learning and achievement? and (3) what is the impact on school organizational and social requirements for an appropriate and effective long term implementation.

Although activity theory (Engestrom, 1987) is being proposed as a way to interpret settings where technology has been introduced, there is still not a well established evaluation framework to track and understand the learning experiences in these kinds of mobile technology supported settings (Traxter, & Kukulska-Hume, 2005). While we have adopted AT as a global perspective to capture people and mediating tools in their context, a mixed approach taking into account quantitative and qualitative techniques is being planned, including questionnaires, a range of interviews, observation of selected tasks, tests, and process and product analysis supported by automatic techniques.

## The Evolution of the System

As has been stated all along in the description of the ENLACE system, a key concept in the evolution of ENLACE is the continuous integration of heterogeneous learning tools into the educational network. However, due to the different configuration needs of each of the learning scenarios, which can include heterogeneous types of devices where the tools and resources have to be deployed in order to carry out the learning activities, the management of the technical infrastructure can in practice become very complex. This task should therefore be better supported to enable this ubiquitous learning framework to be sustainable for a school, above and beyond a pilot study.

A working problem that is being dealt with is developing an intuitive configuration mechanism for the ENLACE educational network that will unburden teachers and pedagogical tutors from the technology issues by enhancing the technical deployment of the learning activities and all the associated learning resources. The goal is not just merely to provide a vehicle for distributing and deploying the needed software resources over different nodes, but to also offer computationally adapted support for a set of learning activities in a particular context for the members of a learning community. This configuration mechanism of the networking infrastructure should model the technological deployment of a scenario design, including a personalization system performing an adapted configuration of resources—tools, data and communication—for each particular type of device and user.

In order to overcome this problem, a distributed technology with AI techniques for personalization and adaptation capabilities would be required. We are exploring an agent-based approach which will deliver a distributed computational layer to the ENLACE system for the personalized configuration of activities in ubiquitous learning scenarios. The basic operations regarding this agent-based layer in the ENLACE architecture are:

- Contextualization—extracting students' environmental information as well as its current situation within a social frame, activity frame, and localization frame context.
- Installation—managing all necessary software tools in an automated way on the collection of devices involved in the current scenario activities.
- Personalization—deciding which set of activities and learning objects are applicable to a specific device-tools environment and

adjust them if necessary for a specific student.

- Selection —extracting the set of necessary learning objects from the LOR and the learning objects that can be considered potentially useful for the generated activities.
- Distribution—deployment of learning objects to the students' devices.
- Configuration—initialization and set up of all the required tools for the scenario, their data and the learning objects for the user's activity with the devices.
- Adaptation—adapting data and interface for the device-network.

A further step for the ENLACE networking architecture is to provide support for tasks that were not previously planned but could emerge from an interesting real time opportunity. During a field trip, occasional events may occur that were not programmed in the activity like seeing an uncommon type of plant or finding some rare animal tracks. The networking infrastructure should react to these kinds of events. The objective is to enrich the experience and enhance the activities by taking advantage of unexpected situations and providing communication mechanisms among the nodes to dynamically reconfigure the current learning activity. The architecture (Celorrio, & Verdejo, 2007) should support the tracking of events from the nodes of the network during the activity in order to take advantage of real-time information, and offer operations for:

- Community Broadcast—depending on the nature of the events, the network can put students in contact or inform particular members of the community about actions that are taking place in the activity.
- Reconfiguration—making the network react to exceptional events that may occur during the development of an activity by reconfiguring the network and redistributing resources amongst users and their devices.

Communication among different nodes (i.e., mobile devices) is still an issue to be solved in order to implement a real-time interaction with p2p operations. The current state of wireless mobile technology is unreliable for guaranteeing full usability in ubiquitous learning scenarios, such as field trips.

## CONCLUSION

This chapter presents a review of the approach, design, and implementation of the ENLACE project to support ubiquitous learning processes. An important issue which has driven our design of the technology from the beginning of the ENLACE system has been to provide a flexible and contextualized composition of services for different learning scenarios. All these services are intended to be general enough to support reusability within diverse learning contexts, in order to fulfill tailored pedagogical requests, and to be modular and interoperable enough to be incorporated within different technological contexts, usually by providing Web services access interfaces.

As mobile technologies are continuously evolving, special effort is made not to compromise our software with a specific type of device or technology. The bulk of our applications are programmed in Java, which is present in the vast majority of mobile devices, and therefore transferring our software models from one device to another would result in few or no changes at all. All these design attitudes very closely match the agenda of the latest initiatives and projects in technology enhanced learning, for instance, the SAIL framework (SAIL, 2007).

## ACKNOWLEDGMENT

The ENLACE Project is partially funded by the Spanish Research Agency, MCYT, grant n. TIN2004–04232. Carlos Celorrio acknowledges PhD grant BES-2005-7178. Agora tool has been developed by former member of the LTCS group Alberto Ruiz Cristina. The authors wish to acknowledge their partners on SEOBirdlife: Federico García and his group of monitors.


## REFERENCES


Avouris, N., Margaritis, M., Komis, V., Saez, A., & Meléndez, R. (2003). Modellingspace: Interaction design and architecture of a collaborative modelling environment. In *Proceedings of the Computer Based Learning in Science 2003*

Branch, J., & Oberg, D. (2004). *Focus on inquiry: A teacher's guide to implementing inquiry-based learning.* Edmonton, AB: Alberta Learning Teaching Resources Branch.

Bote, M. L., Vaquero, L. M., Vega, G., Dimitriadis, Y., Asensio, J. I., Gómez, E., & Hernández, D. (2004). Tailorable collaborative learning system that combines OGSA grid services and IMS – LD scripting. In *Proceedings on the X international workshop on groupware, CRIWG 2004*, Springer-Verlag, LNCS 3198, (pp. 205-321). San Carlos, Costa Rica.

Celorrio, C., & Verdejo, M. F. (2007). Adapted Activity Deployment and Configuration in a Pervasive Learning Framework. In *Proceedings of the Workshop of Pervasive Learning 2007: Design Challenges and Requirements.* Toronto, Ontario, Canada.

Celorrio, C., Verdejo, M. F., & Barros, B. (2006). Un Modelo de Distribución de Repositorios para un Portal de Comunidades de Aprendizaje Colaborativo. *IE Comunicaciones, Revista Iberoamericana de Informática Educativa, 3*, 27-42.

Chen, Y.S., Kao, T.C., & Sheu J.P. (2003). A mobile learning system for scaffolding bird watching learning. *Journal of Computer Assisted Learning, 19*, 347-359.

Crook, C. (1994). *Computers and the collaborative experience of learning.* London: Routledge International Library of Psychology.

De Crom, E. P., & De Jager, A. (2005). The "ME"-learning experience: PDA technology and e-learning in ecotourism at the Tshwane University of Technology (TUT). In *Proceedings of the 4th World Conference in M-Learn 2005.* Cape Town, South Africa

Engestrom, Y. (1987). *Learning by expanding.* Helsinki: Orienta-Konsultit.

Faux, F., McFarlane A., Roche, N., & Facer, K. (2006). *Learning with handheld technologies.* Futurelab Handbook. December 2006. Retrieved July 18, 2007, from http://www.futurelab.org.uk/resources/publications_reports_articles/handbooks/Handbook198

Hoppe, U., Verdejo, M.F., Pinkwart, N., Barros, B., Oelinger, M, Mayorga, J.I., & Zeini, S. (2005). Building bridges within learning communities through ontologies and "thematic bjects." Computer Supported Collaborative Learning, 2005

IMS Tools Interoperability Guidelines. (2006). Retrieved July 18, 2007, from http://www.imsglobal.org/ti

Koschmann, T. (1996). *Paradigm shifts and instructional technology: an introduction.* In CSCL: Theory and practice, (pp. 1-23). Lawrence: Erlbaum Associates.

Larrañaga, M., Rueda, U., Elorriaga, J.A., & Arruarte A. (2002). Using CM-ED for the Generation of Graphical Exercises Based on Concept Maps. In Kinshuk, R. Lewis, K. Akahori, R. Kemp, T. Okamoto, L. Henderson & C.-H. Lee, (Eds.), *Proceedings of ICCE'2002*, (pp. 173-177).

Liu, X. (2002). Using concept mapping for assessing and promoting relational conceptual change in science. *Science Education, 88*(3), 373-396

Mayorga, J.I., Barros, B., Celorrio, C., & Verdejo, M. F. (2006). Accessing a learning object repository through a semantic layer. In *Workshop on 'Learning Object Repositories as Digital Libraries: Current challenges'2006.*

Novak, J. D. (1977). *A theory of education.* Ithaca, NY: Cornell University Press.

O'Malley, C., Vavoula, G., Glew, J. P., Taylor, J., Sharples, M., Lefrere, P. et al. (2005). *WP4 – Pedagogical Methodologies and Paradigms, Guidelines for learning/teaching/tutoring in a mobile environment.* MOBIlearn. Retrieved 18 September, 2006 from www.mobilearn.org/download/results/public_deliverables/

Roschelle, J. (2003). Unlocking the learning value of wireless mobile devices. *Journal of Computer Assisted Learning, 19,* 260-272.

SAIL. (2007). Scalable Architecture for Inquiry Learning. *SAIL home page.* http://www.telscenter.org/confluence/display/SAIL/

Scanlon, E., Jones, A., & Waycott, J. (2005). Mobile technologies: prospects for their use in learning in informal science settings. *Journal of interactive media in education (25)*

Shefton-Green (2004) *Literature review in Informal learning with technology outside school.* Report NESTA Future Labs. (7).

Sharples, M., Taylor, J., & Vavoula, G. (2005). Towards a theory of mobile learning. In *Proceedings of mLearn 2005 Conference.*

Soloway, E., Grant, W., Tinker, R., Roschelle, J., Mills, M., Resnick, M. et al. (1999). Science in the palms of their hands. *Communications of the ACM, 42(8)*, 21-27.

Soloway, E., Norris, C., Blumenfeld, P., Fishman, B., Krajcik, J., & Marx, R. (2001). Log on education: Handheld devices are ready-at-hand. *Communications of the ACM, 44(6)*, 15-20.

Stringer, M., Fitzpatrick, G., Halloran, J., & Hornecker, E. (2005). *Moving Beyond the Application: Design Challenges For Ubiquitous Computing.* Position paper presented at Aarhus2005 workshop 'Ambient Computing in a Critical, Quality of Life Perspective'. Aarhus, Denmark, 21st August .

Tatar, D., Roschelle, J., Vahey, P., & Penuel, W. R. (2003). Handhelds go to school: Lessons Learned. *SRI International. Journal Computer, September*, 30-37

Taylor, J. (2003). A Task-Centred Approach to Evaluating a Mobile Learning Environment for Pedagogical Soundness. In *Presentation at the Mlearn 2004, Birmingham.*

Traxler, J., & Kukulska-Hulme, A. (2005, October 25-28). Evaluating Mobile Learning: Reflections on Current Practice. In *Proceedings of MLEARN 2005.* Cape Town, South Africa.

Vélez. J., Barros, B., & Verdejo, M. F. (2006). A framework to define web based communities. *International Journal of Web Based Communities (IJWBC), 2(*3).

Verdejo, M.F., Celorrio, C., Lorenzo, E., & Sastre, T. (2006a). An Educational Networking Infrastructure Supporting Ubiquitous Learning for School Students. In *Proceedings of 6th IEEE International Conference on Advanced Learning Technologies (ICALT06)*, Kerkrade, The Netherlands.

Verdejo, M. F., Celorrio, C., & Lorenzo, E. (2006b). Improving Learning Object Description Mechanisms to Support an Integrated Framework for Ubiquitous Learning Scenarios. In *Proceedings of the 4$^{th}$ IEEE International Workshop on Wireless, Mobile and Ubiquitous Technologies in Education (WMUTE06).* Athens, Greece.

## ADDITIONAL READING


Bellamy, R.K.E. (1996). Designing educational technology: Computer-mediated change. In B. A. Nardi (Ed.), *Context and consciousness: Activity theory and human-computer-interaction*, (pp. 123-146). Cambridge, MA: MIT Press.

Benford, S. (2005). *Future location-based experiences*. JISC Tech Report (TSW0501).

Chen, Y., Liu, Ch., Yu, M. & Lu, Y. (2005). A study on elementary science classroom learning with wireless response devices-implementing active and experiential learning. In *Proceedings of the IEEE International Workshop on Wireless and Mobile Technologies in Education (WMTE'05).*

Davis, S. (2003). Observations in classrooms using a network of handheld devices. *Journal of Computer Assisted Learning, 19*, 298-307.

DiGiano, C., Yarnall, L., Patton, C., Roschelle, J., Tatar, D., & Manley, M. (2002). Collaboration design patterns: Conceptual tools for planning for the wireless classroom. In *Proceedings of the IEEE International Workshop on Wireless and Mobile Technologies in Education (WMTE'02).*

Jansen, M., Rossmanith, P., Uzun, I., & Hoppe, U. (2005). Integrating Heterogeneous Personal Devices with Public Display-Based Information Services. Paper presented at the *IEEE International Workshop on Wireless and Mobile Technologies in Education (WMTE '05).*

De Jong, T., & Van Joolingen, W.R. (1998). Scientific discovery learning with computer simulations of conceptual domains. *Review of Educational Research, 68*(2), 179-201.

Kurti, A., Spikol, D., Milrad, M., Svensson, M., & Pettersson, M. (2007). Exploring How Pervasive Computing Can Support Situated Learning. In *Proceedings of the Workshop of Pervasive Learning 2007: Design Challenges and Requirements.* Toronto, Ontario, Canada.

Linn, M. C., Davis, E. A., & Bell, P. (2004). Inquiry and technology. In M. C. Linn, E. A. Davis, & P. Bell, (Eds.), *Internet environments for science education* (pp. 3-27). New Jersey: Lawrence Erlbaum Associates.

Rieger, R., & Gay, G. (1997). Using mobile computing to enhance field study. *In Proceedings of CSCL97*, pp. 215-223.

Ryu, H. (2007). Is mobile learning a necessary evil?: The goodness-of-fit of mobile learning situations. In *Proceedings of the Workshop of Pervasive Learning 2007: Design Challenges and Requirements.* Toronto, Ontario, Canada.

Naismith, L., Lonsdale, P., Vavoula, G., & Sharples, M. (2004). Literature Review in Mobile Technologies and Learning: Nesta Futurelab.

Neven, F., & Duval, E. (2002). Reusable learning objects: a survey of Lom-based repositories. *Proceedings of the 10th ACM international Conference on Multimedia* (pp. 291-294). Juan-les-Pins, France.

Milrad, M., Hoppe, U., Gottdenker, J., & Jansen, M. (2004) Exploring the use of mobile devices to facilitate educational interoperability around digitally enhanced experiments. In *Proceedings of the 2nd IEEE International Workshop on Wireless and Mobile Technologies in Education.*

Verdejo, M. F., Barros, B., Mayorga, J. I., & Read, T. (2004). Designing a semantic portal for collaborative learning communities. R. Conejo et al. (Eds.), *CAEPIA-TTIA 2003, LNAI 3040*, (pp. 251-259). Berlin: Springer-Verlag.

Verdejo, M. F., Barros, B., Read, T., & Rodríguez-Artacho, M. (2002). A System for the Specification and Development of an Environment for Distributed CSCL Environments. *Intelligent Tutoring Systems, 6th International Conference, ITS 2002.* LNCS, Springer-Verlag.

Weal, M., Hornecker, E., Cruickshank, D.G., Michaelides, D.T., Millard, D., Halloran, J. et al. (2006). Requirements for in-situ authoring of location based experiences. In *Proceedings of MobileHCI '06,* (pp.121-128). ACM Press.

Zurtia, G., & Nussbaum, M. (2004). Computer supported collaborative learning using wirelessly interconnected handheld computers. *Computers & Education, 42*, 289-314.

## ENDNOTES

[a] ENLACE means *Link*. The project home page: http://enlace.uned.es

[b] SEOBirdlife is a nongovernmental organization dedicated to the study and conservation of birds and nature. SEO stands for Sociedad Española de Ornitología (Spanish Ornithology Society).

# Chapter IX
# Collaboration in Context as a Framework for Designing Innovative Mobile Learning Activities

**Daniel Spikol**
*Växjö University, Sweden*

**Arianit Kurti**
*Växjö University, Sweden*

**Marcelo Milrad**
*Växjö University, Sweden*

## ABSTRACT

*In this chapter we describe our continuing efforts related to the design, implementation and evaluation of innovative educational activities supported by ubiquitous computing in the AMULETS (advanced mobile and ubiquitous learning environments for teachers and students) project. We argue that the design of innovative mobile learning activities should be guided by collaborative learning scenarios in context supported by mobile and ubiquitous technologies. To support this claim, we propose a conceptual framework of collaboration in context that can be used when designing novel mobile learning scenarios. This framework provides the designer with opportunities to tackle the challenges of designing for innovative mobile learning activities. To illustrate our ideas, we present the results of three trials we have conducted with children and adult students since the spring of 2006. These mobile learning activities have been designed and implemented using our proposed framework. Working with the teachers and students gave us the opportunity to design learning activities at authentic settings using meaningful content that has relevance for the school curriculum. The outcome of our efforts suggests that outdoor learning experiences supported by ubiquitous technologies should be combined with learning activities in the classroom to provide learners with meaningful activities.*

## INTRODUCTION

Recent advancements in mobile, wireless, and positioning technologies, combined with contextual computing, are contributing to the development of new mobile applications and services. The rapid adoption of sophisticated mobile devices and applications has created new social tools for people to connect and interact; therefore changing the ways we communicate and collaborate. Educational environments are being subject to these changes, providing an opportunity for curriculum development that can use these socially based mobile devices for supporting different aspects of learning and teaching. Mobility offers new dimensions to support and promote meaningful learning activities that include features such as connectivity, social interactivity and context sensitivity (Klopfer et al., 2002). From this perspective, mobile technologies allow enhancing the learners' context by the creation of embedded ubiquitous environments in authentic settings, thus providing innovative ways of interacting with them. They also present design opportunities for multiple kinds of collaboration to support different aspects of the learning process (Price et al., 2003).

One of the main assumptions we consider as a point of departure for the ideas to be presented in this chapter is the fact that in the coming five years, whether educators would like it or not, more and more students will bring mobile devices with wireless communication into the classroom. These devices can be in the form of tablet PCs, PDAs, cellular phones, smart phones or GPS devices. All these technologies and new forms of mobile communication and collaboration have been adopted by young people and integrated into their everyday lives. Clear indications of this can be found on sites such as www.youtube.com, www.flickr.com, www.blogger.com, and www.facebook.com. Lankshear and Knoble (2006) claim that schools ignore some of these trends and argue that mobile and wireless technologies and new media might be integrated into current school educational activities, as they are transforming and defining new literacies in teaching and learning. Thus, there are a number of challenging questions that deserve further exploration. What are the implications of using mobile computing and wireless communication for supporting teaching and learning? What new scenarios and applications will emerge? Which aspects and processes should be considered while designing new mobile collaborative solutions?

In this chapter we describe our continuing efforts related to the design, implementation and evaluation of innovative educational activities supported by ubiquitous computing in the AMULETS (**a**dvanced **m**obile and **u**biquitous **l**earning **e**nvironments for **t**eachers and **s**tudents) project. We argue that the design of innovative mobile learning activities should be guided by collaborative learning scenarios in context supported by mobile and ubiquitous technologies in authentic settings. To support this claim, we propose a conceptual framework that can be used when designing novel mobile learning scenarios. This framework provides the designer with opportunities to tackle the challenges of designing for mobile computer supported collaborative learning (mCSCL) and mobile-learning (mLearning) environments. To illustrate our ideas, we present the results of three trials we have conducted with children and adult students since the spring of 2006. These mobile learning activities have been designed and implemented using our proposed framework. In the rest of the chapter, we will describe in further details how collaboration in context with mobile support can be used for the theoretical, conceptual and design aspects of our research activities, as well as for evaluating the results. We described the activities in the trials together with a brief explanation of the technology we have developed. We will conclude by discussing the outcomes of the trials in connection to the proposed framework and the challenges facing innovative mobile learning applications.

## COLLABORATIVE LEARNING IN CONTEXT

Learning and collaboration have their roots in many different theories of cognition and development that support different type interactions between peers and experts. Piagetian theories advocate for peer-to-peer interactions between "equals" enabling conversations that can result in cognitive restructuring, while Vygotskian theories can be seen to support a peer-to-mentor interaction where the mentor, the more "able" partner, facilitates the development of knowledge and skills by scaffolding their activity (Price et al., 2003). Both schools advocate social interaction as playing a key role in learning; while Vygotsky's work started a wider theoretical development of "cultural psychology" that together with anthropology and cognitive science formed the basis of situated learning (Littleton & Häkkinen, 1999 citing Lave & Wenger 1991 and Suchman, 1987). Situated learning (Lave & Wenger 1991) is a general theory of knowledge acquisition that is based on the notion that learning (stable, persisting changes in knowledge, skills, and behaviour) occurs in the context of authentic activities.

Learning is a social process. It happens in collaboration between people and together with technology. So, when introducing technology, the view should be shifted from seeing it as a cognitive delivery system to considering it as means to support collaborative conversations about a topic (Brown & Duguid, 2000). The central notion is that learning is enculturation, the process by which learners become collaborative meaning-makers among a group defined by common practices, language, use of tools, values, beliefs, and so on (Hoppe et al., 2005). Our view on collaborative authentic learning activities outside the classroom, which will be presented in the coming sections, has been guided by the ideas of Rogoff and Lave (1984). These authors have suggested that young people and adults learn more efficiently, and perform more competently in realistic settings outside the classroom than they do in many decontextualized environments that school usually provides (see also an elaboration of learning in context by Brown et al., 1989).

Mobile collaboration in context is our suggestion towards a new framework for the design, implementation, and evaluation of innovative mobile learning activities and systems. Both collaboration and context play key roles in the design of mobile learning activities. Mobile devices are prevalent in people's everyday lives and can be easily used in the classroom and in the field, providing more opportunities than computer labs (Vahey & Crawford, 2002). Roschelle and colleagues (2005) suggest that articulating a design framework that spans many mCSCL activities can be a key contribution to further work. Furthermore, collaboration needs to be seen from the perspective of shifting away not only the outcomes and products of collaborative work, but also towards analyzing interactions as means of gaining insights into the processes of collaborative learning (Littleton & Häkkinen, 1999). Dourish (2004) raises the importance of looking at context through interactions focusing on the question; "how and why in the course of their interactions, do people maintain a mutual context for their actions?"

Winters and Price (2005) highlight the importance of the context in which a learning activity is taking place as a crucial component for design. Context plays multiple roles in the interaction between physical and social locations, tasks and activities and the user's situation. In the scope of our research efforts, we define context as *"information and content in use to support a specific activity (being individual or collaborative) in a particular physical environment."* Therefore, our definition of context relies upon a three-axis structure consisting of the following attributes; *location/environment* attributes, *activity/task* attributes and *personal/interpersonal* attributes. The attributes of this structure are interdependent, meaning that information about who the

*Figure 1. A conceptual framework for collaboration in context*

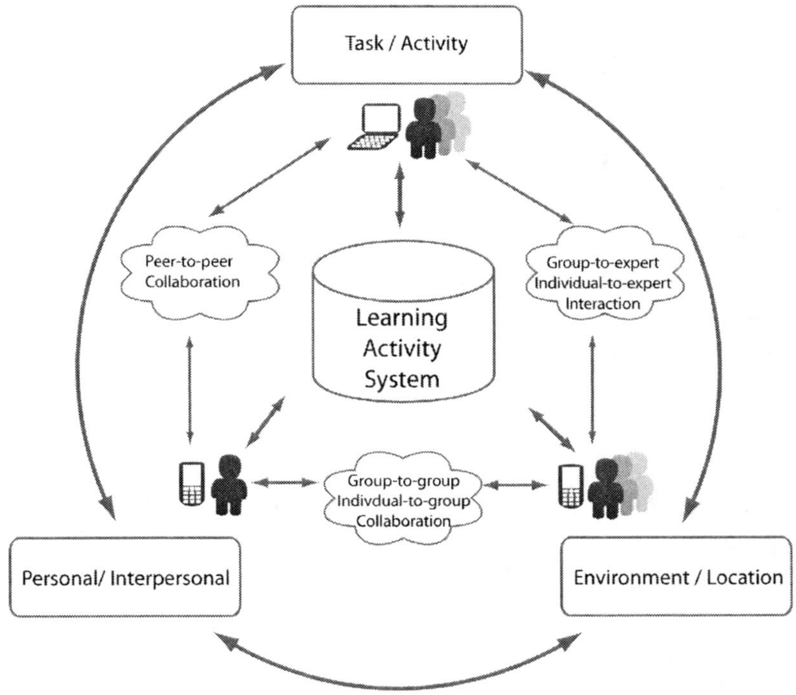


user is; where the user is; what the user is doing and the interplay between these activities need to become valuable inputs to the design process (Kurti, Spikol, & Milrad, 2008).

Figure 1 illustrates the ideas previously described while using them as the central components of a conceptual framework for designing innovative mobile learning activities. A basic component of our framework is the learning activity system (LAS) simply described as a computational system and content repository that provides the technological infrastructure for integrating educational content into the context where the learning activities and collaborations are taking place. The surrounding circle of this conceptual framework defines the context where the activities are happening. The use of this design framework allows for creating engaging active learning activities in which collaboration and context are important components. From a technical perspective, the implementation of the LAS relies upon the use of different software components and ubiquitous technologies, as well as sensors in order to contextually support collaborative activities across locations (Kurti, Spikol, & Milrad, 2007). The learners interact with the LAS and with each other, thus promoting different modes of collaboration. Each one of the three context attributes described in figure 1 can be combined in set of pairs (e.g., task, location; personal-interpersonal/task-activity, etc.) or as a triplet, thus providing the proper context in which the learning activity takes place.

Our research focus is concerned with novel ways of using mobile technologies to support a range of different collaborative learning activities rather than to focus mainly on one form of collaboration around the computer screen. One of the main efforts is on designing new ways of collaboration between learners, and learners with objects in the physical world mediated by different ubiquitous technologies. The main collaboration modes we have explored in our research efforts can be described as follows: peer-to-peer, individual-to-group and individual-to-expert collaboration. Peer-to-peer collaboration happened between students within the same group that needed to discuss and to find solutions for the tasks. The tasks have been designed in the way to encourage collaboration between peers and were mainly direct without technological mediation. Individual-to-group collaboration happened while the learners were solving the tasks and encouraged to collaborate with the other groups outside or inside and thus bridging the location contexts of the same activities. This collaboration was technology mediated and relied upon text (mobile instant messaging) and content (audio/video/picture messages). Individual-to-expert collaboration happened between the children and experts. These collaborations were direct and technology mediated. These modes of collaboration have been used in the design of the different tasks to promote collaborative problem solving. This approach enables learners to interact more freely and engage in a variety of interaction and collaborative modes depending on the different contexts in which learning occurs. By linking collaboration to context our hope is to utilize the fluidity of learners' actions, relations, and locations in a way that further defines collaboration and context in relation to mobility. In the coming section we illustrate a concrete implementation of these ideas by describing a number of activities we have recently conducted with school children and university students.

## THE AMULETS PROJECT

In the AMULETS project we are exploring how teachers can develop and implement novel educational scenarios combining outdoor and indoor activities that use mobile computing technologies together with stationary computers. During 2006 and 2007 we conducted three different trials with children and university students. The first trial took place in June, 2006, in an elementary school while the second trial occurred the following December, in the town square with the same school. The third trial took place between April and May, 2007, and we collaborated with the teacher training program at our university.

For the first two trials, 55 elementary school children performed remote and co-located activities equipped with Smartphones, PDAs, GPS devices and stationary computers in the subjects of natural sciences, history and geography. The educational scenarios consisted of different stages with game like features. At the end of the learning sessions, all these activities have been reconstructed in the classroom using several visualization tools, including among others digital maps. For an elaboration of the results please see the work of Kurti et al. (2007, 2008). These types of activities provide new opportunities for children and teachers to review and to continue the learning experience in the classroom, thus supporting different aspects of learning such as exploration, discussion, negotiation, collaboration and reflection. In the third trial 16 student teachers from an environmental science course at our university used smartphones and stationary computers to explore and to learn about those aspects related to tree morphology. In all three trials we have developed and implemented educational scenarios that were designed together with teachers. In the first two cases the activities were designed to support the regular school curriculum for elementary school children while in the case of the university students the scenarios were designed in collaboration with the instructor to support a module of a university course.

In the first trial the theme of the scenario was learning about "the forest" and in the second trial "the history of the city square through centuries." In the forest scenario conducted in the spring of 2006, 26 $4^{th}$ grade students (10-11 years old) took part working in seven groups. The activities were conducted over a two-day period with only one group performing at a time. The active challenges for the children were based on exploring the physical environment, identifying different types of tree and measuring the height and age of trees. Part of the children's task was to record still images and video clips using the smartphones detailing how they solved the problems. This co-created content automatically encoded with metadata, containing attributes such as GPS coordinates, time stamp, and the phone ID provided rich contextual information for later use in the classroom. Pedagogical coaches supported the children with hands-on techniques describing how to measure the height of trees. Additionally, animated characters delivered content based on a specific location and tag triggered context to the smartphones.

In the city square trial conducted in the fall of 2006, 29 5th grade students (11-12 years old) participated. They worked in three groups; each group was divided into two subgroups of five students. One subgroup worked in the local museum and the second group operated in the field (the square). For this second trial, we introduced collaborative missions in order to provide the children with challenging problems. In order to solve them, children at the museum and in the field were required to collaborate using a number of mobile tools including an instant text messaging system that allowed communication between the smartphones in the field and the stationary computers at the museum. A narrative journey backwards in time relating to the square's history was supported by animated characters and video clips delivered to the smartphones, thus providing the contextual information that was needed in order to accomplish the challenges in the different missions. Children needed to work together in order to complete the tasks including deciphering Roman numerals, finding locations for historical buildings and solving problems in the fields of history and geography.

In the spring of 2007, 16 student teachers (20-35 years old) from our university participated in this activity. The students were divided into four groups and each of these groups split into two subgroups. Again, one group became the field group while the other part became the base group. The field group had two smartphones, with one acting as the communication and messaging device with basecamp and the second phone being used as a camera. The learning activity was on how to teach tree morphology, where the student teachers used a tree key to identify different species of tree by bark, type of buds, and the surrounding environment. The field group task was to locate the trees, send images back to basecamp and collaboratively determine the tree species, as well as to negotiate answers to questions while performing tasks about the environment. After the field group completed two stages (from a total of four) they returned to basecamp and switched roles. This gave the opportunity for all students to experience both field and base work (except for one student with mobility issues who remained in basecamp for both sessions). From observing how the previous trials worked, we introduced this rotation to allow all the students to take part in both activities in order to experience the different roles. The students followed up with a post hoc activity that consisted of an informal quiz about the content they learned during this trial and discussions about the activity. The second part of the trial took place in the following weeks. The students were sent out in pairs into the field to identify two species of plants, mark their choice on a map via GPS and perform general field experiments about the surrounding environment during a three weeks period, in order to understand the impact of microclimate changes on plant growth. The latest component of this activity was a hands-on workshop with the students exploring the design of new mobile learning activities. One of the goals of the AMULETS project is to work closely with teachers and student teachers to help them understand the potential of how new technologies can support their teaching activities. From a design point of view, these activities gave us the opportunity to allow future teachers to act as co-designers (Druin et al., 1998) in the creation of novel learning activities. In the following subsections we describe in more detail the different activities from the three trials.

## Bergunda School Trial

This first trial took place on the outskirts of the Bergunda School (near Växjö, Sweden) in the surrounding natural environment. During the course of these activities, through collaboration (initially with the teacher, later with their peers within the group and at the end with other groups), students learned about different aspects of the forest and basic knowledge that could be used to identify trees in their environment. Once stage one was completed, children were introduced to the field activity including a short hands-on workshop, providing them with the necessary knowledge about how to use the different mobile tools available. The collaboration occurred in peer-to-peer and peer-to-group contexts for the mobile learning activities. For the first task, the children needed to identify a particular sort of tree out of three different kinds of tree located in the surrounding forest. Once they identified the trees and received some additional information on the smartphone, they needed to scan the correct semacode tag (a 2D barcode tag that can be read by a camera-enabled mobile phone for embedding URLs to specific location, see: http://www.semacode.org/) placed on one of the trees. In the case of choosing the incorrect tree the children received additional information describing how to proceed. Upon selecting the correct tree, a video animation was sent to the smartphones in order to give the children the required information to proceed to the next mission.

These exploratory and task-based activities continued by encouraging the children to learn how to measure the height and age of the surrounding trees. In addition, they gained some knowledge about when trees are ready to be processed by the

*Figure 2. Illustrates the flow of the learning activities from the pre-activities, to the mobile learning application, to the post-activities*

forestry industry. During the entire field trial, the children documented their activities by taking photographs and videos that contained automatically generated GPS metadata. As part of the game related aspects of this activity, our system collected the time that it took for each group to accomplish the different tasks as well as the answers to each mission from the different groups. All this data was stored in our repository and we used it for further purposes in the follow up activities. At the end of the two day event, all the children were gathered together in the classroom. The follow up activities took place back in the school where all groups presented and discussed the content created during the trial while this content was tailored to a specific location as explained before. At the end of the activity, the results (times and numbers of points) were presented and the winner was announced. This trial was designed as a pilot test. The learning activity was divided into three stages including a pre-activity (where interaction was primarily group-to-expert, as illustrated in the top left image (a) in figure 2), a field activity (where collaboration was mainly peer-to-peer as illustrated in images (b) and (c) in the figure 2) and a post activity (where knowledge exchange was based on individual-to-group basis as illustrated in section (d) of figure 2).

## Växjö Square Trial

This trial took place at the main square and at the museum of history in the city of Växjö. The overall activity was divided into three sessions over two days. The students were divided in three groups of 10 children. Additionally, each group was divided in two subgroups of five students, where one subgroup was working indoors in the museum, while the other group was outdoors in the city square. In this trial, we introduced several new features that included collaborative learning activities between the indoor and outdoor groups. The indoor and outdoor groups were required to communicate and collaborate across different locations using mobile technologies in order to accomplish a task. The collaboration modes used in this trial were peer-to-peer (between children in the same group) and group-to-group (between children in different groups and in different locations, indoor and outdoor). We relied on the use of jigsaw techniques (Aronson et al., 1978) for the pedagogical design in different locations, thus creating the conditions for our concept of *collaboration in context.* For all these activities we explored how different collaboration modes worked between the children in different locations.

Figure 3 illustrates several of the tasks in this trial that occurred simultaneously, in images (a) and (b) the learners are decoding a roman numeral on the governor's house. Images (c) and (d) illustrate continued negotiations between the groups to determine the age of the governor's house. In images (e) and (f) the outdoor group has just sent a photograph of the square and the indoor group is directing them to where a historical building once stood in order to relate what the square looks like today to what it looked like in the 19$^{th}$ century. The activities were designed around group collaboration to solve the challenges for each task. For the five tasks in this activity, the groups needed to discuss and negotiate, thus switching roles regarding the different actions and decisions to be taken depending on the task. We used mobile instant messaging (IM) and the exchange of digital photos produced by the children to support the discussions and negotiations. The outdoor subgroup was equipped with three smartphones (Nokia 6630) for content delivery, content generation, instant messaging and decoding the semacode tags. The indoor subgroup was equipped with a laptop computer equipped with a GPRS connection and a mobile handset for still photography. Student teachers supervised the groups during the activities. While the outdoor subgroup was in the field, the indoor subgroup

*Figure 3. Växjö square trial activities*

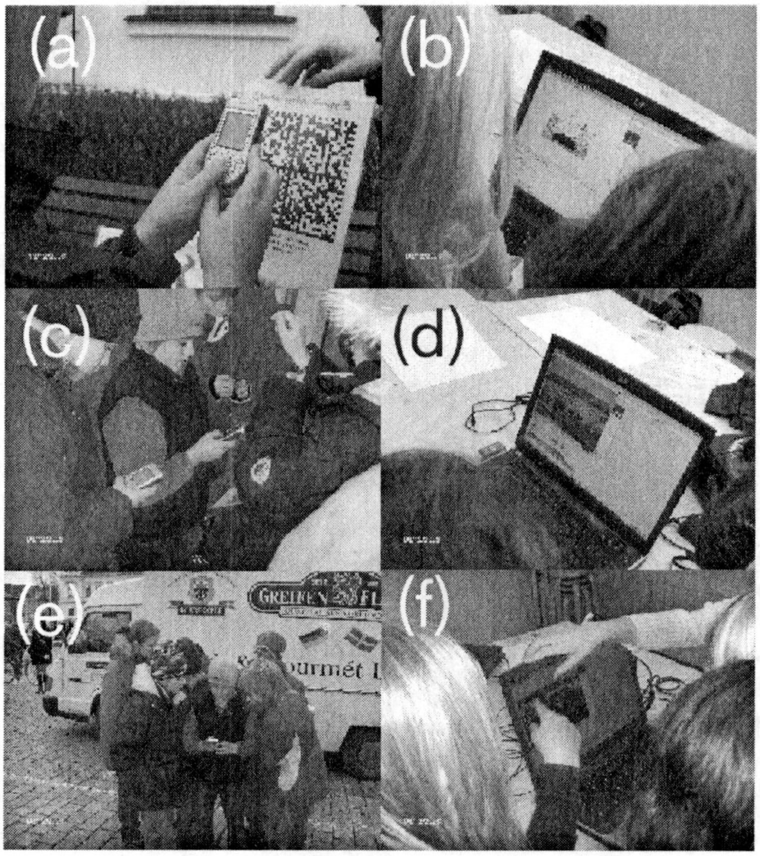


was in the museum. We used animated characters and short video clips to provide information and additional help for the students.

## Student Teacher Trials

This trial took place on campus at Växjö University in the spring of 2007. We worked with 16 student teachers, divided into four groups. Each of these groups was divided into two subgroups. The field groups were equipped with two smart phones, one for game control and information and one for digital documentation. The control smartphone was used with semacodes for the control of the learning activities and for sending messages via a semacode tag, while the second phone automatically delivered the photographs and audio files to base camp once the students took an image or finished recording. The field activities focused around the identification of four different families of trees, where the outdoor group collected data (images, video, and audio files) via the smartphones. The indoor group analyzed the images, audio, and sound in order to determine, with the support of a tree taxonomy instrument, to which family the tree belonged to according to leaf buds, bark colour, and other environmental factors. For this third trial we further refined the learning activity by running simultaneous trials with four groups and splitting the indoor and outdoor ses-

*Figure 4. Teacher student trial activities*

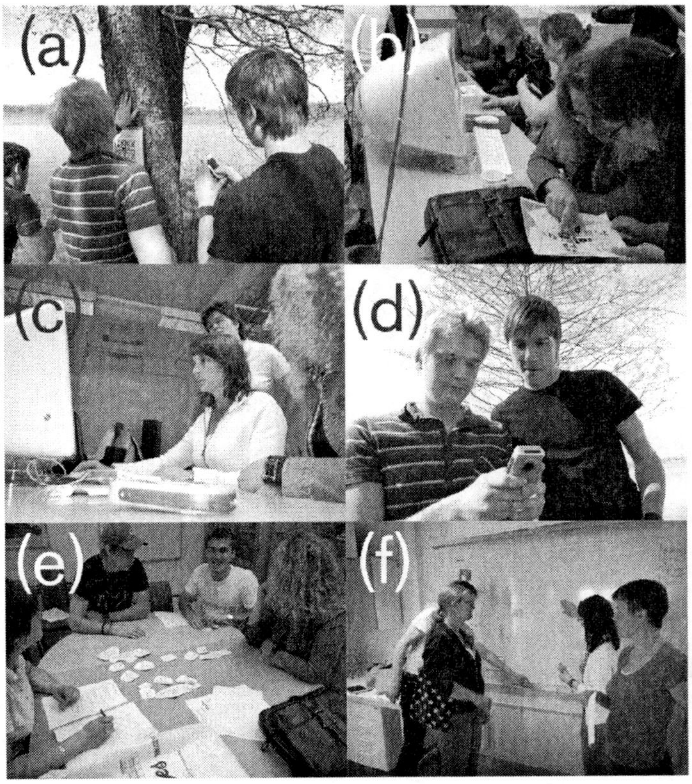


sions between them, enabling all the students to experience the different roles and aspects of the trial. The collaboration modes promoted in this trial were primarily based on peer-to-peer and individual-to-group collaboration. Images (a, b, c, & d) in figure 4 illustrate how the indoor and outdoor groups needed to collaborate to solve the tasks. The images (e & f) show the brainstorming process and idea presentations. In this trial we tried to scale down the number of devices and control the communication to be more effective.

The field students set out for their respective first stations, where after scanning the "startcode" they received an introduction, short audio instructions and a special tree identification form. They used the digital documentation phone to sent photographs and audio recordings to the base station using a special communication semacode. The groups needed to remotely collaborate using the mobile media and a Web interface in order to identify the tree species.

In addition, the field group used the phones to document the environment and to answer additional questions that were designed to inspire them to explore the tree's surrounding environment. Meanwhile, the indoor group compiled additional information for the post activities. For the post activity, the students worked together as a group to reflect over the content generated by the different teams, they participated in an informal quiz and discussed how the trial worked out from a pedagogical point of view.

The second part of the trial comprised a three-week period where the students did traditional fieldwork, collecting environmental data. After this activity, the students participated in a future technology workshop (Vavoula, Sharples, & Rudman, 2002) organized by one of the researchers to flesh out how a mobile learning activity could be designed in this particular domain. The students worked in four groups, brainstorming and testing their ideas, and the workshop ended in a presentation of the best concepts followed up by a general discussion. This trial was designed with two goals in mind. The first goal was to support a module of the environmental science course for the student teachers, thus providing an opportunity for them to introduce a mobile learning component into their course. The second goal of this trial was to actively involve the student teachers in the design process of mobile learning in order for them to become co-designers of new mobile applications. Figure 4 illustrates the outcome of this trial. In the top four images we see the different groups collaborating and the bottom two images show the brainstorming and the idea presentations.

## Applying the Conceptual Framework

Price and colleagues (2003) have suggested a framework for supporting multiple interactions between individuals and groups that provides multiple collaboration opportunities. We hope to address the complexity and fluidity of introducing context in collaborative learning environments using our conceptual framework (Figure 1). In

*Table 1. Applying our conceptual framework*

| *Trial* | *Location / Environment* | *Task / Activity* | *Personal / Interpersonal* | *Collaboration* |
|---|---|---|---|---|
| Bergunda School | The main activity was conducted outdoors and supported with pre and post activities indoors/ School and surrounding forest | Serial tasks about nature and history of the local forest/Co-located | Collaboration between the groups and then group knowledge exchange in the post activity | Peer-to-Peer |
| Växjö Square | The main activity was conducted both in indoor and outdoor settings/ Local museum and town square | Parallel and simultaneous tasks about the history of the square and live in past times/ Co-located Remote | Collaboration between indoor and outdoor groups was mediated using text and content mode | Peer-to-Peer Individual-to-Group |
| University | The main activity was conducted both in indoor and outdoor settings/ University lab and surrounding nature | Parallel and simultaneous and shifting roles. Tasks about tree morphology and the ecosystem/ Co-located Remote | Collaboration between indoor and outdoor groups was mediated using text and content mode | Peer-to-Peer Individual-to-Group |

Table 1, we have tried to categorize the different trials mapped according to the components of our framework based on *location/environment*, *task/activity* and *personal/interpersonal* type of collaboration.

## METHODOLOGICAL CONSIDERATIONS

Design studies are typically conceived as test-beds for innovation. One of the main objectives of this study was to investigate the possibilities for educational improvement by stimulating new forms of learning (Design-Based Research Collective, 2003). We consider our efforts as being an attempt to create innovative, socially-situated exploratory learning experiences through elaborated learning sequences supported by ubiquitous technologies. Within the context of our efforts, the notion of socially-situated extends to the idea of learning activities guided by the context in which they are taking place.

Design-based research is an attempt to combine the intentional design of interactive learning environments (ILE) with the empirical exploration of our understanding of those environments and how they interact with individuals (Hoadley, 2004). The primary aim of Design-Based Research Collective is to develop domain specific theories (Mor & Winters, 2007). Therefore, we consider it as a suitable methodological approach for the field of mobile learning. A recent view regarding the design of ILEs is presented by the Design-Based Research Collective group (2003) who argue that design-based research, which blends empirical educational research with the theory-driven design of learning environments, is an important methodology for understanding how, when, and why educational innovations work in practice. Based on those claims, design is central in efforts to foster learning, create relevant knowledge, and advance theories of learning and teaching in complex settings. According to Edelson (2002), the emerging design-based research paradigm treats design as a strategy for developing and refining theories. Design-based research follows an iterative cycle of designing, implementing, analyzing and modifying.

The research efforts presented in the former sections were conceived and implemented as inspired by the ideas and rationale suggested by this methodology. Such research is more akin to ethnography than to quantitative studies; the emphasis is on design processes as planned, observed and reported in their natural settings. In our particular efforts, the different educational scenarios we developed were created based on prior cognitive, educational and technological research, relevant learning goals and content pedagogy, and knowledge of the specific educational context. Different scenarios were implemented and a variety of data were collected and analyzed to determine the success of the design. Since a successful educational design should operate as an integrated system, the critical elements of the design were identified and their interactions in the educational setting were analyzed. If those elements were not working in the expected way, then the design was modified based on the findings and a revised prototype was implemented. In this sense, design-based research has some of the aspects of a formative evaluation, especially as it informs the next version of the ILE. In the research efforts described in this chapter we have used design-based research as means of exploring how collaboration and context can be used to support the design of innovative learning activities.

### Trial Design Issues

Roschelle and colleagues (2005) have defined co-design as a highly-facilitated, team-based process in which teachers, researchers, and developers work together in defined roles to design an educational innovation, realize the design in one or more prototypes, and evaluate the prototype's significance for addressing concrete educational needs that support our situated learning aims. For the AMULETS trials we have worked in this manner where teachers, researchers, and students engaged in co-design together. Our aim with the overall project is to provide mobile learning tools and methods for teachers to use in different educational settings.

For the first trial at Bergunda School, we developed the technology and worked with iterative design in conjunction with the teachers. This was our initial prototype in terms of how the activities worked out in the pre and post sessions with the children. As shown in Table 1, the first trial took place outdoors with the groups having peer-to-peer collaboration. For the second trial, working with the teachers helped us to broaden the concept of *geographically distributed mobile collaboration* between two groups, providing a space for the children to collaborate and negotiate in order to solve the tasks. The collaboration between the different locations was mediated by the technology of instant messaging and the exchange of photographs between the smartphones and stationary computers. For the third trial, we extended this form of collaboration by having all the students experience both the indoor and outdoor work. Moreover, this activity was designed in a way to support reflection and knowledge sharing using rich media content and via visualizations using a specific web application we developed for this particular purpose.

The trials have been designed in the form of game-based activities. One of the main pedagogical challenges of these activities was to design learning tasks that fostered collaborative problem solving skills within the same subgroup and with their peers. Most of the activities were designed in such a way as to promote the division of labour, fostering collaboration, first within subgroups in the trials and then later across the groups in the second and third trial. During the different stages of the trials, children and students needed to use mathematical and navigational skills, combined with reasoning and argumentation. Strong negotiation skills were needed for the successful accomplishment of the tasks. In addition, group discussions and interactions, as well as collaboration, were also activities that enriched the learning experience. The integration of all these different features into a realistic scenario offered children and students a challenging learning environment.

## THE TECHNOLOGICAL ENVIRONMENT

In order to support the different learning activities we have developed and implemented a number of mobile tools and applications. In the technical architecture presented in Figure 5, we illustrate the three main components of the technical system. The central component is the *learning activity system* (LAS) that is comprised of three main functional blocks, the activity generator, the collaboration tools, and the presentation engine. The activity generator contains the activity control system (ACS) that enables collaboration between users and devices while retrieving and storing the content and it controls the flow of the learning activities. The collect, convert, and send (CCS) component is the content repository and it is used to collect content generated by the different groups and to deliver content to the mobile devices and computers upon request. The educational content delivered to the mobile phones and computers is also stored in this repository. The LAS manages the automatic generation of metadata, storing the tags and the content in the CCS that the two other components, namely the collaboration tools and the presentation engine, create and utilize.

The *collaboration tools* provide the literal bridge between groups outside and inside through instant messaging, images, and audio. For the outdoor activities we have used smartphones and PDAs with GPS capabilities to interact, create, collect, and communicate throughout the learning

*Figure 5. The technical architecture of the learning activity system*

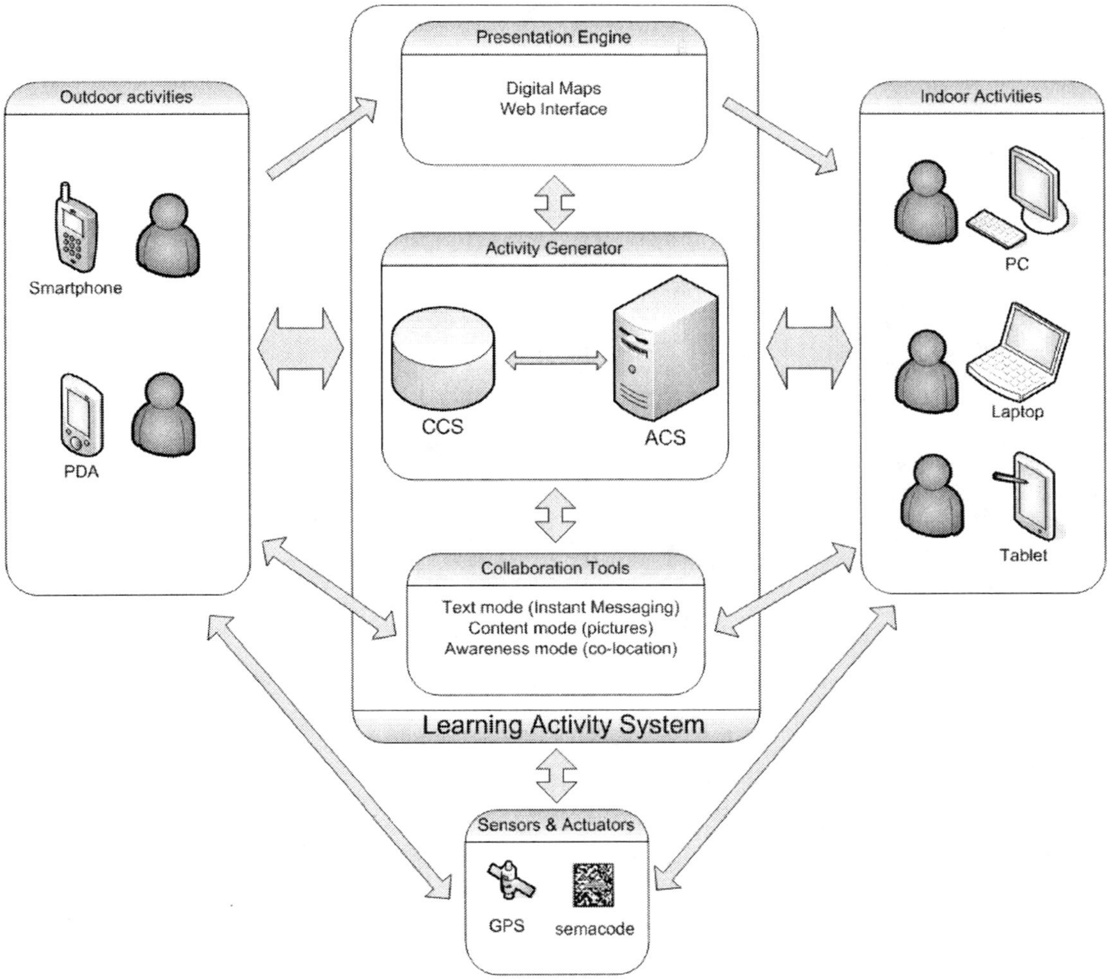


activities. These devices exchange data with the LAS components, retrieving and sending content and information, as well as interacting with the sensors. The *collaboration tools* enable the technology-mediated support for remote groups to work together by providing text, content, and awareness modes. For the text mode collaboration, support was provided by a mobile instant messaging application we developed using instances of the Nokia Raccoon software. Nokia Raccoon has a built-in python script for enabling mobile text communication via instant messages. In addition, we used these features to send photographs, video, and audio files from the mobile phones to the LAS, thus linking the content delivered and created by the group. All the content generated by the learners contained contextual information such as group number, activity type and additional information that was stored in the CCS. The user-generated data was handled by a python application that ran on the smartphones that automatically sent the meta-tagged data to the CSS. Depending on the different learning situations, this content and its associated metadata was available for immediate access to the indoors group via the presentation engine.

The presentation engine provides the visualization tools to support the collaboration during

*Figure 6. Mapping digital content with associated metadata in the 3rd trial*

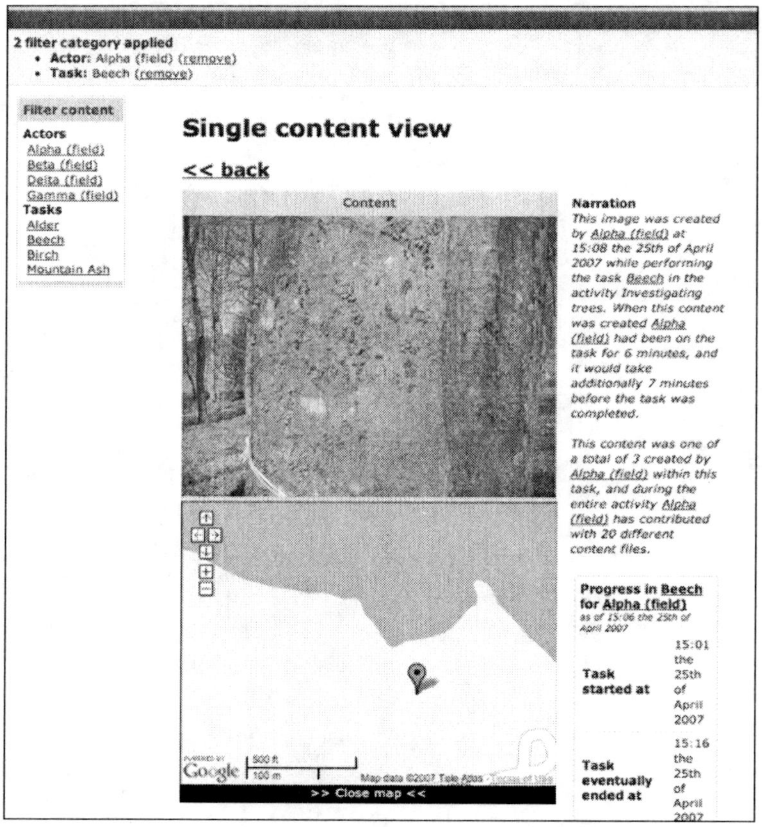


the activities and for reflection in the post activities through the use of metadata and rich media content generated during the group activities. See Figure 6 for an example of how the presentation engine can render the data from specific tasks and actors in the learning scenarios. The fourth block of this architecture consists of the sensors and actuators that support the outdoor activities with location and visual tags (semacodes) to trigger or record events. For the indoor activities the children and students interacted through a web interface linked to the presentation engine, thus providing contextual content and connection to activities performed by the outdoor group

## RESULTS

Assessing the learning processes and outcomes in the type of ILEs presented in this chapter is a very complex task, as there are many variables involved (different tasks, roles, contexts, etc.). Existing and validated methods for pedagogic evaluation for these types of specific applications need to be refined in order to accommodate the properties of mobility and context. In our particular efforts, we used several techniques for data collection including questionnaires and interviews with the children, students, and teachers, as well as observation protocols and stored data files. The questionnaires were mostly used to evaluate usability aspects while the interviews with children, students, and teachers were used more to evaluate the pedagogical related aspects of the trial. The digital content generated during the trials were saved on the CCS repository and we have used those data in different ways in our activities and analysis. In the first trial, the stored data have been used for reconstruction of the field activity in the classroom settings. In the second and third trials, the server log files have been used to trace the messages exchanged between the indoor and outdoor subgroups to investigate the collaboration that occurred between these subgroups. Additionally, in the third trial, data storing techniques were further developed to allow further exploration and visualization of how the activities unfolded by task and group. The main focus of our assessment was to investigate how we could use the conceptual framework presented in section 2 as a basis to explore the technology, pedagogic, and collaborative impact of the mobile learning activities. Taylor (2004) suggests some key points to consider while evaluating mobile learning according to the following:

- The learning opportunities presented by the new mobile technology
- Its (potential) impact on the way people perform learning tasks
- Its (potential) impact on the human social process and interactions
- How these in turn are changed or modified by the technology

We combined Taylor's (2004) task based approach for the evaluation of mobile learning environments with our conceptual framework in order to develop an assessment strategy to analyze the data we collected during the trials. Table 2 presents this strategy, illustrating how context can be used to evaluate the learning activity, describing which key points to look at, and which techniques should be used for data collection.

For the trials we chose to evaluate the learning opportunity related to the *location & environment* based on the nature of the field trip based activities. We used the *task & activity* components to look at how the usability aspects related to the use of mobile technologies that may have some impact on the way the activities here were conducted. For the *personal & interpersonal* we looked at the impact of technology on the collaborative learning aspect and for the *collaboration modes* how the technology may have some influences on the learning outcome. The data collected during the trials and saved on the CCS repository have

*Table 2. Assessment framework*

| Context | Evaluation | Type | Data |
|---|---|---|---|
| Location / Environment | Learning Opportunity | Field Indoor Outdoor Remote & Co-Located | Survey Interviews |
| Task / Activity | Impact on Task | Usability | Survey Observations Data |
| Personal / Interpersonal | Impact on Social Processes | Collaborative | Survey interviews Log files |
| Collaboration modes | Impact of Technology on Collaboration | Learning Outcome | Interviews, Survey, Log files |

been used in different ways in our activities and analysis. The following subsections are used to evaluate the three trials based on the ideas presented in Table 2. Section 6.1 looks at usability issues with the learning activity based on the task, 6.2 on how technology impacted the social and collaborative activities, and 6.3 on the learning outcomes shaped by the technology.

## Impact on Task and Usability

During first two trials, we conducted 55 questionnaires exploring aspects such as perceived ease of use, satisfaction and peer collaboration mediated by ubiquitous technologies. Of the 26 children who attended the first trial, 22 of them described the activities as "very enjoyable" while the other four described them as "enjoyable."

Of the 26 students that participated in the first trial, 16 of them found the usability of ubiquitous technologies as "very easy to use" while the other eight for the same question answered with "easy to use." Only one child described the technology as "not easy to use." In the second trial, the usability questionnaire for the technology was made separately for the indoor and outdoor subgroups since they used different technological tools. From 29 students that participated in this trial, 14 of them belonged to the outdoor subgroups while 15 belonged to the indoor groups. In the outdoor group, eight of the students felt the smartphones were "very easy to use" while the remaining six described then as "easy to use." From these 14, only four of described the use of semacodes as "very easy," nine described them as "easy" and one described the use of semacode as "difficult." Six of students described the instant messaging service used as "very easy;" the other six describe it as "easy" while the last two did not answer this question.

For the third trial, based on the questionnaires of the 17 students who responded about the usability, 13 reported that the technology was easy to use, three felt it was difficult, and one did not answer. Semacode reading was the most difficult technology to use according the responses of 13 of the students. We observed some difficulties in the use of the semacode application. The children and students also requested better possibilities for communication between basecamp and the

Figure 7. Usability

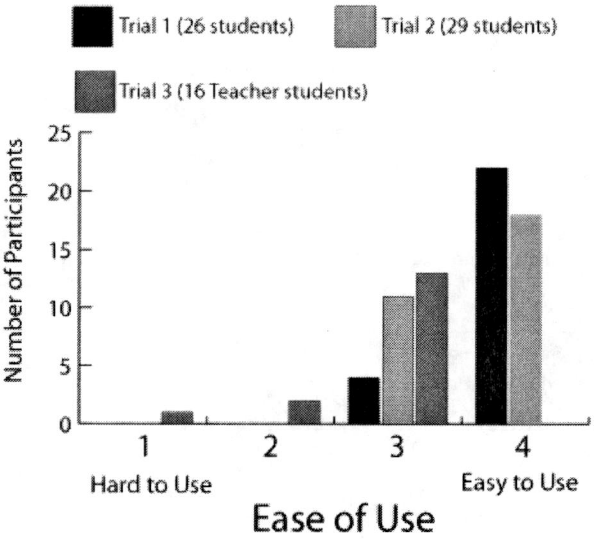

field. Figure 7 illustrates the ease of use of the technology across the three trials.

## Impact on Social and Collaborative Activities

For evaluating the impact of mobile technologies on the social and the collaborative aspects of the different trials we used questionnaires, interviews, and observations. In general, across all the trials, and based on the interviews we conducted, all the participants enjoyed the activities and thought they were fun. From a pedagogical perspective, the teachers believed that the game format and the narrative style of the activity helped the children to concentrate on the task. They both felt that the communication and collaboration tasks the children needed to solve throughout the activity were key components, helping the children to learn social and problem solving skills embedded in the history curriculum. From the instructor's point of view, the social and collaboration task required by the trials and the workshops pushed the teachers' students to discuss more about the learning process than in previous courses.

In the second trial, we used questionnaires to assess the collaboration modes. For the outside groups, when asked about the collaboration with the indoor subgroup, only four of them defined it as "very good," eight of them defined it as being "good" and two described it as "not so good." When asked about the collaboration with the outdoor subgroup, five of them described it as being "very good," seven of them described it as "good" and one child each thought that this collaboration was "not so good " or "bad." Here also, one child did not answer this question. Figure 8 shows the value of collaboration for the students in trial 2, where similar questionnaires regarding collaboration were conducted with 16 student teachers. Based on their responses, we can say that eight preferred the field trials, while seven enjoyed both and two liked the indoor activities. From the 15 children participating in the indoor subgroups, 13 of them described the collaboration between them as "very good," one described it as being "good" and one did not answer this question.

Overall, the students preferred face-to-face collaboration and felt that working together was most efficient. Figure 9 is a breakdown of how the student teachers ranked the importance of collaboration during the experiment by whatever mode they preferred.

*Figure 8. Collaboration in trial 2*

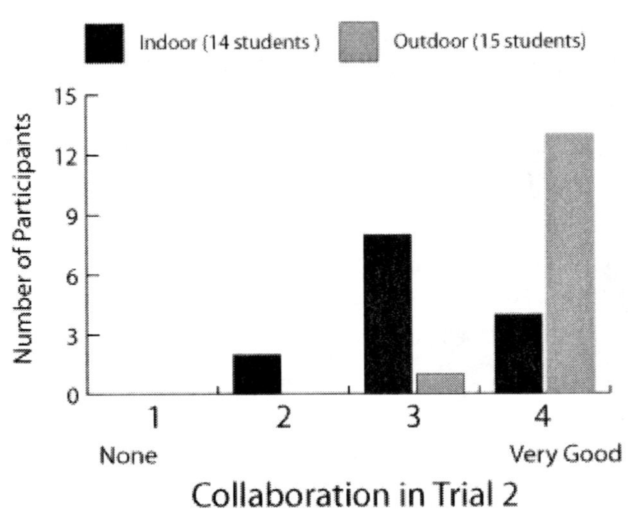

*Figure 9. Value of collaboration in trial 3*

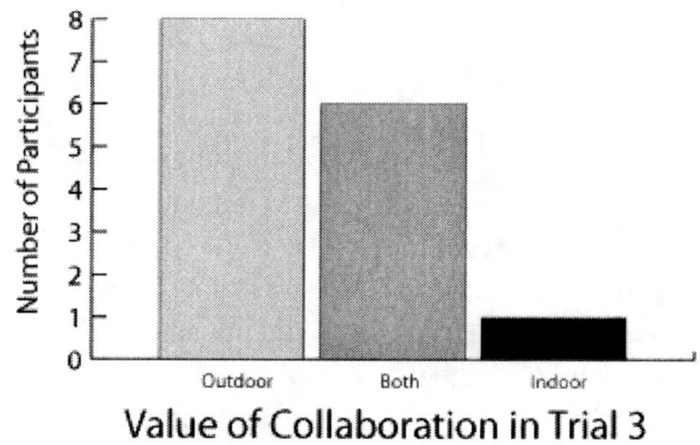


## How Technology Influences the Learning Outcome

In order to assess the impact of the technology on the learning outcome, we used questionnaires, interviews, structured observations, and the stored data files. The interviews with the children in the second trial were conducted some days after the activities. The focus of the interviews was on the overall learning experience. The first question we asked was to the children, *"What did you learn during the activity?"* The main denominator based on their answers was that they believed they learned about what happened in the main square during the different time periods in history, but when reflecting together with them about when things happened in time. It appeared that children had problems to differentiate between events that took place in different time periods.

For the student teachers, we asked similar questions about the activity and similar responses were given. In terms of what they learnt through the activity, they recognized that the different sub groups learnt different things during the trial and in retrospect they realized that they needed to discuss more in the post activity. The students requested better possibilities for communication between the group indoors and the groups in the field. A hands-on workshop was organized around a future technology scenario activity, as a second part of the student teacher trial. The goal of this activity was to explore and brainstorm about new ways of enhancing traditional fieldwork using mobile technologies. The focus of this activity was based on the identification of plants and the measurement of soil temperature, light, and moisture. The purpose of this activity was to encourage teachers to think and act as co-designers of mobile learning activities rather than being only users. One major outcome of this activity was the teachers' concerns about usability aspects of the technology and transparent interactions. In general, they imagined a future mobile learning device that would have more ease of use features than the current crop of smart phones.

## Analysis

The teachers' general impression was that the trials were successful and they both felt that mobile technologies (smartphones in the case of these trials) may help children to become more engaged in the activities When reflecting about novel aspects of this way of learning, the teachers' main concern was the risk of technology potentially overshadowing the learning process. The teachers thought that this game-like-scenario helped the students to focus on the tasks more than traditional learning settings. For the third trial, the instructor felt that the use of mobile technologies helped to involve the student teachers in learning how to teach the subject and presenting them with opportunities to face learning challenges in authentic situations. In terms of what they learnt through the activity, they recognized that the different sub groups learnt different things during the activity and now in retrospect realized that they needed to discuss more in the post activity. Similar to younger students, the additional content delivered to the smartphones was not recognized as important.

For the first two trials with the children we asked the following question, *"Do you think that was an interesting and enjoyable day and do you want to join a similar activity in future?"* From a quantitative perspective, and based on responses from the questionnaires, all of the children (55) said that they are likely to participate in similar activities with different missions in the future and they would like to see more activities of this kind integrated into the daily school activities. The student teachers had more mixed feelings about the learning opportunity; out of the 16 students, nine would consider using mobile technology in their future classrooms if relevant, four were not interested in the technology, since they felt that for very young children, it was not relevant, three were positive towards using the technology, and one did not answer. From the interviews they all expressed concern about how to get access to the technology when they become teachers.

In summary, based on the results presented in the previous sub-sections, we can observe that the mobile learning solutions we developed provided added value to the locations in the different trials. Like all new technologies introduced in educational activities there is always a risk of a technology centric approach away from the learning, but with careful pedagogical planning and the involvement of teachers in the design process this can be avoided. Both the children and the students expressed that face-to-face collaboration was more valuable than the technology during the trials and during the post-activities. Based on our experiences in the second trial, real-life situations that the children encountered during the activities mattered more than the computer generated animated characters delivered to the mobile devices. In general, both children and students feel that the technology can be used for appropriate situations, while the new teachers expressed concern about what tools could help them create future applications easily. From a usability perspective and an interaction point of view, we can clearly see difficulties in using semacodes as the main means for triggering events in the field while working with the smartphone. A key factor in this respect is the need of training sessions to make users become more familiar with this mode of interaction. This fact implies that future efforts should focus on how scaffolding techniques can be integrated into the application. From a social and collaborative perspective, the game-like features worked well for the design of the learning activities and for the overall experience. Both the children and the students expressed enjoyment in working together and felt that the face-to-face collaboration was the most enjoyable. This does raise some issues for the future design of activities in how we may shift the activities to different patterns of collaboration.

## DISCUSSION

Currently, mobile devices are not perceived any more as simple communication tools; they can be seen instead as new social tools to support human collaboration and interaction. In this chapter we have presented our view about the design of innovative mobile learning based on new "social technologies." The hope is that learning in mobile settings can be made more effective by expanding how people collaborate while taking into consideration *context* as a design parameter. The focus of this chapter has been to explore how innovative mobile learning scenarios can be designed by defining new ways of collaboration and interaction between people and devices. We have presented our thoughts in this direction by proposing a design framework based on our view of context and collaboration. The general framework for design and evaluation presented in this chapter offers new possibilities for addressing the challenges for mCSCL and mLearning. Integrating different aspects of the learners′ context into the design of collaborative learning activities can provide new modes of interaction that may help to enhance different aspects of learning. This latest aspect is especially important when it comes to bridging indoor and outdoor learning activities. Our framework is open enough to provide an underlying foundation for future research efforts and open to different theories. By providing mobility to the learners we can offer more authentic learning opportunities. According to Jonassen, Peck and Wilson (2000) meaningful learning will take place when learners are engaged in real world activities. This approach to technology-enhanced learning may contribute to a richer, more authentic grounded experience than conventional learning activities conducted in classroom settings using traditional material such as textbooks or demonstrations of experiments.

The use of design-based research provided us with a methodology that combines the practice and theory of learning. Using this approach combined with working together with the teachers and the students gave us the opportunity to design learning activities in authentic locations using meaningful content that has relevance for the school curriculum. The co-design activity together with the student teachers aimed to address the needs of creating a simple toolbox for both students and teachers. The outcome of our efforts suggests that outdoor learning experiences supported by ubiquitous technologies should be combined with learning activities in the classroom to provide learners with meaningful activities in order to:

- Learn and to explore a topic in authentic settings,
- Collaborate in order to construct common knowledge,
- Reason and to argument in order to come to the solution of a problem,
- Reflect upon things and to support abstract thinking

Our explorations into collaboration in context evolved over the three trials presented in this chapter. In the Bergunda School trial, the students worked in groups in the field. In the second trial at the Växjö Square, we introduced the co-located subgroups collaborating in different location scenarios while the collaboration was mediated by mobile technologies. In the third trial, the student teachers rotated between the outside and the inside activities, thus providing all students with learning experience at the different locations. Based on the assessment of these trials and the post activities, we have learned that the users placed high value on the collaboration aspects of the learning activities and the need to develop easier forms of communication for collaboration in context using mobile and ubiquitous technologies.

Mobile and ubiquitous technologies offer the potential for a new phase in the evolution of technology-enhanced learning, marked by a continuity of the learning experience across different learning contexts. Chan and colleagues (2006) use the term *"seamless learning"* to describe these new situations. Seamless learning implies that students can learn whenever they are curious in a variety of scenarios and that they can switch from one scenario to another easily and quickly using their personal mobile device as a mediator. These scenarios include learning individually, with another student, a small group, or a large online community, with possible involvement of teachers, relatives, experts and members of other supportive communities, face-to-face or in different modes of interaction and at a distance in places such as classrooms, outdoors, parks and museums. Seamless learning spaces refer to the collection of the various learning scenarios supported by personal (and also collaborative) mobile technologies. In the different cases illustrated in our trials we have presented several examples in which we have implemented seamless learning spaces by *augmenting physical spaces with information exchanges* as well as using *geospatial mappings* between the mobile device and the real-world that facilitate navigation and context-aware applications. According to Pea and Maldonado (2006) these last two features play an important role in designing mobile applications with an emphasis on inquiry processes, social constructivist theories, and distributed cognition designs.

## FUTURE RESEARCH DIRECTIONS

These experiments have provided us with some ideas to develop a set of recommendations for the design of new trials in the AMULETS project and others efforts. In order to support the design of innovative educational practices it is necessary to take an integrative perspective to technology-enhanced learning where pedagogy and learning theory are the driving forces rather than mobile technologies. From this perspective, mobile technologies can be used as collaborative mindtools (Hoppe et al., 2005) that help learners (in both formal and informal settings) to conduct activities and accomplish results that are impossible to achieve without these technologies. Thus, it might be beneficial to continue to elaborate this framework to help designers to identify educational situations and requirements in which mobile technologies fill a unique role while trying to support innovative educational practice. By looking at the application of our conceptual framework (Table 1) and the assessment framework (Table 2) we can observe how mobile collaboration in context can be used to guide future work. Further development and implementation of these ideas can result in guidelines that can be used for the design of technology-enhanced learning environments using mobile technologies to support innovative educational practices.

In our future work, we plan to improve and to modify our existing activities and technical solutions in order to increase the validity of the learning situations, as well as providing post-activities for fostering reasoning, argumentation and reflection combining mobile media, positioning techniques and digital maps. From a technical point of view, we will explore how to integrate RFID tags and Near Field Communication technology in our technical platform. By doing that, we want to asses if this technology facilitates the ways people interact with mobile phones and the objects compared with the visual tag solutions we have been using until now. We will also try to develop new ways for promoting collaboration, since the students and the teachers identified the issue of collaborative problem solving as one of the most appreciated things during these trials. Another research direction we will be exploring is how contextual information and positioning technologies can be used to deploy mobile and ubiquitous learning environments that will be responsive to the learners and the context in which the learning is taking place. We are starting to explore how intelligent support techniques can be integrated in these environments and we are involved in an EU initiative together with colleagues from the UK, Finland, and Holland where we are investigating new methods and technologies for efficient context-aware collaborative learning for workplace learning situations.

## ACKNOWLEDGMENT

This work has been partially supported by the Swedish *KK-foundation* and Växjö University under the ICT and Teacher Training program, project *Young Communication.* Additional funding has been provided by the *Internet Infrastructure Foundation* of Sweden, project MeMiMo. We thank all our team members, especially Martin Svensson and Oskar Pettersson for the technical support and Jonas Wereén, Gerd Ouchterlony, and Louise Roth for the pedagogical concepts and content.

## REFERENCES

Aronson, E., Blaney, N., Stephin, C., Sikes, J., & Snapp, M. (1978). *The jigsaw classroom.* Beverly Hills, CA: Sage Publishing Company.

Brown, J.S., Collins, A., & Duguid, P. (1989). Situated cognition and the culture of learning. *Educational Researcher, 18*(1), 32-42.

Brown, J.S., & Duguid, P. (2000). *The social life of information.* Cambridge, MA: Harvard Business School.

Design-Based Research Collective. (2003). Design-based research: An emerging paradigm for educational inquiry. *Educational Researcher, 32*(1), 5-8.

Dourish, P. (2004). What we talk about when we talk about context. *Personal Ubiquitous Computing,8*(1), 19-30.

Druin, A. (1998). *The design of children's technology.* San Francisco, CA: Morgan Kaufmann Publishers.

Edelson D. (2002). Design research: What we learn when we engage in design. *The Journal of the Learning Sciences, 11*(1), 105-121.

Hoadley, C. (2004). Methodological alignment in design-based research. *Educational Psychologist, 39*(4), 203-212.

Hoppe, H.U., Pinkwart, N., Oelinger, M., Zeini, S., Verdejo, F., Barros, B. et al. (2005). Building Bridges within Learning Communities through Ontologies and Thematic Objects. Proceedings of the *International Conference on Computer Supported Collaborative Learning* (CSCL2005). Taiwan.

Jonassen, D., Peck, K., & Wilson, B. (2000). *Learning with technology: A constructivist approach.* Upper Saddle River, NJ: Prentice Hall.

Klopfer, E., Squire, K., & Jenkins, H. (2002). Environmental detectives: PDAs as a window into a virtual simulated world. *In the proceedings of the International Workshop on Wireless and Mobile Technologies in Education.* Växjö, Sweden, (pp. 95-98).

Kurti, A., Milrad, M., & Spikol, D. (2007, July 18-20). Designing Innovative Learning Activities Using Ubiquitous Computing. *Proceedings of the 7th IEEE International Conference on Advanced Learning Technologies* on, 2007 in Niigata, Japan, (pp. 386-390).

Kurti, A., Spikol, D., & Milrad, M. (2008). Bridging Outdoors and Indoors Educational Activities in Schools with the Support of Mobile and Positioning Technologies. Accepted for publication at the *International Journal of Mobile Learning and Organization,2*(2), 166-168.

Lankshear, C., & Knobel, M. (2006). *New literacies: Everyday practices and classroom learning.* Berkshire, UK: Open University Press.

Lave J., & Wenger E. (1991). *Situated learning: Legitimate peripheral participation,* Cambridge, UK: University of Cambridge Press.

Littleton, K., & Hakkinen, P. (1999). Learning together: Understanding the processes of computer-based collaborative learning. In P. Dillenbourg (Ed.) *Collaborative learning* (pp. 20-30). Oxford: Pergamon.

Milrad, M., Chan, T-W., Roschelle, J., Hsi, S. et al. (2006). One-to-one technology-enhanced learning: An opportunity for global research collaboration. *Research and Practice in Technology Enhanced Learning Journal, 1*(1), 3-29.

Mor, Y., & Winters N. (2007). Design approaches in technology-enhanced learning. *Interactive Learning Environments, 15*(1), 61 – 75.

Pea, R., and Maldonado, H. (2006). WILD for learning: Interacting through new computing devices anytime, anywhere. In K. Sawyer (Eds.), *Cambridge handbook of the learning sciences* (pp. 427 - 442). New York: Cambridge University Press.

Price, S., Rogers, Y., Stanton, D., & Smith, H. (2003, June 14-18). A new conceptual framework for CSCL: Supporting diverse forms of reflection through multiple interactions. *In Proceedings of the International Conference on CSCL'03* (pp. 513-522). Bergen, Norway. Kluwer Academic Publishers.

Roschelle, J., Rosas, R., & Nussbaum, M. (2005, May 30 - June 04). Towards a design framework for mobile computer-supported collaborative learning. *In Proceedings of The 2005 Conference on Computer Support For Collaborative Learning: Learning 2005: the Next 10 Years!* (pp. 520-524). Taipei, Taiwan: Computer Support for Collaborative Learning, International Society of the Learning Sciences.

Taylor, J. (2004). A task-centered approach to evaluating a mobile learning environment for pedagogical soundness. *Learning with mobile devices: research and development* (pp. 167-171). London, UK: Learning and Skills Development Agency.

Vahey, P., & Crawford, V. (2002). Palm Education Pioneers Program: Final Evaluation Report, Menlo Park, CA: *SRI International.*

Vavoula, G., Sharples, M., & Rudman, P. (2002, Aug 28–29). Developing the 'Future Technology Workshop' method. In M. Bekker, P. Markopoulos & M. Kersten-Tsikalkina (Eds.), *Proceedings of the International Workshop on Interaction Design and Children (IDC2002)* (pp. 65–72). Eindhoven, The Netherlands: Shaker Publishing.

Winters, N., & Price, S. (2005, September 19-22). Mobile HCI and the learning context. *The 7th International Conference on Human Computer Interaction with Mobile Devices and Services.* Austria: University of Salzburg.

## ADDITIONAL READINGS

Benford, S. (2005). Future location-based experiences. *JISC Tech Report* (TSW0501).

Milrad, M. (2007). How should learning activities using mobile technologies be designed to support innovative educational practices? In M. Sharples (Ed.), *Big issues in mobile learning: Report of a orkshop by the Kaleidoscope Network of Excellence Mobile Learning Initiative.* Nottingham: Learning Sciences Research Institute.

Nova, N., Girardin, F., & Dillenbourg, P. (2005). Location is not enough! an Empirical Study of Location—Awareness in mobile collaboration, In Proceedings of *3rd IEEE International Workshop on Wireless and Mobile Technologies in Education* (pp. 21-28). Tokushima, Japan.

Price, S., Rogers, Y., Stanton, D., & Smith, H. (2003). A New Conceptual Framework for CSCL: Supporting Diverse Forms of Reflection through Multiple Interactions, In B. Wasson, S. Ludvigsen, & U. Hoppe (editors), *Designing for Change in Networked Learning Environments. Proceedings of the International Conference on Computer Supported Collaborative Learning* 2003.

Raessens, J. (Forthcoming). Playing history. Reflections on mobile and location-based learning. In T. Hug (Ed.), *Didactics of microlearning. Concepts, discourses, and examples.* Münster: Waxmann Verlag.

Rogers, Y., Price, S., Randel, C., Frases, D.S, Weal, M., & Fitzpatrick, G. (2005). Ubi learning integrates indoor and outdoor experiences. *Communications of the ACM*, 18(1), 55-59.

Rogoff, B., & Lave, J. (1984). *Everyday cognition: Its development in social context.* Cambridge, MA: Harvard University Press.

Sharples, M., Taylor, J., & Vavoula, G. (2007). A theory of learning for the mobile age. In R. Andrews & C. Haythornthwaite (eds.), *The sage handbook of e-learning research* (pp. 221-47). London: Sage.

Sharples, M., Milrad, M., Arnedillo Sánchez, I., & Vavoula, G. (In Press). Mobile learning: Small devices, big issues. In Balacheff N., Ludvigsen S., de Jong, T., Lazonder, A., & Montandon, L. (Eds.), *Technology-enhanced learning. principles and products*. Springer-Verlag.

Stenton, S.P., Hull, R., Goddi, P.M., Reid, J.E., Clayton, B.J.C., Melamed, T.J. et al. 2007). Mediascapes: Context-aware multimedia experiences. *IEEE MultiMedia*, *14*(3), 98-105.

Uden, L. (2007). Activity theory for designing mobile learning. *International Journal of Mobile Learning and Organisation*, *1*(1), 81–102.

# Chapter X
# Participatory Simulation for Collaborative Learning Experiences

**Chengjiu Yin**
*University of Kyushu, Japan*

**Hiroaki Ogata**
*University of Tokushima, Japan*

**Yoneo Yano**
*University of Tokushima, Japan*

## ABSTRACT

*In traditional education, most of the communication is one-way, where the teacher teaches and the students listen. The students are therefore less motivated to learn. In this chapter, we use the scaffolding technique to design an interactive simulation framework called SPS (scaffolding participatory simulation) for collaborative learning. Based on the SPS framework, a system called PLASPS (PDA-based learning algorithm system using participatory simulation) was developed to help students learn about sorting algorithms. Using this system, the teacher can assign tasks to their students and ask them to sort a list of numbers according to a certain algorithm. Learners then collaborate together to complete the task. The system checks the result and provides feedback to the students if there is a mistake with the positions of the numbers. The learners then correct the number positions and send the result back to the system. The collaborative activities and discussions, along with information about any errors, help the students to understand the sorting algorithm.*

## INTRODUCTION

In traditional education, most of the communication is one-way, where the teacher teaches and the students listen. With knowledge transmission from teacher to student based on instruction, students are not required to be active learners and can be only passive recipients: all the information or knowledge related to learning is automatically given through a teacher irrespective of the students' needs or problems, even if they are in their classrooms. In such situations, learners are less motivated to learn. In contrast, if learners solve their own problems for their own sake, they will try to actively acquire the required knowledge and skills (Gerhard & Masanori, 2006). Traditional education is passive, lacks interaction, and makes complicated content hard to teach and understand. Poor communication between students and teachers is one of the major problems of mass lectures. However, mobile devices can help to address this problem and improve interactive communication to increase the motivation of students (Kopf et al., 2005). Innovative educational projects are evolving in response to the new opportunities that are becoming available by integrating advanced technologies (Barak, Harward, & Lerman, 2007).

To encourage more active participation and interaction among the learners, we designed a research framework to support collaborative learning. A prototype of a mobile learning object on Java programming for the PDA was developed to help the students learn sorting algorithms with PLASPS (PDA-based learning algorithm system using participatory simulation). The research framework is applied to the PLASPS. This system was then implemented and evaluated.

## BACKGROUND

In the near future, we believe, with the evolution of improved wireless telecommunications capabilities, open networks, continued increases in computing power, battery technology, and the emergence of flexible software architectures, that these technologies can be commonly used in mobile learning. The world becomes more fluid, networked and complex, and organizations will continue to become more mobile than ever before (Liebowitz, 2007). Mobile technology is associated with any device that is designed to provide access to information in any location, or while on the move. Specifically, this includes, but is not limited to: wireless notebooks, tablet computers, mobile phones and personal digital assistants (PDA). One important field where mobile technology can make significant contributions is in education (Barak et al., 2007).

Mobile devices (e.g., PDAs or notebooks) provide different services aiming at the improvement of interactivity and creating additional, computer-moderated channels of communication between the learner and the teacher. Mobile devices can therefore reduce the problem of lack of communication between students and teachers, and by improving interactive communication we may help to increase the motivation of the students (Papert & Harel, 1991).

Further, mobile handhelds can easily be used in any classroom or field site; hence they can be used more often than computer labs (Vahey & Crawford, 2002).

Using mobile devices for supported collaborative learning is known as MCSCL (mobile computer supported collaborative learning). There has been an increasing amount of research about MCSCL in order to enhance learning and teaching (Chen et al., 2002; Okada et al., 2003). Many studies have examined the use of wireless mobile devices in learning. According to Roschelle (2003), "90 percent of teachers in a study of 100 palm-equipped classrooms reported that the handheld was an effective instructional tool with the potential to impact learning positively across curricular topics and instructional activities."

There are many projects using MCSCL in Tokushima University, such as CLUE (Ogata & Yano, 2003a, b) and JAPELAS (Yin, Ogata, & Yano, 2004). CLUE is a prototype system for embedding knowledge awareness maps, and facilitates the sharing of individual knowledge and learning through collaboration. Learners provide their own knowledge about language learning in their everyday lives, and share and discuss them with others. JAPELAS also addresses language learning. It is very difficult for overseas students to learn Japanese polite expressions because these expressions change in a complicated manner according to context, for instance, hyponymy, social distance, and the formality of the conversational situation. JAPELAS is intended to help foreigners to learn Japanese polite expressions using a PDA, anywhere and anytime.

Types of MCSCL may be classified as follows (Roschelle, 2003):

**Classroom response systems**: This is a system where learners can answer the teacher's question immediately with a mobile device, and the system will then display the overall result. According to this result, the teacher can know the level of understanding of each learner about the course content. The system turns every learner into an active learner. For instance, "ClassTalk" is a classroom response system that immediately shows teachers the statistics about learners' answers in the classroom.

**Collaborative data gathering**: The mobile learning environment changes the monotonous way of teaching in the classroom whereby the learners are only listeners. It lets the learners gain experience from real life, and understand more deeply what they have learned. The learner touches and feels the actual object. For example, the bird watching assistance system (Chen et al., 2002) was developed for this purpose to support the learning environment.

**Participatory simulations**: This is to help the learner understand the course content better through participation/practice. The learner uses a mobile device to take part in a common participatory simulation. Through this active participation, the learner can discuss and get the correct answer, consequently better understanding what they have learned. For example, the Virus Game was developed at MIT (Colella, 2000) to explain the process of how a virus is spread. Another project, JAPELAS (Yin et al., 2004), allows participatory simulation where learners play roles in different kinds of social situations.

**Others applications**: A number of researchers have developed university-based systems using PDAs to support study activities. For example, Tokushima University has a project called BSUL (Basic Support for Ubiquitous Learning) (Saito, Ogata, Paredes et al., 2005) to aid study activity. Another project called Perkam (El-Bishouty, Ogata, & Yano, 2006) allows the learners to seek for and share knowledge, interact, collaborate, and exchange individual experiences.

Our approach uses the scaffolding technique to design a participatory simulation framework to support collaborative learning. The framework is called SPS (scaffolding participatory simulation) to support student-centered learning. Making use of this framework, the learner plays the role of data in the simulation of logic to visualize the data flow of computers in the real world. This framework uses participatory simulation to encourage more active participation and interaction among the learners. Students are not simply "active" in a participatory simulation, they are *enactive* and this stands to improve both student motivation and understanding (Wilensky & Stroup, 2000).

Based on the SPS framework, we have implemented and evaluated a support system to learn sorting algorithms called PLASPS (PDA-based learning algorithm system using participatory simulation). In the following sections, we describe how the system uses a participatory simulation environment for learning about sorting algorithms, and how we used the scaffolding technique to develop this system. We also describe the implementation of the PLASPS, the evaluation of the system and plans for future work.

## PARTICIPATORY SIMULATION

During the past 10 years, computers have been used increasingly as simulation machines. The widespread popularity of game software like SimCity and SimEarth gives a clear indication of the extent to which simulation has permeated popular culture. Participatory simulations grew out of an age-old tradition of using role-playing to help people develop personally meaningful understandings of complex or nuanced situations (Colella, 2000). Students engaged in participatory simulations act out the roles of individual system elements and then see how the behavior of the system as a whole can emerge from these individual behaviors (Wilensky, 1999).

Participatory simulations are learning games where players play an active role in the simulation of a system or process. Thinking Tags are an example of wearable computing (Colella, 2000). They are small computational devices about the size of a name badge that have been used to involve people in participatory simulations, by wearing small, communicating computers (the Thinking Tags), students are transformed into players in a large-scale micro world. Like classic micro worlds, participatory simulations create a scenario, mediated by a set of underlying rules, that enables inquiry and experimentation. In addition, these new activities allow students to "dive into" a learning environment and directly engage with the complex system at hand. Simulations of this type have recently come to the attention of educators through the work of Colella (2000). Students participate in an active way, analyze information, make decisions and see the outcome of their actions. This increases the motivation and the learning success improves.

Role-playing activities have traditionally been used in social studies classrooms, but have infrequently been used in science and mathematics classrooms (Wilensky, 1999). Our use of the term participatory simulations is intended to refer to such role-playing activities used primarily in science and mathematics classrooms to help students understand computer science theories more easily and clearly.

The SPS framework focuses on using PDA participatory simulations to learn a concept (or rule) through doing it. PDA participatory simulations use Palm OS handheld computers (for now only Palm OS is supported) to embed people inside simulations. Interactions between players in the game are mediated by peer-to-peer beaming. Using an SPS-based system, the teacher can assign tasks to learners who collaborate together to perform them. The system supports the learner to do the task step by step, checking every step and providing feedback to the learner. Learners can understand the concept (or rule) through trial and error. When the learner masters the concept (or rule) on a certain level, the system will reduce the help function gradually and more responsibility is shifted to the learner.

PLASPS (PDA-based learning algorithm system using participatory simulation) is a specific implementation based on the SPS framework that is intended to help teach sorting algorithms. In this system, all the learners stand in a line with PDAs, and the teacher can assign tasks to the learners and ask them to do the task by using the system. Learners receive these tasks, collaborate together and exchange their physical positions according to the algorithm. The system helps the learner to understand the features of the various sorting algorithms.

This research is advocated by pedagogical theories such as hands-on learning, authentic learning, active learning and discovery learning. Brown, Collins, and Duguid (1989) define authentic learning as coherent, meaningful, and purposeful activities. When classroom activities are related to the real world, learners receive greater academic rewards. There are four types of learning to ensure authentic learning: action learning, situated learning, incidental learning, and experimental learning (Ogata & Yano, 2003, 2004). The SPS framework employs two forms of authentic learning; action learning and experimental learning based on face-to-face communication. The SPS framework brings the learners to learn in the 'real world.' The SPS framework also employs interactive learning. Interactive learning involves interactions, either with other learners, teachers, the environment, or the learning material. It also uses active learning and discovery learning. Active learning is an umbrella term that refers to several models of instruction that focus the responsibility of learning on learners. Bonwell and Eison (1991) popularized this approach to instruction. According to Mayer (2004) strategies like "active learning" developed out of the work of an earlier group of theorists, those promoting discovery learning.

Discovery learning is a method of inquiry-based instruction. Discovery learning is a constructivist based approach to education. Jerome Bruner is thought to have originated discovery learning in the 1960s. Bruner argues that "Practice in discovering for oneself teaches one to acquire information in a way that makes that information more readily viable in problem solving" (Bruner, 1961). This philosophy later became the discovery learning movement of the 1960s. The mantra of this philosophical movement suggests that we 'learn by doing.' Discovery learning takes place in problem solving situations where the learner draws on his own experience and prior knowledge. Discovery learning is a method of instruction through which students interact with their environment by exploring and manipulating objects, wrestling with questions and controversies, or performing experiments.

## SPS FRAMEWORK

This system is like a scaffold. Scaffolding is a great technique that can help the students to understand the sorting algorithm. At the beginning, this system supports the students by providing some instructions, and later the fading process is started where the students have to practice independently. There are two parts to this system; one is system-driven, which uses a scaffolding technique, and the other is learner-driven, which allows the student to work independently.

The SPS framework comes from traditional education and can be seen as an extension of traditional education. In the traditional education system, teachers give priority to the learners, and learners are normally very passive. For example, in learning computer science theory, when the teacher teaches some complicated concept or algorithm, they only explain it literally but it is difficult to make it clear to the learners. Making use of this framework, a mobile system to help students to understand concepts more easily and clearly is developed. In addition, it improves interactive communication, and increases the motivation and enthusiasm of the learners.

### What is Scaffolding?

Before moving on to the framework, we would like to explain the term 'scaffolding.' The term 'scaffolding' comes from the works of Wood, Bruner and Ross (Wood et al., 1976). In the paper "Building Scaffolds in Your Classroom" (Feldman, 2003), the 'scaffolding' has been explained with the following sample of learning: To ride a bicycle gives children a wonderful sense of accomplishment. Many of us can recall running alongside a child, holding the bike steadily as the child gained speed and then wobbled off independently, unaware that he or she was pedaling without support. Scaffolding, as provided by human tutors, has been well established as an effective means of supporting learning (Soloway et al., 2001).

SPS was designed using the scaffolding and fading concept, so the framework will give the right advice at the right time to the students. In the area of educational research, the term 'scaffolding' has been used to describe various instructional techniques for supporting learning activities that reflect authentic task situations. Scaffolding enables the learners to engage in out-of-reach activities; having a 'knowledgeable other' or 'more capable peer' to support the learners and 'share the cognitive load' (Jackson et al., 1996).

The scaffolding concept has also been used in other researches such as the BWAS (bird watching assistance system) (Chen et al., 2002) for collaborative data gathering to support the learning environment. This system offers a mobile learning system, which supports the learners learning through scaffolding. The BWAS system contains a testing subsystem that evaluates each learner's level of learning. It has a structured-assistance learning system; more assistance is given for beginners, and as they become more proficient, less assistance is given. Learners in the system use mobile learning devices to acquire and learn through a wireless transmission tool anytime and anywhere. Evaluation from their experiences was taken and it showed that the learners benefited substantially even though the scaffolding support was gradually withdrawn. It was also found that the learners who used the bird watching system improved their learning; above and beyond what they would normally be expected to learn.

The timing of the scaffolding is very important. When children are just learning to ride a bicycle, it is necessary to hold the bike steady to give them a sense of security. However, when they can pedal without support, if we still hold the bicycle, it will prevent them from achieving more speed. Therefore we should begin the process of "fading," or the gradual removal of the scaffolding, which allows the learner to work independently.

## Design of the SPS Framework with the Scaffolding Technique

The design of the SPS framework is divided into four sequential parts (see Figure 1), the first one is the initial process, the second one is the system-driven part, which is the process of scaffolding to support task execution, and the third is the learner-driven part, which is the process of "fading" to train the learners to think by themselves, and the last part is the reflection.

During the initial process of learning and problem solving, the teacher will provide general ideas, concepts, rules, examples, and so forth to the learners to guide them and to make sure they have a basic idea of what has to be learned. Following that, the system-driven component will assist the learners by providing some instructions, the learner will gain experience from the aid and help messages provided by the system to tackle the problems presented in the PDA. After some practice using the PDA to solve problems, the learner will gradually learn and understand the methods and techniques to solve the problem. When the learners become more experienced with the concept (or rule), the fading process is started where the learners use the system to practice independently. At this stage, the system will reduce help messages gradually and more responsibility is shifted to the learner. Lastly, the learners will be able to solve the problem themselves without the help of the system. At the end, learners can reflect with the history record.

### Initial Process

The teacher explains the concept (or rule), and points out the major emphasis, and then explains to the learners how to use the system. The final explanation will be conducted using the examples on the help page. This is very useful to help learners to understand.

### System-Driven

In the system-driven part, the system guides the learner to do the task step by step. The system-driven part acts like a bridge, to enable learners to master the concept (or rule). All the learners do the task using PDAs, with the system will guiding them on how to do the task. A learner can discuss and compare with the other learners by indication, before exchanging the position. Here, mistakes are expected from the learners, but the system will provide the learners with some information, which points out the position if it is incorrect and how to correct it for the learners to be able

*Figure 1. Design of SPS framework*

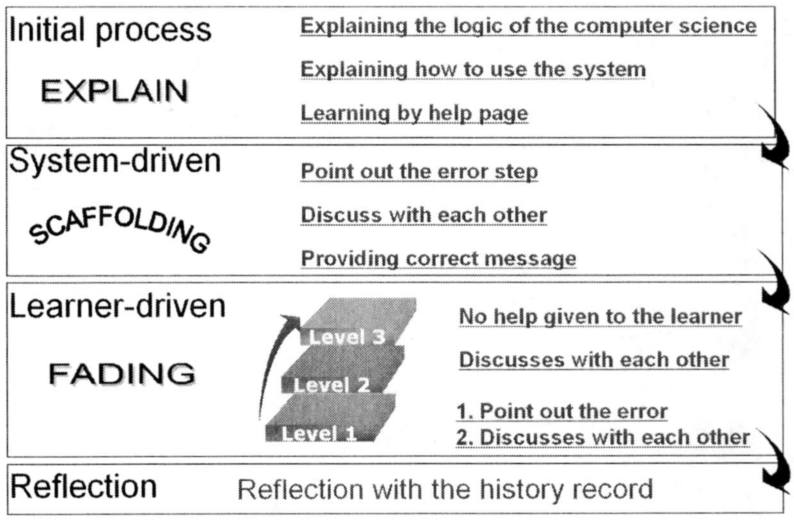


to achieve the task. The learner comprehends through participation in the discussion and help from the system to understand the feature of this concept (or rule). There are three characteristics to support learners in system-driven:

1) Discussion and help with each other.
2) Pointing out the error position.
3) Providing messages to correct the error.

## Learner-Driven

The learners use the learner-driven system to practice doing the task by themselves. When the learner masters the concept (or rule) on a certain level, the process of fading begins. The teacher will judge according to the level of understanding of the learners and reduce the system's help function gradually.

We design the learner-driven part into three levels depending on the process of fading as in Figure 1:

**Level 1** points out the error, but the method of how to correct it must be completed by the learners, and they can discuss and compare with their neighboring learner, before attempting to correct the error.

**Level 2** does not point out the error, and the learners have to correct it by themselves. They can discuss and compare with their neighboring learners before attempting to correct the error.

**Level 3** (see Figure 2) lets everyone do the task with the PDA. For example, there are 6 learners with PDAs named "Student A," "S1," "S2," "S3," "S4" and "S5." In this example, the task is to perform a sorting algorithm. S1 represents number 18, S2 represents number 3, S3 represents number 12, S4 represent number 5 and S5 represents number 10. Student A will have to rearrange the numbers according to the task. Student A stands before them indicating how to do switch positions and do the task. Facing the other learners, student A orders them to exchange their position without discussing (Figure 2) and he will ask S1 and S2 to exchange their positions. If Student A passes the third level, we can say he can complete the task independently. When he finishes the task, if there are some mistakes, the other students can point out the error and correct him.

*Figure 2. Overview of Level3*

## Reflection

Every step is stored in the history record. After finishing the task, the learners can see the history of each step, which points out the wrong positions for the learners and the corrections. With the history record the learner can reflect on the algorithm. The mobile device is used to communicate with the server and to store every step of the process. The stored steps can allow reviewing or restarting at any time.

## Characteristics of the SPS Framework Relating to Educational Scaffolding

Here are the characteristics of the SPS framework relating to educational scaffolding as below (Jamie, 1999):

1. **Provision of clear directions.** The SPS framework offers step-by-step directions to explain just what learners must do in order to meet the expectations for the learning activity. A message about how to do the task will be provided to support every step.
2. **Clarification of purpose.** The SPS framework keeps purpose and motivation at the forefront. Scaffolding aspires to meaning and worth. At first the teacher should explain the concept of the algorithm, and point out the major emphasis including what is the target or result to be achieved. The SPS framework will help the learner to do the task based on the target of every step.
3. **Keeping learners on task.** By providing a pathway or route for the learner, each time a learner or team of learners is asked to move along a path, the steps are outlined extensively. The SPS framework checks every step, and the right step will be shown to the learners.
4. **Assessment to clarify expectations.** From the very start, the SPS framework provides

examples. By these examples the learners can know what result is expected at the end. Right from the beginning, learners are shown rubrics and standards that define excellence. In traditional school research, learners were often kept in the dark until the product was completed. Without clearly stated criteria, it was difficult to know what constituted quality work.

## Characteristics of the SPS Framework

There are three main characteristics in this framework:

(1) The framework uses participatory simulation, and enables students to dive into a simulation just like in a game. Active participations and discussions in the system also help to increase their motivation to learn.
(2) The framework was designed using the scaffolding concept, providing the right advice at the right time to the students.
(3) Using the history record in this framework, reflection can be realized for better understanding. Also, every step will be saved allowing the system to be restarted at any stage.

## PLASPS

When learning computer science theory, it is essential to learn sorting algorithms, but it is not easy to understand the concept of the different sorting algorithms. A prototype of a learning object on Java programming for the PDA was developed to help the student to learning algorithms with the name PLASPS (PDA-based learning algorithm system using participatory simulation). We use the SPS to develop the PLASPS system, which was modeled following the standard specifications of Java 2 Platform, Standard Edition (J2SE), and using Apache Struts. We used a wireless LAN (IEEE 802.11b), and tomcat 5.0 as the server running on a desktop computer, and learners ran this system on Pocket PC2003, and an Access database.

### Architecture

There are six sorting algorithms in this PDA participatory simulation system. They are bubble sort, insertion sort, selection sort, quick sort, heap sort, and shell sort. There are three modules in the system driven part (see Figure 3).

#### Server Module

The server will send messages to the teacher or the learners. Each time after sorting, the server will receive data records from the learner module and the server will save these data records automatically. With these records, the server will validate the sorting whether it is correct or not, and send messages to the learners or teacher.

#### Teacher Module

The main job of the teacher module is to select the sorting algorithm and set the number of the learners. After the random data is generated, the teacher sends this data to learners. The teacher can use a desktop personal computer or a PDA for this purpose.

#### Student Module

The learners will get the data to be arranged from the server according to the ID of the learners. Then the learners have to analyze, compare, discuss, and swap the obtained data. The result will then be sent to the server and the server will make a comparison of the correctness of the result. At the same time, the teacher can view the result and also measure the understanding of the learner and revise any new ways to explain the compilation of data.

*Figure 3. Architecture of system*

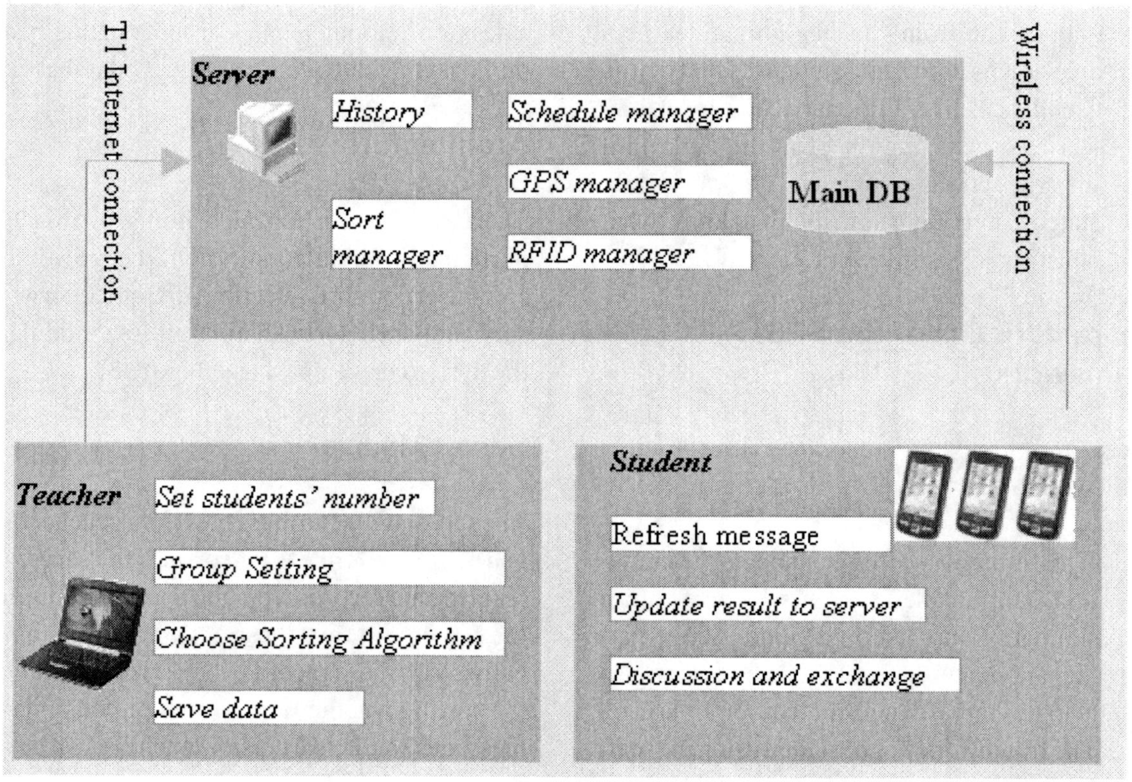


## User Interface

We have designed the system as a central server, with four interfaces, the first one is a user interface for login, the second one is a learner interface, the third one is a teacher interface, and last one is a help interface.

There are two types of users, the teacher and the learner. The system can detect the user type from his account and display the correct interface accordingly.

### Help Interface

Before using the system, the learners should know the concept of the sorting-algorithms, and how to sort these algorithms.

## Teacher Interface

As shown in Figure 4, there are three check options in this interface; an error checking option, a help message to correct the error, and a choice for either ascending or descending sort. When the error checking option and help message option is on, it is in the system-driven mode, if one or all of them are off, this system is in the learner-driven mode.

The teacher logs in as a teacher user, and the teacher window will appear as shown in (A). In this window, the teacher can select the sorting algorithms and set the number of learners. If the selection is ok, the numbers and first-time position will appear as in screen (B). The system will generate a list of random of numbers and send this list and the position numbers to the learners. P1, P2, P3...PN are the positions of the number and this data will be saved in the database, these are all the settings for the teacher. Then, when the teacher clicks 'next,' the sorting will be started. The system will assign one number in this list to each learner. The system also will assign one number in this list to each PDA.

## Student Interface

As shown in Figure 4, when learners login with their user ID, the (C) 'Initialization for student' window will be shown. When it is refreshed, the number of the user will appear as in the (D) 'Step for student' window. There are also messages from the server and the position of each number. The

*Figure 4. Interface of system*

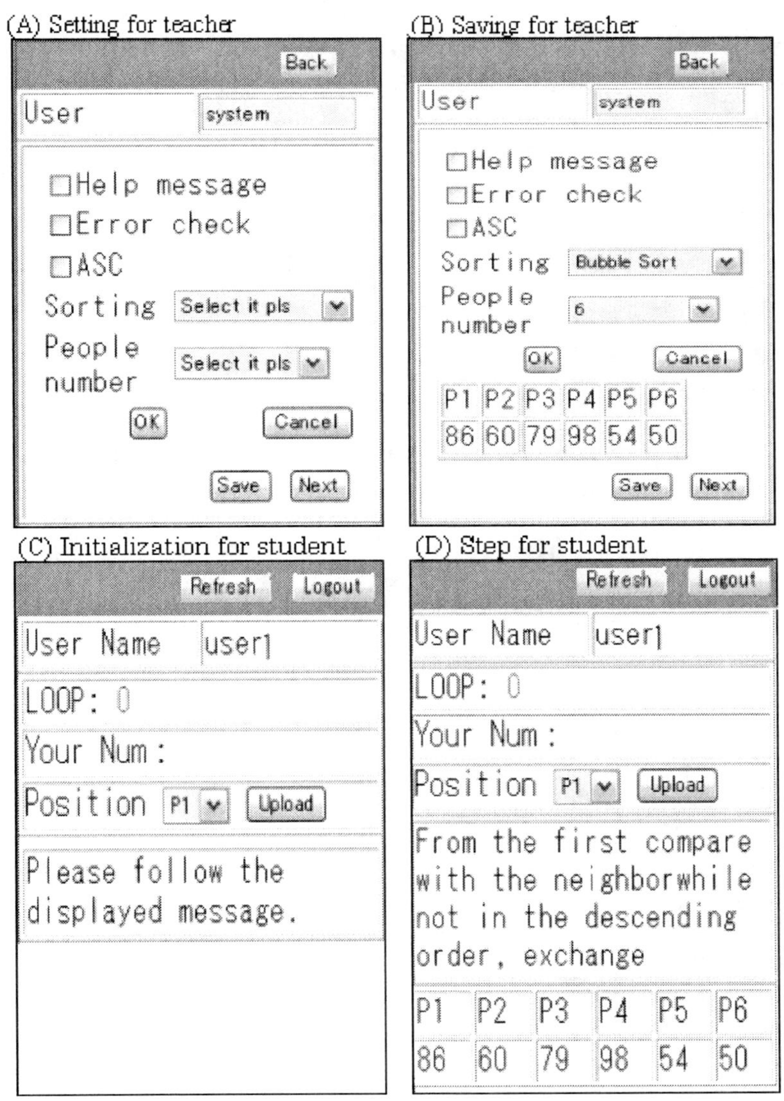

learner will change his location and the number position according to the sorting algorithm. The 'upload' button is used to send the result to the server after every step. Students can also review the sorting algorithm here and also the number of loops for the algorithm.

## Scenario

This is a scenario about using the system. The teacher assigns an array of numbers to his learners and asks them to sort these numbers according to a certain algorithm. Learners receive these tasks, collaborate together to do the task and the result of each step is sent to the server. The system will check it and feedback the learner with the positions of the numbers if there is a mistake. The learners will correct the number positions and send their new positions again to the server. Learners can understand the algorithm through discussions and their errors.

Figure 5 is a sample of using the system to learn the bubble sort algorithm. In the evaluation conducted in a sorting algorithm class, there were altogether 30 students and a teacher. The students were divided into three groups and were instructed to use this system at the same time. In this case, the teacher chose the bubble-sorting algorithm. The system generates the data randomly and sends them to the learners. In the example shown, the array list "84, 94, 43, 93, 96, 83, 14, 93, 99, 58" was to be sorted in descending order. Learners initially stand in a line with a PDA, which displays their numbers and positions.

At the beginning the system will give hints and instructions to help solve the problem. In loop 1, an initial help message like this will appear: "The first person (in this case P1) has to compare and change his position with the neighbor if it is not in descending order." After the comparison, the new position is uploaded to the server. The students will also change their physical standing position in the line. The server will evaluate the

*Figure 5. Bubble sort*

| | P1 | P2 | P3 | P4 | P5 | P6 | P7 | P8 | P9 | P10 |
|---|---|---|---|---|---|---|---|---|---|---|
| Start | 84 | 94 | 43 | 93 | 96 | 83 | 14 | 93 | 99 | 58 |
| Loop1 | 94 | 84 | 93 | 96 | 83 | 43 | 93 | 99 | 58 | 14 |
| Loop2 | 94 | 93 | 96 | 84 | 83 | 93 | 99 | 58 | 43 | 14 |
| Loop3 | 94 | 96 | 93 | 84 | 93 | 99 | 83 | 58 | 43 | 14 |
| Loop4 | 96 | 94 | 93 | 93 | 99 | 84 | 83 | 58 | 43 | 14 |
| Loop5 | 96 | 94 | 93 | 99 | 93 | 84 | 83 | 58 | 43 | 14 |
| Loop6 | 96 | 94 | 99 | 93 | 93 | 84 | 83 | 58 | 43 | 14 |
| Loop7 | 96 | 99 | 94 | 93 | 93 | 84 | 83 | 58 | 43 | 14 |
| End | 99 | 96 | 94 | 93 | 93 | 84 | 83 | 58 | 43 | 14 |

change of positions and send an error message if the change was done incorrectly.

In case of any mistakes in the change of positions, the message will then ask the learners to correct it such as: "Numbers 84 and 94 are not in descending order, please change your positions." Conversely, if the positions were changed correctly, the server will generate a message like this: "P2 and P3, please compare your order and upload the new positions." As in the case of P1 and P2, if the change was incorrect, the server will generate a message to aid the learners again and vice versa if the change was correct. This process is also carried out for the other learners until P10.

As we move on to Loop 2, each learner will discuss and compare with his neighbor learners according to the messages from server as in Loop 1 and this process goes on for a few loops depending on the problem until the whole array is sorted. In each loop, circled numbers are correct and fixed while the underlined numbers indicate that these positions require changes in next loops.

## EVALUATION AND EXPERIMENTAL RESULTS

### Single Group

In order to evaluate the usability and get some feedback for this system, we asked 10 learners to evaluate the system twice. Ten PDAs were used in this system too. All the participants were Master's students who had learned sorting algorithms about three years previously when they were undergraduates. Each student holds a PDA with a number and stands in a line to sort (see Figure 6). Most of them have not used the sorting algorithm for a long time so they may have forgotten the rules of the sorting algorithms. We are proposing that they can use the system to review the algorithm again.

We chose learners who have not used sorting algorithm for a long time to get their feedback and the efficiency of re-learning the sorting algorithm again using the system.

Most of the learners commented that they could learn sorting algorithms using this system, and

*Figure 6. View*

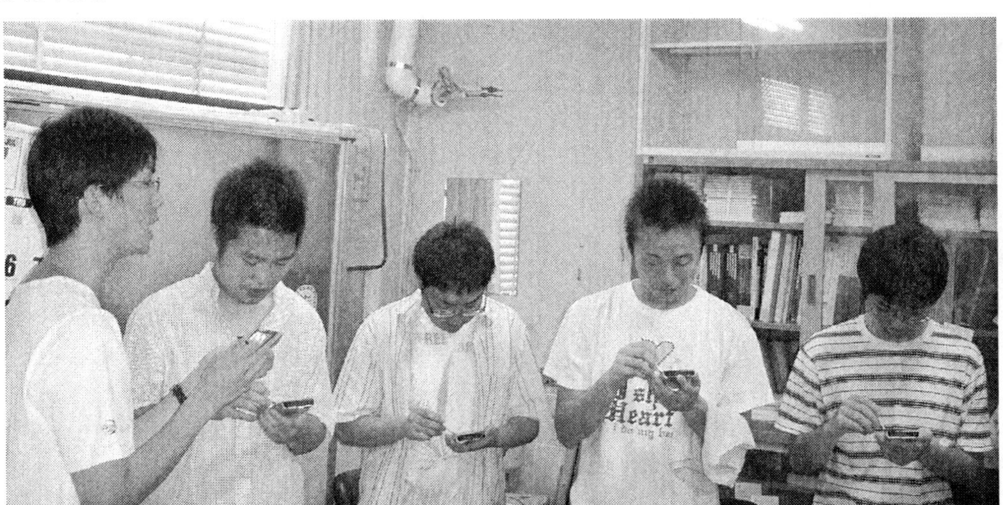

*Table 1. Comments from single group evaluation*

| | *What do you think about this learning method?* |
|---|---|
| 1 | It is very easy to be understood using this way of teaching algorithm. |
| 2 | We can exchange our views on the solution of the algorithm and get the best way to solve the problem. |
| 3 | In this way I can get help from the other learner |
| 4 | Studying in this way is quite easy to understand, just like studying in games. |
| 5 | Learning from using the system is more interesting than learning from teaching material in a normal class setting. |
| 6 | We can do revision on the algorithm learned in class with this system. |

***Table 2.*** *Questionnaire used for multiple group test*

| | *Questionnaire.* |
|---|---|
| Q1 | Do you think this system is helpful for learning algorithms? |
| Q2 | The system checks each step and gives help information; do you think it is helpful? |
| Q3 | Do you think this system is easy to use? |
| Q4 | Do you want to keep using this kind of system to assist learning? |
| Q5 | Can you understand the algorithm better after making mistakes? |
| Q6 | Can you understand more deeply when you get help from the other learners? |
| Q7 | Can you explain well when you help someone else? |
| Q8 | Do you understand more deeply and are enlightened by the discussion? |
| Q9 | Do you think it is interesting to study while discussing with other learners? |
| Q10 | Do you think it is good to use the history record to reflect on the process of the sorting? |

it is a good way to explain the algorithm by the participatory simulation. This system gives the opportunity to the learners to see in a practical way how the sorting algorithms are done.

After completing this experiment, we got comments from these learners for improving the interface, such as "please change the position of the buttons." After we improved the system, we evaluated system again. In the experiment, we obtained some comments as shown in Table 1. According to the comments, we concluded that it is a good method to study by this system, which is like to playing games in class. It is more interesting than just learning using books only.

## Multiple Groups

Multiple groups can use this system at the same time. For the third experiment, we asked 30 learners to evaluate the system and divided them into three groups.

After completing the experiment, a system evaluation questionnaire was given out (Table 2).

The learners evaluated the system by rating each of 10 questions on a scale from 1 (the lowest) to 5 (the highest). The scale labels were; 1: totally disagree, 2: partially disagree, 3: Neither agree nor disagree, 4: partially agree, 5: totally agree). The average of the points was 4.0. Figure 7 shows the results of the evaluations by the questionnaires.

According to the results for question (1), the system is helpful for learning algorithms. For question (2), about the system checking each step and giving helpful information, the result was not as good because they said there some messages did not point out the error position. For question (3), the evaluation was not as good because this was the first evaluation and the interface of system was not user-friendly enough. Modifications will be made gradually from time to time to improve the user-friendliness of the system. We also need more explanation for the learners on how to use it. Results for question (4) show they liked using this kind of system to help learning.

From the results of question (5), they can understand the algorithm better after making mistakes. For questions (6) and (7) using this method, they appear to benefit form helping each other. Questions (8) and (9) show that they could understand the algorithm more deeply and were enlightened by the discussion. Thus, we are also thinking about how to enlighten and help to improve the understanding of learners learning sorting algorithms. In the discussion, they can tell each other what they have learned, and they liked studying by discussion. Question (10) asked about using the history record to reflect on the process of the sorting. Some learners commented that by the history record they can know the whole process, and they can ponder and reflect on the algorithm from the history records.

Most of the learners commented that they could learn sorting algorithms using this system, and participatory simulation is a good way to explain the algorithms. This system gives the opportunity to the learners to see in a practical way how the sorting algorithms are done, but the current system is not easy to use.

From the evaluation, we can observe that the students were able to learn better and enjoy the learning process in such group settings. Each group would compete with each other to be the fastest to sort the algorithm.

## FUTURE RESEARCH DIRECTIONS

We noted from the student feedback that the system is still not easy to use, so we are planning to improve the interface and ease of use in the future. We will then perform further evaluation of this system to get more in-depth feedback from the learners in order improve it to achieve the goal of using it in class.

Other future work will include a comparative experiment with alternative approaches. The next stages of the evaluation will include students who have no background of algorithm and will also compare these outcomes with those when the mobile device is not used.

## CONCLUSION

This chapter describes the SPS framework, which was designed using the scaffolding technique. Making use of SPS, the learner plays the active role of data in the simulation of logic to visualize sorting algorithms. This encourages more active participation and interaction among the learners. Active participation and discussion while using the system also help to increase the motivation to learn. The SPS framework uses the participatory simulation method to realize authenticity, and using the history record in this framework, reflection can be realized for better understanding. The framework was designed using the scaffolding concept, providing the right advice at the right time to the students. Using the system based on the SPS, students not only listen and read but also study through discussion with each other and doing practical tasks.

Based on this SPS framework, as a sample system, the PLASPS system was implemented and then evaluated. The teacher can use this sys-

Figure 7. Mean ratings on the questionnaires (min.: 1, max.: 5)

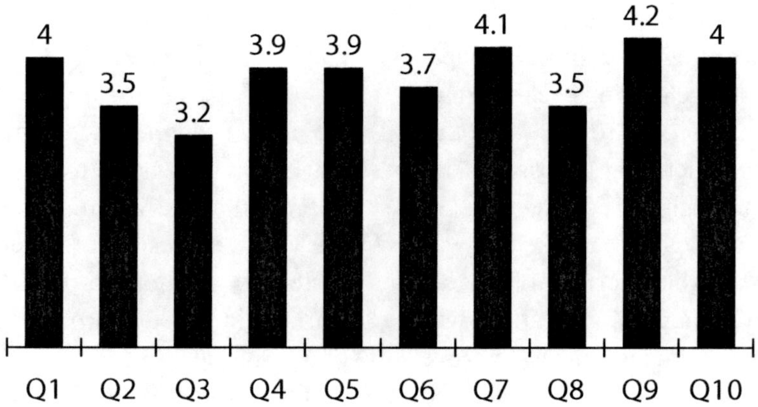


tem to help the learner to understand algorithms more deeply. From the evaluation, we conclude that by using the PLASPS, participants can learn effectively through their discussions, trials and errors, and the system is useful for learning algorithms. The communication and discussion is always an essential part of the learning process. Most of the learners commented that they could learn sort algorithms using this system, and that it is a good and interesting way to explain the algorithms using participatory simulation. We found that it was very easy for learners to understand and the main goal is achieved, whereby each learner understood the algorithm deeply, and this framework encourages more active participation and interaction among the learners.

The PLASPS system can help to teach algorithms and the scaffolding technique is suitable to support the teachers in interactive lectures. A major advantage of participatory simulations is the fact that it is easier for learners to see patterns and understand coherences. We believe that the emerging field of mobile interactive services and participatory simulations can improve the learning experience especially in the case of complex problems.

## REFERENCES

Barak, M., Harward, J., & Lerman, S. (2007). *Studio-based learning via wireless notebooks: a case of a Java programming course. International Journal Mobile Learning and Organisation, 1(1), 15–29.*

*Bonwell, C., & Eison, J. (1991). Active Learning: Creating Excitement in the Classroom AEHE-ERIC Higher Education Report No. 1. Washington, DC: Jossey-Bass.*

Brown, J. S., Collins, A., & Duguid, P. (1989). Situated cognition and the culture of learning. *Educational Researcher, Jan.-Feb.*, 32-42.

*Bruner, J. S. (1961). The act of discovery. Harvard Educational Review, 31(1), 21–32.*

Chen, Y.S., Kao, T.C., Sheu, J.P., & Chiang, C.Y. (2002). A mobile scaffolding-aid-based bird - watching learning system, Paper presented at *the IEEE International Workshop on Wireless and Mobile Technologies in Education* (WMTE'02) (pp. 15-22). Växjö, Sweden

Colella, V. (2000). Participatory simulations: building collaborative understanding through impressive dynamic modeling. *Journal of the Learning Sciences, 9*(4), 471-500.

El-Bishouty, M., Ogata, H., &Yano, Y. (2006). Personalized knowledge awareness map in computer supported ubiquitous learning, *Paper presented at the International Conference on Advanced Learning Technologies (ICALT2006)*, Amsterdam, The Netherlands.

Feldman, S. (2003). *Building scaffolds in your classroom*. Teaching Pre K-8, pp. 6.

Gerhard, F., & Masanori, S. (2006). Supporting self-directed learners and learning communities with socio technical environments. *Research and Practice in Technology Enhanced Leaning, 1*(1), 31-64.

Jackson, S.L., Stratford, S.J., Krajcik, J., & Soloway, E. (1996). *Model-It: A Case Study of Learner-centered Software Design for Supporting Model Building*. Report No. MVK50700. Michigan University, Ann Arbor, MI. (ERIC Document Reproduction Service No. ED 446903)

Liebowitz, J. (2007). Developing knowledge and learning strategies in mobile organisations. *International Journal of Mobile Learning and Organisation, 1*(1), 5–14

McKenzie, J. (1999). Scaffolding for success. From now on. *The Educational Technology Journal, 9*(4), Retrieved August 27, 2005, from http://www.fno.org/dec99/scaffold.html

*.Mayer, R. (2004). Should there be a three-strikes rule against pure discovery learning? The case for guided methods of instruction. American Psychologist, 59(1), 14-19.*

Ogata, H., &Yano, Y. (2003a). *How ubiquitous computing can support language learning*.Proc. of KEST 2003, (pp.1–6).

Ogata, H., &Yano, Y. (2003b). *Supporting knowledge awareness for a ubiquitous CSCL*, Proc. of E-Learn 2003, (pp.2362–2369).

Ogata, H., & Yano, Y. (2004). *Context-Aware Support for Computer-Supported Ubiquitous Learning*. Proc. of WMTE 2004, IEEE Computer Society Press, (pp.27-34).

Okada, M., Yamada, A., Tarumi, H., Yoshida, M., & Morita, K. (2003). *Digital-EE II: RV-augumented interface design for networked collaborative environmental learning*, Proc. of CSCL 2003, (pp. 265-274).

Papert, S., & Harel, I. (1991). *Constructionism, chapter Situating Constructionism*. Ablex Publishing Corporation, Norwood, NJ.

Roschelle, J. (2003). Unlocking the learning value of wireless mobile devices, *Journal of Computer Assisted Learning, 19*, 260-272.

Saito, N., Ogata, H., Paredes, R., & Yano, Y., & Ayala, G. (2005). *Supporting Classroom Activities with the BSUL Environment*. Proc. of WMTE 2005, IEEE Computer Society Press, (pp. 243-250).

Kopf, S., Scheele, N., Winschel, L., & Effelsberg, W. (2005). *Improving Activity and Motivation of Students with Innovative Teaching and Learning Technologies*, Methods and Technologies for Learning, pp. 551-556.

Vahey, P., & Crawford, V. (2002). *Palm education pioneers program: Final evaluation report*. Menlo Park, CA: SRI International.

Wilensky, U. (1999). *Learning through participatory simulations: Network-based design for systems learning in classrooms*. Computer Support for Collaborative Learning, (pp. 667-676).

Wilensky, U., & Stroup, W. (2000). *Networked gridlock: Students enacting complex dynamic phenomena with the HubNet architecture*. Proc. ICLS 2000, Lawrence Erlbaum Associates, (pp. 282-289).

Wood, D., Bruner, J.S., & Ross, G. (1976). The role of tutoring in problem solving. *Journal of Child Psychology and Psychiatry, 17*, 89-100.

Yin, C., Ogata, H., & Yano, Y. (2004). Supporting Japanese polite expressions learning using PDA towards ubiquitous learning. *The Journal of Information and Systems in Education, 3*(1), 33-39.

## ADDITIONAL SOURCES OF PARTICIPATORY SIMULATION

http://education.mit.edu/pda/

http://www.ncrel.org/sdrs/areas/issues/students/learning/lr1scaf.htm

http://condor.admin.ccny.cuny.edu/~group4/

http://www-yano.is.tokushima-u.ac.jp/ogata/clue/index.html

http://www.mobilearn.org/

http://l3d.cs.colorado.edu/clever/

http://gseacademic.harvard.edu/~hdul/

http://www.pebbles.hcii.cmu.edu/

http://www.m-learning.org/

http://www.pjb.co.uk/m-learning.htm

# Section IV
# Enhancing Situated Learning Experiences

# Chapter XI
# Situated Learning with SketchMap

**Sosuke Miura**
*University of Tokyo, Japan*

**Pamela Ravasio**
*University of Tokyo, Japan*

**Masanori Sugimoto**
*University of Tokyo, Japan*

## ABSTRACT

*This chapter presents the SketchMap system, which supports children's situated learning by their experience of creating maps. In an outdoor environment, each child creates a map in the region of his or her school using a SketchMap client. The map is uploaded to the SketchMap server to be shared with other children who have created maps of different areas. Children can add new information to the maps or can edit them in their classrooms or in their homes. The goal of the SketchMap project is to investigate whether it the integration of outdoor and classroom activities, and the sharing of the children's experiences through the maps, can actually promote collaborative learning. This system has been used in the classes "safety map" and "nature exploration" in a Japanese elementary school, and an evaluation of the system has also been performed. Some issues that were identified during the educational activities are also described here.*

## INTRODUCTION

In this chapter, the SketchMap system (Enjoji, Ravasio, & Sugimoto, 2006; Ravasio, Tscherter, & Sugimoto, 2006; Sugimoto, Ravasio, & Enjoji, 2006) that supports children's situated learning activities through the creation of maps is presented. A key function of SketchMap is the enhancement of the children's learning experiences (Sugimoto, 2005). To make children's learning situated and contextual, SketchMap supports their outdoor fieldwork by providing a technologically enhanced learning environment. It also integrates the children's fieldwork in an outdoor environment and advances their reflection and further learning processes in a classroom, which is important for reaching a deeper understanding (Ackermann, 1996). SketchMap is used (1) to allow children to easily record what they have found during their fieldwork, (2) to reflect their learning experiences and enhance further learning while using their maps, and (3) to support knowledge sharing and construction processes by sharing individual children's experiences through the maps.

In SketchMap, children are asked to create a map of the neighborhood of their school. They use a tablet PC to retain the natural feeling of a sketch with pen and paper, while avoiding the difficulties of pen and paper based sketches. For example, using a tablet PC simplifies changing, or modifying, the scale of a map. The child draws a street with a stylus pen or places an icon, representing a landmark, by selecting from the icon list. A SketchMap tablet PC is augmented by a GPS receiver and a USB camera. The child can capture an image, a sound, or a video using the attached camera, and easily position it anywhere on the tablet PC display. During the map creation task, all the child's activities on the tablet PC, with time and location data captured through the GPS receiver, are logged. As the tablet PC is always connected to the Internet via its wireless Internet card, a map being created by a child can be uploaded automatically to the SketchMap server. After the children have returned to their schools or their homes, they can access the server and browse the uploaded maps. Then the children can edit their own maps or add new information to other children's maps.

Even while the SketchMap project was underway, evaluation of the use of the system was conducted in collaboration with a Japanese elementary school. This chapter describes some findings gained, lessons learned through this practice, and suggests some future work.

## RELATED WORK

The use of a mobile device, such as a PDA, to individualize tuition in educational environments has rapidly led to the transfer of widely known, effective pedagogical methods to this novel medium. Examples are the MapIt and PicoMap applications for concept mapping (Chan & Sharples, 2002) and Cooties (Soloway, Norris, Blumenfeld et al., 2001) projects for science classes. A mobile butterfly-watching system (Chen, Kao, Yu, & Sheu, 2004) uses content-based image retrieval and allows individual learners to retrieve information on a butterfly whose image is captured using a camera built into the handheld device in an outdoor environment.

The range of existing collaborative educational systems covers topics as diverse as participatory simulations (Benford, Rowland, Flintham et al., 2005; Colella, 2000; Wilensky & Stroup, 1999), classroom communication systems (Dufresne et al., 1996), and problem based learning (Cuartielles, Malmborg, & Schlaucher, 2003).

Individual and collaborative practices can be integrated into a single system. Ishizuka et al. (2004) compared the achievements of pupils using Web-based training systems in combination with PDAs in the "integrated study" classroom, outdoors, and in a social education facility. Deguchi et al. (2006) described a system for use in an integrated study class, whose aim was to make pupils collaborate in the design of a town and experience the relevant environmental implications. Yatani, Sugimoto, & kusonoki (2004) developed a system in which pairs of pupils collaborated to solve a knowledge quiz about the individual insights they gained from the science exhibition they were visiting, while Takenaka et al. (2006) proposed a system that allows elementary school children to collaboratively create and share a virtual botanical garden by using a mobile phone equipped with a camera. Finally, the purpose of the Ad Hoc Classroom project (Chang & Sheu,

2004) is to develop a wireless platform so that teacher and students can establish a classroom dynamically, irrespective of location and time limitations (that is, anytime and anywhere).

There have been proposals for systems that aim to enhance the use and application of geo-spatial data and tools in the context of other knowledge domains. An example is the e-MapScholar project (Jones et al., 2003), which, together with others, provides teaching case studies from varying areas of expertise, such as archaeology, civil engineering, or forestry. In the WebPark project (Dias et al., 2004), a mobile information system that provides visitors to natural areas and parks with information about their surroundings has been constructed, using smart phones and a GPS that is equipped to access the information in the system.

Environmental detectives (Squire & Klopfer, 2007) is for learning about potential contamination of a local water supply using a PDA that provides learners with location-sensitive visualization and for supporting them in their socially-situated scientific practices. JAPELAS and TANGO (Ogata & Yano, 2004) are context-aware language learning support systems, especially for assisting foreign students in Japan to learn polite Japanese expressions through interactions with real world objects. DIGITAL-EE (Okada et al., 2003) is for supporting learners conducting fieldwork in an outdoor environment and for those in a classroom to share the virtually augmented fieldwork and enhance their collaboration. CyberTracker (2000) is a PDA application designed for conservation biology data collection and used in the BioKIDS curriculum (Parr, Jones, & Songer, 2002) to enhance inquiry learning in an outdoor environment. The Ambient Wood project (Rogers et al., 2005) uses pervasive technologies to augment woodland digitally in a contextually relevant manner and to enhance the physical experiences of children exploring the outdoor world.

Although there are similarities between these predecessors and SketchMap, it differs from them as follows.

- The basic objective of SketchMap is that children's experiences are augmented by their articulation and recognition of the real world, and by expressing them as a map.
- SketchMap enhances children's conventional fieldwork by using a tablet PC that retains the features of pen-based interaction.

*Figure 1. An overview of the SketchMap system*

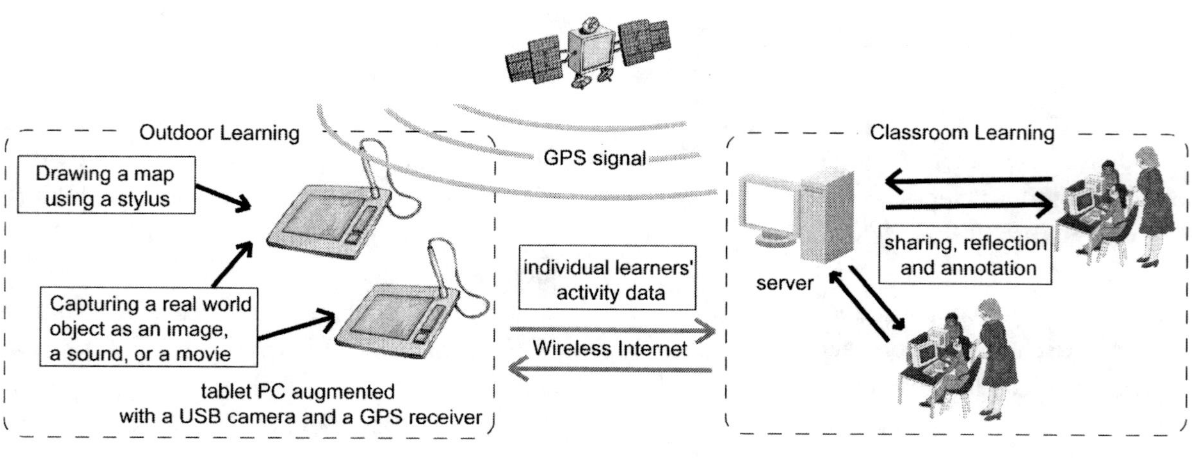

- In order to share individual children's outdoor experiences, enhance their reflections, and use them for their further learning, SketchMap allows the children to edit and annotate the maps through the server.

## CONFIGURATION OF SKETCHMAP

### Overview

As shown in Figure 1, SketchMap comprises multiple client computers and a server. The client computer is used to support the children in creating maps in outdoor environments, and the server computer is chiefly used to support their individual/collaborative learning, using the maps in their classrooms or in their homes.

### SketchMap Client

When we began designing the SketchMap client, we investigated numerous commercial software packages that support children in creating maps. Our investigations showed that most of the packages prepare a blank map and ask children to complete it by placing landmark icons on it (e.g., a police station or a fire department). However, let us assume that we create a map on a blank piece of paper. We first have to decide what should be placed/drawn on the paper. From an epistemological point of view, this is closely related to our articulation process, namely, how we recognize the three-dimensional real world, how we select a limited number of objects, such as roads, buildings, and shops, from an unlimited number of objects, and how we represent and arrange the objects on the paper as a two-dimensional map. This map creation task requires greater cognitive capability or effort than tasks that require children simply to place a ready-made icon corresponding to a real world object (because the world is already "articulated"). As one of the main goals of the SketchMap project is to enhance learners' experiences and reproduce them for their further learning, a map creation task that requires a higher level of articulation, based on the individual child's viewpoint, is necessary. Therefore, we decided to design a SketchMap client that allows the child to create a map on a blank canvas while receiving moderate assistance.

*Figure 2. A SketchMap client*

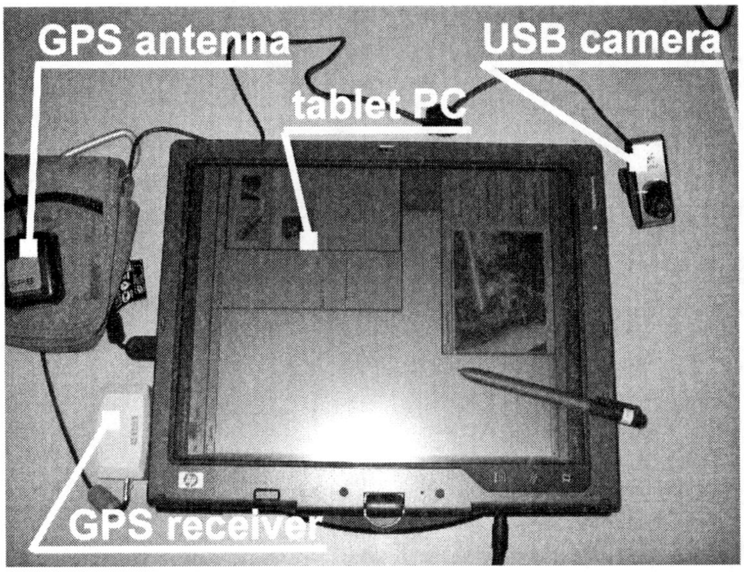

A SketchMap client comprises a tablet PC (HP Compaq tc4400), a USB camera (Logitech Qcam for Notebooks Pro), and a GPS receiver (IO DATA CFGPS2) as shown in Figure 2. The child carries the client and creates a map with a stylus pen. The user interface of the client is shown in Figure 3. When the child draws a street or places an icon, such as a hospital or a municipal office, on the canvas (Figure 3 (A)), he/she taps the menu in the upper part of the user interface (Figure 3 (B)) and selects the corresponding item. A USB camera attached to the tablet PC is manipulated through buttons on the popup window (Figure 3 (C)) and children can capture an image, a sound or a video with the camera. When this has been done, an icon that represents the captured image (thumbnail icon), sound (microphone icon), or video (thumbnail icon) is automatically added to the palette, as shown in Figure 3 (D). Then the child can drag the icon from the palette to anywhere they like on the map. They can easily enlarge/shrink an image or a video, or replay a sound by tapping the corresponding icon.

While children are creating maps in an outdoor environment, the SketchMap client receives GPS signals. Children's activity and location data, as given by the GPS, are automatically logged in the following format: time, latitude, longitude,

*Figure 3. A user interface of a SketchMap client (A Japanese word is translated into English in each parenthesis.)*

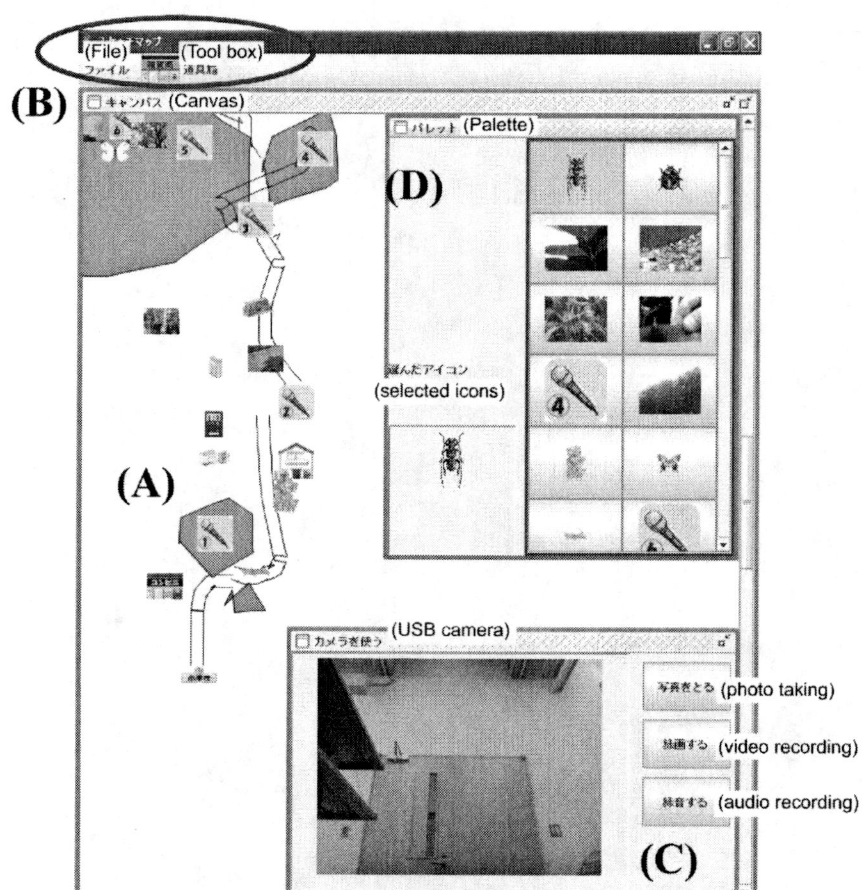

user manipulation, and item information. The user manipulation represents manipulations by children using the SketchMap clients. As the client makes pen strokes on the tablet PC, it can record actions such as photographing, video/audio recording, icon dragging, item drawing/erasing (a street, park, etc.), and canvas scrolling. The item information is written in the SVG/ XML format that describes a type of item (e.g., a street, park, or icons) and its graphical information (e.g., shape, or points of x–y coordinate values). These log data are uploaded to the server through the wireless Internet and managed by the server for sharing and annotation, as described in the following section.

## SketchMap Server

To use the maps created by children in an outdoor environment for collaborative learning in a classroom, the SketchMap system on the server provides the children with the following functions, which are available through a Web browser.

(A) A function for supporting reflection

Children repeatedly sketch real world objects on their tablet PC and capture them using the camera in an outdoor environment, after which they finalize their own map. Although children can share their own final maps with other children and use them for further learning in their classrooms and homes, they cannot completely remember the map creation processes from the final maps. If individual children are effectively reminded of when, where, and how they discovered an object described on the map, or why they drew or captured it, they can more actively discuss and totally understand their own and other children's maps.

SketchMap allows children to replay their map creation processes using the log data described previously. Figure 4 shows the user interface of the replay module, with the slider at the bottom of the figure representing the time that elapses while creating the map. Children can change the position of the slider and visualize the map that was created at the corresponding elapsed time. By using the replay module, children can easily discover how they created the map; for example, they perhaps first drew a street and then interviewed a person walking by. As the person told of a relevant location and pointed to it, the child took a picture of the place.

(B) A function for annotating maps

SketchMap allows children to add new information or experiences, related to what they have discovered during their outdoor activities, to their maps. As each child can view maps created by the other children or information annotated by them, the children can collaboratively share information and knowledge about neighboring areas in the vicinity of their school.

A Web interface for editing and annotating a map is shown in Figure 5(a). When the children click a hand-drawn map on the left hand section, a comment box with images, videos, or sounds recorded in an outdoor environment appears on the right.

The Web server is accessible in their classrooms and in their homes, or in any other place where a public Internet service is available. This allows children to add photographs, sounds, or videos that have been captured while they commute, and also involves their parents in collaborative knowledge construction and sharing processes through their participation in the map editing and annotation.

Figure 5(b) is an example of the Google Earth based interface that also shows results of children's map creation process. A red icon represents where children recorded an image, a sound, or a video in an outdoor environment, which is placed on the photomap based on the GPS data. Children click the icon, and write a comment on the popup window with a photo, a sound, or a

*Figure 4. A user interface for replaying the map creation processes*

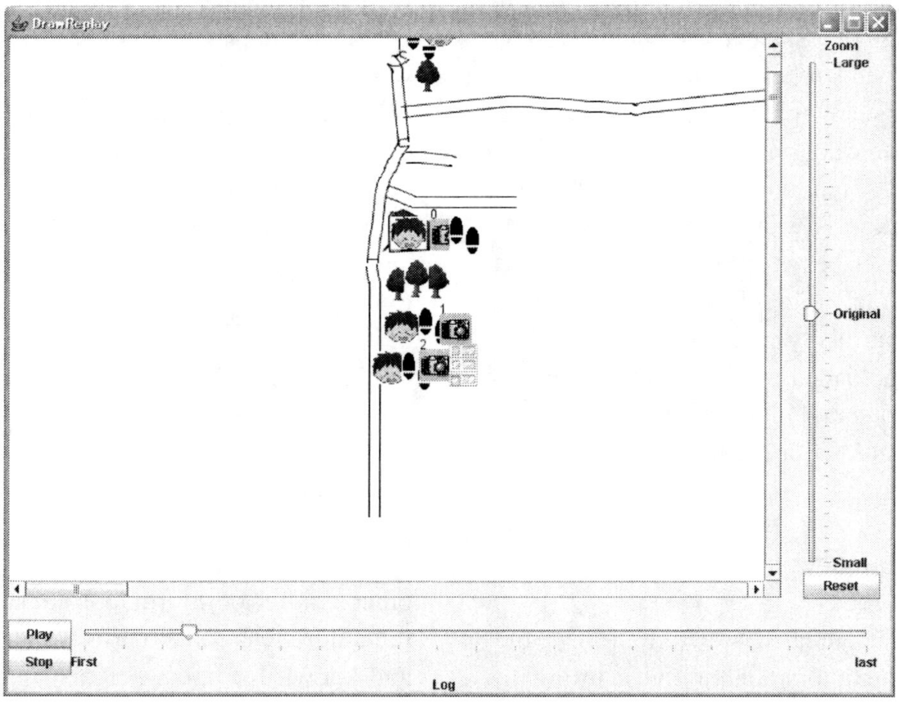


movie. As a map created by individual children is often inaccurate in its scale or orientation, it might be difficult for the other children, teachers, or parents to understand the map. If only the child who created it can understand a map, active discussion or participation in annotating the other children's maps does not happen. As the Google Earth interface is a photo-based map, it is anticipated that it will allow children, their teachers, and parents to more accurately identify information about the surrounding environment where children conducted their fieldwork.

The current version of SketchMap has been developed using the Java language (J2SE5.0), an XML database server (eXist), and a Web server (apache), and is executable on both Microsoft Windows XP and Red Hat Linux.

## EVALULATIONS THROUGH EDUCATIONAL PRACTICES

### Overview of Educational Practices Using SketchMap

Since February 2006, we have undertaken educational activities using the SketchMap system in collaboration with a local elementary school (Chiba prefecture, Japan), where the system has been used for "safety map" class and "nature exploration" class. In these classes, evaluations of the SketchMap system have been conducted three times during 2006 (February, June, and October) and twice during 2007 (February and June). To date, more than 200 children have used the SketchMap system.

*Figure 5. Web interfaces for editing and annotating maps*

*(a)*

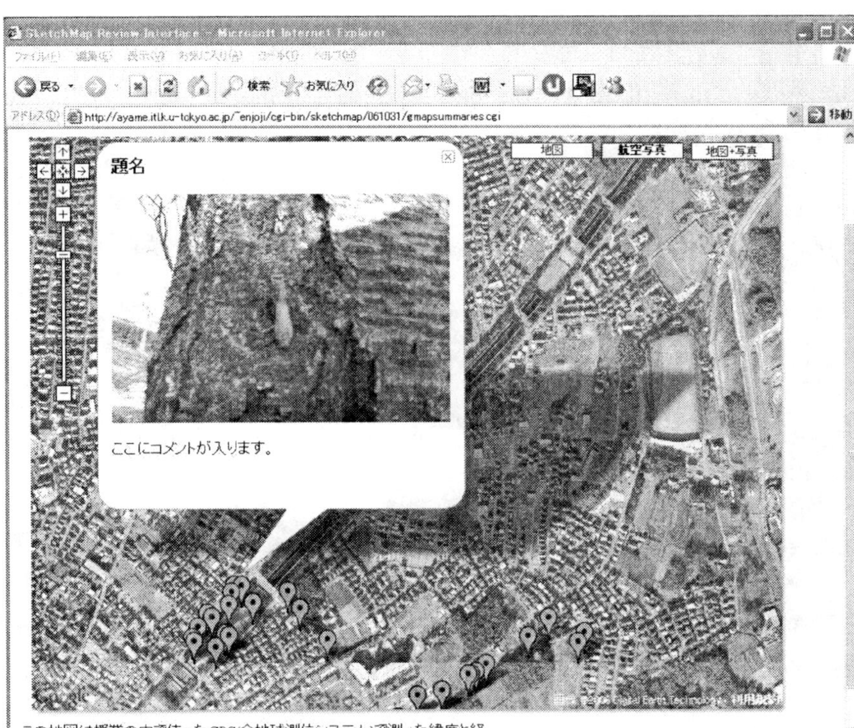


*(b)*

(1) Safety map class

Recently, the number of brutal crimes against children has increased, and ensuring their safety while commuting is one of the most critical issues in Japanese elementary schools (Japanese Ministry of Education, Culture, Sports, Science and Technology, 2006a). With this in mind, one of the measures for protecting children from such crimes is to create a safety map from a child's viewpoint. Here, children create a map using pen and paper, record photographs with a digital camera, and complete the map after they return to their classrooms (Japanese Ministry of Education, Culture, Sports, Science and Technology ,2006b). By using SketchMap, which supports pen-based interaction with a tablet PC enhanced by a camera and a GPS receiver, it appears possible to identify differences from, and advantages over, the conventional paper-and-pen based method. A map created by children is a useful resource for ensuring their safety and raising their level of awareness about danger, as well as focusing the concerns of the children, their teachers, and their parents. Therefore, it is expected that children, their teachers, and their parents will access the map, annotate it, and discuss it voluntarily.

In the safety map class, as shown in Figure 6, sixth grade children (age 11 to 12 years) were divided into groups of five. Each group was given one SketchMap client, and was asked to create a map of a specified area around their school, by drawing a street, taking a photo, and by placing an icon that represented an "empty street" or a "blind spot" (these icons denote a dangerous area), as well as by recording interviews with residents. The task took approximately 90 minutes. After the children returned to their classroom, using their map uploaded onto the SketchMap server, each group presented to the other groups what they had found during their outdoor activities.

(2) Nature exploration class

The purpose of a nature exploration class (Figure 7) is to interest children in living things, such as insects and plants, by creating a map around their school. A child's experiences in an outdoor environment can be a motivation for further learning. When they are back in the classroom, children share their maps and discuss or question what they found during the fieldwork. Moreover, as teachers and parents participate in creating the map in the classroom or at a home, they can easily assess the level of their children's interest in nature and provide them with more relevant information through discussion or annotation of their maps.

In this class, fourth grade children (age 9 to 10 years) were divided into groups of five. Each group was asked to create a map of a specified area around their school using the SketchMap system. The task lasted approximately 60 minutes. In the same manner as in the safety map class, when they returned to their classroom, using their map, children presented what they had found to the other children.

In this chapter, experimental results from the safety map class are shown. This class was conducted from the second week of June to the first week of July in 2007 with the following schedule.

- One day for outdoor fieldwork using a SketchMap client
- Three successive weeks for reflection and collaborative map annotation using the SketchMap server

Seventy-five sixth graders participated in this educational practice, in which the purposes of the evaluations were to investigate (a) the usability of the SketchMap client for children outdoors and (b) how outdoor activities motivate children to undertake collaborative learning and knowl-

*Figure 6. Children in the safety map class. Three of the five children in one group appear in the photograph. The girl on the left is taking a photograph to capture a potentially dangerous spot. The girl on the right is drawing a map on the tablet PC. The boy in the center is managing the individual children's tasks.*

*Figure 7. Children in the nature exploration class. Three of the five children in one group appear in the photograph. The boy sitting in the lower right of the figure is directing a USB camera to a plant. The girl holding a tablet PC in the center is taking a photograph and placing a thumbnail icon on their map. Another girl (second from the right) is coordinating the other two children's tasks. A student supporting the children in their field work is at the left of the figure.*

*Figure 8. Results of post-experiment interviews*

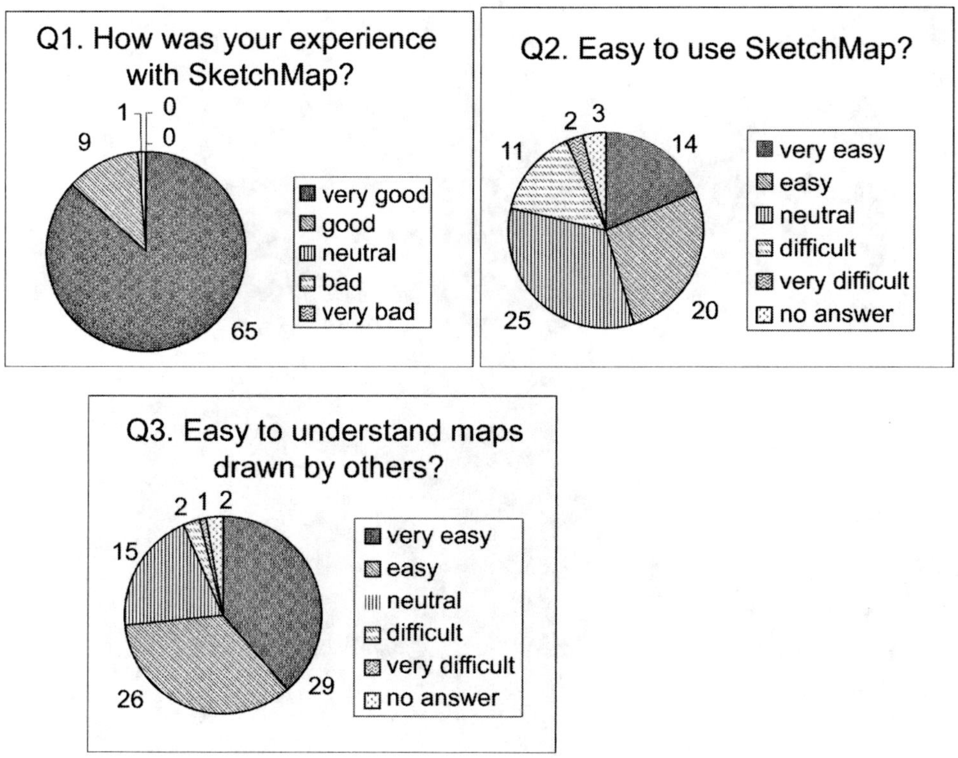


edge sharing/construction in their classrooms or homes.

In this class, maps were not uploaded using a wireless network. Instead, the maps were uploaded immediately after the children returned to their school. The children's activities outdoors and in the classroom were videotaped and post-experimental interviews and questionnaires with children and their teachers were conducted.

## Experimental Results and Discussions

(1) Usability of SketchMap client

Figure 8 shows the results of the post-experimental interviews conducted at the end of the one-day fieldwork. The children's responses were positive overall, showing that they enjoyed creating a map by using SketchMap, and that its usability was satisfactory.

Video analyses during the children's outdoor activities clarified several issues related to the advantages and disadvantages of the SketchMap client, some of which are summarized as follows.

- The five children in each group voluntarily discussed and allocated their individual roles, for example, capturing with a camera, drawing a map, discovering an object to be captured, and managing individual tasks, and changed them halfway through their task, although they were not asked to do this beforehand. This division of roles allows children to conduct their map creation smoothly.

- Children could use the SketchMap system easily. During the fieldwork, to record what they were interested in, or what they thought important for creating their safety map, they frequently used the camera to take a photograph, for example of a bush, to record a sound, for example, an interview with a resident, or to make a video, for example, of a barking dog.

The remarkable difference between SketchMap and the conventional map creation method is that children could not only draw a street, but also immediately put a photograph, a sound, or a video on the map. For example, if they did not like a photograph on their map, they could immediately take a new photograph and replace it easily. In contrast, it would have been impossible for children to place photographs, videos, or sounds on the map and then confirm if they were acceptable during the fieldwork, when using pen and paper. The comments from the teachers showed that the children seemed to be more motivated and engaged in the tasks when using SketchMap than when using the conventional method.

- Weather sometimes made it difficult for children to use a tablet PC outdoors. During the safety map class, the weather was cloudy, and the children could easily distinguish what was shown on their tablet PC display. However, in one of the nature exploration classes on a different day, the weather was sunny, and the children could not read the displayed information and draw the map easily. To solve this problem, it might be better to use a tablet PC with a low reflection display.

(2) How outdoor activities motivate children for collaborative learning and knowledge sharing/construction

Figure 9 shows the number of comments made by children on the SketchMap Web server. The total number during the three weeks was 29, which appeared relatively small. One of the main reasons was that the school teachers did not use the SketchMap server in their class. The teachers expected their children to access the server voluntarily, and to edit or provide comments on their maps whenever and wherever they liked, such as

*Figure 9. Number of comments during the three-week experiments*

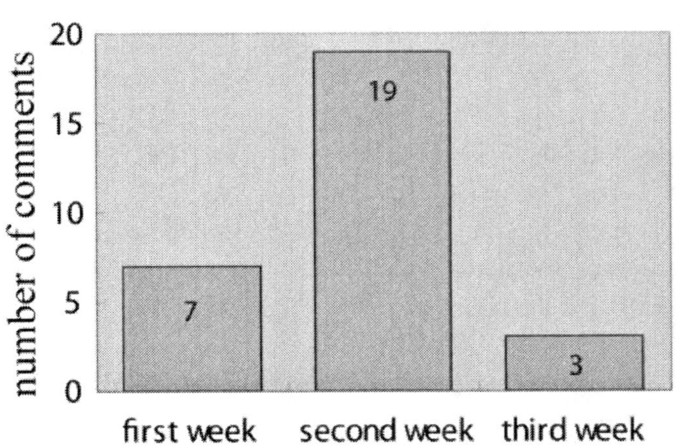

in their free time before, between, or after classes in their school, or while they were at home.

The following is an example of a series of comments made by four children (C1, C2, C3, C4) and one teacher (T1) on the SketchMap server. They tried to enumerate potentially dangerous places by referring to a certain part of the map and photographs on the server. (Phrases and sentences within the brackets have been added by the authors to clarify the meaning of the comments).

C1: This parking lot seems dangerous, because suspicious individuals can easily hide behind cars. We might be attacked while commuting.

T1: A location behind the step of the apartment house [next to the parking lot] is also dangerous, because it is dark and difficult to recognize.

C1: There are a bush and a hut [around the parking lot] that allow suspicious individuals to hide.

C2: Suspicious individuals can easily enter these places, and we cannot recognize them when it becomes dark.

T1: This photograph [on the map] shows the parking lot taken from a distant location.

C3: I think it is very dangerous because cars parked there make it difficult to identify suspicious individuals.

C4: That location [behind the steps of the apartment] seems dangerous because it is dark and [the wall of the apartment is] covered with graffiti.

In this case, C2 and C4 walked by the parking lot and found information that they had not found during the fieldwork. Active discussion among the children happened because they revisited a map created in their fieldwork and, therefore, had a certain level of knowledge about the place.

The questionnaires after the three-week experiments proved that (1) children who accessed the SketchMap server and added comments on the map could increase their level of understanding of the map and (2) could share more information and knowledge with the other children.

The experiments also clarified several issues to be investigated in our future work.

(1) More participation by the children is necessary.

In this experiment, the teachers neither used the SketchMap server in class nor strongly suggested that the children use it in their schools or homes, which reduced the number of children involved with the annotation. In future experiments, a curriculum that utilizes the SketchMap server will be designed to make children actively participate in the annotation and relate their classroom activities to their outdoor activities.

(2) Role of teachers

The role of teachers was very important for discussions about the SketchMap server. Teachers provided suggestions or clues to activate the children's discussions and often moderated them by deleting inappropriate comments, such as defamation of other children. Every time a comment was added to the map, an email notifying a teacher that children had annotated the map on the SketchMap server was automatically sent. Although some teachers told us that such notification emails were useful for easily and promptly checking the children's comments, others said that they were sometimes very distracting.

(3) User interface of the SketchMap server

Before the practical experiments, the school teachers and the authors thought that children might not be able to understand a map created by other children (Figure 5(a)) and therefore prepared a photomap (Figure 5(b)) that is more accurate than the hand-drawn map. However, the children did not use the photomap frequently, apparently because they had certain levels of knowledge about the vicinity of their school, and photographs and icons attached by children to the hand drawn map helped the other children to identify their physical locations.

(4) Parents' participation

In this experiment, contrary to the teachers' and authors' expectations, the children's parents did not participate in annotating maps on the server. The reason for this is still to be investigated thoroughly, but the following two issues can be suggested: (a) parents were not well informed about the experiment or the server's URL, as they only knew about the experiment from their children, and (b) parents hesitated to add comments because they did not want to make their comments visible to the other children and parents. A method for effectively promoting parents' participation must be explored.

(5) Reflection on outdoor activities

Using the replay module reminded the children of what they had done using SketchMap, for example, drawing a street, taking a photo, or putting an icon on their tablet PC. Some children found clues for starting discussions or were motivated to revisit a place after their fieldwork through reflecting on their outdoor activities.

## CONCLUSION AND FUTURE RESEARCH DIRECTIONS

This chapter describes a system called SketchMap that integrates outdoor and classroom activities to enhance children's learning experiences. Children use a tablet PC augmented with a GPS receiver and a USB camera to create their maps in outdoor environments. The SketchMap system allows children to share and annotate the maps for collaborative learning and has been used in classes in a Japanese elementary school. The evaluations of SketchMap proved that children's responses were positive and indicated that it could promote children's collaborative knowledge sharing and construction through their outdoor learning experiences. However, as the number of children and parents who participated in the discussions on the SketchMap server was small, more investigations are necessary. Consequently, in rhe future it will be important for us to undertake further educational practices using the SketchMap system, to derive more conclusive results to prove its effectiveness.

## ACKNOWLEDGMENT

This research project has been supported by (1) the Japanese Ministry of Education, Science, Culture, Sports and Technology under a Grant-in-Aid for Scientific Research and (2) the Hewlett Packard Company. The authors thank teachers in Nishihara Elementary School in Kashiwa, Chiba prefecture, for their support and collaboration.

## REFERENCES

Ackermann, E. (1996). Perspective-taking and object construction. In Y. Kafai, & M. Resnick (Ed.), *Construction in practice: Designing, thinking, and learning in a digital world* (pp. 25–35). Marwah, NJ: Lawrence Erlbaum.

Benford, S., Rowland, D., Flintham, M., Drozd, A., Hull, R., Reid, J. et al. (2005). Life on the edge: Supporting collaboration in location-based experiences. In *Proceedings of CHI2005* (pp. 721–730), Portland, OR.

Chan, T., & Sharples, M. (2002). A Concept Mapping Tool for Pocket PC Computers. *In Proceedings of IEEE WMTE2002* (pp. 163–164), Vaxjo, Sweden.

Chang, C., & Sheu. J. (2002). Design and Implementation of Ad Hoc Classroom and eSchoolbag Systems for Ubiquitous Learning. In *Proceedings of IEEE WMTE2002* (pp. 8–14), Vaxjo, Sweden.

Chen, Y., Kao, Y., Yu, G., & Sheu, J. (2004). A Mobile Butterfly-Watching Learning System for Supporting Independent Learning. In *Proceedings of IEEE WMTE2004* (pp. 11–18), Taoyuan, Taiwan.

Colella, V. (2000). Participatory simulations: Building collaborative understanding through immersive dynamic modeling. *Journal of the Learning Sciences, 9*(4) 471–500.

Cuartielles, D., Malmborg, L., & Schlaucher, P. (2003). Components for a Ubiquitous Learning Environment. In *Proceedings of HCII2003*, Crete, Greece.

CyberTracker (2000). *CyberTracker.* Retrieved June 2007, from http://www.cybertracker.co.za/

Deguchi, A, Yamaguchi, E., Inagaki, S., Sugimoto, M., & Kusunoki, F. (2006). CarettaKids: A System for Supporting Children's Face-to-Face Collaborative Learning by Integrating Personal and Shared Spaces. In *Proceedings of IDC2006* (pp. 45–48), Tampere, Finland.

Dias, E., Beinat, E., Rhin, C., & Scholten, H. (2004). Location aware ICT in addressing protected areas' goals. In *Research on computing Science 11, Special Edition on e-Environment*, 273–289.

Dufresne, R., Geracce, W., Leonard, W., Mestre, J., & Wenk, L. (1996). Classtalk: A classroom communications system for active learning. *Journal of Computing in Higher Education, 7*, 3–47.

Enjoji, H., Ravasio, P., & Sugimoto, M. (2006). A Collaboration Support Tool for Cartography Learning in the Open Air. In *Proceedings of CollabTech2006* (pp.90–95), Tsukuba, Japan.

Ishizuka, T., Horita, T., Takada, K., Ishihara, K., Ogawa, M., Moriya, K. et al. (2004). Development and Practical Study on Mobile Learning Environment using Personal Digital Assistants for Elementary School Children. In *Proceedings of ICCE 2004* (pp. 969–981), Melbourne, Australia.

Japanese Ministry of Education, Culture, Sports, Science and Technology (2006a). Retrieved September 2006, from http://www.mext.go.jp/b_menu/houdou/17/12/05120900.htm (in Japanese)

Japanese Ministry of Education, Culture, Sports, Science and Technology (2006b). Retrieved September 2006, from http://www.mext.go.jp/b_menu/houdou/17/12/05120900/007/{005,020,021,024}(in Japanese)

Jones, A., Blake, C., Davies, C., & Scanlon, E. (2004). Digital maps for learning: review and prospects. *Computers and Education, 43*, 91–107.

Ogata, H., & Yano, Y. (2004). Context-Aware Support for Computer-Supported Ubiquitous Learning. In *Proceedings of IEEE WMTE2004* (pp. 27–34). Taoyuan, Taiwan.

Okada, M., Yamada, A, Tarumi, H., Yoshida, M. & Moriya, K (2003). DigitalEE II: RV-Augmented Interface Design for Networked Collaborative Environmental Learning. In *Designing for Change in Networked Learning Environments* (pp. 265–274). Kluwer Academic Publishers.

Parr, C. S., Jones, T., & Songer, N. B. (2002). CyberTracker in BioKIDS: Customization of a PDA-Based Scientific Data Collection Application for Inquiry Learning. In *Proceedings of ICLS2002* (pp. 574–575), Seattle WA.

Ravasio, P., Tscherter, V., & Sugimoto, M. (2006). SketchMap: A Collaborative Tool to Learn Basic Cartographical Concepts. In *Proceedings of COOP'06* (pp.115–122), Carry-le-Rouet, Province, France.

Rogers, Y., Price, S., Randell, C., Fraser, D. S., Weal, M., & Fitzpatrick, G. (2004). Ubi-learning Integrates Indoor and Outdoor Experiences. *Communications of the ACM, 48*(1), 55–59.

Soloway, E., Norris, C., Blumenfeld, P., Fishman, B., Krajcik, J. & Marx, R. (2001). Handheld devices are ready-at-hand. *Communications of the ACM, 44*(6), 15-20.

Squire, K., & Klopfer, E. (2007). Augmented reality simulations on handheld computers. *The Journal of the Learning Sciences, 16*(3), 371–413.

Sugimoto, M. (2005). A support for learning by enhancing interactions with physical worlds. *Journal of Japanese Society for Information and Systems in Education, 22*(4), 161–168 (in Japanese).

Sugimoto, M., Ravasio, P., & Enjoji, H. (2006). SketchMap: A System for Supporting Outdoor Collaborative Learning by Enhancing and Sharing Learners' Experiences, In *Proceedings of ICCE Workshop on Design and Experiments of Mobile and Ubiquitous Learning Environments* (pp. 1– 8), Beijing, China.

Takenaka, M., Inagaki, S., Kuroda, H., Deguchi, A., & Ohkubo, M. (2006). Fieldwork Support System Using Mobile Phones: Evaluations of Information Sharing in the Second Grade's Life Environment Study. In *Proceedings of ED-MEDIA2006* (pp. 1325–1331), Chesapeake, VA.

Yatani, K., Sugimoto, M., & Kusonoki, F. (2004). Musex: A System for Supporting Children's Collaborative Learning in a Museum with PDAs. In *Proceedings of IEEE WMTE2004* (pp. 109–113), Taoyuan, Taiwan.

Wilensky, U., & Stroup, W. (1999). Learning Through Participatory Simulations: Network-based Design for Systems Learning in Classrooms. In *Proceedings of CSCL'99* (pp. 667–676), Stanford, CA.

# Chapter XII
# An Architecture for a Personalized Mobile Environment to Facilitate Contextual Lifelong Learning

**Dionisios N. Dimakopoulos**
*London Knowledge Lab, UK*

**George D. Magoulas**
*Birkbeck College, University of London, UK*

## ABSTRACT

*In response to the ever-increasing need of private and public organizations for lifelong learning, research in technology-enhanced learning has been geared towards the provision of systems that will better support those who pursue education at a number of different stages in their life, move between institutions while pursuing a course of study, or make use of resources and facilities from more than one institution/organization. The new generation of mobile phones can support applications that would facilitate contextual learning through life. This chapter presents our approach to designing a mobile application for contextual lifelong learning and the prototype created. The application assists learners to access, compose and manage their learning in a range of institutional, informal and work-based settings by keeping them connected with content that is relevant to their studies, and its use is demonstrated in three lifelong learning scenarios.*

## INTRODUCTION

Lifelong learners increasingly look for more intelligent services and support in order to better meet their learning goals and needs (Baajour, Magoulas, & Poulovassilis, 2007). As Sharples (2000) points out, in lifelong learning it is important to make learning contextual, that is:

- Facilitate learning at different times and places, when a person tries to solve a problem, share an idea, gain an understanding or reflect on the current situation; and
- Allow people to continually enhance their knowledge to address immediate problems and to participate in the process of professional development.

In everyday life, a lifelong learner tries to solve problems, share ideas, or understand situations. In order to accomplish their goals, learners need information to clarify the current situation, or a way to communicate with other members of their community in order to share ideas. Furthermore, they might want to enhance their knowledge by attending courses, which follow specific schedules that have to match with their busy lives. This generates new requirements for delivering learning content, accessing information, learning on demand, and for support at the workplace, in the classroom or in the current situation.

Mobile learning appears to be a promising approach to enable learning in context and analysis of real world problems with the use of mobile devices and applications to better support the learning process. In this vein, Miertschin and Willis (2004) used tablet PCs in a freshman introductory computing course. Although a tablet PC might be a good choice regarding the goals of the project presented by Miertschin and Willis, its size and dependability on wireless networks, mainly on campus, does not fulfill the requirements for a type of learning that is not restricted to specific places. On the same wavelength, Valiquette, Williams, Kutner, and Seeger (2005) proposed a mobile application that requires a laptop PC as the mobile device and is also restricted to the campus. These approaches serve their purpose well, but do not accomplish the features of contextual lifelong learning, which according to Sharples (2000) requires applications to be highly portable, personalized, unobtrusive, available anywhere, persistent, useful and easy to use.

These requirements make us revisit mobile phones as the medium for lifelong learning education. While it was not very long ago that mobile phones were used just to make calls, the advent of the 3G standards and technology has provided additional features which may transform mobile phones into communication facilitators suitable for contextual lifelong learning.

In particular, with mobile phones being owned by most of the students (Lindquist, Denning, Kelly, et al., 2007), the creation of mobile applications that present and manage the knowledge needed by lifelong learners seems apparent. A mobile phone by definition is highly portable, personal, available anywhere and unobtrusive. A thick mobile application can be created that adapts to the user needs, is persistent, useful and easy to use. Lastly, mobile phones also support context awareness, a highly desirable feature for the learning process (Lonsdale, Baber, & Sharples, 2004).

This chapter presents our approach to the design of a personalized mobile learning environment for lifelong learning and our prototype, demonstrating its use through usage scenarios. The environment interacts with a personal learning space, giving back information to the user, either specific to the situation she is in, or relative to the courses she is attending (Dimakopoulos & Magoulas, 2006). It comprises server side middleware that performs the intensive calculations that could not be executed on a mobile phone, and connects the information in the database with the mobile client that runs a mobile application. Building on our previous work (Dimakopoulos & Magoulas, 2006), here we extensively discuss design considerations for mobile learning systems, proposing solutions to the usability issues of small mobile phone screens and we demonstrate a design that adopts hierarchical menus and lists of items to map the tasks and activities that a lifelong learner can do. The tasks and activities can be formed as categories, subcategories and items in the personal learning space of a student. The issue of the limitations on processing power, memory and storage has been addressed by the Java Micro Edition (Java ME) that can be used to develop a thick mobile client. Lastly, for dealing with bandwidth and connection performance, an approach is adopted that is based on the idea of employing a connection adaptation mechanism as proposed in the literature (Alvarez-Cavazos, Garcia-Sanchez, Garza-Sanchez, et al., 2005).

The chapter is organized as follows. The next section presents previous work in the area of mobile applications for learning. After giving an overview of our approach and discussing the design considerations that led to the development of our prototype, we present the architecture of the system. Then we give a user's view of the prototype environment by demonstrating its use through lifelong learning scenarios. The chapter ends with future research directions.

## BACKGROUND

The literature (Kukulska-Hulme, 2002; Waycott, Scanlon, & Jones, 2002) suggests that mobile learning should be seen as an extension of traditional learning and not a replacement. This is because not all kinds of activities and content are appropriate for mobile devices (Keegan, 2001). According to Trifonova (2003), a good approach for using mobile devices as learning tools is to create learning modules that are short, no more than 10 minutes long, to fill the gaps of the highly fragmented attention of the user while on the move. There are, however, several technical and user-related issues that need to be addressed before mobile learning applications reach their full potential.

From a technical point of view, there are issues that concern the architectural aspects of mobile learning systems that need to be addressed. A potentially useful approach has been proposed by Alvarez-Cavazos et al. (2005) for accessing digital libraries on the move. Their architecture includes a mobile thick client written in Java ME, which is designed to cope with the limitations of the mobile environment. It communicates with a middleware application on the server using the XML-RPC protocol. The mobile client offers a subset of the functionality that a desktop client offers. This mobile client performs functions to deal with the limitations of the mobile environment and includes:

- A local storage mechanism, which stores documents in the device for later offline usage;
- A connection adaptation mechanism that provides a constant response time regardless of the connection variability, and
- User interaction support that allows the user to manage his or her personal digital library.

The mobile client reads the results received from the middleware and displays them on the screen. The mobile connection middleware stands between the database that holds the digital library and the mobile devices. This is because accessing the database is a resource-intensive procedure that mobile clients might not be able to deal with. The accessing of the databases is done by the middleware and the results are cached in order to be served to the users the next time someone asks for them. Finally, the middleware performs adaptation of the content so that it fits the display of the devices. The architecture of our application, which is presented later in this chapter, adopts several of the ideas of Alvarez-Cavazos et al. (2005), such as the local storage mechanism, the connection adaptation mechanism, the user interaction support and the accessing and adaptation of the content with the use of middleware.

Another architectural approach, called eBag, has been proposed by Brodersen, Christensen, Grnbk, et al. (2005) for nomadic learning. Their approach targets the school domain and it is designed to support pupils working in a project-oriented and nomadic learning way, inside or outside the school buildings. It includes a discovery mechanism that is based on Bluetooth proximity sensors, which can be used to obtain contextual information and to inform about the existence of a device within a specific range. In our architecture we adopt the idea of incorporating a discovery mechanism, as described in that work, in order to inform remote servers about the existence of a mobile device.

In an attempt to better support context learning, Berri, Benlamri, and Atif (2006) have proposed an architecture that employs ontologies, which are rule-based and are driven by a learner profile, and a search agent that searches a set of distributed learning objects repositories for lightweight learning objects. This approach takes into account the operational environment and the special requirements of mobile learners. Compared to previous approaches, this architecture puts more emphasis on the user needs through the use of a user profile as a set of rules on an ontology that helps to contextualize learning content. In this way, the extracted conceptual knowledge from the source ontology is tailored specifically to the needs of the learner. Along the same lines is also the work of Lonsdale et al. (2004) that uses context information to provide filtered content appropriate to users' goals, settings, and resources. Their resources include services and various options, such as collaboration tools and interaction preferences, on top of the actual learning content. This is done using a subsystem that selects content reflecting the requirements of the user. Thus, this approach again combines context awareness with user related information. A user profile is included in our architecture too. It holds information about the user, and is used to tailor the content delivered to the learner.

The importance of understanding the learner's needs in mobile learning application development has of course been recognized long before this work. For example, Sharples (2000) was among the first to investigate the general requirements for personal tools to support contextual lifelong learning. He identified a set of requirements, such as portability, personalization, unobtrusiveness, anytime/anyplace availability, persistence, usefulness, and ease of use. He also described a mobile system for elementary school children. The system has been built as a Java mobile application that runs on a tablet PC using a metaphor (a rabbit avatar) to communicate the different actions to users. The system comprises a personal learning resource and several tools and activities that create new resources. These tools are drawing, writing, and communicating, while the actions connected with the avatar are capturing images, browsing Web pages, and creating and annotating images, notes and drawings. However, during user testing, the children couldn't handle the size of the tablet PC as it was too big for them and therefore it was not serving the unobtrusive requirement set by Sharples.

Lastly, with the users in mind, Holzinger, Nischelwitzer, and Meisenberger (2005) have presented a multimedia-based application for mobiles, the so-called mobile learning engine. Their application is constructed so that it is intuitive and users can use it without online help. It can serve different didactical and pedagogical strategies, such as problem-solving, exploratory learning and situated learning.

In the following sections, we describe the approach proposed in this work. The next section presents our design approach that addresses technical and user related issues, such as the small screen problem, mobility, location awareness, and discovery. Furthermore, our approach includes user profiling to personalize the content, and the use of collaboration and communication tools. In contrast to many existing mobile learning applications, which provide learners with learning content and resources that relate to a particular stage of their lifelong learning, our approach aims to create an extensible personal information space for lifelong learners that will allow them to manage their personal information beyond their desktop or a single institution, facilitating access, composition and management of learning under varying circumstances and settings, such as cross-institutional, informal and work-based situations.

## DESIGN CHOICES FOR A PERSONALIZED MOBILE LEARNING ENVIRONMENT

Usability considerations play a vital role in promoting mobile applications. One of the big challenges in this area is how to fit the information onto the small screens of mobile phones (Ziefle, Schroeder, Strenk, & Michel, 2007). A variety of approaches have already been proposed for browsing Web sites using mobiles (Baluja, 2006; Blekas, Garofalakis, & Stefanis, 2006) but there are no guidelines for designing a user interface (UI) that would present all the information in one screen. Instead, application specific approaches have so far been proposed when it comes to screen space real estate (Holtzblatt, 2005; Newcomb, Pashley, & Stasko, 2003).

In the context of mobile learning, Berri et al. (2006) explain that it is not applicable to simply port existing learning resources to mobiles from desktop based learning. Also, Parsons and Ryu (2006) point out that we should not simply apply known design requirements from e-learning into the mobile learning context, because of the special limitations and benefits of the mobile phones. In particular, Parsons and Ryu (2006) look at the design of mobile learning applications from four perspectives: the mobile environment, the learning context, the learning experiences, and the learning objectives. Regarding mobile environment issues, they argue that features of the mobile phones like mobility and limitations like small screen, poor input methods, and limited battery life dictate the creation of a user interface that should meet users' needs without overloading them or operating too slowly. They propose the use of rich media objects that do not distract the attention of the learners from their main activities. In order not to distract the learners, the objects have to be in short pieces and in several media types. They mention that a widely noted feature of mobile phones is communication support, and they argue that it can contribute to alternative models of mobile learning. Regarding the learning context, they argue that mobile learning is quite different from traditional learning, and they propose design requirements by reviewing the features of m-learning, based on Wang's six dimensions of the mobile learning context. Particularly relevant to our approach are the identities of the learners, which can be valuable for the development of personalized learning experiences, the styles of learners related to their learning experience, the collaboration that can take place between a group of learners, and the spatio-temporal dimension that refers to time and location awareness. Regarding the learning experiences and objectives, they refer to Preece's claims that application designers should consider not only usability goals for their products, but also user experience goals to assess how enjoyable, satisfying and motivating their product is. They propose two metaphors, the cinematic and the game metaphor. The cinematic metaphor has content that comprises of narrative or story elements that the learners can easily follow, whilst the game metaphor helps to engage learners by leveraging the popularity of games.

Drawing on the preceding discussions, Table 1 lists the design considerations used for the creation of our prototype personal learning space, discussed in the next section, whilst the rest of this section provides a more detailed account of the main choices.

### Overcoming the Small Screen Problem

The small screen of mobile phones is typically cited as being restrictive for the creation of a usable interface. Beyond handling and visibility problems, the restricted screen space allows only

*Table 1. Design guidelines*

| Design consideration | Reason | Example |
|---|---|---|
| Tasks and activities modeled as a directory (Krikos et al., 2005) | Imposes hierarchy and relevance | Tasks and activities represented as categories and items |
| Hierarchical menus (Luchini et al., 2004) | Task-based workspaces | Each main task is a category |
| Reordering of items in the menu (Smyth and Cotter, 2002) | Easier access to most used tasks | Most used tasks appear first |
| Use of rich media objects | Overcome limitations of the screen | Audio or video content |
| Audio as content in taxonomy (Ractham and Zhang, 2006) | More information provided | Audio lectures as directory items |
| Video as content in taxonomy (Liu and Choudary, 2004) | More information provided | Video lectures as directory items |
| Use of RSS to summarize content (Blekas et al., 2006) | Facilitate categorization | Easier creation of directory items |
| Small modules of information | Fit the limited amounts of time within the fragmented user's time | Small slices of theory, examples, exercises etc |
| Possibility to post or save quiz answers | Use of fragmented time to reflect on course material | Submission of quiz answers using text or voice |
| Grouping of search results (Kules et al., 2006) | Categorization of results into meaningful directory categories | List of percentages that categories participate in the results |
| Organizers and timetables | Remind user for important deadlines | Calendar and reminder services |
| Ability to communicate with other users | Reinforcement and feedback | Discussion board |
| User profile visibility | Knowledge of the personal information held and fine tuning | UI that allows the user to change the user profile easily |
| Location awareness (Aalto et al., 2004) | Discovery of adjacent personalized services | Bluetooth sensor that informs remote servers about its presence |
| User profile sharing (Chatfield et al., 2005) | Personalization by systems that don't have knowledge of the user. | Transmission of the user profile to a remote server for personalization |
| Saving user profile on the phone (Brar and Kay, 2005) | Portability | Access to the user profile anywhere |
| Connection with the base terminal | Synchronization of the user profile | Send and receive the user profile to/ from the base system |
| Social adaptive navigation support (Brusilovsky et al., 2004) | Indication of useful material | Number of students that accessed each category |

a little information to be displayed at a time. Because of this, memory load is increased, (Ziefle et al., 2007), particularly for older users whose cognitive abilities (verbal memory and spatial ability) are less pronounced. Users with high verbal memory capacity might have a better orientation within technical menus. Nevertheless, orientation in menus might still be complicated for some users, particularly those who cannot experience how the menu can be spatially structured (Ziefle et al., 2007).

*Figure 1. A hierarchical taxonomy*

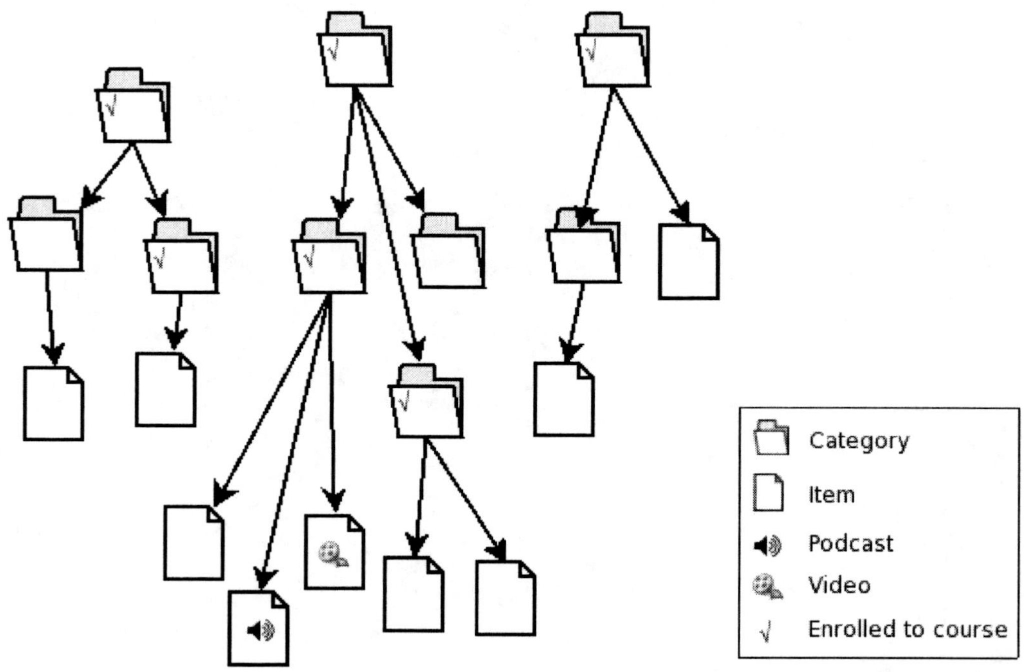


Our approach to solve the problem of designing the UI was based on decomposing the activities the user is called on to accomplish when using the personal information space. In lifelong learning, learners need to be able to access information relevant to tasks and activities that they need to accomplish, when they are not at their "base," this being their home, office or university lab desktop computer. Such activities are information about the courses that they are currently taking, information about other learning opportunities offered, and access to digital resources in order to find information relevant to the task they are performing and communicating with peers and tutors.

The activities that are provided to learners in our approach are modeled as root categories of a taxonomy. The taxonomy includes all the information that is relevant to their courses and status as lifelong learners. Figure 1 shows a snapshot of the hierarchical taxonomy used in our approach. In our implementation, the root categories are activities such as viewing learner's current courses, searching a digital library, participating in a discussion board, and so forth. Taxonomies are used in the Open Directory Project, and other online information directories. In our approach, the application relies on a main taxonomy that includes all the information, and is common for all learners (see Figure 1). Different users make different uses of the taxonomy, visiting only the parts they are interested in, creating a "personally ordered, edited version" of the taxonomy, (Lave, 1988), that would match their personal preferences (Nardi, 1996). Figure 1 shows how the learner is initially registered to be able to see specific parts of the category, a step done by her institution when she enrolls.

Each category of the taxonomy in our system is represented as a menu item. By selecting the menu item, a list of items and subcategories contained in that category are shown. The idea of using menu-hierarchies to let a user navigate through the different menu items that represent the functionalities of a mobile phone has been investigated frequently in the literature (Buchanan, Jones, Thimbleby, et al., 2001; Ziefle et al., 2007). Since the users are already familiar with the use of hierarchies (Ziefle et al., 2007), the UI design of our application follows the same standard but in our case the menu items represent tasks and activities the user needs to have access to when she is mobile. This approach provides a feasible way to guide the user in navigation (Brusilovsky & Rizzo, 2002). Ziefle et al. (2007) also report that older users, who are an important group in lifelong learning programs of study, become more disoriented when using hypertext structures interconnected by links than when using hierarchical menus. In addition, our proposed visualization of each category/task follows the design guideline proposed by Luchini et al. (2004) that suggests task-based workspaces. Moreover, Smyth and Cotter (2002) note that the reordering of the items in the menu, according to the user preferences, helps the users find what they want more readily. Hence, our application reorders the items of the menu according to the way the user likes to see them, or depending on the usage of each item, placing frequently used items at the top.

The items in our prototype personal learning space can represent text or multimedia, e.g. podcasts or instructional videos. For example, an item can link to a digital repository, facilitating the searching for information in books and journals for students on the move and on the field of study, lecture notes and manuscripts in text, audio or video formats. Lindquist et al. (2007) proposed a system where the lecturer can post online slides that are viewable by almost any computing device, in any order, at any time. Such an approach is beneficial for a lifelong learning application, because the user will be able to review specific slides in short amounts of time.

Thus, in a real life scenario, the items of the categories that are related with teaching material, such as text or video, are expected to reside on the tutors' Web sites. Such a Web site could be a mix of Web pages, where each Web page could be for one lecture, or a group of lectures. Such an arrangement of Web pages would not fit well in the personal learning space, and would be difficult to browse and manipulate using a mobile device. One way to alleviate this situation would be to use RSS feeds in order to get essential information from a Web site as Blekas et al. (2006) suggest. An RSS feed delivers a set of items that are chunks of information, with a title, a description and a link. These chunks are used to describe items in the personal learning space instead of the actual Web page that contains this information and in our approach can be used to populate menu items. This might make browsing and searching of this information easier for mobile users.

With regards to offering multimedia content, which has been proposed in the literature as a means to overcome limitations due to the lack of output screens, Lee and Benbasat (2003) propose to convert some parts of the content into audio format, so that output space can be saved on the mobile screen. Converting text to audio will minimize the use of screen for reading that is tiring due to the continuous scrolling needed for long texts. Liu and Choudary (2004) created a system that broadcasts lecture videos to mobile devices by using a modified encoding of the video, optimized for mobile devices. Ractham and Zhang (2006) propose the use of podcasting to deliver information to the students. With the use of podcasting the students would be able to download and listen to the lectures, or other information uploaded by the tutors, on the move. Recent generations of mobiles frequently feature an mp3 player, which could be used to playback the podcasts. Of course, there are two related drawbacks of multimedia over text. One is high cost and the other is slow download time. Therefore, the podcasts should be small in size, and contain only the information that is needed. Since the audio would be solely speech, a small bit rate and frequency when encoding the podcast would help. The problem faced though is the extraction of only the information needed by an audio file. The user should be able to download and listen to only the part of the podcast that would be relevant to what she is searching. This means that large podcasts, like a whole lecture, should be sliced into small parts that are meaningful and tags could be used to make the information available for searching (Chang, Zhong, & Kumar, 2001).

## Facilitating Navigation in the Information Space

Providing users with a large taxonomy and letting them navigate to the parts they are interested in would be a waste of time and bandwidth for them, as they would always need to navigate only to particular parts of the information. There is a need for providing views of the taxonomy that are adapted to users' needs, allowing users to navigate a personalized subset of the main taxonomy. Initially, the personal space is set up by the institution or the organization providing a service that matches the profile of the student, for example, depending on the courses she enrolled in, but gradually the user is able to customize the content and enrich the information that is linked to her personal information space. Searching for information within the personal space is assisted by a mechanism of automatic categorization of the search results into categories of the personal learning space, as proposed by (Kules et al., 2006). This kind of categorization creates a directed situated navigation, which helps the users understand the context of the results they obtain (Tabard, Mackay, Roussel, & Letondal, 2007).

The personalization of the system requires the creation of a user profile. The user profile is a collection of attributes that will depict the preferences of the user regarding various aspects of the application and also their usage of it. Beale and Lonsdale (2004) have noted that context should not be considered as static but as a dynamic process, which changes through users' interactions with the learning materials and the surrounding world over time. The context features should be used either as excluders or as filters. Therefore, the system needs a mechanism to fine tune the user profile according to the interaction of the user with the mobile phone, in order to have an up to date view of the user stored in the user profile. The user should be able to review and change the user profile, so that it matches her way of working and thinking. The user should also be able to let the system modify the user profile automatically, according to the observed use of the mobile application, and, if needed, to tweak the values to correctly depict her choices. Consequently, the system has a screen that the user can review the current user profile and change it at will.

## Facilitating Communication and Information Sharing

Communication between the learners for sharing practices and reflections on course concepts could provide additional motivation and help learner's engagement with mobile learning (Davis, 2002). In this direction, a discussion board is used in our application to enable users, mobile or otherwise, to communicate and share their experiences and questions.

Along with the above, in our system a category/item can be annotated with information about how many times other members of the community accessed it. This can indicate what parts of the overall taxonomy other people from their class or group found interesting. This is in line with previous work by Brusilovsky and Rizzo (2004) who proposed a Web-based system that creates a map of resources and indicates the times other members of a group have used a specific resource by the saturation of the color of the cell.

## Supporting Mobility, Location Awareness and Discovery

Of course, mobile users access content from a variety of locations, and so the situations they are involved in may change (Weißenberg, Voisard, & Gartmann, 2004). For example, one case could be that the user enters a public library to search for information about a specific topic. We assume that the topic is relevant to the program of study or to an assignment that the learner has to write; therefore her user model will contain keywords that describe the topic. It would be of a great help for a user in such a situation if she can receive a personalized view of the library catalogue that would present only results that match her profile, instead of searching the whole catalogue. For this to happen, the library system should somehow get hold of certain attributes stored in the user profile. This of course imposes many privacy issues. First of all, the user might not want to share personal information, or share only some part of the information. Previous work has found that users are willing to share personal information if they perceive that they will benefit from the personalization that will be performed, depending on whether they trust the system regarding the way their personal information will be used (Chatfield et al., 2005). This approach was also used in a system proposed by Brar and Kay (2005), where the user profile was saved on the user's mobile device, and could be released in exchange for personalized services. Users can also create partial profiles, called personas, allowing only some of the elements of the personal information to be released, depending on the level of privacy that the user wants. To this end, Brar and Kay (2005) adapted the W3C specification for P3P (W3C, 2006b) in order to inform the users regarding the use of their personal information from the system, and APPEL (W3C, 2006a) for the users to specify their privacy preferences.

Our approach is centered around systems that use Bluetooth, which is common nowadays on mobiles, to get information about users' location and to receive offerings for personalization from other systems (Aalto et al., 2004). Ackerman and Cranor (1999) found that users prefer to be informed when an exchange of personal data will take place, and confirm it. Therefore, when the user is inside the range of a Bluetooth network in a place like a public library, which could offer personalized content, she is informed about the services the library system provides, the way her details will be used and for how long they will be stored, as described by the P3P policy. Then, she has the option to either send some of her personal data to the system, or decline the offer, and use the system without personalization. In order for the systems, the mobile device and the service provider to exchange information, they should be able to communicate using a standard way of exchanging information, such as using P3P and APPEL, or some other XML based protocol.

Our prototype makes use of the above ideas, which have been summarized in the design guidelines of Table 1, and will be described in the next sections.

# THE ARCHITECTURE OF THE MOBILE LEARNING ENVIRONMENT

The system architecture that appears in Figure 2 shows the connections between the main components of the system:

- The mobile client
- The Web service that acts as middleware between the mobile client and the database
- The database

The connection between the mobile client and the Web service is done using the Internet and specifically the HTTP protocol. This approach was followed in order to make the mobile application lightweight, taking into consideration the limitations of the mobile phones. All the logic and

*Figure 2. High level architectural description of the system*

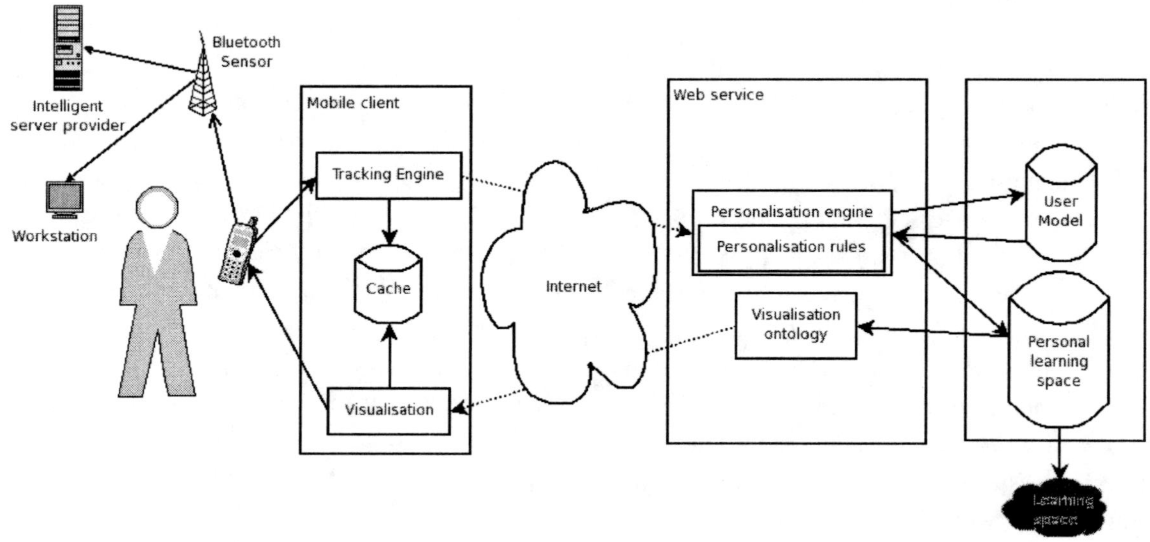


expensive calculations take place on the server, and the results are provided via Web services, which are presented by the mobile application on the device.

The mobile client has two main functions. First, it tracks user actions, passing them to the Web service, where calculations take place on the server. It then reads the result of the calculations and displays the data taken from the Web service. A cache of the actions tracked and the corresponding results from the Web service are held (cached) on the mobile, so that in the event of loss of connectivity the system could remain usable, at least for the actions already performed by the user. This is done by having a small database in the mobile client. The second function of the client is to act as device that can be discovered by remote sensor systems, where personalized services are offered.

The Web service component collects the data from the mobile client and decides on what the user sees next. It uses a user profile to track the user preferences. This user profile can be changed either directly by the user, through a screen on the mobile device, or be adapted by the system, through the personalization engine, according to the use of the mobile application by the user.

The data that are passed from the mobile phone to the Web service are a list of events, which show how the user has interacted with the application, and also the request made by the user, regarding the part of the personal learning space she wants to see. The data passed from the Web service to the mobile application include the content the user requested, the way the content should be shown, the way they should be visualized in the application, and finally links to other content that might be available and is linked by the user to the content to be viewed. Table 2 shows the attributes of the data passed from the mobile to the Web service and vice versa.

The information space incorporates a hierarchical taxonomy similar to Web directories like the Open Directory Project. The space is available to all students and holds the information from all the courses that are available, and also links to other resources like a digital library or a discussion board. Its contents can be text, audio or video. Each student visits only the parts that are interesting to her, creating a model of the taxonomy that is personally ordered and edited (Lave, 1988). The default personal learning space for each student is formed by the categories that are relevant to the courses she enrolled in, so the personal learning space is a subset of the learning space provided, but the student has the ability to explore the whole taxonomy and expand the personal learning space. Furthermore, the student can expand the personal learning space by creating new categories that do not exist in the initial learning space and by importing into them new or existing categories and items from the learning space. This gives flexibility to the student to expand and order the personal learning space in a way meaningful to her. The student can also create connections between existing categories that do not exist, but are relevant according to the student's view.

The user can select between saving the user profile on a base system, such as a desktop PC at home or at the university, or on a university server that runs the Web service, or on a mobile device and only release a specific amount of personal information for each interaction with the server. If the user profile is saved securely in a personal computer, then the mobile phone could download the user profile by using Bluetooth to build a direct connection. If the user profile is stored on the university server, then it can be downloaded using HTTP.

The system tracks the interactions of the user with the mobile client, using a tracking engine, and feeds the interactions into the tracking engine of the server. These events are passed to the user profile, either directly, if the user selected to have a static user profile, or after being processed intelligently, to decide whether the user profile should be changed or not. The user profile on the server either contains all the information or only the information that the user agreed to. The event captured is then passed to the personalization engine where a set of rules will run to personalize the data the user asked to see. These data will come from the personal learning space of the user, where they will either reside being cached from previous calls, or they will be fetched from the learning space and be personalized on the fly and stored in the personal learning space. These data will be passed to a visualization ontology that will decide on the way the data should be presented to the user. Then the data together with the visualization technique will be passed back to the mobile client to be shown to the user.

Beyond the communication with the server component, the mobile device is able to communicate with remote servers available in a public

*Table 2. Attributes of data transferred*

| Mobile application to web service | Web service to mobile application |
|---|---|
| Timestamp, event | Name of category<br>Subcategory names<br>Links to categories<br>Visualization |
| 1191919191, view 1 | "My courses"<br>Dissertation, Database design 2 (35), Database design 1 (14), HTML Design (37)<br>401, 204, 104, 101<br>Category |
| 1191919200, view 204 | "Database design 2"<br>Relationships (37), Queries (33), ...<br>2041,2042,...<br>Category |

library or a museum. The communication with these servers is done both by the use of Bluetooth technology and HTTP. First, using Bluetooth, the phone is discovered by a remote server and an offering is sent regarding the services that are provided, together with the parts of the user profile that are needed and the URL where the Web services of the server can be located. If the user accepts the offering, then an HTTP connection is created with the remote server to send the portion of the user profile needed, and to use the services provided. The services provided by remote servers appear in the personal learning space as new categories at the top level. The full contents of the services are available while the system is in range of the mobile client, but a cached version of documents that were retrieved can be held on the mobile device as proposed by Cohen, Herscovici, Petruschka, et al. (2002).

## PRESENTATION OF THE ENVIRONMENT

The design of the UI was created by decomposing the activities the learner is able to perform when using the application. These activities are relevant to the learning subjects she is currently enrolled in, and also include services that are important to lifelong learning that can be used on the move. These activities are the root categories of the personal learning space, and appear on the main screen of the application (Figure 3). The categories shown in Figure 3 are the pre-initialized values. These values were created by the institution when the learner enrolled. They map the learning subjects and services relevant to the courses she enrolled in. The learner can browse the courses she has already taken; search digital libraries or post comments and questions on a discussion board, where she can communicate with other students or teachers. The user has access to the whole taxonomy, but with the use of a user profile, she initially gets only the information relevant to the context she is currently in. If desired, she can access other parts of the taxonomy that are not shown in the beginning. There is the option to inspect the user preferences, where the user can either change them manually or fine tune the automatically generated preferences that are created by the system, so that they better reflect the learner. There is also the option to transfer the user model from the mobile to the base system. This can be done by using Bluetooth technology to create a peer to peer network with the base system.

In the following, we present three scenarios in order to demonstrate the use of the system in lifelong learning.

### Scenarios

Martina is a lifelong learner who has enrolled for several courses at her local Lifelong Learning Network (Gaeta, Ritrovato, & Salerno, 2005; LLN, 2007). She enrolled for a diploma that will help her improve her professional skills. The courses she has enrolled in are relevant to her current job. She is currently working full time in a company, and travels 60 minutes every day by bus to and from work. She is keen on using her mobile phone for messaging and games during the traveling, and has enrolled on the mobile learning application helper of the university. She plans to use the application to reflect on or understand concepts presented in recent lectures and also to organize coursework according to the existing deadlines.

Here we will present three scenarios where Martina:

- Talks about an idea she has about some coursework she would like to do as part of her diploma, and tries to find colleagues that would like to join the project,
- Tries to resolve a problem at her work using the reference material of the courses she is attending and

*Figure 3. The main screen of the system*

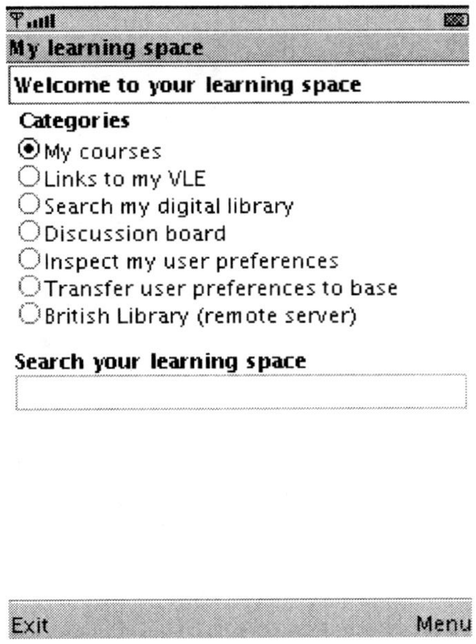


- Tries to understand how to put to use the knowledge she got from the courses to help her with a situation at work.

## Scenario 1

Martina is very keen with databases. She wants to become a database administrator, so she wants to have a good knowledge of the way databases work and how they should be designed. She is working as a data analyst and uses databases constantly, so she thought to do a dissertation on the creation of a database that would hold data similar to what she is using at work, and also to create a Web front end that will make it possible to edit the contents of the database from it. She logs on to the discussion board of the mobile application (Figure 4a), through the option that is available on the main screen of the application (see Figure 3), and she creates a new personal message to send to her tutor (Figure 4b).The tutor replies that the subject is quite big for a dissertation and suggests that she should share the topic with another student. She goes back to the discussion board (Figure 4a) and posts the request publicly (Figure 4c, 4d) so that other students of her diploma can read it and reply accordingly. In parallel, she is doing some research about relevant work that exists in the course notes, by following the favorite categories of her fellow students (Figure 5), and searching her personal space (Figure 6) using the search facility that appears at the bottom of the main screen (Figure 3). Figure 5 shows the favorite subcategories of the learner. Here she can get a quick pick of the most used subcategories within a specific category. In that view she can find out how many of her group members visited each category. This can help her find out which are the popular categories among the students of her class, helping her to create a reading list contributed by the learners. Figure 6 shows the search results screen, where the user is provided with a break down of the results into the categories of the taxonomy. A percentage is shown beside each of the top level categories that appeared in the results, explaining which categories contributed most on the results. This helps the user understand and choose the closest context to the one she is searching for, in case a search term is relevant to two or more subjects being studied.

## Scenario 2

Martina has a problem at her work regarding the update of a database. When traveling back home from work, she uses her mobile to access course materials on database queries that are relevant to the problem she is facing. After accessing the courses category (Figure 7a) and then the course items (Figure 7b), she selects the category with the podcasts (Figure 7c) that are relevant to database updating. The course items of Figure 7c are received from tutors' Web sites using RSS, and the multimedia files are small chunks of information; no more than 10 minutes in length. The user can either download the files to listen them immediately or save them in order to listen

*Figure 4. Different views of the discussion board*

*(a) The discussion board menu, while creating a new personal message*

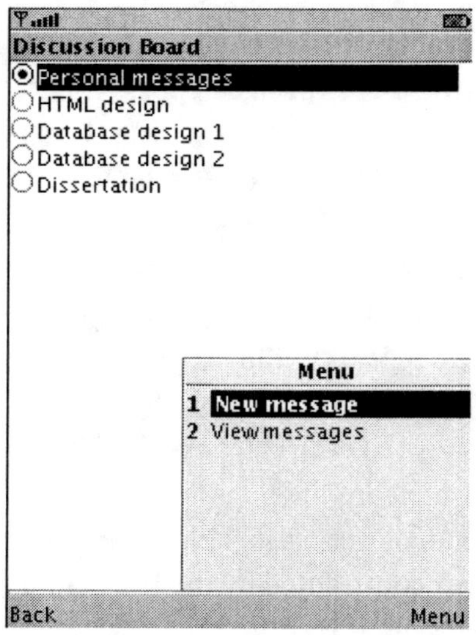


*(b) Writing a new personal message*

*(c) Creating a new public message under the "Dissertation" discussion board*

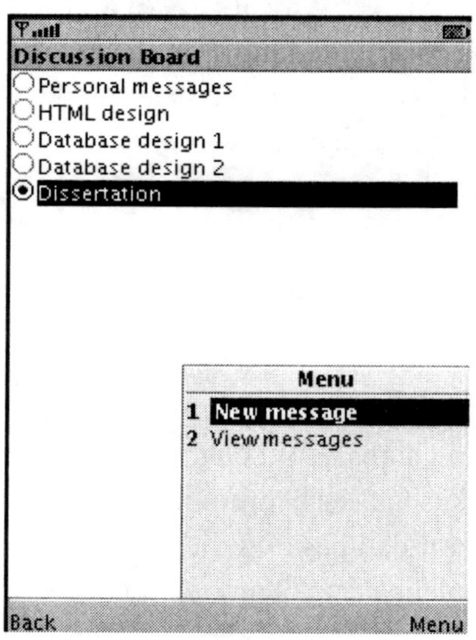


*(d) Writing a new message for the "Dissertation" discussion board.*

*Figure 5. Favorite subcategories and popularity index*

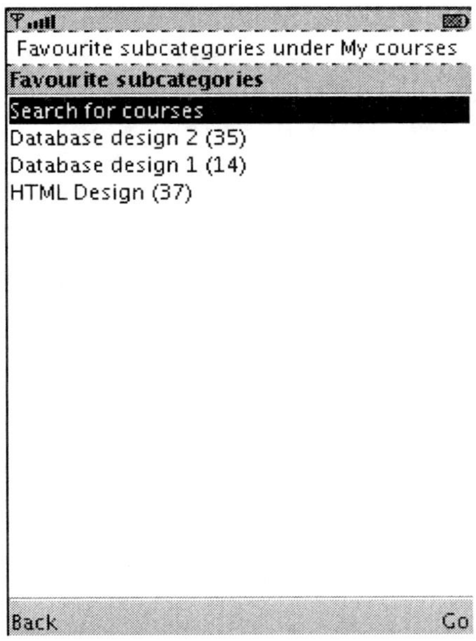


*Figure 6. Search results*

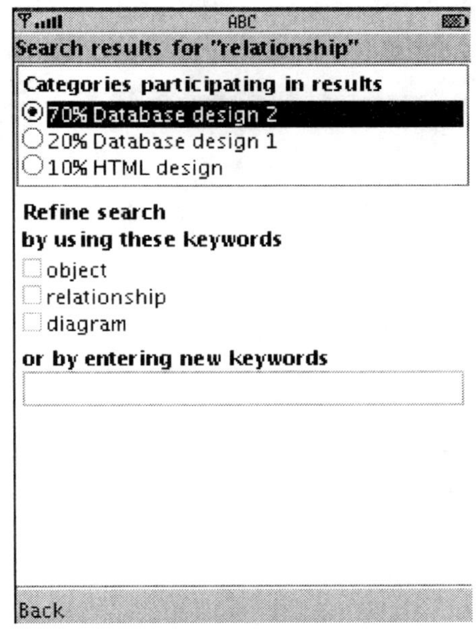


them later, subject to the available memory of her mobile phone.

Martina decides to download the part on "Queries…Update" that seems relevant to the situation at her work (Figure 7d) and listens to it. She understands the concept, but she has some questions regarding the way it works in practice. She follows the videos category (Figure 7c) and chooses to watch a video demonstrating the updating of a database (Figure 7e). The material covers the subject superficially, so she still has a question about updating records. She posts a topic on the discussion board (see Figure 4a), hoping that some of her fellow students have already faced this problem and know how to solve it.

## Scenario 3

Martina is using databases extensively, but she is following the same procedure every time, making her work tiring because of repetition. In the last databases lecture, the tutor demonstrated how to create scripts that automate the procedures done on a database, and she thinks it is a good idea to put that in practice in her everyday job. Traveling to work she logs on to the system and downloads the category of examples to look at some of the examples provided. She first watches a video of the part of the lecture where the tutor explains the procedure (Figure 7c), and then reads a sample script (Figure 8a). Finally, she attempts to answer some multiple choice questions (Figure 8b) to test her knowledge on the new subject. She can submit the answers directly for marking or save them on the mobile phone to finish the test later. She can also export the solved questions to the "base" computer, making it possible for her to complete a part of the test on the move with her mobile phone, and another part at her "base" system.

Lastly, she adds a new category at the main application screen to place the information she

*Figure 7. Screens of the application*

*(a) My courses*

*(b) Course items*

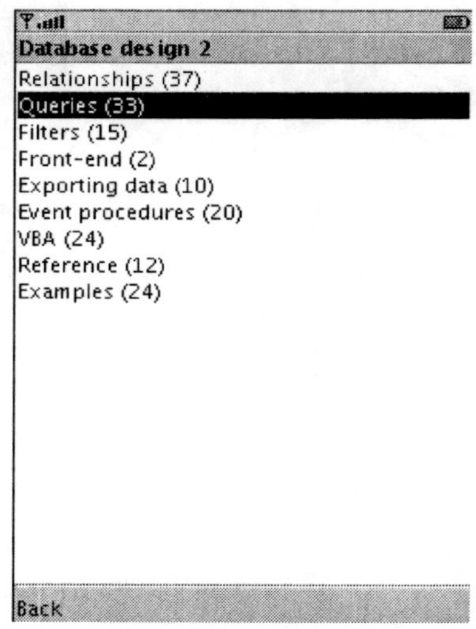


*(c) Item categories*

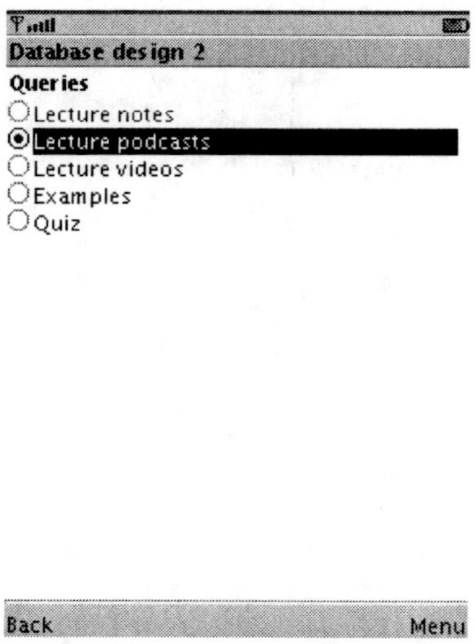


*(d) Podcasts items*

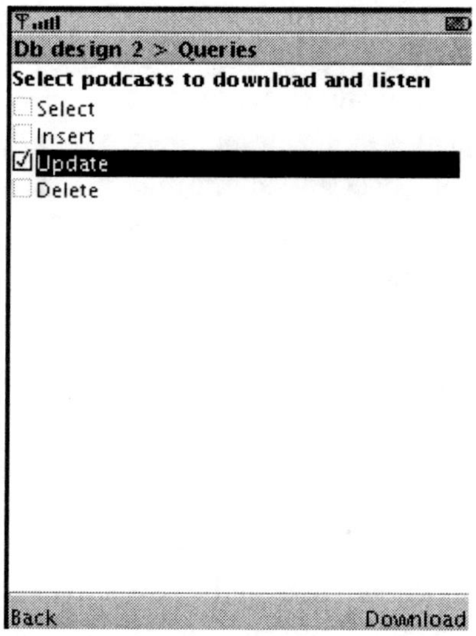


*(e) Video items*

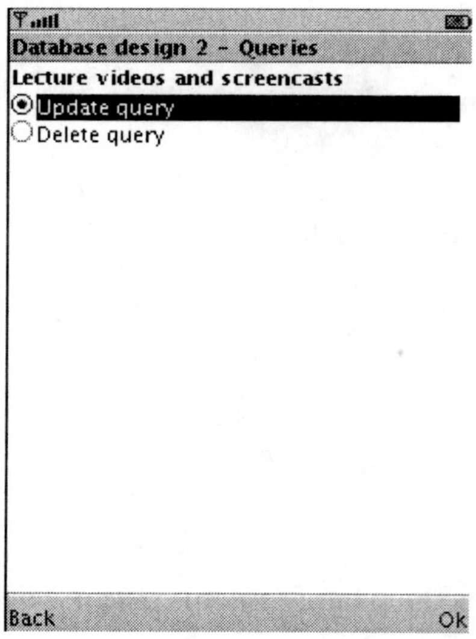

*Figure 8. Screens of the application*

*(a) A list of examples*

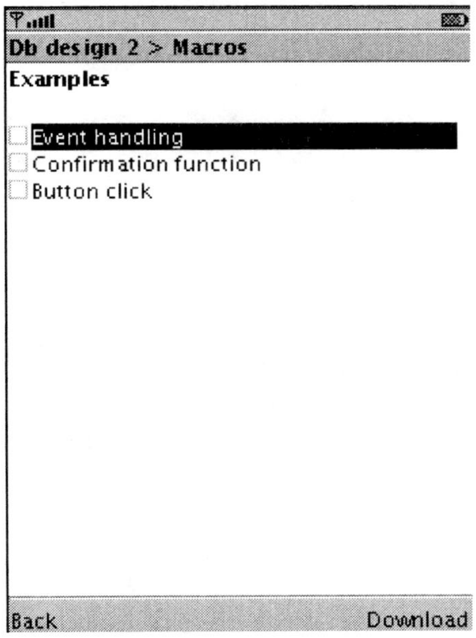


*(b) Multiple choice questions*

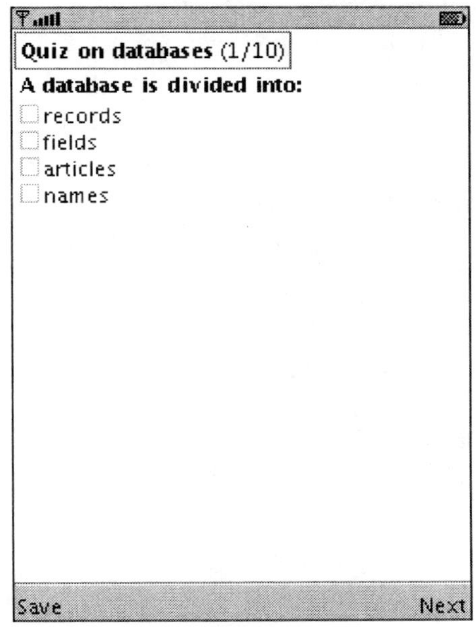


*(c) Add category*

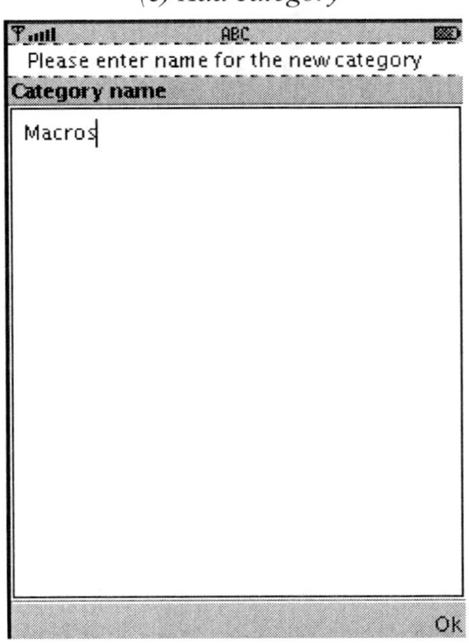

found in one easy to access category (Figure 8c). Generally the user is free to create new categories and assign content from other categories of the taxonomy, or even link content of one category with content of another, creating links that carry semantic meaning for the individual. The system allows user to add a category at any level in order to create meaningful semantic links between categories in a personalized way. These links provide information about the context in which specific categories are viewed, with regard to the categories the links are connected to.

## CONCLUSION AND FUTURE RESEARCH DIRECTIONS

Lifelong learners form a rather diverse student population with a variety of backgrounds, needs and accessibility requirements. Making contextual lifelong learning happen in practice requires learners to become more aware of their own studying and thinking processes, along with tools and guidance that support the planning of learning. To this end, technology should be used effectively to assist lifelong learners to access, compose and manage their learning under varying circumstances and settings, such as institutional, informal and work-based.

We proposed design guidelines and an architecture that can help create useful applications to facilitate lifelong learning activities. Nowadays, various devices exist that have big differences when it comes to their capabilities, which impacts the way mobile learning environment can be created.

The emergence of Lifelong Learning Networks and University Federations, which are envisaged as groups of institutions that come together across a city, area or region to offer new progression opportunities for lifelong learners, creates new drivers for change to accommodate the needs of learners who will move between various groups and institutions. Much of this has been made possible by the new possibilities offered by the Internet and the new approaches in systems design. However, lifelong learners may belong to radically different groups; they come from different backgrounds, and are very much concerned with the accessibility of the various systems within Lifelong Learning Networks or federations, the transferring of their personal information for cross-institutional e-learning, and the flexibility of the curricula offered. Thus, new approaches are needed for managing and exchanging performance or competency data between different systems, assembling or manipulating fragmented user data that were stored in different distributed systems. In the future the new generation of phones will provide more features that will overcome current limitations and provide more opportunities to create mobile learning environments in this context.

## REFERENCES


Aalto, L., Göthlin, N., Korhonen, J., & Ojala, T. (2004). Bluetooth and WAP Push based location-aware mobile advertising system. In *Proc. Second International Conference on Mobile Systems, Applications and Services*, (pp. 49 - 58). Boston, MA.

Ackerman, M. S., & Cranor, L. (1999). Privacy critics: UI components to safeguard users' privacy. In *Proc. Computer Human Interaction Conference (CHI '99) Extended Abstracts on Human Factors in Computing Systems*, (pp 258-259). Pittsburgh, Pennsylvania

Alvarez-Cavazos, F., Garcia-Sanchez, R., Garza-Salazar, D.A., Lavariega, J.C., Gomez-Martinez, L.G., & Sordia, M. (2005, October 17-20). Universal access architecture for digital libraries. In *Proceedings of the 2005 conference of the Centre for Advanced Studies on Collaborative Research*, (pp. 12-28). Toronto, Ontario, Canada

Baajour H., Magoulas G. D., & Poulovassilis A. (2007, June 25-29). Designing services-enabled personalisation for planning of lifelong learning based on individual and group characteristics. *In* P. Brusilovsky, M. Grigoriadou, K. Papanikolaou (eds.) *Proc. of Workshop on Personalisation in E-Learning Environments at Individual and Group Level*, 11th International Conference on User Modeling (UM 2007), (pp. 8-15). Corfu, Greece.

Baluja, S. (2006, May 23 - 26). Browsing on small screens: recasting Web-page segmentation into an efficient machine learning framework. In *Proceedings of the 15th International Conference on World Wide Web*, (pp. 33-42), Edinburgh, Scotland WWW '06. New York, NY: ACM Press.

Bealle, R., & Lonsdale, P. (2004). Mobile context aware systems: the intelligence to support tasks and effectively utilise resources. In *Proceedings of Mobile HCI 2004*, (pp. 240-251), Glasgow, UK.

Berri, J., Benlamri, R., & Atif, Y. (2006, July 03 - 06). Ontology-based framework for context-aware mobile learning. In *Proceedings of the 2006 International Conference on Communications and Mobile Computing*, (pp. 1307-1310), Vancouver, British Columbia, Canada.

Blekas, A., Garofalakis, J., & Stefanis, V. (2006, May 22 - 22). Use of RSS feeds for content adaptation in mobile Web browsing. In *Proceedings of the 2006 international Cross-Disciplinary Workshop on Web Accessibility (W4a): Building the Mobile Web: Rediscovering Accessibility?* (vol. 134) (pp. 79-85), Edinburgh, U.K, W4A. New York, NY: ACM Press.

Brar, A., & Kay, J. (2005). *Privacy and security in ubiquitous personalized applications* (Tech. Rep. No. 561), School of Information Technologies, University of Sydney.

Brodersen, C., Christensen, B. G., Grnbk, K., Dindler, C., & Sundararajah, B. (2005). Ebag: a ubiquitous Web infrastructure for nomadic learning. In *Proceedings of the 14th international conference on World Wide Web (WWW '05)*, (pp. 298-306), New York, NY:ACM Press.

Brusilovsky, P., & Rizzo, R. (2002). Map-Based Access to Multiple Educational On-Line Resources from Mobile Wireless Devices. In *Proceedings of the 4th international Symposium on Mobile Human-Computer interaction*, (pp. 404-408), Pisa, Italy

Brusilovsky, P., Chavan, G., & Farzan, R. (2004, August 23-26) Social Adaptive Navigation Support for Open Corpus Electronic Textbooks. In: W. Nejdl, & P. De Bra, (Eds.), *Proceedings of the Third International Conference on Adaptive Hypermedia and Adaptive Web-Based Systems*, (pp. 24-33), Eindhoven, the Netherlands.

Buchanan G, Jones M., Thimbleby H., Farrant S., & Pazzani M. (2001, May) Improving mobile Internet usability. In *Proceedings of the 10th International Conference on World Wide Web*, (pp. 673-680), Hong Kong: ACM Press.

Chang, S.F., Zhong, D., & Kumar, R. (2001, December). Real-Time Content-Based Adaptive Streaming of Sports Video. In *Proceedings of the IEEE Workshop on Content-Based Access to Video/Image Library*, (p.139), Hawaii, Los Alamitos.

Chatfield, C., Carmichael, D., Hexel, R., Kay, J., & Kummerfeld, B. (2005, November 21 - 25). Personalisation in intelligent environments: managing the information flow. In *Proceedings of the 19th Conference of the Computer-Human Interaction Special Interest Group (CHISIG) of Australia on Computer-Human interaction: Citizens online: Considerations For Today and the Future*, (pp. 1-10), Canberra, Australia. ACM International Conference Proceeding Series, (vol. 122). Narrabundah, Australia: Computer-Human Interaction Special Interest Group (CHISIG) of Australia.

Cohen, D., Herscovici, M., Petruschka, Y., Maarek, Y. S., & Soffer, A. (2002, May 07 - 11). Personalized pocket directories for mobile devices. In *Proceedings of the 11th international Conference on World Wide Web*, (pp. 627-638), Honolulu, Hawaii.

Davis, S. M., (2002, August). Research to industry four years of observations in classrooms using a network of handheld devices. In *Proc. of IEEE International Workshop on Wireless and Mobile Technologies in Education (WMTE 2002)*, (pp. 31-38), Vaxjo, Sweden.

Dimakopoulos, D.N., & Magoulas, G.D. (2006, October 5-8). A personalised mobile environment for lifelong learners. In *Proc. of IADIS International Conference on WWW/Internet 2006*, (pp. 31-38). Murcia, Spain.

Holtzblatt, K. (2005). Customer-centered design for mobile applications. *Personal Ubiquitous Computer, 9*(4), 227-237.

Holzinger, A., Nischelwitzer, A., & Meisenberger, M. (2005). Lifelong-learning support by m-learning: example scenarios. *eLearn 2005, 11*(Nov. 2005), 2.

Gaeta, M., Ritrovato, P., & Salerno, S. (2005). Making e-learning a service oriented utility: the European learning grid infrastructure project. In P. Ritrovato, C.Allison, S.A. Cerri, T. Dimitrakos, M. Gaeta & S. Salerno (Eds.), *Towards the learning grid: advances in human learning services, frontiers in artificial intelligence and applications,* (vol. 127) pp. 63-76.

Keegan D. (2001). *The future of learning: From eLearning to m Learning.* Retrieved July 31, 2007, from http://learning.ericsson.net/mlearning2/project_one/book.html

Krikos, V., Stamou, S., Kokosis, P., Ntoulas, A., & Christodoulakis, D. (2005). DirectoryRank: ordering pages in Web directories. In *Proceedings of the 7th Annual ACM international Workshop on Web information and Data Management*, (pp. 17-22), Bremen, Germany.

Kukulska-Hulme A. (2002, June). Cognitive, Ergonomic and Affective Aspects of PDA Use for Learning. In *Proceedings of the European Workshop on Mobile and Contextual Learning*, (pp. 32-33), Birmingham, UK.

Kules, B., Kustanowitz, J., & Shneiderman, B. (2006). Categorizing Web search results into meaningful and stable categories using fast-feature techniques. In *Proceedings of the 6th ACM/IEEE-CS Joint Conference on Digital Libraries*, (pp. 210-219). Chapel Hill, USA.

Lave, J. (1988). *Cognition in practice.* Cambridge: Cambridge University Press.

Lee, Y. E., & Benbasat, I. (2003). Interface design for mobile commerce. *Communications of the ACM, 46(12)*, 48-52, 2003.

Lindquist, D., Denning, T., Kelly, M., Malani, R., Griswold, W. G., & Simon, B. (2007, March 7-10). Exploring the potential of mobile phones for active learning in the classroom. In *Proc. of SIGCSE'07*, (pp. 384-388). Covington, Kentucky.

Liu, T., & Choudary, C. (2004). Real-time content analysis and adaptive transmission of lecture videos for mobile applications. In *Proc. of the 12th Annual ACM international Conference on Multimedia*, (pp. 400-403), New York.

LLN (2007). Lifelong Learning Networks, Higher Education Funding Council for England (HEFCE). Retrieved from http://www.hefce.ac.uk/widen/lln/

Lonsdale, P., Baber, C., & Sharples, M. (2004) A Context Awareness Architecture for Facilitating Mobile Learning. In J. Attewell & C. Savill-Smith (Eds.) *Learning with Mobile Devices: Research and Development. London. Learning and Skills Development Agency*, (pp. 79-85).

Luchini, K., Quintana, C., & Soloway, E. (2004). Design guidelines for learner-centered handheld tools. In *Proc. of the SIGCHI Conference on Human Factors in Computing Systems*, (pp. 135-142), Vienna, Austria.

Miertschin, S. L., & Willis, C. L. (2004). Mobile computing in the freshman computer literacy course what impact? In *Proc. of the 5th Conference on information Technology Education*, (pp. 149-152), Salt Lake City.

Mikleia, E. (2006, June 27 - July 01). Mobile devices to be applied as supporting tools in Research Method class for undergraduate students. In *Proc. of the 7th international Conference on Learning Sciences*, (pp. 918-919), Bloomington, IN: International Society of the Learning Sciences.

Nardi, B. A. (1996). *Studying context: A comparison of activity theory, situated action models, and distributed cognition. Context and consciousness*. MA: The MIT Press.

Newcomb, E., Pashley, T., & Stasko, J. (2003). Mobile computing in the retail arena. In *Proc. of the SIGCHI Conference on Human Factors in Computing Systems*, (pp. 337-344). Ft. Lauderdale, USA.

Parsons, D., Ryu, H., & Cranshaw, M. (2006). A Study of Design Requirements for Mobile Learning Environments. In *Proc. of the 6th IEEE International Conference on Advanced Learning Technologies*, (pp. 96-100). Washington, DC: IEEE Computer Society.

Ractham, P., & Zhang, X. (2006). Podcasting in academia: a new knowledge management paradigm within academic settings. In *Proc. of the 2006 ACM SIGMIS CPR Conference on Computer Personnel Research*, (pp. 314-317), Claremont, USA.

Sharples, M. (2000). The design of personal mobile technologies for lifelong learning. *Comput. Educ., 34(3-4)*, 177-193.

Smyth, B., & Cotter, P. (2002). Personalized adaptive navigation for mobile portals. In *Proc. of the 15th European Conference on Artificial Intelligence*, (pp. 608-612), Lyon, France.

Tabard, A., Mackay, W., Roussel, N., & Letondal, C. (2007). PageLinker: Integrating contextual bookmarks within a browser. In *Proc. of the SIGCHI conference on Human factors in computing systems (CHI '07)*, (pp. 337-346), San Jose, California, USA.

Trifonova, A. (2003). *Mobile Learning: Review of the literature*. (Technical Report #DIT-03-009).

Valiquette, P., Williams, M., Kutner, N., & Seeger, E. (2005). Rensselaer's mobile computing program a whale of an idea! In *Proc. of the 33rd Annual ACM SIGUCCS Conference on User Services*, (pp. 405-410), Monterey, USA.

W3C, A P3P Preference Exchange Language 1.0 (APPEL1.0). (2006). Retrieved July 31, 2007, from http://www.w3.org/TR/P3P-preferences/

W3C, Platform for Privacy Preferences (P3P) Project, (2006), Retrieved July 31, 2007, from http://www.w3.org/P3P/

Waycott, J.,Scanlon, E., & Jones, A. (2002, June). Evaluating the Use of PDAs as Learning and Workplace Tools: An Activity Theory Perspective. In *Proc. of the European Workshop on Mobile and Contextual Learning*, (pp. 34-35), Birmingham, UK.

Weiβenberg, N., Voisard, A., & Gartmann, R. (2004). Using ontologies in personalized mobile applications. In *Proc. of the 12th Annual ACM international Workshop on Geographic information Systems*, (pp. 2-11), Washington DC.

Ziefle, M., Schroeder, U., Strenk, J., & Michel, T. (2007, April 28 - May 03). How younger and older adults master the usage of hyperlinks in small screen devices. In *Proc. of the SIGCHI Conference on Human Factors in Computing Systems*, (pp. 307-316), San Jose, California.

## ADDITIONAL READING

Hawkey, R. (2004). *Learning with digital technologies in museums, science centres and galleries* (Report 9). Futurelab. London: King's College.

Naismith, L., Lonsdale, P., Vavoula, G., & Sharples, M. (2004). *Literature Review in Mobile Technologies and Learning* (Report 11). Futurelab, University of Birmingham.

Rodden, T., Cheverst, K., Davies, N., & Dix A. (1998, May). *Exploiting Context in HCI design for Mobile Systems*. C. Johnson (Ed.) Proceedings 1st Workshop on HCI with Mobile Devices (pp. 12-17).. Glasgow, Scotland: University of Glasgow.

Sefton-Green, J., (2004). *Literature Review in Informal Learning with Technology Outside School* (Report 7), Futurelab, WAC Performing Arts and Media College.

Smailagic, A., & Siewiorek, D. (2002). Application design for wearable and context-aware computers. *IEEE Pervasive Computing, 1(4), October-December 2002*, 20-29.

# Chapter XIII
# Designing Situated Learning Experiences

**Hokyoung Ryu**
*Massey University, New Zealand*

## ABSTRACT

*To ensure the success of future mobile learning environments, it is essential to develop affordable and effective applications that are well matched to the needs of the users. Depending on their unique requirements, effective mobile learning applications should keep up with their learning activities rather than simply providing them with conventional course materials on mobile devices. As an early exploration of this line of study, this chapter designs and evaluates a location-aware learning organizer that helps university students to manage their learning activities on campus. We confirmed that this situated learning support experience could lead to markedly different in-depth learning activities. Empirical testing of the learning organizer also revealed some potential opportunities for the learners to be more engaged in further mobile learning activities.*

## INTRODUCTION

Mobile technologies now allow us to explore new opportunities around "m-" neologisms such as m-office, m-government, m-commerce, m-health, m-learning, and so forth. The prevalent mobile paradigm inevitably continues to change the ways in which people fulfill their lives. For instance, several government-funded mobile-based research projects presented increasing possibilities to make some societal changes with relevant mobile technologies, e.g., anti-alcohol education (Demirjian & David, 1995), quit-smoking program (Horwood, 2003) and HIV/AIDS education programs in African countries (Swedish International Development Agency started this program in 2002 and since). These mobile-based education programs exploited the penetration of mobile devices and technologies as the medium of delivery of the learning contents, which might indicate effective learning (or education) experiences from the learner's perspective. These mobile learning programs are still evolving with the widespread support of pervasive computing environments, capturing more direct contextual information and providing appropriate learning contents at the right time and the right place.

Regardless of one's intentions, all of us can learn something from our daily lives, such as from museum visits or consultation with doctors, which is what is principally meant by the premises of pervasive or ubiquitous learning environments. Atkins (2005) coined this particular type of learning activity as '*informal learning*', normally a very situation-oriented and more goal and task-oriented activity, compared to the formal learning experience offered by most educational institutions. In particular, as the cost of mobile computing devices has fallen well below that of conventional desktop machines, we have been designing new learning activities employing novel mobile or wireless technologies, enabling access to pedagogical applications anywhere and anytime (Becerra-Fernandez, Cousins, & Weber, 2007; Eschenbrenner & Nah, 2007; Kukulska-Hulme, 2007).

This dominant mobile paradigm has changed the underlying pedagogical experiences of what we have hitherto taken for granted, namely traditional face-to-face or distance learning (or e-learning) experiences from educational institutions. Several researchers (e.g., Kukulska-Hulme & Traxler, 2005; Rogers et al., 2005) have claimed that the most important change mobile learning (m-learning) can make, which traditional learning experiences cannot achieve, is that learners can be hooked into the situations where learning actually occurs (e.g., the binge drinkers in the pub or the smokers outside), proactively putting themselves in control of what and how they choose to learn. An early example quoted in one of the Pan-European m-learning projects is thus worth returning to:

*"If you walk as a tourist around an archaeological site in Greece, you may be interested in what this heritage is about, how long it took to build, what historical accounts are, and et cetera, But it is very unlikely that you will take notes in order to undertake further research when you get back home..."*

This seems to imply a significant loss of potential learning opportunities associated with the place where the learning could happen. That is, *situated learning* is not just for the classroom or intentional learning activities, but for the world outside classroom doors, which would be a better learning arena in many cases.

We should, however, be cautious about applying these anecdotally plausible stories to other learning activities, especially those occurring in formal education sectors. Indeed, people, institutions, universities, and businesses have suffered enough from the prophecies of futurologists who promise and project that "mobile *X*" would be the virtue of everything in the near future. In effect, what we need to do is to take the measure of when and how this kind of situated learning activity would best serve pedagogical purposes in traditional learning arenas.

There are, therefore, a number of concerns regarding the outcomes from this new type of learning activity. Kukulska-Hulme and Traxler (2005) stated that m-learning alone may not find it easy to attract university learners, to engage them with educational activities, and keep their motivation during learning activities, thanks to its un-intentionality or distraction (Ryu & Parsons, 2007). Also, chapter V in this book shows the "one-size-fits-all" fallacy that m-learning researchers may find themselves trapped by. The primary issue to be addressed in the development of future m-learning systems therefore may be how to associate (or bridge) these new learning experiences with more conventional ones in order to ensure their usefulness. Indeed, there have been a number of studies that have tried to bridge the gap between m-learning and conventional learning environments. One example is the *'Ambient Wood'* project, carried out by Sussex university (Rogers et al., 2005), which demonstrated a playful learning experience where primary school students could explore and reflect upon a physical environment that had been augmented with a medley of digital devices. A field trip 'with a context' was created, where children discovered, hypothesized about and experimented with ecological processes taking place within a realistic physical environment, and these experiences were discussed later back in the classroom. This project was developed with a view to creating future m-learning environments that may provide more constructive learning outcomes at the primary school level, that is, enhancing individual learning outcomes coupled with the contextual learning experience.

Yet, whilst the *Ambient Wood* project demonstrated a number of features that we may find useful in building future m-learning applications, it is unlikely that they would have the same benefits for those (e.g., professionals or university students) whose overall motivation is, to a large extent, different from primary school students. It might be argued that the learning style of the *'Ambient Wood'* project would be perceived as being less valuable for professionals or university students, and the time-consuming nature of the learning activity might be de-motivating to this audience.

This chapter therefore addresses the specific context of the university learner and associated user profiles in order to explore and model some design features that one should pay attention to when it comes to designing a situated m-learning environment for this category of learner. The following section begins with a brief account of which features of m-learning environments and theoretical foundations are closely related to our understanding from early studies, so that m-learning researchers may be aware of the design issues involved in creating effective m-learning applications for university students. The next section covers the learner-centered design practices (including modeling situated learning experiences). Usability testing with the intended user group in their learning context is then outlined to demonstrate the benefits of the situated learning experience. Finally, a practical implication of this study is described.

## THEORETICAL FOUNDATIONS ON DESIGNING SITUATED LEARNING ACTIVITIES

As previously discussed, it has been widely proposed that mobile technologies would provide excellent opportunities to develop novel education or learning environments, extending the current educational context into a versatile "anytime and anywhere" learning experience. It has thus been suggested that mobile technologies could be used as a bridge between various learning activities, thereby breaking the boundaries of known learning contexts such as formal learning in schools (see more details in chapter IV), life-long learning (chapter XII), real-world problem solving activities (chapter VIII), and establishing one learning platform to cover the different learning contexts. Yet, there has been little discussion about how the new mobile technologies can impact on current educational activities and relationships, which are now discussed in many chapters in this book. To understand the diverse roles of educational activities that m-learning (or any learning applications) can support, we first focused on theoretical foundations that could effectively assist the design of the system proposed in this chapter.

### Situated Learning and Activity Theory

To a large extent, the literature on mobile or pervasive computing has emphasized the context as the most important feature it must have (Dix, Finlay, Abowd, & Beale, 2005), defining the context feature as any information that can be used to characterize the situation of an entity. Traditional learning applications, especially desktop-based distance learning applications, have considered only a fairly limited set of situational cognitive concerns (Brusilovsky, Schwarz, & Weber, 1996). These might include user characteristics (i.e., the level of understanding of the learning contents, their skill level and so forth). In contrast, many m-learning applications augment the approach of these conventional learning applications with a wide variety of contextual and situational information from users' frequently changing circumstances and mobility (Tarasewich, 2003). It is thus the author's belief that both *situated learning theory* (Lave & Wenger, 1991) and *activity theory* (Bendy, Seglin, & Meister, 2000; Engeström, 1999; Leont'ev, 1978) offer us powerful theoretical foundations for eliciting and modeling situational requirements, and designing more useful m-learning applications as a result.

*Situated learning theory* (Lave & Wenger, 1991) is a general theory of knowledge acquisition that is based on the notion that learning occurs in the context of activities that typically involve a problem or task, other persons and an environment or culture (Kurti, Spikol, Milrad, & Svensson, 2007), emphasizing highly interactive and productive roles for the skills that are acquired through the situated learning process. Perhaps many mobile learning applications have been implicitly or explicitly built upon this theory, for instance, chapters VIII and IX in this book clearly embody this belief, and relate learning activities to a situated learning process.

The main principle of this theory clearly pinpoints that the learner is not gaining abstract knowledge that they will then transfer and reapply in later context. Instead, they acquire the skill to perform by actually engaging in the process, under the legitimate peripheral participation, in which the relevant situational settings are dictating actions and learning outcomes. Many research outcomes (e.g., Brown, Collins, & Duguid, 1989) have increasingly supported this learning theory, that the failures of students to apply concepts learnt in different contexts are in many cases due to too much abstraction and de-contextualization of learning. The very nature of situated learning theory is to revisit the contextual knowledge that provides a barometer to guide and support the learner. Yet, designing or modeling this situated learning activity is challenging, since in many cases there are many features that contribute to capturing the right contexts or situations. This chapter focuses on the location context of the learner, to reflect the principal implications of situated learning activities (Lyytinen & Yoo, 2003).

To model situated learning activities, we also took another theoretical framework, the *activity theory* (AT) (Bendy et al., 2000; Engeström, 1999; Leont'ev, 1978). AT can effectively model situational and contextual circumstances that are linked to one's learning activity (Bertelsen & Bødker, 2003). Furthermore, many previous studies (e.g., Kurti, Spikol, & Milrad, 2007; Sharples, Taylor, & Vavoula, 2005; Uden, 2007) showed some related evidence that it could provide a broad conceptual framework to describe the structure, development, and context of tasks that are supported by a mobile learning system. For instance, the author has recently modeled a system that helps extramural students conduct their course projects on a group basis with AT as shown in Figure 1.

The main objective of this system was to help extramural students who had enrolled in the "user-centered design" course at Massey University, to conduct three group-based practical project assignments through this system. This course necessitates this group-based practical project experience to apply the concepts and techniques from the online lecture for a more realistic information systems design exercise, which is clearly represented as the learning outcome.

*Figure 1. Activity is an intertwined system, if one entity changes, the whole system becomes unstable and must develop to obtain renewed stability. This is a part of modeling the relation between the six entities from extramural students' perspective.*

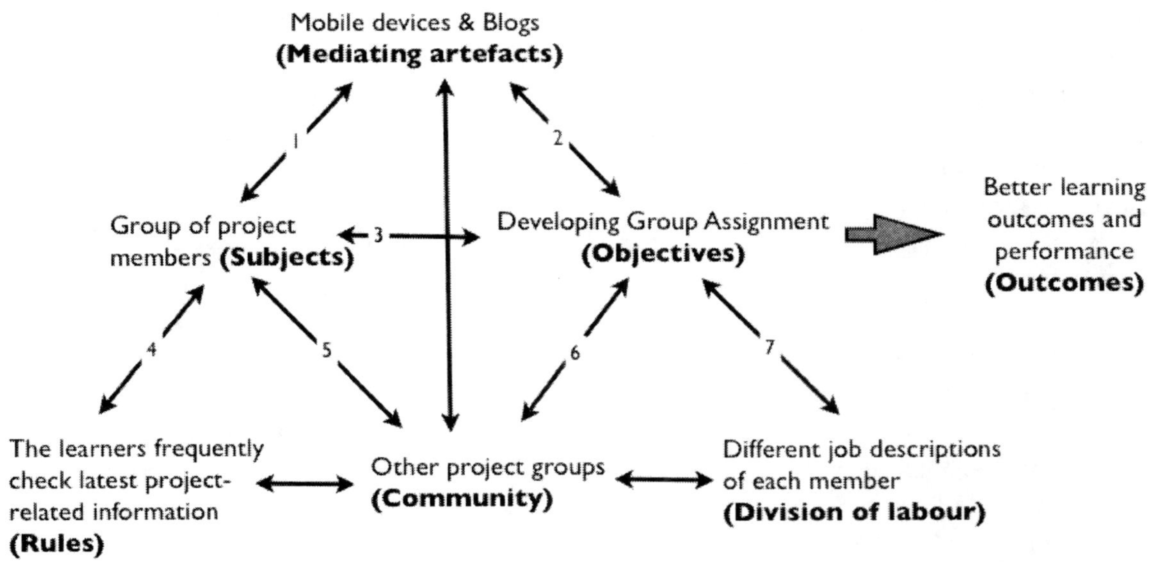


Principally, the group of extramural students is forced to use only two mediating artifacts (numbered as "1" in the figure), text messages on their own mobile phones and Web-based blog systems. These mediating artifact-producing activities include the intrinsic communication activities among the group members. Using these two artifacts the students must complete their group assignments (arrow 2), which indicate the object activity, that is, three assignments. In effect, to carry out these three assignments, the students should be in contact with one another using text messages and further motivate their collaboration with their own project blogs (arrow 3). This group assignment basically needed a lot of interviews, a literature review, and focus group studies (arrow 7), the responsibilities for which were to be split between the members of the group. The group members were asked to regularly report their own work-in-progress amongst themselves via the blog or text messages (arrow 4), which are rule-producing activities. The present status of the other group projects is also available through the course blogs (arrow 5), which is of great value to motivate (or even regulate) their own project performance in the following project assignments (arrow 6). The activities drawn in Figure 1 would be constantly developing to match new needs or requirements. For instance, if the lecturer would like to be informed about the current status of the assignments, she or he can add one rule-producing activity in the corner of the triangle to re-create new activity systems, consequently the changing part of the systematic structure naturally generates relevant requirements as needed.

As can be seen in Figure 1, the situational and contextual requirements in a mobile learning system can be modeled with the useful mechanism that activity theory provides. In particular, the three entities—community, rules and division of labor—that denote the situated social context allow us to consider the situational and contextual requirements in an effective way. See Bertelsen & Bødker (2003) and Uden (2007) for more details.

By modeling these situated requirements through the two theoretical foundations discussed above to contextualize them, we can increase the richness of learning activities and make it possible to produce more useful learning experience at the university level. The following section describes the designing experience of the mobile learning system to offer this opportunity in terms of how the learner's locations in the learning space can be further associated with the students' learning activities.

## DESIGNING SITUATED LEARNING EXPERIENCE IN THE UNIVERSITY

One practical example of developing a situated m-learning system for university students is Scornavacca, Huff and Marshall (2007)'s interactive SMS (short text messaging service) classroom system. They specifically noted a lecturing situation which involved large numbers of first-year university students in a big lecture theatre. Clearly, this situation is not easy either for the lecturer or the students. In most cases, the lecturer would like to know whether the students understand the contents that have had delivered to them, but the students are unlikely to express themselves or ask the lecturer any questions in this context.

To facilitate a certain level of engagement with the class, the students were allowed to ask any questions via text messages with their own mobile phones, so the lecturer could interactively capture important questions from the students and effectively respond to them. Indeed, this form of educational technology is not new, and has already been commercialized, for example, 'Clickers™' (Gilbert, 2005), but Scornavacca et al.'s SMS classroom system demonstrated that common low-tech mobile phones could be very useful at university level without the purchase of high-end devices.

*Figure 2. Learning organizer's time manager and course manager screenshots, reprinted from Corlett et al. (2005)*

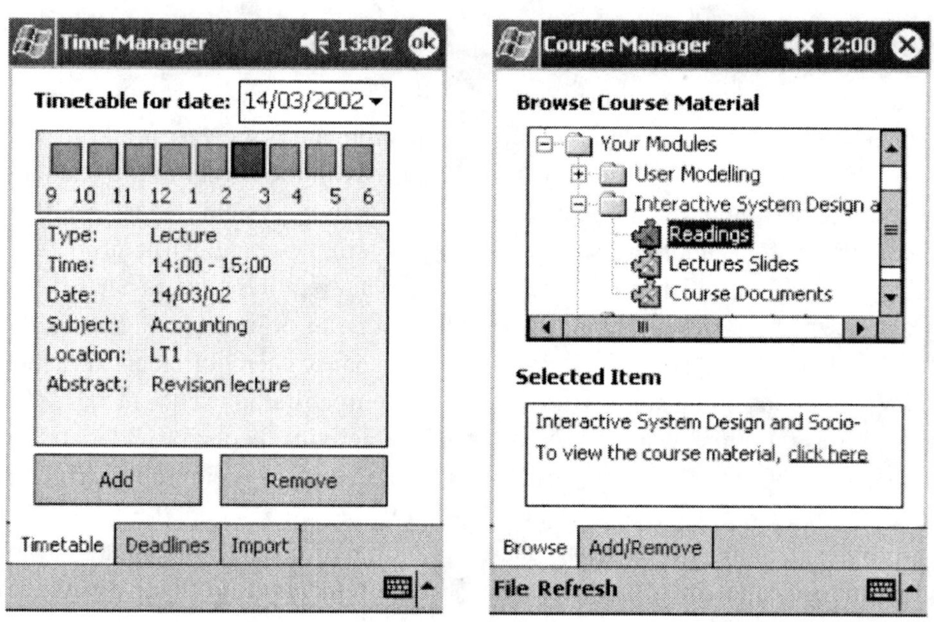

They found, however, that the learning outcomes had not been significantly enhanced by the system. This result takes into account that this system does not aim to provide a bridge between the m-learning system and other learning activities, but simply to enhance the level of engagement.

Supporting the notion of enriching the level of engagement by mobile devices, Sharples et al. (2005) developed a learning organizer on PDAs (personal digital assistants) that could assist university students in some aspects of managing their studies. This was a kind of event calendar system that could present basic information about lectures (i.e., time-management) or provide access to some of the course content, as shown in Figure 2. A usability study of the system, however, provided no conclusive evidence as to its usefulness to the students' learning processes (Corlett et al., 2005), though it hinted that more contextual information (e.g., reminders of course deadlines around the due dates) might be of use to the students. In this kind of physical learning space, the contextual information seems to be very relevant to the locations in which the learners are. For instance, if they are in the lecture room, more preferable and desirable information at that time might be the course related information.

Several attempts to include the student's location information in m-learning applications have also been made. For instance, the Portable Help Desk (PHD: Smailagic & Kogan, 2002) demonstrated the usefulness of both context and location awareness. The campus-wide, sensor-based location awareness system provided some useful contextual information (e.g., which printer has the shortest print queue or the location of their friends on the campus to set up a meeting time and so forth) to support university students' campus life. However, an early test of PHD did not show any clear benefits to the students' learning outcomes, illustrating only minor potential for university-based m-learning applications.

Having identified some areas of challenge in this line of research, what we intend to explore in this chapter is how to extend the university student's learning experience with situated help or location information, by reviewing university students' contexts and requirements. It was assumed that the two theoretical foundations, situated learning and activity theory, could elicit the relevant contextual requirements of this learning system for the university students. We also assumed that a critical success factor for an m-learning application for university students would be to provide two types of enhancements at the same time: learning experience and learning outcomes. PHD (Smailagic & Kogan, 2002) and the SMS-based classroom interaction system (Scornavacca et al., 2007) enhanced the learning experience of the university student. In contrast, the learning organizer (Corlett et al., 2005) significantly improved some learning outcomes. This chapter takes account of these dual objectives, providing a location-aware learning organizer to ensure that the system meets university student's requirements within their learning contexts. In particular, we would also like to see if this learning application enhances the learning outcomes of the university students, and whether it can hook them into subsequent learning activities. Usability testing was therefore undertaken to attempt to measure these factors.

## Eliciting Requirements and Modeling Situated Learning Activities

There have been a wide variety of methodologies proposed for eliciting, presenting, and validating requirements. Specifically, the fundamental approaches for collecting requirements are shifting away from the data to be processed by the system and moving more towards human-computer interaction or social and organizational factors. Sommerville (1996) claimed that social and organizational factors affect the requirements analysis process and proposed the use of ethnography on which social analysis must be based for examining the real everyday needs of an organization. However, ethnography analysis requires a substantially long period of time to perform observation activities, which poses severe restrictions on performing ethnography analysis in practice.

We attempted, instead, to collect the university students' learning experiences by using the critical incident method (CIM) (Flannigan, 1954, 1962) which is used to identify critical elements from previous experience. We firstly collected as many critical and practical incidents as possible from 20 university students, asking what types of contexts, tasks, information and design features would suit a university student's expectations. The interview questions used were as follows:

- Tell me about an incident that you remember which was an example of an ineffective learning experience;
- What were the general circumstances leading up to the incident?
- Tell me exactly what it was about the learning experience that was so ineffective at the time
- When did the incident occur?

*Table 1. Interview data summaries. Frequency indicates the number of interviewees from each user profile which gave the incidents for each category.*

| Category | Frequency: New students (8) | Frequency: Senior students (12) | Sample excerpts |
|---|---|---|---|
| Mobility (or location) issues | 7 | 1 | *I am a new student, so sometimes…, I have no idea of where QA2 is, I don't even know what QA2 means. That means I missed the first lecture in my first ever school life<br>* The lecturers never told us where to go for the tutorials and lab sessions. I was looking for the labs immediately after the lecture, but it wasn't so easy to find the place which is in the other precinct in five minutes.<br>** This … course has changed the lecture room in Week 5. But I haven't got any notice of this room change *(interviewer asked why)* Actually, I missed the Week 4 lecture, so I also missed the Week 5 lecture. |
| Learning (or contents) issues | 2 | 10 | * I want the university to help me easily find assignments, examinations, library information, room changes, and so on, cos' the lecturers never kept the course outlines, so I have some difficulties to find the correct information.<br>** When I arrived to the campus in the morning, I liked to set up today's itinerary, says where I have to get now, and then, and then, when or where I can have a lunch and so on. Also, if I have some time off between lectures, I want to stay at the library to read some books and do some assignments. However, sometimes, the library is full of students, so I cannot find a place to study.<br>** The lecturer told us that we had quizzes next week, but I didn't know what I had to prepare for this, so I had been to his office, but he was not in the office. And then I found his telephone number from my mobile phone, and left a message in his voice mailbox. But he never returned me at all.<br>** I requested an inter-library loan book. … I didn't collect this book for some time, because I didn't know I could. I had no calls or emails from the library, and they charged me $2 fine. |

* *the excerpts from new students;* ** *the excerpts from senior students*

The 20 interviewees were all Massey University students, aged between 18 and 32, randomly selected from the user-centered design course that the author was teaching in 2007. Twelve of them were male, and eight of them were first-year students. Eleven of them were international students. Each interview took around fifteen minutes.

After transcribing the interview data, we decided upon a coding scheme containing two categories for handling the interview data: mobility issues and learning issues, as depicted in Table 1. In the first category, the interviewees presented some requirements relating to location issues around the campus. In the second category, the interviewees presented several ineffective learning experiences that they had at university. The frequency column in Table 1 gives the number of interviewees who contributed to each category. Note that one asterisk (*) indicates excerpts from the first-year students, and two asterisks (**) indicates excerpts from those who are senior students (i.e., second and third year students).

Interestingly, the frequencies in Table 1, as derived from the protocols, appeared to show certain differences between the user profiles (new students vs. senior students). That is, the senior students had more concerns about learning issues, whereas the new students reported more mobility issues.

These interview data served to specifically identify the design requirements of the m-learning application that we were developing in terms of three perspectives. Firstly, the interview data revealed that most of the new students (seven out of eight students) had little idea of where the classrooms and laboratory facilities were located in a widely distributed campus (in fact there are three separate precincts at Massey University in Auckland). Because of this, new students are often unsure of where their next meeting or lecture is to take place. In contrast, many of the senior students seem to be already aware of this type of information, but require more in-depth information about their personal studies (learning issues). Their concerns during "campus life" revolve around the organization of their studies, such as being aware of assessments, while also being up to date with messages and resources from lecturers and/or other students involved in their program of study, which were given little consideration in Corlett et al.'s learning organizer (2005). The new students' contexts and requirements were thus relevant to aspects of interface design, such as map support, while the requirements of the senior students related to the personalization of their learning environment.

The next consideration was the students' learning context, which emphasizes the needs and intended outcomes of specific learning activities. In particular, the interview found that spatial awareness is very important for the junior students, but temporal issues were more relevant to the senior students, for example relating the time of day to their study schedules. One of the interview data showed that when a senior student arrives on the University campus, the contextual information that they are searching for may be related to their course schedule for the day, room changes or important messages (e.g., assessment information, library records, and so forth). These different activities define different design requirements. As noted previously, the junior students want more navigational support on the campus, but the senior students require more contextual knowledge and social interaction with lecturers or tutors.

The key point to note from this requirement elicitation process is that the success of any m-learning application relies on taking into account different contexts of use for different learner groups, who may have different expected learning experiences, as we have identified from the responses above. Figures 3 and 4 show the part of the activity systems used to visualize the requirements of the different learning activities.

*Figure 3. Activity system for junior students. It characterizes the subject, object, and mediating tools*

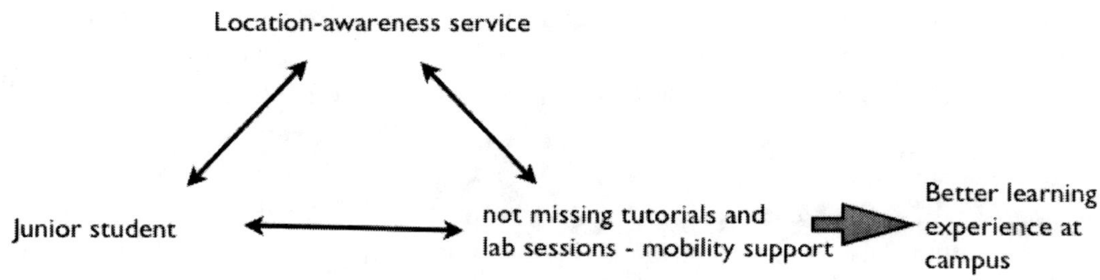


*Figure 4. Activity system for senior students, situating the senior students' system in use.*

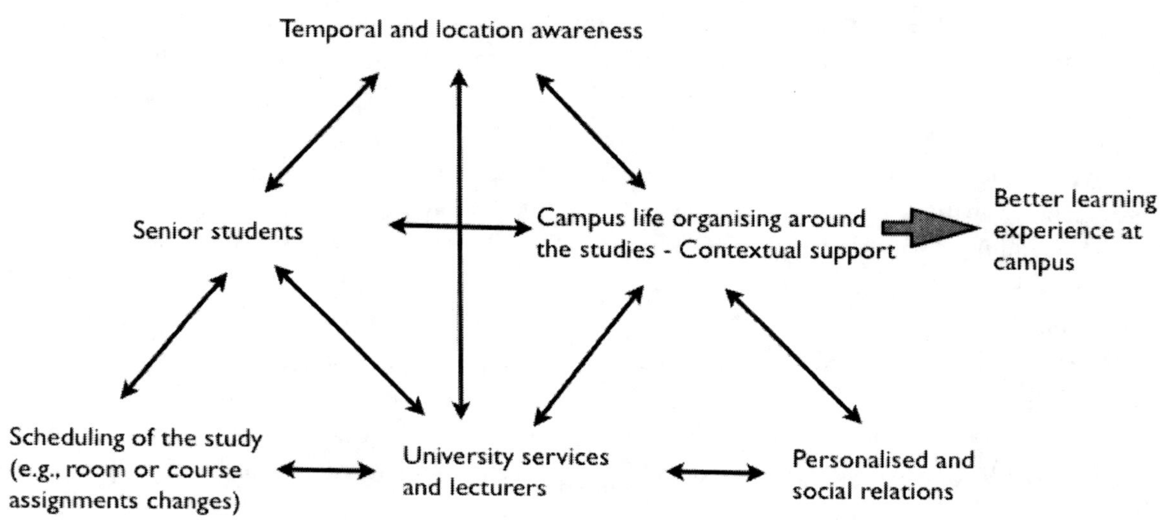


## Design of a Location-Based Learning Reminder

To address these design requirements, location support was thought to be an essential function to be implemented, as depicted in Figure 3 and 4, with contextual information linked to the location-aware service. Indeed, the location in everyday life regularly tends to present a particular circumstance or stimulus of our activity. According to Suchman (1987), the nature of our activities or plans of actions would be highly associated with particular social and physical circumstances pertaining at the time. To account for this under the university context, for instance, as a student is passing by the library, she or he may think about something that is related to the place (e.g., return a book requested by another borrower, or collect an inter-library loan book). Similarly, the lecture room where the student is present would be likely to raise awareness of the course-related information (e.g., assessments or assignment due dates and so on.) This location-based reminder, which was not considered in Corlett et al.'s learn-

*Figure 5. Apparatus used in the system development, GlobalSat BT-338 GPS (Courtesy of GlobalSat) and HP-iPAQ 6300 (Courtesy of Hewlett-Packard)*

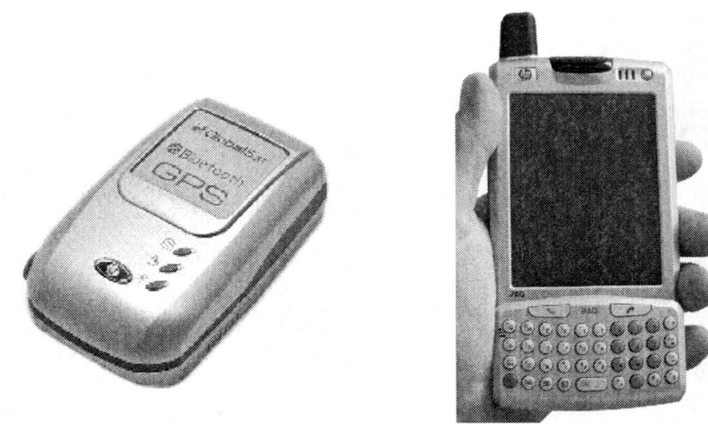

ing organizer (2005), would be central to the m-learning application that is described in this chapter. It is the author's belief that this application allows the learner to be hooked more seriously into subsequent learning activities.

We developed this location-aware learning organizer with a PDA phone (HP IPAQ™ 6300) with a separate Bluetooth GPS device (See Figure 5). It is able to locate the current position of each learner on the university map, and to access the server using a mobile carrier network (Vodafone New Zealand). The aerial photos of the campus were obtained from the North Shore City Council, and calibrated for use with a GPS device by constructing software for the task. Microsoft SQL server was used to relate the contextual information and the location data. Each component of the software was built using Microsoft Visual Studio 2005.

The assumptions of typical users and the physical environment for the development of this application were that science or engineering students were concurrently studying four or more courses and that activities related to their daily study routine included accessing learning content and collecting or returning library books. In terms of their coursework, most courses include lectures, tutorials, and/or laboratory sessions. In terms of the physical environment, the Massey University campus consists of three separate areas, presenting spatial-temporal issues related to a student's location context. For instance, most lectures take place in theatres in the upper precinct of the campus, while many of the tutorials and

*Figure 6. Screenshot of navigation software for a user who wants to get to the Quadrangle building for her or his next lecture*

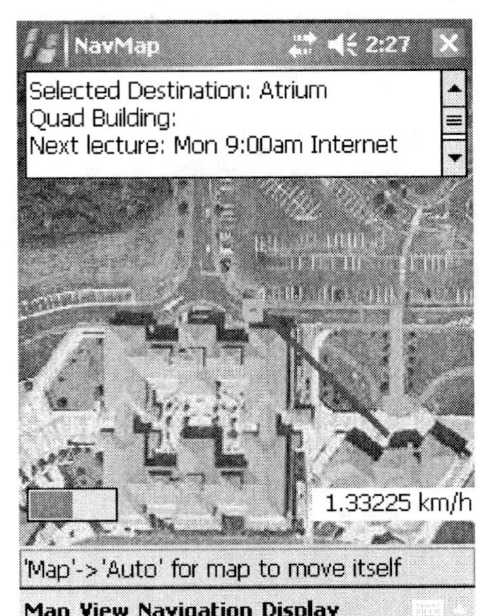

laboratory sessions are some distance away in the lower precincts. Therefore, students need to be constantly aware of the spatial contexts of their learning activities. Figure 6 shows a screenshot from the location-aware system. There are indicators representing the user's current position, and a path to their selected location. The map is synchronized with the users' location, moving and rotating in line with their movements. This type of feedback provides the user with the necessary navigational information. In addition, the user can select their next destination from the course calendar, and the path to the location appears on the map.

Figure 7(a) shows the location-aware learning organizer's system architecture. It automatically sends the user's current location to the PDA server. The server builds the query for the SQL server database (as shown in Figure 7(b)), which retrieves the client's information. The PDA server returns the information to the user's PDA. As users move through an area, the PDA server delivers contextual information relating the current position based on the predefined user's profile. Therefore, the user can request information to provide new knowledge and assist their activities related to study at the university. The major challenge was the data set we should put into the database, because of the privacy protocols that the university applies. We therefore used the student records of course enrolments with the consent of each student and the lecturers' voluntary support for each class.

In terms of user-driven information, a user could utilize this system to find what books have been placed on reference in relation to their course of study. As the user walks past the library, they can automatically see this type of contextual information. Figure 8 demonstrates some contextual information a user may receive while walking close to a lecture theatre. In addition, if the user clicks the 'Assignment 1 available' link to their learning environment, they can see what the assignment is and its relevant information from the library. In addition, they can access the whole m-learning module as shown in Figure 9.

Central to this research was to explore whether this location-aware learning organizer could effectively support tertiary students' learning experiences without any significant difficulty-

*Figure 7. System architecture (a). The system receives the user's query and matches it to an appropriate contextual help, accessing the SQL server database (b)*

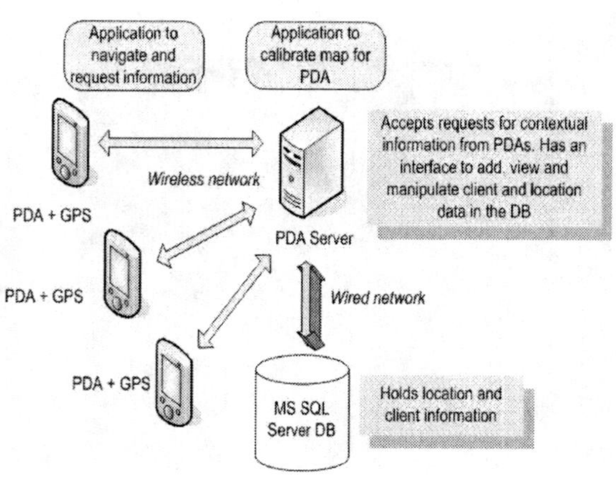


(a)

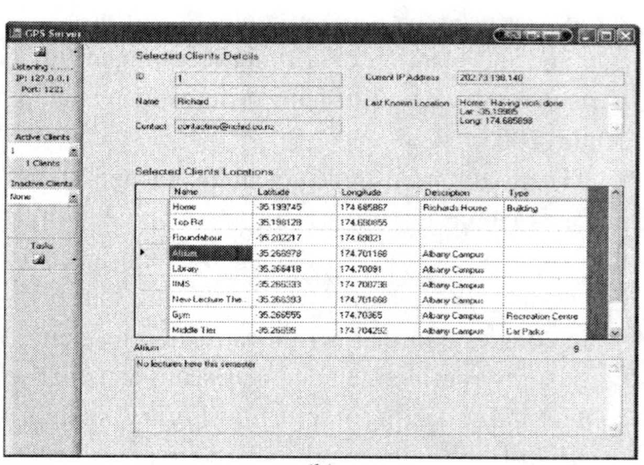

(b)

*Figure 8. Contextual information presented to the user. If a user clicks 'Assignment 1 available' on the system, one can access the m-learning module*

*Figure 9. Learning module design for the learning organizer. It provides their current location and additional learning module students can download*

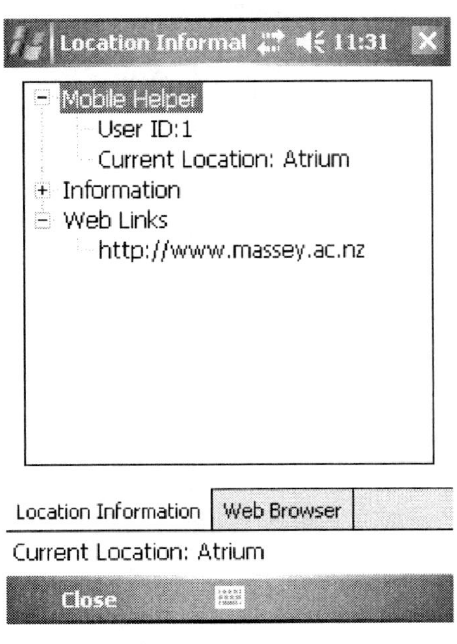


performance tradeoffs. Also, it was intended to confirm if this type of m-learning application could hook the learner into subsequent learning activities. To do this we examined the perceived usefulness of the location-aware learning organizer and its likely support of the increased motivation with other learning activities, which we describe in the following.

## OBSERVING THE LOCATION-AWARE LEARNING ORGANIZER IN USE

To explore the perceived usefulness of the location-aware learning organizer, we obtained the participants' perceptions of their experience of the system. This approach has been validated through many usability studies for predicting user acceptance and explaining the use of newly developing systems (Davis, 1986, 1989; Mathieson, 1991).

The deployment of this m-learning application involved exposing 20 student participants (10 new and 10 senior) to the system, and then asking them to give seven ratings on four questions. All the participants (18 males and two females) were Massey University students who had enrolled in the author's undergraduate course in 2007. For the contextual help information, participants' profiles such as their interests, coursework, and library information were collected before the evaluation. Based on these profiles, contextual information for each participant was added to the system database.

To evaluate the system in use, the participants roamed the campus for 15 minutes, where they experienced the two major functions of the location-aware learning organizer, (i) finding the location of their classrooms, and (ii) contextual help. Each user was first asked to find the location of their next lecture with this system, and then get to that location. As they approached the location, relevant contextual information appeared.

*Figure 10. The questionnaire about perceived usefulness and ease-of-use*

1. I think this device would be very useful in my university life
2. The system made me more organized in my campus life
3. I would like to use this system again
4. The system made me more engaged with my studies

*Figure 11. Mean ratings for scale 1 'I think this device would be very useful in my university life'*

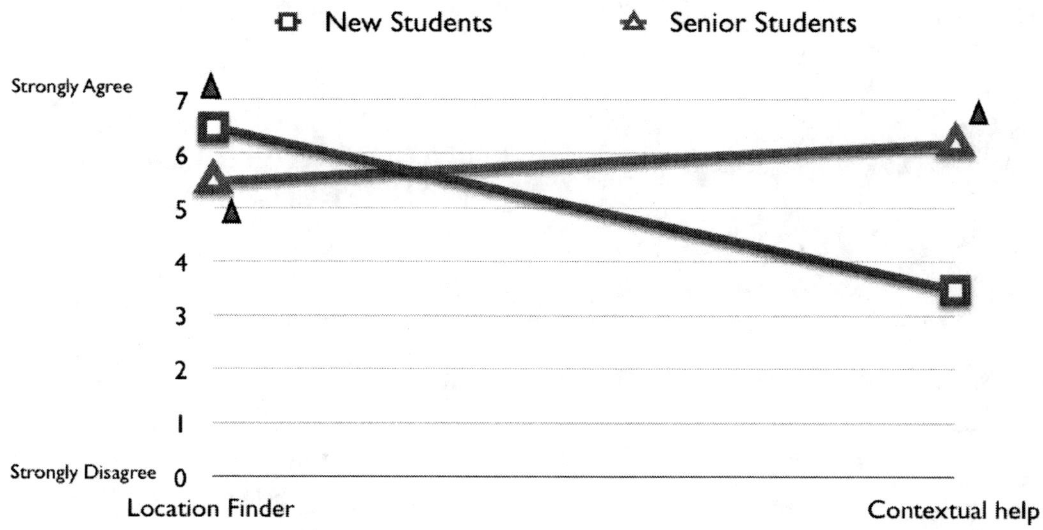


For instance, if the students were passing the library, the system proactively displayed their library information, for example if there was a new book based on their interests or books due for return. After using each function, they were asked to rate the usefulness of the system on a seven-point Likert scale. In this way, we were able to examine how much the system was perceived to be useful and useable by the different user groups, simulating a realistic learning context at university level.

Figure 10 depicts the questions used in this usability study. The purpose of this evaluation was to determine if this sample of participants (10 first year students and 10 second or third year students) found the system useful. To answer this question the ratings were analyzed in terms of their absolute level, that is, whether the mean rating was significantly higher (agreement) or lower (disagreement) than the neutral value 4.00.

Figure 11 gives mean ratings for Scale 1 'I think the device would be very useful in my university life.' A rating of 1 indicates that the participant strongly disagrees with the statement, a rating of 7 that they strongly agree. Arrows indicate that a mean is significantly different from 4, at 0.05 level, by one sample t-tests.

Looking at the separate lines in Figure 11, it can be seen that there is a reverse effect of the functions. The location finder functions were perceived as being very useful by new students, while in contrast, senior students give a higher rating to the contextual help. This tells us that the expected learning experiences from the learning organizer vary. One-sample two-tailed t-tests against the neutral value 4.00 showed that the

*Table 2. Mean ratings for Scale 2 and 3. Scale 2 'The system made me more organized in my campus life:' Scale 3 'I would like to use this system again.' (mean/s.d)*

| | New students | Senior students |
|---|---|---|
| Scale 2 | 6.60 (0.84) | 6.40 (0.97) |
| Scale 3 | 7.00 (0) | 6.80 (0.63) |

*Table 3. Mean ratings for Scale 4 'The system made me more engaged with my studies.' (mean/s.d)*

| | New students | Senior students |
|---|---|---|
| Location finder | 1.30 (0.48) | 1.50 (0.71) |
| Contextual help | 3.90 (1.20) | 3.70 (2.00) |

new students efficiently utilized the system to discover new locations while feeling confident in themselves in using the system. The senior students, however, were attracted to the contextual information.

The ratings for Scale 2 and 3 were around 7.00, indicating that most of the participants gave the maximum rating for both functions (so the ratings of both functions were averaged out, see Table 2). These results show very little variance for these scales. Because of this overall ceiling effect, the small differences between means are not interpretable, so they made it hard to identify any behavioral differences between the new students and the senior students. However, the face values of Table 2 were broadly in support of the usefulness of the location-aware learning organizer in the university context.

Finally, for Scale 4, the ratings for the location finder function were uniformly low, but the ratings for the contextual help were relatively high, even though they were not statistically different between the new students and the senior students (see Table 3).

There are a number of possible explanations for these results. One is the novelty of the technology. The participants who use this location-aware learning organizer might have been attracted by the new technology, and enjoyed it more as a result.

To explicitly see how the students would interact with the system, we further assessed the system using the think-aloud technique. Ten additional students (five new students and five senior students) were further recruited from the campus, and the same task procedure was applied. Verbal protocol results concurred with the usability study, placing emphasis on ease of use and the ability to learn from contextual information. All the participants were able to complete the tasks in a straightforward manner. They would create an objective to either find a location, acquire some contextual information while moving around, or a combination of the two. In each occurrence the participants interpreted new contextual information, assisting in their purpose of studying at Massey. Also, they appeared to find the contextual information both useful and informative in relation to their goals of being at the location, such as "library books due" messages when they were near the library. The participants with specific information appeared to feel more satisfied with the system, and its abilities. Furthermore, we witnessed that six out of the ten students who participated in this study expressed that they were motivated to access the relevant

Web sites (e.g., the library Web site or the course Web page) just after they saw the contextual help in the readout. Therefore they were more likely to have a positive attitude to, and respect for, the system, learning more and potentially hooked into subsequent learning activities.

## CONCLUSION AND FUTURE RESEARCH DIRECTIONS

There have been a number of repeated questions raised about the usefulness of m-learning from a pedagogical perspective, such as, is it really possible to learn with such small handheld devices? What kinds of people would use mobile devices for teaching and learning? What sorts of subjects and situations are appropriate for mobile learning?

Following from these concerns, we assumed that relatively informal and situated (more specifically location-aware) learning activity support, such as the location-aware learning organizer described in this chapter, can be more situated, experienced, and contextualized within specific domains.

### Limitations of the Study

Though this study seems to provide us with some guidance for the potential of future m-learning applications within the university context, it would be difficult to say whether this location-aware learning organizer would encourage the uptake of m-learning. Most importantly, we did not directly assess the most important function, that is, the content of the course, which can bridge the m-learning application with the conventional learning experiences. Indeed, we did implement this function, but the very slow connection to the content through the GPRS wireless network did not allow us to continue this functional test with our participants. Future studies will therefore need to use the faster HSDPA (high speed downlink packet access) network. Also, the effort required to collate the large amount required to implement a large scale version of this system would be also a significant challenge, and may inhibit the uptake of this type of m-learning system. As a practical approach to address this issue, a student management system that is currently up-and-running at Massey has also been approached to identify some working scenarios that would be available for future development ideas.

### Future Work

This study was designed to discover whether university students would gain significant benefits from contextual help, and how m-learning processes might be integrated with current learning experiences. Also, we briefly reviewed the two theoretical foundations that would work with the development of situated and contextual learning experiences. We partially reached these objectives, identifying the importance of the contextual help, especially outdoor location data. However, there is no conclusive evidence regarding integration of current learning activities, though the location-aware learning organizer was perceived as being useful.

The results of this study suggest several opportunities that could be pursued in future research, in particular, we found that the location-based reminder would be a useful function to support students' learning activities, which has not been demonstrated empirically before. Another would be that the indoor location data can be highly related to students' learning activities.

Indeed, this chapter is not the final result of this activity; the main objective here is to foster a wider discussion of m-learning design practices, based on an analysis of current research and future challenges.

## REFERENCES

Atkins, C. (2005). *Adult education and social reform.* Lecture notes in the distant lifelong education of the University of Nottingham. Hong Kong.

Becerra-Fernandez, I., Cousins, K. C., & Weber, R. O. (2007). Nomadic context-aware knowledge management systems: Applications, challenges and research problems. *International Journal of Mobile Learning and Organization, 1*(2), 103-121.

Bendy, G. Z., Seglin, M. H., & Meister, D. (2000). Activity theory: history, research and application. *Theoretical Issues in Ergonomic Sciences, 1*(2), 168-206.

Bertelsen, O. W., & Bødker, S. (2003). Activity theory. In J. M. Carroll (Ed.), *HCI models, theories, and frameworks: Toward a multidisciplinary science* (pp. 291-324). San Francisco, CA: Morgan Kaufmann.

Brown, J. S., Collins, A., & Duguid, P. (1989). Situated cognition and culture of learning. *Educational Technology, 18*(1), 32-42.

Brusilovsky, P., Schwarz, E., & Weber, G. (1996). A tool for developing adaptive electronic textbooks on www., *The Webnet 96 Conference of the Web Society.* San Francisco, CA.

Corlett, D., Sharples, M., Bull, S., & Chan, T. (2005). Evaluation of a mobile learning organizer for university students. *Journal of Computer Assisted Learning, 21*, 162-170.

Davis, F. (1986). *A technology acceptance model for empirically testing new end-user information systems: theory and results*. Unpublished Doctoral dissertation, MIT.

Davis, F. (1989). Perceived usefulness, perceived ease of use, and user acceptance of information technology. *MIS Quarterly, 13*(3), 319-340.

Demirjian, A., & David, B. (1995). Learning medical and dental sciences through interactive multi-media. *Medinfo, 8*, 1705.

Dix, A., Finlay, J., Abowd, G., & Beale, R. (2005). *Human-computer interaction* (3rd ed.). London: Prentice Hall.

Engeström, Y. (1999). Activity theory and individual and social transformation. In Y. Engestrom & R. Miettinen & R.-L. Punamaki (Eds.), *Perspectives on activity theory* (pp. 19-38). New York: Cambridge University Press.

Eschenbrenner, B., & Nah, F. F. (2007). Mobile technology in education: Uses and benefits. *International Journal of Mobile Learning and Organization, 1*(2), 159-183.

Flannigan, J. C. (1954). The critical incident technique. *Psychological Bulletin, 51*(4), 327-358.

Flannigan, J. C. (1962). *Measuring human performance*. Pittsburgh, PA: American Institutes for Research.

Gilbert, A. (2005). New for back-to-school: 'Clickers.' CNet news.com.

Horwood, A. (2003, 01.03.2003). Txt 2 help u quit smoking. New Zealand Herald.

Kukulska-Hulme, A. (2007). Hooked into learning: Designing for the Unexpected. In *W. o. P. L.* 2007 (Ed.). Toronto, Canada.

Kukulska-Hulme, A., & Traxler, J. (Eds.). (2005). *Mobile learning: A handbook for educators and trainers*. London: Routledge.

Kurti, A., Spikol, D., & Milrad, M. (2007). Bridging outdoors and indoors educational activities in schools with the support of mobile and positioning technologies. *International Journal of Mobile Learning and Organization.*

Kurti, A., Spikol, D., Milrad, M., & Svensson, M. (2007). *Exploring how pervasive computing can support situated learning, Pervasive Learning.* Toronto, Canada: Centre for Mobile Computing, Massey University.

Lave, J., & Wenger, E. (1991). *Situated learning: Legitimate peripheral participation*. Cambridge: University of Cambridge Press. .

Leont'ev, A. N. (1978). *Activity, consciousness, personality*. Englewood Cliffs, NJ: Prentice Hall.

Lyytinen, K., & Yoo, Y. (2003). Issues and challenges in ubiquitous computing. *Communications of ACM, 45*(12), 62-65.

Mathieson, K. (1991). Predicting user intentions: comparing the technology acceptance model with the theory of planned behaviour. *Information Systems Research, 2*(3), 173-191.

Rogers, Y., Price, S., Randell, C., Fraser, D. S., Weal, M., & Fizpatrick, G. (2005). Ubi-learning integrates indoor and outdoor experiences. *Communications of the ACM, 48*(1), 55-59.

Ryu, H., & Parsons, D. (2007). A learner-centred design of a location-aware learning reminder. *International Journal of Mobile Learning and Organization*.

Scornavacca, E., Huff, S. L., & Marshall, S. (2007). *Developing a SMS-based classroom interaction, mobile learning technologies and applications*. Auckland, New Zealand.

Sharples, M., Corlett, D., Bull, S., Chan, T., & Rudman, P. (2005). The student learning organizer. In A. Kukulska-Hulme & J. Traxler (Eds.), *Mobile learning: A handbook for educators and trainers*. London: Routledge.

Sharples, M., Taylor, J., & Vavoula, G. (2005). *Theory of learning for the Mobile age, MLEARN*. Cape Town, South Africa.

Smailagic, A., & Kogan, D. (2002). Location sensing and privacy in a context-aware computing environment. *IEEE Wireless Communications, 9*(5), 10-17.

Sommerville, I. (1996). *Software engineering*. London: Addison-Wesley.

Suchman, L. A. (1987). *Plans and situated actions: The problem of human machine communication*. Cambridge, U.K.: Cambridge University Press.

Tarasewich, P. (2003). Designing mobile commerce applications. *Communications of the ACM, 46*, 57-60.

Uden, L. (2007). Activity theory for designing mobile learning. *International Journal of Mobile Learning and Organization, 1*(1), 81-102.

# Chapter XIV
# Developing a Mobile Learning Platform for a Professional Environment

**Ana Dzartevska**
*Sandfield Information Systems, New Zealand*

## ABSTRACT

*This chapter describes some conceptual and practical issues in the development of a professional mobile learning system. The main focus of the chapter is the implementation of a mobile learning solution for the requirements of a specific working environment. The study proposes a conceptual framework for mobile learning applications, based on multiple studies of mobile learning design requirements, and uses it as a design tool for the development of the new system based on situated learning in the workplace. This approach is illustrated through the development life cycle of a prototype mobile learning solution. Finally, the chapter reviews and evaluates the success of the prototype and the utility of the framework as a design tool, identifies potential issues and outlines ideas for further development and research.*

## INTRODUCTION

### The Demographics of Professional Learning

If we consider today's business environment there have been some significant changes in workforce demographics and the technologies that support this environment. The 'millennials' is the term used for generation Y, people born between 1979 and 1997, who already account for 15 percent of the workforce (Hignite, 2007). While the 'baby boomers' (born between 1946 and 1964) currently outnumber the millennials by three to one in the workplace, it is estimated that within 10 years this proportion will change as more baby boomers retire and younger millennials reach their working age (Hignite, 2007). The differences between the new generation of employees and the current workforce are significant across multiple levels. Deloitte Consulting summarize the profile of generation Y as (Leah, 2006):

- Characteristics: techno-savvy, connected 24/7, independent, self-reliant;
- Beliefs: education is good, integrity is admirable;
- Technology: grew up online, never experienced life without computers, all information is a click away, expect information on the go, use instant messaging, have carried cell phones since high school (or earlier);
- Work attitude: work well with friends and in teams, collaborative thinkers, comfortable with speed and change, impatient, love a challenge.

These changes in the workforce create challenges for organizations to attract and also to retain new employees. The millennials are the most education-oriented generation in history. Furthermore, they are motivated and willing to constantly learn and gain new skills. They see an organization as a hub of resources that will provide training, learning, and relationship opportunities. They are highly mobile and technology literate and expect all information to be at their fingertips. In the light of these changes, businesses are faced with the challenge to not only create a good learning and training program within the organization but also to rethink the way learning is conducted and delivered amongst their employees (Koschembahr & Sagrott, 2005).

## The Mobile Professional

In terms of the technology that supports the current business environment, major emphasis is given to mobile devices and wireless communication of information. The market penetration of mobile device technology in the business professional environment has been tremendous over the last decade. Mobile devices such as PDAs and smartphones have grown to become an absolute necessity for employees in order keep up with their busy schedules and manage their work load in a highly dynamic, competitive, and time critical business environment. This significant push of mobile device usage has been supported by considerable advances in wireless technology. The rich functionality list featured in the mobile devices with their increasingly user friendly design, wider network coverage, higher bandwidth, and faster and more stable connections, make mobile devices not only practical to use but also a virtual requirement for many professional users to deliver their objectives, meet targets and keep up with the competition. This sophistication of mobile technology allows for more complex utilization opportunities that go far beyond simply making a phone call, sending an email or updating an organizer.

## Professional Mobile Learning

Companies are facing a serious challenge to revolutionize the way they deliver and implement professional learning among their employees. Training in a conventional manner results in significant loss of time which leads to potential profit loss for the organization. Every hour spent in a classroom means missed sales opportunities for many employees on the road. Thus, learning needs to be flexible in terms of time and place, available, and ad hoc in order to fit into the new workforce demographics and business environment.

Professional mobile learning is a way to address these issues. Mobile learning enables employees to choose what they want to learn, in their personalized format, at a time convenient to them and wherever they are. Learners can use their 'down time' to extend their professional skills or product knowledge. Learning can be done during breaks, before meetings with clients or in traveling time. '*Mobile learning is a solution that opens new possibilities to already existing technologies*' (Koschembahr, 2005a). Companies already have their field employees armed with wireless mobile devices, with which they are very familiar and comfortable. People find mobile devices personal and convenient. They currently have their phones, calendars, inboxes, contacts, organizers and conference environments in their pockets. Since they carry almost the entire office around with them, a classroom should easily fit into that same pocket as well.

The research described in this chapter focuses on the mobile learning process in a commercial context. The concept of learning in this kind of environment is significantly different from conventional academic learning. It requires 'just-in-time' (Koschembahr, 2005b) and 'byte sized' learning that will be available on-demand and that could easily be incorporated into the organizational day to day workflow (McGee, 2003). In many cases professional mobile learning is more content focused than the type of learning offered by educational institutions, and is often focused on performance support (Kukulska-Hulme & Traxler, 2005). However, a professional mobile learning concept needs to provide more than just a support for day to day operational activities. While the same platform can merely be used for obtaining time critical information, there is much greater potential for the context of true professional development of the employees within an organization. Therefore in this chapter we do not simply address down time didactic learning. Instead, the focus here is on situated and collaborative learning.

## Overview of the Problem Domain

This research targets a dynamic professional user who is already using a mobile device to perform day to day operational activities. The study described here was conducted with the help of a participating organization and a prototype application was developed using a relevant context from the business environment. The participating organization is a major player in the supply chain industry in New Zealand and Australia and the research focused on the logistics management component of their business activities. The organization already has formal classroom based training in place. It is a requirement that every warehouse employee undergoes five days of induction training. This training consists of several components:

### Induction and Customer Service Training

This section covers the history of the company, the company's culture and standards, health and safety issues and training on customer service.

### Overview of the Warehouse Management Information System (WMS)

This part gives an overview of the technology used across the business and focuses on the warehouse management information system that is being used across all warehouses in Australia and New Zealand.

Forkhoist training
Practical on site training.
Warehouse operations training

The employees are introduced to all types of stock management operations that occur in a warehouse on a daily basis. Every operation is broken down into a series of steps that need to be verified or completed in order for the operation to be accurately and successfully accomplished. All employees are trained on the standard procedures of handling inwards, orders, picking, packing, pricing, and inventory management and control

### Dangerous Goods Training

As the company specializes in transportation and storage of hazardous goods, every employee is trained on all different classes of dangerous goods and the standard health and safety procedures for their handling.

The mobility of the organization's employees is driven by the common operations in supply chain logistics management such as picking, packing, placement and inventory management. The employees already have access to mobile devices (Symbol Windows Mobile Pocket PCs) which they use to fulfill some of the logistics management processes. The mobile devices connect to a Wi-Fi network that is set up across the warehouse. This enables the communication between the central WMS and employees' handheld devices.

When goods arrive in the warehouse, the list of products, product descriptions and receiving quantities are sent from the WMS to the Pocket PC of the employee assigned to that job. The employee confirms the arrival of the goods by scanning each product, entering product characteristics, quantities and other information such as expiry dates, batch numbers and pack types. This information is then sent back to the WMS and the goods become operational in the system.

Similarly, when an order is received in the WMS, the goods need to be picked, packed, and sent to the client's address. The information about the requested products, product characteristics, quantities and picking location is again sent from the WMS to the Pocket PC of the employee assigned to the job. The employee uses this information to locate the goods in the warehouse, to make sure the picked products match the details in the order and to record any discrepancies such as product shortfalls, grade mismatch, and so forth. Once the goods are picked, confirmation information is sent to the WMS and the order is ready to be packed and dispatched.

For the purposes of the research, a prototype was developed to partially mobilize the learning content of the warehouse operation training and the dangerous goods component. The idea of the research was not to develop a mobile solution that would replace the classroom based training. Instead the application was intended to provide ongoing support for the users in terms of product and process information as well as provide the basis for skill improvement and development. In this sense, it was designed as a combination of blended and situated learning, reinforcing more traditional training delivery with context relevant experiential learning.

## Research Objectives

The main objective of this research project was to propose a suitable framework for mobile learning design requirements that provides a solid and well-structured basis for implementation of professional mobile learning applications. It integrates relevant factors and design requirements for mobile learning that have been identified by previous studies in this area. The framework also recommends how mobile applications should be organized and designed in order to meet these requirements. The study then puts the proposed framework into practice by using it as a design tool in the implementation of a mobile learning application for a specific professional environment.

The content delivered via this application needs to be relevant and valuable for the professional development of the employees of the collaborating organization. It needs to provide an engaging context for learning, which will keep the user focused and interested. This application is designed for mobile devices that can connect to a wireless network. The challenge is to utilize the obvious advantages that come with this technology such as ubiquity, personalization, and convenience but also to deal with some potential pitfalls, such as limited real estate, for presenting the content on a mobile device, inconsistent connectivity, network coverage, and so forth. Finally, the proposed solution should conceptualize a solid and useful platform for professional mobile learning that brings in measurable results.

The research questions addressed in this study may be summarized as follows:

1. From previous work, what design frameworks can usefully be adapted to develop an effective mobile learning system for a professional context?
2. What enhancements might be made to such frameworks for the requirements of this particular implementation?
3. What practical implementation issues have to be addressed in order to produce a working prototype?
4. To what extend does the implementation, driven by the design framework, validate the presumptions of that framework, as measured by user evaluation?

The prototype developed from the research questions and requirements was evaluated against the collaborating organization's objectives in the mobile learning context: improving professional skills and knowledge in operational activities and client's products; enhancing employee's loyalty and providing ongoing support for mobile employees.

### Chapter Structure

The next section reviews the conceptual and physical design of the system, reviewing design tools and approaches along with academic literature relevant to professional mobile learning. It introduces the design framework and the methodology that are used in this research. The following section covers the development of the prototype system, providing a description of the iterative development cycle, its phases, and its milestones. The next section covers the user evaluation of the system, using a group of participants from different areas of expertise. The final section summarizes the results in the context of the initial research proposal. It draws some lessons to be learned from the development experience and outlines some ideas and suggestion for further research relevant to the developed system.

## CONCEPTUAL AND PHYSICAL DESIGN

In order to create a suitable design for the professional mobile learning system, it was necessary to address a number of areas of research and development. First, there are issues related to the physical mobile device platform for mobile learning. Second, there are the specific requirements of the application context. Third, there are conceptual issues around the components of an effective mobile learning application, which may be summarized in a suitable conceptual framework. This framework may then be used to support the proposed system architecture. Finally, an appropriate research methodology must be chosen. This section addresses these issues in the context of relevant academic and commercial literature.

### Mobile Platform Issues

A successful m-learning application gains from the benefits of a mobile device but is also aware of its limitations. In the context of the technology platform, the application needs to provide a learning context relevant to the user's requirements; present it in an interesting and engaging manner and deliver a specific learning objective. Thus, mobile learning design must address both the characteristics of the mobile environment and the learning requirements of the application.

A major decision therefore is the selection of appropriate implementation technologies for the application and the type of client that will be developed. According to the system's requirements the client architecture can range from a thin client that requires a browser application running on the mobile device to a smart client developed in, for example, Microsoft .NET Compact Framework

*Table 1. Features of thin, thick and smart client architectures*

| Feature | Thin client | Thick client | Smart client |
|---|---|---|---|
| Provides a rich user interface | Yes, but difficult to develop, test, and debug. Generally ties the application to a single browser. | Yes. Easier to develop, test, and debug. | Yes. Easier to develop, test, and debug. |
| Can take advantage of hardware resources on local computer | Yes, but only through additional components. | Yes | Yes |
| Can interact with other local applications | No | Yes | Yes |
| Can interact with server side resources | Yes | No | Yes |
| Can be multithreaded | No | Yes | Yes |
| Can function offline | No | Yes | Yes |
| Can perform well in low bandwidth and high latency environments | No | Yes | Yes |
| Easy to deploy | Yes | Varies. Difficulty depends on application requirements. | Varies. Difficulty depends on application requirements. |
| Low maintenance and change management costs | Yes | Varies. Updating the application may be problematical. | Varies. Costs depend on application requirements. |
| Can be deployed to a wide variety of clients with varying capabilities | Yes, although more complex thin clients may require a single browser. | Depends on the selected implementation technology, e.g. A thick client developed in .NET will have more limited deployment options than one developed in Java ME, though even here there are issues with versions and libraries. | As for thick client, depends on the selected implementation technology. Relies on suitable communication mechanisms being enabled on the device |

or J2ME (Java 2 Micro Edition). A third type of client that can be developed is a thick client. This is also called a heavy client and unlike the thin or the smart client it does not need a network connection to be fully functional. The main aspects that need to be considered in selecting between a thin, thick, or smart client architecture are summarized in Table 1.

## Thin Client

Thin clients rely on mobile browser technology to render pages using mark-up languages. However in the mobile context even thin clients are not without their difficulties. A number of different mark-up languages are used such as WML (wireless markup language), XHTML-MP (extensible hypertext markup language—mobile profile) or cHTML (compact hypertext markup language), Devices also have different interface characteristics (different screen sizes and orientation, color displays, etc.), different network connectivity (cellular connection, wireless LAN) and may have other differences in capacity (graphical support, plug-in support) so developing web applications that will be accessed by wireless browsers is quite challenging (Microsoft, 2007). However there are a number of server side frameworks available that can generate adaptive markup. At execution time, the mobile control generates the suitable mark up for the device that makes the request (Kumar, 2007).

### Thick Client

A mobile thick client has the benefits of a locally installed application. It uses local resources and can provide a richer user experience. It takes advantage of the local CPU, local memory, and other devices connected to the client such as bar code readers, and so forth. It also can use and interact with other software applications installed on the device, for example the client can launch a media player in order to play a video file. It is off line capable so it can be used independent of wireless networks and it is easily deployed and configured using a simple file copy or a download. However, it has no network connectivity and therefore cannot update itself or its data using a network connection.

### Smart Client

A mobile smart client combines aspects from both local applications (thick client) and Web applications (thin client). It benefits from local resources and is offline capable, but is also a connected application that exchanges data on the Internet or an enterprise network, enabling it to be updated wirelessly and cache data locally (Diws, 2005). A smart client usually talks to Web services that provide access to a central database which enables the data communication flow to and from the device in a standard XML format (Poddar, 2003).

## Mobile Learning Application Context

The previous section identified some design requirements and challenges that arise from the characteristics of the mobile environment. The focus now shifts to the mobile learning application context, with an overview of a number of system specific design requirements. These requirements are reviewed from the following perspectives: user and context analysis, interface design, application structuring and organization, learning experiences, and learning objectives.

### User and Context Analysis

Appropriate user and context awareness is crucial in designing applications that are suitable for a variety of learners, given that they may have very different skills and motivations to learn in varying contexts (Syuvanen et al., 2005). As the user has a central role in the learning process the user analysis is a fundamental element in system development. The analysis process should identify the characteristics of the user; the user's familiarity with the mobile device; their motivation and interest in learning using a mobile device and their expectations from the mobile learning process. Different contexts and situations in which the application would be used should also be considered and catered for.

### Interface Design

Regardless of the type of client, interface design decisions need to be made based on the characteristics of the mobile device the application will be deployed on. Grasso and Roselli (2005) suggest several guidelines for usable interface design; correspondence to the hardware characteristics and functions, rationalized content, and simple, handy usage. They also suggest interface design guidelines from a pedagogical perspective, including the role of help and feedback on learning tasks, personalization of the application such as adding comments or notes for future review, and integrating previously learned content into new content.

## Application Structuring and Organization

Meisenberger and Nischelwitzer (2005) outline four elements that should be fulfilled by the learning application to achieve greater learning richness; exposition, exploration, construction, and communication.

Exposition defines the characteristics of the mobile learning object (MLO). In a mobile environment, these should be limited in size, presented in a sequence of slides and concluded with a set of questions. Where possible, different media should be applied to compensate for the limited amount of continuous text used. Generally, mobile learning objects should not contain as much information as other types of learning objects (i.e., e-learning objects). Due to the nature of the mobile environment, the content should be split across multiple MLOs. The MLO should be highly cohesive but also logically connected with other MLOs. Another similar concept is that the content should be 'byte sized' that will be available on-demand and that could easily be incorporated into the organizational day to day workflow (McGee, 2003).

Exploration allows the user to explore the learning content on their own without having to follow a predefined learning path. This puts the users in charge and motivates them to pursue the learning at their convenience. Hyperlinks can be used to support this concept though these can lead to the 'lost-in-hyperspace' phenomenon which may lead to negative experiences (Meisenberger & Nischelwitzer, 2005).

Construction gives the user the opportunity to 'learn by doing,' for example by using simulations and models to create a virtual environment that reflects a real life situation that the user follows or participates in.

Communication utilizes the benefits of the mobile environment to create a dynamic and collaborative learning experience. The right utilization of synchronous or asynchronous communication models leads to building an engaging and motivating system by enabling competition, challenge and social interaction among the users (Parsons, Ryu & Cranshaw, 2006).

## Learning Experience Metaphors

The application needs to provide an engaging context for learning which will keep the user focus and interested. Two useful metaphors are suggested for delivering a satisfying and enjoyable mobile learning experience: the cinematic metaphor and the game metaphor (Parsons et al., 2006). By using the cinematic metaphor, the content is presented in a narrative manner that is easy for the user to follow. For professional mobile learning it is important to try to contextualize learning into real-life situations. A narrative technique can be utilized to illustrate such an example. The technique also allows the one-way narration to be extended with some input from the user in such a way that they can decide the flow and direction of the presented content (Walker, 2004).

The game metaphor incorporates the use of games to increase the degree of engagement and interest in the application. In a mobile learning context, a game can be used to test/confirm the knowledge gained after a series of mobile learning objects have been reviewed. Schwabe and Goth (2005) identify several elements that create an engaging context:

- Rules — define and organize the game. Should be unambiguous, intuitive and easy to follow;

- Goals and objectives — that the users try to achieve. The user will be more engaged and interested to pursue a game if the objective is clearly outlined and known from the start;
- Outcome and feedback — evaluating results against the goal. The user should know how they did and what they achieved after they finished the game;
- Conflict, competition, challenge, opposition and interaction — these elements increase the motivation of the user and create a dynamic and unpredictable environment;
- Representation of the story — in the mobile learning context this is the way the learning content is incorporated and presented in the game.

## Learning Objectives

The learning objectives can range from improving personal skills or gaining new skills to improving team building and social interaction within the company (Parsons et al., 2006). The specific learning objectives are defined in the context of the developing application. Different types of design approaches and techniques can then be selected, utilized, and evaluated for the delivery of the desired set of objectives.

*Figure 1. A framework for M-learning design requirements (from Parsons et al., 2007)*

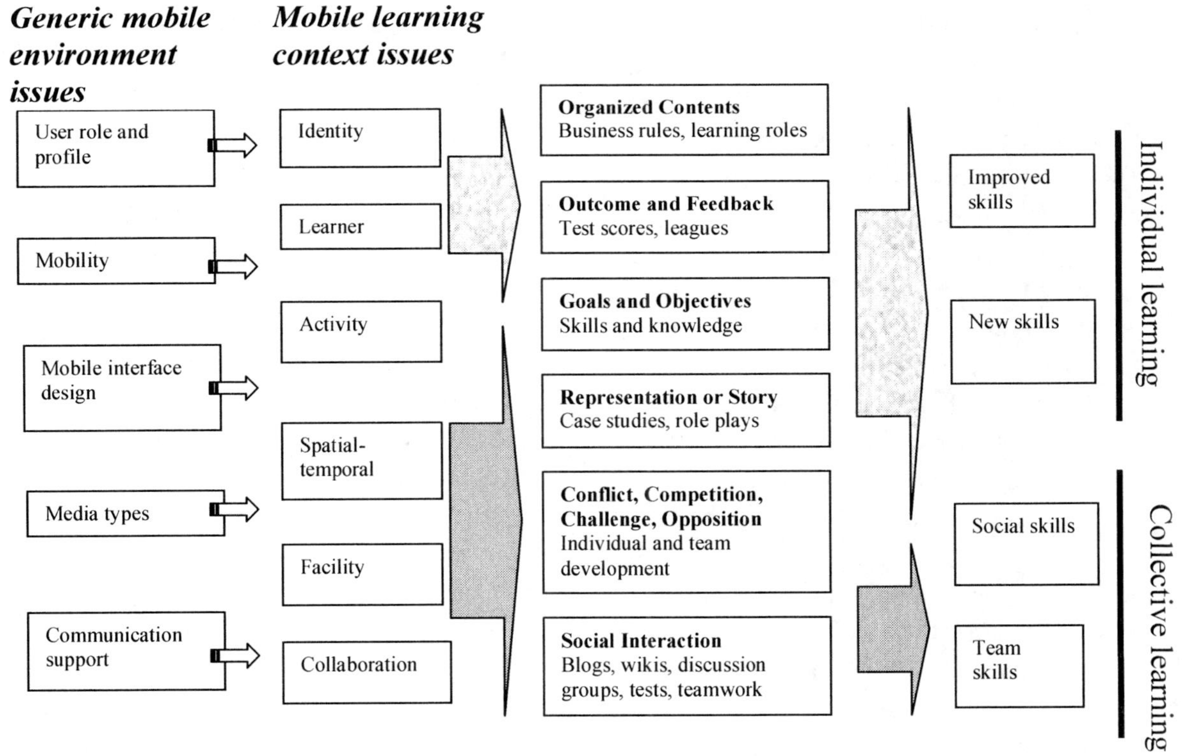

## Deriving a Framework for Mobile Learning Design Requirements

So far, a number of environment and application specific design requirements have been outlined and grouped into categories based on the context from the mobile learning area they target. Parsons, Ryu and Cranshaw (2007), propose a conceptual framework for mobile learning applications that combines most of the design categories mentioned in the previous section. Figure 1 shows the four pillars for mobile learning design requirements that compose this framework: generic mobile environment issues, mobile learning context issues, learning experience and learning objectives, and the interaction between them.

As an example that illustrates the interaction between the design requirements and the logic flow in the framework, consider an organization whose learning objective is the development of social and team skills. The development of these skills is supported by learning experiences such as: social interaction, competition, challenge, opposition and conflict. To deliver these learning experiences, the application needs to implement some kind of communication and interaction module. The design of this module needs to be supported by an adequate analysis of the user's activities; the user's collaboration in those activities as well as their spatial-temporal dimension. These requirements can then be mapped against some relevant mobile environment issues such as network connectivity; type of mobile device that will be used or media that needs to be supported.

Further analysis of the framework points out a view of the design requirements from a broader perspective. The left hand side of the framework defines the design requirements: learning objectives and learning experiences. These can be seen as expected outcomes from the learning application, requirements that are set to 'work towards.' As illustrated in Figure 1, in order for a mobile learning application to encourage development of social and team skills it needs to provide experiences such as social interaction, competition, challenge, and so forth. The developed application is expected to deliver these learning experiences to meet the required learning objectives.

On the other hand, in the framework, design requirements relevant to the learning context and the technical issues in mobile learning environment are defined. These design concerns arise from the characteristics that are specific to the business environment the mobile application is developed for and the mobile environment itself, that is, appropriate analysis of the business environment identifies the different types of learners the application should be designed for and defines the business activities that determine the content of the learning application. Therefore, they can be seen as input requirements or requirements that are given to 'work with.'

Another practical study in this area undertaken by Meisenberger and Nischelwitzer (2005) refers to the four categories of mobile learning application structuring and organization: exposition, exploration, construction, and communication. Their study gives an overview of how those categories can be practically realized within a mobile learning application, and the definition of these four categories is used as a design tool in the development of the mobile learning engine, a computer-aided and multimedia-based learning application created for mobile phones (Meisenberger & Nischelwitzer, 2005). Compared to the previous study that concentrated on the specifics and the expectations of environment, this one puts the focus on the application being developed and provides some guidelines and ideas for appropriate content organization and design that delivers different levels of learning-richness.

Based on the two frameworks previously introduced, Figure 2 shows a new framework of mobile learning design requirements that is proposed by combining key features of these two complementary frameworks.

*Figure 2. Proposed framework for mobile learning design requirements*

| **Mobile Environment Issues and Interface Design** | **User and Context Analysis** | | **Application Structuring and Organization** | | **Learning Experience** | **Learning Objectives** |
|---|---|---|---|---|---|---|
| Implementation Technology | User roles and profiles | ⇨ | Exposition | ⇦ | Organized Content; Goals and Objectives | Improved Skills, New Skills |
| Interface Design | User identity and categorization | ⇨ ⇨ | | ⇦ | Outcome and Feedback | |
| Media supported | User activity definition | ⇨ | Exploration | ⇦ | Motivation; User in Charge | |
| Mobile Device | User activity Spatial-temporal dimension | ⇨ ⇨ | Construction | ⇦ ⇦ | Representation or Story; User's Input | |
| Connectivity | User activity collaboration | ⇨ ⇨ ⇨ ⇨ ⇨ ⇨ | Communication | ⇦ ⇦ ⇦ ⇦ ⇦ ⇦ | Social interaction: Competition, Challenge, Opposition, Conflict | Social Skills, Team Skills |

The proposed framework incorporates the four categories of mobile learning application structuring and organization; exposition, exploration, construction, and communication. These categories make up the fifth pillar of mobile learning design requirements, 'application structuring and organization,' which sits in the middle of the framework. The idea is that both the 'work with' and 'work towards' design requirements feed information into this part of the framework which puts these requirements into the perspective of actual application design. This new column actually sits like a bridge between the two parts of the framework and provides a good starting point for implementation of any mobile learning application as it directs the developer to the specific element of the system that requires work and investigation in order for those requirements to be met, that is, if an organization requires development of new skills or improvement of existing skills among the users, then the application should feature one or more of the following elements: exposition, exploration and construction. As shown in Table 2, these elements deliver learning experiences that support the desired learning objectives, such as: logically organized and easy to follow content; delivering outcome and feedback back to the user; clearly outlined goals and objectives that motivate the user and help them engage; using a cinematic metaphor for story representation and story construction; putting the user in charge of their learning process. The development of these elements of the system needs to be supported by analysis and categorization of different user profiles and roles and their activities. Subsequently, this will be followed by investigation and analysis of relevant mobile environment issues and design requirements.

One of the research objectives of this project was to put the proposed framework into practice by using it as a design tool in the implementation of a mobile learning application for a specific pro-

*Table 2. Organization's learning objectives*

| Learning objective | Description |
|---|---|
| Improved operations execution | The aim is to accurately follow the standardized company's procedures to minimize error rate and provide higher standards of inventory management. |
| Obtaining new knowledge in operational activities | The mobile learning application should extend the classroom based training for new operational activities. |
| Support | The application needs to provide critical, just in time information for mobile employees such as product information, dangerous goods qualification, handling procedures etc. |
| Collective learning | Users should be able to share their comments, feedback or questions with each other, and participate in open discussions and forums. The aim is to increase social skills and provide an environment that supports and encourages collective learning. |

*Table 3. Learning experiences*

<table>
<tr><th>Learning objective</th><th>Learning Experience</th></tr>
<tr><td>Improved operations execution</td><td rowspan="2">Organized Content; Representation or story; Goals and Objectives; User in Charge; Outcome and Feedback</td></tr>
<tr><td>Obtaining new knowledge in operational activities</td></tr>
<tr><td>Support</td><td>User's Input; Outcome and Feedback</td></tr>
<tr><td>Collective learning</td><td>Social interaction, Communication, Feedback, Information Sharing</td></tr>
</table>

fessional environment. Following this framework, an initial outline of the system architecture and design can be made at this stage.

## Proposed System Architecture

Following the information gathered from the literature review with regards to the design requirements for a professional mobile learning platform, an informal proposal describing the learning objectives for the system is summarized in Table 2. This table was compiled from the initial requirements analysis and was used as a starting point for the system development.

To support the learning objectives outlined in Table 2 the mobile learning application needed to deliver appropriate learning experiences. Table 3 links each of the previously specified learning objectives with a corresponding set of learning experiences from the framework.

The learning content is organized into a sequence of logically connected learning objects, each designed to deliver a specific knowledge objective. The way the flow of the learning objects is organized and delivered to the user defines the content representation or story. This can be a one way representation with the possibility to add on an interaction element where the user defines the progression and direction of the story/learning sequence. This puts the user in charge of the learning process and they can adapt the delivered content to a specific scenario or requirement. Each set of learning objects should be concluded with a set of questions and exercises. This helps to provide an appropriate outcome and feedback to the users that will summarize their effort and achievements in completion of this objective.

As mentioned in the learning objectives section, the application needs to provide critical, just-in-time information for mobile employees such as product information, dangerous goods qualification, handling procedures. In order to support this objective, the application needs to provide a way of capturing the user's input, interpreting it and returning relevant information, that is, the user should be able to find the definition of a certain industry term, abbreviation or process. Also, the user should have access to important product information by entering a product code or scanning a bar code. Finally a mobile learning feature that enables immediate sharing of information, comments, questions or feedback with other employees or trainers provides an environment for successful collective learning.

In terms of application structuring and organization, for exposition, the cinematic metaphor was adopted. A section of training material was selected and separated into series of MLOs, logically and sequentially connected with each other but each representing a cohesive and well defined item on its own. Every MLO was presented by a series of slides, controlled by navigational keys, which defined the tasks that needed to be performed in order for the activity presented by the MLO to be successfully completed. For exploration an index table was created for frequently used keywords in the industry. The user could use this facility as a quick reference for the meaning of abbreviations, keywords, and jargon terms. Similarly, the application would accept user's input in the form of a product code and return important information such as product description, handling procedures, and so forth. To cater for the construction element of the system a walk through approach was implemented where a user could select a direction of execution at decision points according to a real life scenario. Finally for communication, a game metaphor was applied in the form of an interactive quiz. The questions were derived from the content of the MLOs presented in the exposition part of the system. According to the posted answers, appropriate feedback was delivered back to the user via the application's interface. To encourage collective learning and

*Table 4. User and context analysis*

| Criteria | Description |
|---|---|
| Familiarity with device | As the employees have been using the mobile devices in their day-to-day activities, their familiarity with the device is very good. There would be no need for additional training. |
| User's interest in m-learning | Generally users expressed interest in the concept of delivering learning material on their mobile devices. Different patterns of learning styles and preferences have been identified among the users in this step. |
| User's expectations | The application will provide support for their day-to-day activities;<br>The application will provide opportunities for future skill development and improvement. |
| User categorization | The users are separated into teams according to their business activity. Every team has a defined set of activities that they perform, i.e. placement, picking, packing and control teams. |
| Time dimension | The application should be able to run while executing an operational activity. It should also be available for use during down time periods. |
| Place dimension | The application should be available on the company's premises. |

information sharing a form of mobile forum/discussion board was developed where users could post their comments and questions. The findings from the initial user and context analysis are summarized in Table 4.

In terms of interface design, the application followed a similar look and feel to the other mobile web applications that are currently in place within the participation organization. Special attention was paid to personalization by recognizing the user and loading content according to predefined setup or preference. The user was able to make some changes to the content or add notes which were available only to that specific user. The application in general provided feedback to user's actions and allowed the user to review, cancel or modify an action.

For the implementation technology, as the employees are already using a mobile device in their normal operational activities it was an obvious decision that the prototype application should be running on the same device, a Symbol Windows Mobile Pocket PC. Since the mobile devices are chosen as a company standard, an assumption could safely be made that the application does not have to cater for multiple platforms or investigate compatibility issues. To cater for some requirements like communication, collaboration or personalization the application needed to be connected to a network and potentially query a central database.

Given the set of requirements gathered for the system, one clear requirement was that the system should have network connectivity. Therefore (from Table 1) the client would need to be either a thin or smart client. However, (also from Table 1) the limitations of thin clients in terms of user interface and high latency environments, coupled with the known target platform, reducing issues of multi-platform, deployment, suggested that a smart client solution would be the most appropriate for this system. The prototype was therefore developed as a mobile smart client using Microsoft .NET Compact framework. However, the initial iterations in fact comprised a thick client implementation because those modules that required network connectivity were only implemented in later iterations.

## Research Methodology

The methodology applied in this study was design research (also knows as design science). The underlying idea of design research is the concept of knowledge—process interaction (Thompson et al., 2004). Basically the research and the design part of the project are not considered as two separate processes with a single serial execution. They are iterative processes that provide feedback to each other, thus refining and improving one another with the outcomes of each step. From one hand, the knowledge is utilized to build the process (design). However, the outcomes of the processes, their evaluation and implementation generate new ideas and clarifications for the knowledge that has been put into practice. Each iteration goes through several stages: Awareness of problem, suggestion, development, evaluation and conclusion (Vaishnavi & Kuechler, 2004/5).

Although the developer has a main goal in mind, the research and the knowledge gained from each cycle shape the way the next iteration is conducted, that is, every milestone is fragmented into a series of tasks. The execution of a task depends on the outcome and the findings from the previous task. This means that a new task can be set up on the fly, or removed if it is no longer required or relevant to the context.

Each of the process steps can also be explained through the expected outputs; a proposal, a tentative design, an artifact (a formal proof of the proposal and the initial design), performance measures and results. These results can generate new knowledge or an intriguing topic which can be used as an idea for future design research.

In the specific context of developing a framework for a professional mobile learning platform, the design research methodology accommodates the complexity and dynamics of the environment of this project. The iterative nature of this approach caters for the uncertainties of the environment and constantly encourages suggestions for new requirements, improvements or modifications. In fact it has much in common with commercial approaches to iterative software engineering. However, "*Routine design is the application of existing knowledge to organizational problems...design-science research addresses important unsolved problems in unique or innovative ways or solved problems in more effective or efficient ways*" (Hevner, March, Park, & Ram, 2004, p. 81).

Van den Akker, (1999) suggests several strategies within the design research methodology to deal with the complexity of the design research process. These strategies are primarily linked to the requirements gathering and evaluation stages of the process. In this study, the requirements gathering stage comprised interviews with a domain expert in the area of professional learning from the collaborating organization. A prototype was developed as an initial proof of concept, to gather some feedback and model the requirements moving forward. The evaluation stage comprised meetings with an expert from the collaborating organization followed by a formal usability study conducted among the potential users of the application.

## SYSTEM DESIGN AND PROTOTYPE DEVELOPMENT

The development of the mobile learning application was executed in three phases. Table 5 summarizes these development phases along with the structural elements of the application the particular iteration was focused on. The last column outlines the deliverables of each phase. Each deliverable is a piece of working software which represents a logical module of the application.

The next section follows the implementation of the mobile learning application in a structure that reflects the iterative development approach applied.

### Phase One

Following the initial set of requirements and the proposed system architecture, a first cut of a working mobile learning application was developed. During this initial phase, the emphasis was on project and client setup; interface design; the exposition and exploration element of the application.

#### Implementation

Following the company's standards and its preferred implementation technology, the mobile

*Table 5. Application development phases: Focus and deliverables*

| Phase | Structural Element | Output |
|---|---|---|
| One | Exposition, Exploration | Logistics Chain of Events, Index Search, Product Information |
| Two | Exposition, Construction | Forkhoist Training Videos, Step by Step |
| Three | Exposition, Communication | The Challenger, Message Board |

*Figure 3. A typical MLO structure in the mobile learning system*

learning application was developed using Microsoft .NET Compact framework. In this initial stage of the system implementation, a thick client was developed, that is, the client was stand alone and no network connectivity was required to be able to use the application. The application used only the resources of the device it was deployed on for its operation.

## Interface Design

The interface design of the initial prototype was kept as simple as possible, following some of the previously mentioned guidelines for designing useable interfaces from both technical and pedagogical perspectives. Because of the limited real estate on the mobile device to display learning content, tab controls were used to rationalize space utilization. Where appropriate, the application provides feedback to notify the user of an invalid action, mistake or completion of a certain task.

## Exposition

The exposition element of the system defined the format of the learning content delivered to the user. Linking back to Grasso and Rosseli's (2005) technical and pedagogical guidelines for designing useable applications, some text exposition standards were defined at this stage.

Following the available classroom training material and the teaching style applied so far, it was easy to break down the main learning content into a set of operations that were connected between each other in some logical order. Every operation could be broken down into a chain of events. Depending on the particular scenario, the events could follow a different order and one operation might have multiple execution paths. An event, on the other hand, can be further broken down to individual tasks which are cohesive on their own but together describe the event in greater detail. The task was utilized as a building block for the mobile learning object.

A learning object delivers the details of one task. It is logically connected to other learning objects and together they make up a larger learning deliverable; an event. The MLO is limited in size in order to fit the real estate on a mobile device screen, but also to keep the focus of the user. As can be seen on the screen shots in Figure 3, the user can use the navigation buttons to go through the learning objects following their predefined sequence. Graphical representation was also used to illustrate the logical connection between individual events within one operation as well as between operations that occur in the day to day business within the warehouse. This structure allows the user to grasp the big picture of the business processes, while giving the opportunity for more in depth detailed exploration. This is shown in Figure 4.

*Figure 4. Example of designing logical connection between events*

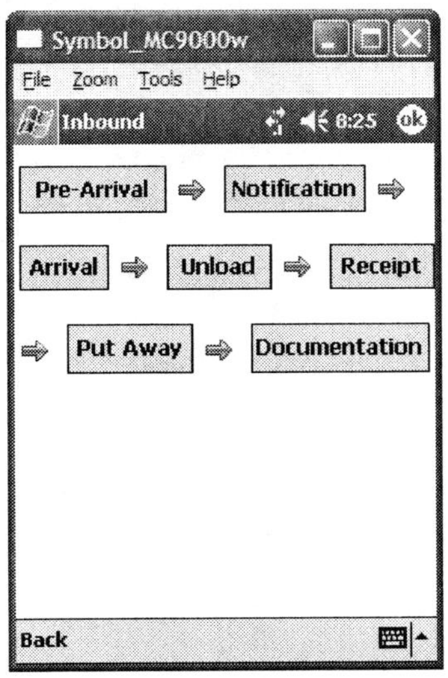


*Figure 5. A simple index search screen*

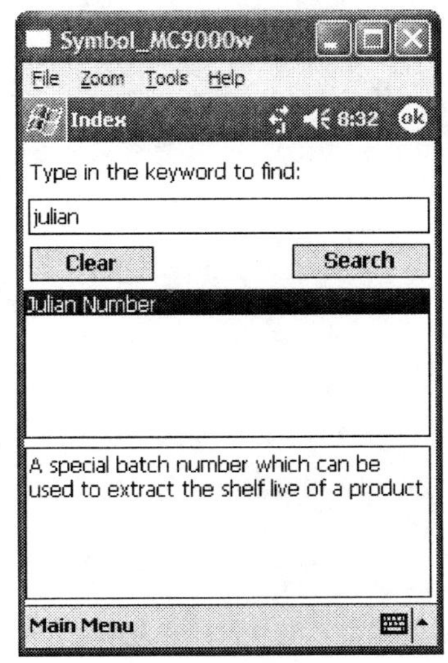


*Figure 6. Screenshots of a product search screen*

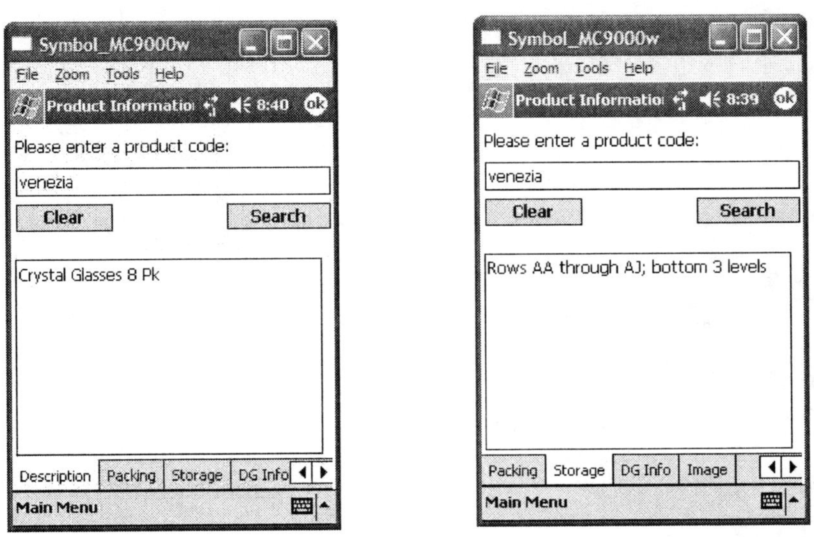

## Exploration

The exploration element of the application gives the opportunity to the users to take control of their learning process and empowers them to choose the content and the delivery style of the learning material. Initially, a quick search screen was developed where the user can just partially enter a term from the business, an abbreviation or a keyword. The search engine returns all valid matches to the supplied entry and when the user selects a particular one, its definition or description is displayed in the bottom pane of the screen. Figure 5 demonstrates an example of an index search screen in the mobile learning application:

Providing support for the mobile employees in terms of delivering time critical information in relation to products, dangerous goods and handling instructions was mentioned as an important learning objective of the participating organization. A warehouse employee comes in contact with thousands of different products on a daily basis, some of which are of a hazardous nature and have quite specific handling and storage procedures. It is practically impossible to deliver this important information to the employees via one off conventional classroom based training, or even to expect that it will be correctly memorized. A product detail screen was developed as part of the initial development phase that conceptualizes the basis of this requirement. An employee can simply enter a product code in the product search screen and the system will display important details such as description, handling and storage information, and so forth. As the employees are already using the mobile device in their daily operations, this really crucial information is right at their fingertips and can be accessed whenever such a situation arises. Figure 6 shows some screen shots of the product screen developed at this stage.

The initial phase of the development provided the basis for the mobile learning solution, defined a typical structure of a textual MLO used in the exposition element of the solution and delivered an example of exploration functionality. In the following phase, alternative ways of delivering content to the user were examined and a functionality that supports the construction element of the system was designed and developed.

*Figure 7. Audio-bar for playback of an audio file*

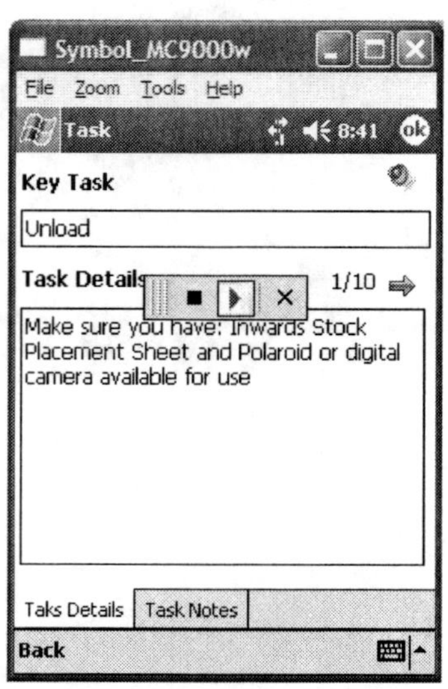


*Figure 8. Using images for delivering content*

## Phase Two

### Exposition

As previously mentioned, the application's exposition element is all about the way the content is delivered to the user. During the first development iteration, only text, (a very traditional medium), was used for these purposes. When it comes to delivering learning material there is some content that is too complex or simply unsuitable to be delivered in a textual format. Therefore, alternative means of presentation needed to be explored.

Since the initial application was developed as a thick client, it could utilize the resources from the local machine. This enabled audio objects to be added to some existing MLOs to further emphasize an important note or just to deliver some additional information associated with a particular event. The user can play the audio file with a click on a button and control it via an audio bar. An example of this is shown in Figure 7.

Figure 8 shows the integration of an image within a MLO as another example of delivering content in a different format.

In this particular example, the image is used on the product details screen and it provides an additional safeguard for ensuring correct management of a specific product.

Operating a forkhoist is a very important dimension and is an integral part of every warehouse operation. Every employee is required to undergo a forkhoist course that involves a one day, practical, on-site training. When considering this element of the training process, video content was selected as the most suitable media for delivering learning material. Short video files containing some common errors when operating a forkhoist were made available through the application. For the purposes of the non-commercial prototype, videos were taken from publicly available samples (IVES Training, 2007). The user can select from the list of available videos and by clicking on the file name, the application launches Windows Media Player which plays the file. Going back to Grasso and Roselli's (2005) principles for interface design with regards to the use of media objects, the selected files are concise and summarize one concrete point. Figure 9 demonstrates the use of video media objects within the mobile learning application.

*Figure 9. Video media objects within the mobile learning application (IVES Training 2007)*

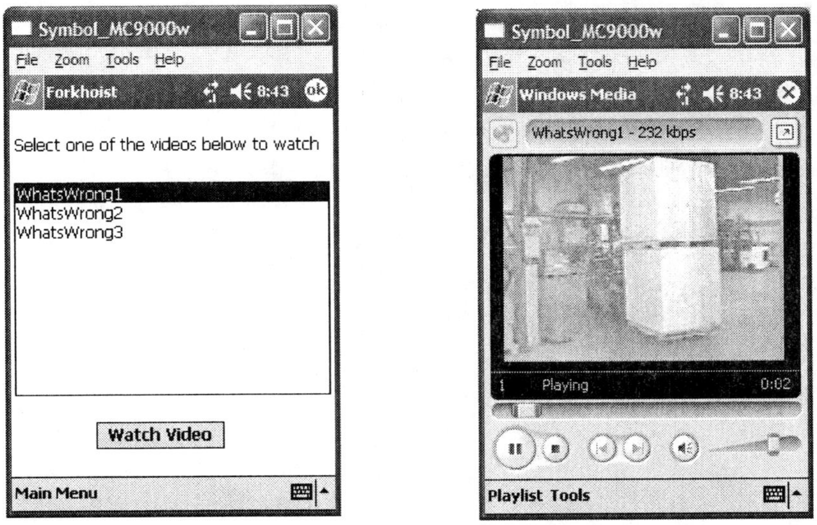

*Figure 10. An example of a construction (decision point) element*

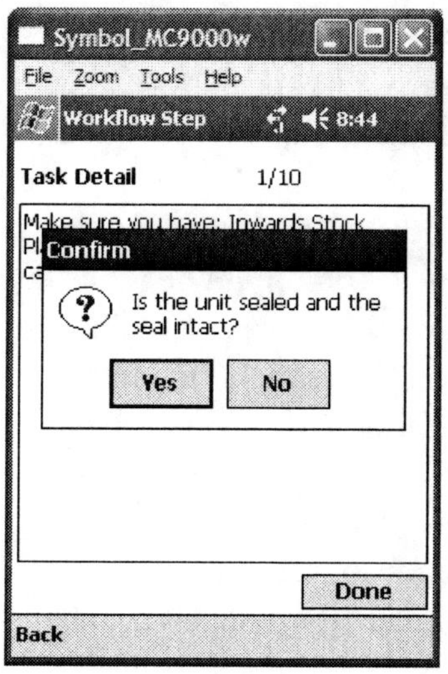


## Construction

The construction element of the system allows the user to incorporate their learning into a real life scenario. Operations do not always follow a predefined and static execution path. Very often the chain of events can take a particular direction based on a condition that occurs at a decision point of the execution of the operation. To cater for this uncertainty a walk through approach was implemented. This feature enables the user to control the execution of an operation based on a real life scenario. From a programmatic point of view, the flow through the MLOs halts when it reaches a decision point. When the user enters their selection, the execution continues in the selected direction. Figure 10 illustrates a decision point in middle of a MLO flow.

The second phase of the system development was focused on building a media rich exposition element of the mobile application with the implementation of various ways of delivering content to the users. At the same time, functionality that supports the construction element of the application was developed. The final phase of the implementation focuses on the fourth structural element of a mobile learning application, communication, and it introduces a slightly different form of exposition in the system.

## Phase Three

The final phase of system development introduced the communications element to the application and evolved the application from a thick client to a smart client. These features in the system mainly target the social and team skills of the users and provide for improved collaborative learning and inter team interaction. From a programmatic perspective, this final development iteration included the building of a web service component that enabled the communication of data between a central database and the mobile device.

### Implementation

As mentioned in the proposed system architecture section the system was intended to be a mobile smart client using the Microsoft .NET Compact Framework. The functionality developed thus far did not require any network connection as all information was maintained in a local database file deployed on the mobile device itself. However, developing an application as a stand alone, thick client in the context of professional learning environment has several weaknesses. The application has no capability for accessing real time data, which makes it difficult to justify the relevance of the support module of the system. The resources of the mobile device are also quite limited, so the volume of the data that can be stored locally cannot possibly cater for all information that employees would require to access. With no connectivity it is also impossible to develop any module that will assist social or collaborative learning or design for delivering user tailored content.

In order to demonstrate a possible solution to overcome the issues mentioned above, in this final phase of the system development, a web service was developed to acquire information from a central database where real time logistics data is maintained for the business. The mobile device uses a GPRS connection to connect to the internet and access the web service. Use of GPRS in the prototype rather than Wi-Fi was primarily to enable off-site testing and demonstration. Information is exchanged between the web service and the mobile device in XML format using HTTP via the internet.

## Communication

To demonstrate the use of the mobile learning application in a connected environment, a message board feature was developed. The idea of this module was that employees could post questions, comments, ideas, suggestions or messages under a certain topic that would be immediately visible to other users. All messages are maintained in a central database and shared between the users. The application accesses web service methods via the internet to get messages that are entered under a specific topic but also to save a new posting a user wishes to share. This module enables the employees to communicate with each other, exchange information and improve their collaboration skills. Figure 11 shows an example of a list of topics and messages entered in the message board.

## Exposition

So far, text, images, audio, and video files have been described for delivering learning content to the user. Learning material can also be delivered in a more interactive manner, where the information is not just simply presented to the users, but the users are expected to recognize, fill in or respond to this content. To illustrate this concept, an interactive quiz was developed containing questions in a multi choice and true/false format. The answers to these questions are analyzed by the system and the users get a hint if they make a mistake. This hint lets the users rethink their answer and encourages them to try to get to the correct solution on their own.

*Figure 11. Example of users' posts to the message board*

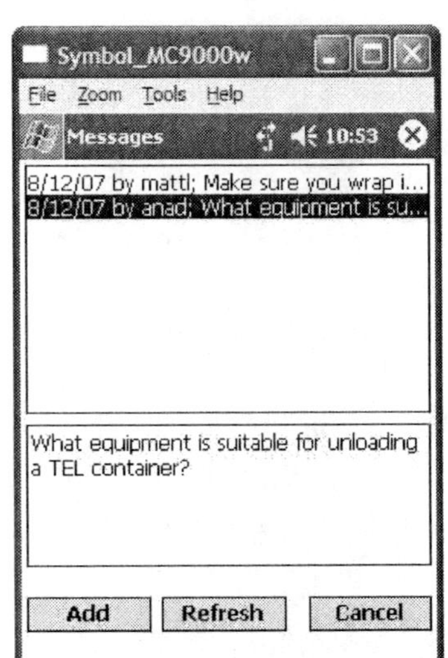

## SYSTEM EVALUATION

The final phase of the prototype development was followed by a formal evaluation process in order to get a critical overview of the working system. A group of participants from different fields of expertise in the area was selected to use, evaluate and provide feedback on the developed system. Cooperative evaluation, a variation of the think aloud evaluation technique was used for these purposes. Think aloud is an observational evaluation technique where the user is asked to talk through the actions they are performing while using the system (Dix & Finlay, 2004), that is, the user is usually asked to execute some predetermined task, talk about what they are doing, what they expect and what they believe is happening. This technique gives an insight into the decision making process of the user and their attitude towards the presented solution, and can also identify potential issues with the interface or logic design of the system. Cooperative evaluation is a more relaxed approach where the user is encouraged to actively participate in the evaluation process instead of just acting as an experimental participant (Dix & Finlay, 2004). In this scenario the evaluator can ask additional questions during the observation. The users are less constrained so they can provide additional explanations, views and critical opinions of a certain issues.

Three participants were selected to take part in the evaluation process: a logistics training specialist, a mobile application development expert and a logistics operations specialist. Tasks from all four elements of the system, exposition, exploration, construction and communication, were given to the participants to work through. Video recording was used to capture the user's feedback and behavior. The results of the cooperative evaluation are summarized below, broken down into four subsections, one for each structural element of the developed application.

### Exposition

To evaluate the exposition element of the application, the participants were asked to perform two tasks: take the challenger quiz and watch a forkhoist training video.

For the first task, the evaluators were asked to answer a series of quiz questions. This task was evaluated by the logistics training specialist. She found the functionality straightforward and easy to use and outlined several benefits she considered important from this module and areas in the business where she saw this functionality being useful. In particular, she saw the potential for monitoring the learning process and targeting learning activities according to learner profiles and work activities. However support for richer media objects was suggested, and the inclusion of warehouse documents as learning objects. The second task was given to the warehouse operations specialist, and it involved going to the forkhoist training screen and launching one of the provided videos. The participant was happy with the format and the content of the video, but expressed concern about the quality of the video, especially when played in a real warehouse environment due to extraneous noise in the environment.

### Exploration

To evaluate the exploration element of the application, the participants were given a task to search for details for the 'Venezia' product. This task involved getting to the product search screen, typing in the required product code and going through the returned information.

The first participant, the warehouse operations specialist, found data input on the search screen difficult using the on-screen key pad. It was suggested that being able to scan a product code instead of typing it in would be a potential solution. The same task was then given to the mobile application development expert who had similar issues with the data input method. He suggested an alternative approach where the user could type in the first couple of characters and the system could return a list of matching codes, letting the user select the required one. Once the product information was displayed on the screen, the participant commented on the fact that in a production environment this module should be retrieving real time information from a centralized database. Several guidelines were pointed out by the mobile application development specialist here, regarding the choice between Wi-Fi networks, which are cheaper and more secure, or mobile phone networks that would support wide mobility beyond the warehouse environment.

## Construction

For the construction element of the application, the participants in the evaluation were given a task to use the step by step functionality for the unload process. The initial impression of the warehouse operations specialist was that this module is quite similar to the logistics chain of events functionality (which is part of the exposition element of the system), but designed in a more sophisticated and smarter way with the addition of decision points and considering the selected choice by the user. In terms of the interface design of the module, the user preferred the more graphical representation of the tasks within one operation instead of a simple list of tasks. Overall, the user was happy with the idea behind the module and could see the benefit of it in a real life scenario. The participant made a comment that the system should provide more feedback to the user to make them aware of what they accomplished and highlight the next task that logically follows. Another important thing that was pointed out in the evaluation process was the necessity to repeat this process in a real environment and examine the behavior of the system when running in parallel with other applications on the same device.

## Communication

The communications element of the application was evaluated through the message board functionality. The participants were given a task to view all messages under the topic "Unload" in the message board and post a new message under the same topic. The first participant to evaluate this functionality was the mobile application development specialist. Initially, his comments were around the performance issues experienced when the system tried to retrieve information by connecting to a GPRS network. As the cooperative evaluation technique encourages the participants to express their ideas, comments or suggestions regarding the system being examined, the mobile application specialist outlined some industry standards for improving performance and availability in a mobile environment, including smart caching. The participant was not happy with the way the system handles the process of entering messages in the message board. He found it quite tedious and frustrating to enter a whole message using the key pad of the mobile device. Next, the task was given to the logistics training specialist. The first impression of the whole idea behind the message board was quite positive as a means of communication and broadcasting. Along these lines it was suggested that it would be quite useful if the message board was visible to the office staff and training people who are not based on the floor and not using the mobile devices. When the participant clicked on the "Unload" topic to get all related messages, a comment was made regarding the time required to download this information from the central database. Another issue highlighted was the lack of appropriate feedback to the user. If the system is waiting for the information to download, a notification needs to be displayed to the user that loading is in progress which might take some time. Again, as with the previous participant, the logistics training specialist was not happy with the input methods of the system.

### Discussion of Results

The evaluation process highlighted several areas in the developed mobile learning application that require further work or investigation. More flexible input mechanisms, a wider spread of rich media types (in particular in the message board context) and more feedback need to be added. Learner profiles are also need to identify different learning patterns among the users of the system, that is, some learners have a more structured learning style where they simply follow logically organized content which explains a certain topic. Others prefer learning through exploration or collaboration, and so forth. The application needs to follow the learning progress of the user and correspondently update the content being delivered in a manner that is suitable to the learning style and current status of the user.

With applications that rely heavily on network connectivity, availability and performance are potentially problematical. The results from the evaluation highlighted concerns around the speed of the message downloading and uploading process in the interaction board. Evaluation of caching approaches and the performance of different type of network needs to be undertaken under realistic conditions. System integration is also important, integration the mobile learning platform with other systems such as human resources.

So far the system has not been evaluated in actual deployment. One of the characteristics of a good mobile learning solution is that it should be easily incorporated into the organizational day to day workflow (McGee, 2003). The only way to evaluate this is to put the learning solution in practice and examine how well it fits with the working routine.

## CONCLUSION AND FUTURE RESEARCH DIRECTIONS

The main objective of this research project was to propose a conceptual framework for professional mobile learning design and use it to guide the implementation of a mobile learning application for a specific professional environment. The developed prototype was then evaluated against the collaborating organization's objectives in supporting their learning program.

The framework identified relevant factors and design requirements in the mobile learning environment and also guided how mobile applications should be organized and designed to meet these requirements. Figure 12 shows how the prototype may be mapped to the framework. On the one hand in the table we have design requirements and issues that arise from the business environment the application was developed for, and the mobile environment itself. Thus, the first two columns summarize the results from the user and context analysis, and document the implementation choices made in the development process in terms of the technology that was used, the type of mobile device the application was deployed on, media types the application supports and type of network connection the systems uses. The other side of the framework defines the expected learning objectives as defined by the participating organization. These objectives are mapped to relevant learning experiences the system delivers in order to support and fulfill the learning objectives. The column in the middle puts these two sets of design requirements in the perspective of actual application design. It defines the different structural elements developed in the application and outlines the names of the specific modules that were built to illustrate these characteristics of the system.

The experience gained from the implementation process of the prototype application demonstrated that the framework provides a good starting point for the development of mobile

*Figure 12. Analysis of the prototype using the design framework*

| **Mobile Environment Issues and Interface Design** | **User and Context Analysis** | **Application Structuring and Organization** | **Learning Experience** | **Learning Objectives** |
|---|---|---|---|---|
| Implementation Technology: *Mobile smart client using Microsoft .Net Compact Framework.* | User categorization: *Users belong to different warehouse operational teams* | Exposition: *Logistics chain; Forkhist video; The Challenger quiz* | Organized Content; Goals and Objectives; *content based on professional responsibilities, assessment based on achieving core competencies* | Improved Skills, New Skills *Improved operations execution;* |
| Interface Design: *Consistent design; Avoid information overload* | User activity definition: *Users' tasks are defined based on the team they belong to.* | | Outcome and Feedback; *contextual learning tasks assessed with interactive responses and quizzes* | *Obtaining new knowledge in operational activities;* |
| Media Supported: *Text, image, audio, video* | Spatial- temporal: *Available within warehouse premises; able to run while executing daily tasks as well as during downtime.* | Exploration: *Index search; Product search* | Motivation; User in Charge, *professionally relevant content driven by user requirements in context* | |
| Mobile Device: *Windows Pocket PC* | Familiarity with device: *Experienced users* | Construction: *Step By Step* | Representation or Story; *User's input affects narrative flow of learning experience* | *Support* |
| Connectivity: *using GPRS to connect to the internet* | Collaboration Activity: *Sharing information, provide support and feedback.* | Communication: *Message Board* | Social Interaction: Competition, Challenge, Opposition, Conflict: *Shared information resource* | Social Skills, Team Skills *Collective learning* |

learning applications as it outlines and clearly formulizes all relevant design issues that need to be analyzed and considered. This transparency easily defined the direction of the actual design and development of the application. The introduction of the middle pillar of the design framework, 'Application structuring and organization,' was especially helpful across multiple levels. Initially, it focused the development on relevant elements of the application that need to be developed in order to support the expected learning objectives and experiences. Furthermore, the definition of each structural element as explained by Meisenberger and Nischelwitzer (2005), suggested multiple development ideas that were utilized to design and implement many of the modules of the application. Finally, this structure and organization of the application was found useful in the evaluation process as it provided a good guideline for different aspects and characteristics of the system that needed to be evaluated and it formulated and conceptualized the evaluation process itself.

The results from the evaluation outlined multiple areas of the system that require further work and investigation such as: usability design, media use, user recognition and learning content customization. However, it has been identified that these issues can again be mapped back to the framework. This means that they have been identified in the framework and are known to the developer, but perhaps more work needs to be done or a different approach needs to be taken in order for those issues to be resolved.

The prototype application was designed for the specific requirements of a collaborating professional organization. Therefore, the delivered learning content is specific and limited to the area of interest of this organization. However, this approach to developing mobile learning applications can easily be applied across different business learning environments. It has also been identified that the success and the acceptance of a learning solution depends greatly on the creativity and design of the learning objects that are delivered to the users. The creators of the learning content in the application need to have a good understanding of the business itself, but also be able to transform that information into learning material with good pedagogical values that will deliver the learning objectives while keeping the users interested and engaged.

As mentioned in the beginning of this chapter, the developed prototype is by no means a complete and final solution. It is used as a proof of concept that helps to evaluate the framework proposed in the study. The evaluation results brought up numerous points that can be considered as an awareness of new problems and thus used as an input into a new design research cycles. Therefore, it is recognized that the framework has not been completely assessed. Further work will include completing the application as per the requirements of the collaborating organization, and testing the system in a real life environment.

## REFERENCES

Diws. (2004). Building Smart Client using .NET. Retrieved 17 January, 2008, from http://www.codeproject.com/dotnet/DotNetBuildSmClnts.asp

Dix, A., J. Finlay, et al. (2004). *Human-computer interaction*. Harlow, Pearson Education Limited.

Grasso, A., & Roselli, T. (2005, November). Guidelines for Designing and Developing Contents for Mobile Learning. *3rd IEEE International Workshop on Wireless and Mobile Technologies in Education* (WMTE'05), (pp. 28-30). Tokushima, Japan.

Hevner, A., March, S., Park. J., & Ram, S. (2004). Design science in information systems research. *MIS Quarterly, 28*(1), 75-105.

Hignite, K. (2007). The word on generational diversity. *HR Horizons, 2*(3), July 2007.

IVES Training (2007). What's Wrong With This? Retrieved 17 January, 2008, from http://www.ivestraining.com/forklift_training_aids/training-videos.php

Koschembahr, C. (2005a). Optimizing Your Sales Workforce through Mobile Learning. Learning Circuits. Retrieved 17 January 2008 from http://www.learningcircuits.org/2005/apr2005/vonKoschembahr.htm

Koschembahr, C. (2005b). Mobile Learning: The Next Evolution in Education. Retrieved 17 January 2008 from http://www.clomedia.com/content/templates/clo_article.asp?articleid=849&zoneid=71

Koschembahr, C., & Sagrott S. (2005). The future of learning at IBM: Empowering employees through mobile learning. In A. Kukulska-Hulme & J. Traxler (Eds.), *Mobile learning: A handbook for educators and trainers* (pp. 164-173). Routledge.

Kukulska-Hulme, A., & Traxler, J. (2005). Mobile teaching and learning. In A. Kukulska-Hulme & J. Traxler (Eds.), *Mobile learning: A handbook for educators and trainers* (pp. 25-44). Routledge.

Kumar A. (2005). How Mobile Web Applications Work. *The Code Project.* Retrieved 17 January 2008 from http://www.codeproject.com/dotnet/mobilewebappsecret.asp

Leah, A. R. (2005). Who are the millennials? *Deloitte Consulting.* Retrieved 17 January, 2008, from http://www.deloitte.com/dtt/cda/doc/content/us_consulting_millennialfactsheet_080606.pdf

McGee, M. K. (2003). E-learning on the fly. *Information Week*, November 3rd 2003. Retrieved 17 January 2008 from http://www.informationweek.com/story/showArticle.jhtml?articleID=15800505

Meisenberger, M., & Nischelwitzer, A. (2005). The mobile learning engine (MLE) – a mobile, computer-aided, multimedia based learning application. Retrieved 20 April 2007 from http://drei.fh-joanneum.at/mle/docs/Matthias_Meisenberger_MApEC_Paper_mLearning.pdf

Microsoft (2007). ASP.NET Mobile Controls Overview. Retrieved 17 January 2008 from http://www.asp.net/mobile/intro.aspx?tabindex=3&tabID=44

Parsons D., Ryu H., & Cranshaw M. (2006, July 5-7). A Study of Design Requirements for Mobile Learning Environments. Sixth IEEE International Conference on Advanced Learning Technologies, 2006, Kerkrade, The Netherlands.

Parsons, D., Ryu, H., & Cranshaw, M. (2007). A design requirements framework for mobile learning environments.*Journal of Computers,* **2**(4), 1-8.

Poddar, S. (2003). Introduction to .NET Web Services. Retrieved 10 October, 2007, from http://www.codeproject.com/dotnet/intro2websvc.asp

Schwabe, G., & Goth, C. (2005). Mobile learning with a mobile game: Design and motivational effects. *Journal of Computer Assisted Learning, 21*, 204-216.

Syuvanen, A., Beale, R., Sharples, M., Ahonen, M., & Lonsdale, P. (2005, November 28-30). *Supporting Pervasive Learning Environments: Adaptability and Context Awareness in Mobile Learning.* In proceedings of the 3rd IEEE International Workshop on Wireless and Mobile Technologies in Education (WMTE'05).Tokushima, Japan.

Thompson, T., Heer, D., Brown, S., Traylor, R., & Fiez, T.S. (2004 October 20–23). *Educational Design, Evaluation & Development of Platforms for Learning.* Proceedings of the 34th ASEE/IEEE Frontiers in Education Conference, Savannah, GA, USA.

Vaishnavi, V., & Kuechler, W. (2004/5). Design Research in Information Systems. Retrieved 17

January, 2008, from http://www.isworld.org/Researchdesign/drisISworld.htm

Van den Akker, J. (1999). Principles and methods of development research. In J. Van den Akker et al. (Eds.), *Design approaches and tools in education and training* (pp. 1-14). Kluwer Academic Publishers

Walker, K. (2004, October15). *Learning on Location with Cinematic Narratives.* Proceedings of the 1st ACM Workshop on Story Representation, Mechanism and Context (pp. 55-58). New York.

# Section V
# Challenges and Future Mobile Learning

# Chapter XV
# Handheld Educational Applications: A Review of the Research

**Yanjie Song**
*University of Hong Kong, Hong Kong*

## ABSTRACT

*This chapter reviews and discusses research on applications of handheld devices in education. The author classifies these applications into six categories based on their functions: educational communication, managing, multimedia access, games and simulations, data collection, and context-aware applications. Each category is examined individually and the consequences arising from these applications are analyzed. The author posits that handheld educational applications, in many cases, have duplicated traditional practice, and that these uses do not seem to have much impact on educational practices. Continued research is needed to focus on longitudinal studies of how students' perceive and use handheld devices for their own needs in contexts that change their educational practices.*

## INTRODUCTION

Mobile technologies have been increasingly used in education. Though laptop computers (Gay et al., 2001; Nicol & MacLeod, 2005), tablet PCs (Corlett & Sharples, Bull, & Chan 2004; Shih & Mills, 2007) and other portable devices like iPods (Belanger, 2005) have been tested in various educational contexts, according to some research reports on mobile technology applications, it appears that the most widely used forms of mobile technology in education are mobile phones and Palms/PDAs (personal digital assistants) (Sharples & Beale, 2003), which are together referred to as handheld devices in this chapter. Due to the rapid development of handheld technologies, more and more educational applications have been developed, designed, and used in education to "ultimately support a lifetime of personal and social enrichment" (Roschelle, Sharples, & Chan, 2005, p. 161). Many researchers assume that the use of these devices will exert considerable impact on educational practices (Lai & Wu, 2006). However, to date, it appears that handheld educational practices have not been sustainable and transferable. The impact that educational applications of handheld devices have brought about does not yet seem apparent (Song & Fox, 2007). What consequences have handheld applications brought about? In predicting the potential consequences of new information communication technology (ICT) use, Sproull and Kiesler (1995) proposed "two-level effects" (p. 4). "First-level effects" result from planned uses of ICT, and are often associated with improved efficiency, but go hand in hand with offsetting consequences. For example, delivery of learning content to learners' handheld devices in universities may create more opportunities for learners to learn without time and space constraints (e.g., Thornton & Houser, 2005). However, the side effects may be that the small screen size and limited computational power of held devices are likely to prevent learners from effective use of the learning content. "Second-level effects" result from unpredicted and uncontrolled ICT use, and are often associated with changes in social practices and the contexts in which social practices take place. For instance, students use handheld devices to collect data in the field, write annotations for the data collected and transmit the data to the teacher who is not in the field but in need of the data (e.g., Lai et al., 2007). This practice in the specific context cannot be easily duplicated using other devices such as desktop PCs. Many researchers have taken context into consideration in their research work on the use of handheld technology. Uden (2007) posits, 'Context plays a crucial role in the understanding and development of mobile learning applications' (p. 82). Context consists of multidimensional overlapping and interacting factors in which handheld educational practices are embedded (Gay & Hembrooke, 2004). This indicates that handheld educational uses cannot be studied without considering the immediate contexts in which they occur (e.g., Abowd & Mynatt, 2000). In the meantime, the immediate context cannot be separated from broader contexts such as family, institution and community (e.g., Lund, 2006; Uden, 2007), which are woven together. Sproull and Kiesler (1995) posited that new technology use is not simply a matter of add-on experience. It leads to new ways of thinking, and thereby changes fundamentally how people work and interact within a given context.

Handheld educational applications have been reviewed by different researchers from different perspectives. However, there are very few comprehensive reviews based on handheld research reports published in journal articles, conference proceedings or as book chapters. In addition, the consequences brought about by handheld educational applications have rarely been analyzed and reflected upon. Understanding the resultant effects from these practices is critical for future research directions. The purpose of this review on handheld educational research is thus to understand the consequences arising from the educational practices and their implications.

## BACKGROUND

Since the year 2000, handheld educational application programs have sprung up in European countries as well as the United States (Keegan, 2002; Vahey & Crawford, 2002). Subsequently, these applications have spread throughout the world. Three important reviews on the use of handheld devices in education have been published in recent years, one in the *Journal of Computer Assisted Learning* (Roschelle, 2003), one in the *Futurelab* series (Naismith, Lonsdale, Vavoula, & Sharples, 2004), and one in *Computers & Education* (Patten, Arnedillo-Sanchez, & Tangney, 2006). Roschelle (2003), focusing on the "key [classroom] communication issues" (p. 262) of mobile technologies that may affect pedagogic practices, classified the applications into three types. They are classroom response systems (e.g., Davis, 2003), participatory simulations (e.g., Facer, Joiner, Stanton, et al., 2004) and collaborative data gathering (e.g., Vahey & Crawford, 2002). Roschelle (2003) points out that handheld pedagogical applications are often associated with 'complex views of technology and simplistic views of social practices' (p. 260), and the lack of standardization of handheld technologies for education hinders the large-scale adoption of handheld applications. However, this literature review is limited to addressing a narrow scope of classroom educational practices. Naismith et al. (2004) claim that most reviews of handheld applications have been concerned with the use of the technologies to address specific areas of educational practice. Therefore, Naismith et al. (2004), from an activity-centered perspective, review the literature of handheld technology applications in six theory-based categories of activities: behaviorist, constructivist, situated, collaborative, informal and lifelong learning, and learning and teaching support. Though this review summarizes the pedagogical applications of handheld devices in a wider scope, it tends to be descriptive without critical comments on the effects brought about by the handheld devices. Recently, Patten et al. (2006), based on previous reviews on mobile technology functional applications and pedagogic practices, developed a framework that they termed "pedagogical underpinning" (p. 300), merging the two perspectives of functionality and pedagogy into one. The authors use pedagogical underpinnings to analyze the functional breakdown of applications within the functional framework consisting administration, referential, interactive, microworld, data collection, location aware and collaborative, and draw a conclusion that the functional applications of administration, interactive, location aware and data collection involve little pedagogy. This review combines handheld functional uses with pedagogical underpinnings. However, the potential of handheld technology cannot simply "be disassembled into predictable cause-and-effect descriptors that can be counted and correlated" without considering the context in which they are used (Laurillard, Stratfold, Luckin et al., 2000, p. 12). Although all these reviews provide useful overviews of handheld applications in education from different perspectives, there is an emerging need to make a more analytical study of these applications under a more comprehensive framework to obtain a better understanding of the technology use and its consequences.

Handheld devices are considered by many researchers to have the unique characteristics of "mobility or portability," and "connectivity" (e.g., Corlett et al., 2005; Markett, Arnedillo Sanchez, Weber, & Tangney, 2006; Swan, van't Hooft, Kratcoski, & Unger, 2005), which ultimately contribute to other characteristics such as convenience, expediency, immediacy, accessibility, individuality and interactivity (e.g., Chen & Kinshuk, 2005; Cui & Bull, 2005; Markett et al., 2006). These characteristics make handheld educational applications in relation to functions distinctive from other technologies such as desktops or laptops. This technology use is highly dependent on the context in which handheld practices happen. In the light of this, the chapter reviews uses of handheld devices to examine the effects that these practices have brought about on education in the hope of providing some references for future research work.

## HANDHELD EDUCATIONAL APPLICATIONS AND DISCUSSION

This section reviews and discusses handheld educational applications in six categories with sub-categories (see Table 1).

*Table 1. Categories of handheld applications in education*

| Categories | Sub-categories | Characteristics |
|---|---|---|
| Educational communication | Pushing | One-way or two-way communication among students, or between students and the teacher that is achieved via handheld devices for educational uses (Liang et al., 2005).. |
| | Posting | |
| | Response and feedback | |
| | Messaging | |
| | File-exchanging | |
| | Classroom communication systems | |
| Managing | Personal management | Personal arrangements for learning, social activities and administrative work in terms of managing assessments via handheld devices. |
| | Administrative work | |
| Multimedia access | | The information or resources that are downloaded and stored in handheld devices, or can be accessed via Internet using handheld devices for learners to seek useful information and utilize it (Churchill & Churchill, in press). |
| Games and simulations | | Providing models of real world settings in handheld devices for the learners to construct their knowledge through active participation in learning activities (Patten et al., 2006). |
| Data collection | | Using handheld devices to record and transmit data usually in the field or in certain environments (Parr, Jones, & Songer, 2004). |
| Context-awareness | Personalization | A system using context to provide relevant information and/or services to the user depending on the user's tasks (Dey, 2001). |
| | Active context-awareness | |
| | Passive context-awareness | |

## Educational Communication Applications and Discussion

### Educational Communication Applications

Liang et al. (2005) list a set of communication mechanisms governing classroom learning. They are response-collecting, posting, pushing, controlling, file-exchanging, and instant messaging mechanisms. Borrowing some of the terms from this list, a new classification has been derived, termed "*educational communication*," which refers to one-way or two-way communication between students, or between students and the teacher that is achieved via handheld devices for educational uses. Educational communication is composed of *pushing, posting, response and feedback, messaging, file-exchanging and classroom communication systems* (See Figure 1), which are expounded as follows:

### Pushing

This category of applications refers to course content, assignments, and quiz and assessment materials delivered to the students' handheld devices to support their learning without time and

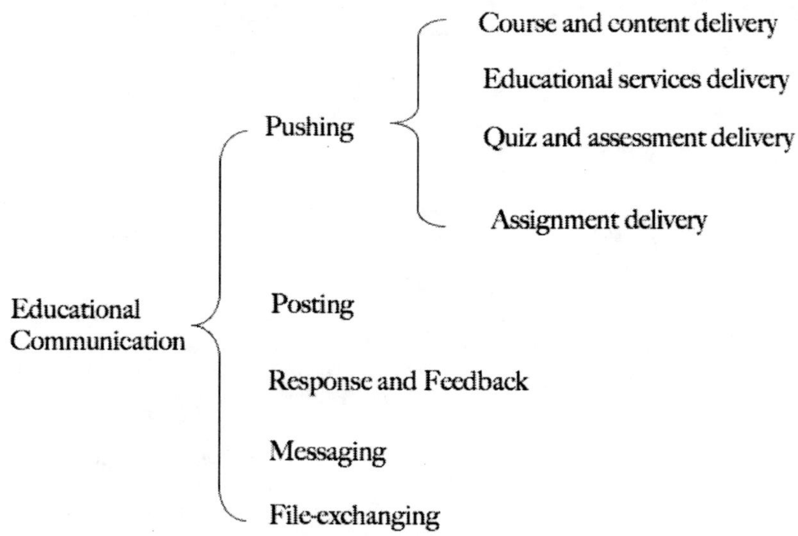


*Figure 1. Educational communication*

place constraints. *Course and content delivery, assignment delivery, quiz and assessment delivery,* and *educational services delivery* are included in this application. One obvious characteristic of the handheld uses in the *pushing* category is to deliver learning material, assignments, quizzes and assessments, or educational services to the learners' handheld devices by taking advantage of the portability and connectivity of handheld devices for anytime and anywhere learning. However, access to content using handheld devices seems simply another kind of resource access similar to the access using laptop or desktop computers (Corlett et al., 2005).

*Course and content delivery* — The low-threshold technology SMS (short message service) has been recognized as having great potential for course and content delivery. Levy and Kennedy (2005) reported an Italian vocabulary learning program delivered via SMS for university learners. Likewise, e-mails were also tested in delivering English vocabulary lessons to the mobile phones of the university students in order to improve their EFL (English as a foreign language) learning (Thornton & Houser, 2005). Some content development tools are developed to deliver learning content adaptively to a wide range of devices such as different types of handheld devices, desktops or laptops for ubiquitous access (Ally, Lin, McGreal, & Woo, 2005). By using compatible applications, handheld devices extend the capabilities of existing tools, such as desktops and laptops. They allow course material to be delivered across heterogeneous computing platforms.

*Assignment delivery* — Some teachers use the handheld device to beam the nightly homework assignments to students' handheld devices (Pfeifer & Robb, 2001). Also, students can submit their assignments via beaming, Bluetooth or e-mail using Wi-Fi networks to their teacher, and the teacher can read the submitted work whenever appropriate (Lai & Wu, 2006).

*Quiz and assessment delivery* — van't Hooft, Diaz, and Swan (2004) envisage that handheld devices have a great potential for self-assessments with low stakes. Such assessments can encourage students to track their progress over time. PDAs have been trialed to deliver quizzes in a number of projects (Segall, Doolen, & Porter, 2005; van't

Hooft et al., 2004). However, such quiz and assessment types are limited to closed questions instead of open-ended questions owing to the limitation of the text input (Segall et al., 2005).

*Educational services delivery*—This application tends to create mobile learning platforms on handheld devices to provide various educational services for both learners and teachers. Such services include news, help, classified ads, academic, financial, library and account management (Chen & Kinshuk, 2005).

## Posting

Within this application, information presentation, dissemination, and annotation via handheld devices can be achieved, usually in fixed physical settings together with other technologies. The learners are encouraged to reflect on and add a brief explanation to what they have read or seen using their handheld devices, and then publish it by sending it to the handheld devices of both students and the teacher or shared devices. An example of this is an approach to extending co-constructive modeling and discussion environments using PDAs realized by an annotation tool (Pinkwart, Hoppe, Milrad, & Perez, 2003). This application provided the possibility for the learners to make private annotations to public slides developed by the teacher, and allowed the presentation of students' annotations on the whiteboard to initiate certain collaborative tasks among the student group following the presentations. Lai, et al. (2007) described a study on experiential learning in a real garden using PDAs at an elementary school, in which students were encouraged to write annotations during the observation in the garden and share them with the teacher. This provided important references for the learners to understand what they had seen. Other examples such as students annotating lecture notes during the class using handheld devices (West, Foster, & Clayton, 2005) exist. This application supports the learners to make just-in-time comments and reflections, allowing those learners who are unwilling to speak publicly to express their views, leading to active participation and collaboration in learning activities.

## Response and Feedback

Within this application, quizzes, and assessments are a type of instant two-way communication that can be delivered by handheld devices. A few interactive quizzes have been tested on handheld devices by both school and university students to enhance their learning (e.g., Liang et al., 2005). For example, multiple-choice questions on handheld devices were reported to have increased communication among students, and between students and the teacher by Liang et al. (2005). One scenario ran like this: The teacher delivered multiple-choice questions to the handheld devices of all the students in class, and asked them to answer the questions by pressing certain buttons. After all students had responded, the teacher could show their answers in a statistical chart, which was used to reflect their performance and facilitate their discussion. All these handheld applications, taking advantage of the opportunities that handheld devices can offer, provide a new experience for the learners to do the quizzes or assessment anywhere, anytime. However, without this "mobility" feature, this form of learning may easily be experienced using other technologies, like desktops or laptops.

## Messaging

SMS messaging is used mainly in this application to allow, via handheld devices, instant exchange of short messages between students and the teacher or among students. SMS messages are used for notifying students that their feedback is available (Seppälä & Alamäki, 2003), enhancing students' awareness of participating in collaborative learning activities (Liu et al., 2005), and directing the learner to tutor support (Kadyte, 2004). In addition, SMS Message Loop via handheld devices was used to increase interactivity in a project (Markett et al., 2006). It allowed for low-cost pedagogic implementation of real-time, text-based interactions across multi-levels, that is, student-student, student-lecturer, student-content levels. These short messages usually function as "alerts" or "reminders" for the learners to participate in collaborative learning or social activities, check feedback or learning material posted periodically, and do their assignments. This application generally support small group learning, and provide educational services for individuals.

### File-Exchanging

File exchanging refers to the files that are exchanged and shared anywhere, anytime via the handheld devices between students and a teacher, or among peers. This is also a two-way communication. A teacher can send exercises via his or her device to the students' devices (Liang et al., 2005). After students finish writing the exercises, they will send them back to the teacher's device for review. The review can be collected later by the students via their handheld devices (Liang et al., 2005), the supervisors and teacher trainees can discuss teaching issues via the handheld devices using digital pictures transmitted to the handheld devices as a part of the supervising process (Seppälä & Alamäki, 2003). In this process, the learners can share the exchanged information anytime, anywhere. This encourages collaboration and communication among students.

### Classroom Communication Systems

Short, asynchronous messages sent via handheld devices, are aggregated using a display board or a presentation tool in classroom settings (Markett et al., 2003). Interaction achieved this way is limited, and face-to-face interactions are diminished to some extent.

## Discussion

Among all the categories of handheld educational uses, the category of *educational communication* includes the largest number of sub-categories and uses. These practices, at first sight, appear to be innovative. By *pushing*, students can receive the learning content anywhere, anytime; by *posting*, students can make their private annotations published to the whole class (e.g., Pinkwart et al., 2003), enhance their understanding in experiential learning (e.g., Lai et al., 2007), or comment on other students' performances (e.g., Hackemer & Peterson, 2005); *response and feedback* via interactive quizzes can enhance student learning; *messaging* can promote instant communication (e.g., Liang et al., 2005), *file-exchanging* and *classroom communication systems* can increase interactivity (e.g., Markett et al., 2006). However, things are not as satisfactory as they may at first appear. These practices often go together with their offsetting consequences (Sproull & Kiesler, 1995). According to Roschelle's (2003) survey, handheld educational applications take an overcomplicated view of technology and a simplistic view of the social practices surrounding these applications, in particular the social practices that are critical to the success of the application. This situation is still prevalent in current educational practices using handheld devices.

It appears that many uses have fallen into the sub-category of '*pushing*' in *education communication*. This means that many of these applications have made use of the handheld devices as content delivery and presentation tools to do drill-and-practice or simple tutorials. These applications center around "pushing" content to the learners' handheld devices, and sometimes to the desktops, at the same time without the awareness that handheld devices have not been ready to be the only platform for learning (Ketamo, 2003). Students are not particularly interested in these content materials as generally the same thing can be performed better on a desktop or laptop due to the severe limitations of screen size of the handheld devices and other factors (Crowe & van 't Hooft, 2006). This indicates that handheld devices should not be used as the sole delivery vehicles of learning materials. Jonassen, Howland, Moore, and Marra (2003) posit that technology should not be used as delivery vehicles. However, this does not mean that pushing applications have no educational value. Whether it is useful depends on how this technology is used in pedagogic practices. A few studies have tested integrating mobile technology into other technologies to support learning. In these studies, mobile working adults can learn English vocabulary anywhere, anytime using SMS technology via handheld devices, and get detailed instruction on Web-based resources using a computer (e.g., Song, in press); students can get just-in-time scaffolding when they encounter English vocabulary difficulties or want to learn culture knowledge while watching English films on interactive TV (e.g., Fallahkhair, Pemberton, & Griffiths, 2007). In these studies, handheld devices have been used to meet students' learning needs.

Next, *response and feedback* is similar to that of pushing for assessment delivery and presentation but feedback can be triggered after the submission of the assessment. In addition, *messaging* has been used for quick alerting notices to remind students to take part in social or learning activities. This does not seem, however, to have exerted much impact on educational practices. As for *posting, file-exchanging* and *classroom communication systems*, it seems that these handheld educational uses are more interactive, collaborative, constructive, and contextualized. In fact, a majority of these communications via handheld devices have been achieved in classroom settings. Most often, handheld devices have been integrated with other technologies to form a learning system, in which handheld devices have been used as an educational communication tool for posting, messaging, file-exchanging, response, and feedback. To increase interactions with the students in classroom, an annotation tool on handheld devices has been developed and used to post students' private annotations onto public slides. However, this way of interaction does not seem to encourage students to make participate verbally, and the mobile devices have been used in the fixed physical setting of the classroom. Ultimately, the use of the handheld device to increase communication has resulted in silent and elusive communications in the face-to-face learning environment. Teacher-controlled instruction is evident. Also, it may distract student attention from the lecture while the student is inputting the annotations using the handheld device. The value of interactive handheld devices used in face-to-face classroom for silent asynchronous interaction is therefore questionable. These practices have involved complicated design for the learning systems in fixed physical settings, in which handheld devices have been used for limited communication purposes. Ironically, the opportunities for the learners to make oral interactions in face-to-face settings are probably reduced, and asynchronous short messages may be encouraged but are controlled by the teacher. In his early review of classroom use of handheld devices, Roschelle (2003) pointed out that many handheld applications focused on teacher-controlled messaging applications among participants in face-to-face settings. The communication, via messaging, was usually in the form of short, asynchronous structured data messages, which were aggregated via a shared public display that was used to make these aggregates available for discussion. These features are probably still prevalent across handheld educational uses for communication in classroom settings. These practices do not seem to bring about much social impact. Handheld technology is distinct from other computer technologies in terms of its 'mobility' and 'connectivity' (Corlett et al., 2005; Markett et al., 2006). However, many research studies do not seem to have taken full advantage of the 'mobility' factor. In contrast a recent study carried out by Wei, Chen, Wang, and Li (2007) makes exceptional use of handheld technology. In this study, a mobile knowledge forum is set up for students to access via their handheld devices. Students who encounter learning difficulties can ask and record their questions using MMS, then e-mail the captured data to a teaching assistant who will post the message to the forum for help. Other students who want to offer help can record their answers and use the same system to post their messages to the forum. Students can also post recorded discussion files and photos of the learning content to the forum for sharing and comments from other students. Students can solve their learning difficulties and share information without time and place constraints by simply 'pushing' and 'posting.' The handheld technology is used in an uncontrolled and unexpected manner based on the learning needs of the students. Such a use changes the way students learn, and will lead to 'second-level effects.'

## MANAGING APPLICATIONS AND DISCUSSION

### Managing applications

In this category, handheld devices are used to schedule personal arrangements for learning and social activities. They have also been adopted as test administration tools in schools and higher education. For instance, calendar and timetabling features have been widely used by students to manage their own study and life (Chen & Kinshuk, 2005; Corlett et al., 2005). Students' basic organizational skills have been improved, and they have a strong sense of personal empowerment for scheduling various activities and taking notes (Pfeifer & Robb, 2001). In addition, handheld devices have been found useful to reduce the instructors' workload in grading, and enrich students' experience in just-in-time learning at both universities and schools (Segall et al., 2005; van't Hooft et al., 2004). In general, applications in this category help learners foster organizational skills, and handheld devices are thought of as a robust educational test administration option (Norris & Soloway, 2004; van 't Hooft et al., 2004).

### Discussion

Applications in the *managing* category have also replicated the traditional practices of using handheld devices to organize various levels of administrative work. This personal managing application seems to enable the learners to organize their activities more efficiently, but the knock-on effect is that they have been more controlled by the handheld devices. It seems more challenging for assessments, because delivery measures have to be taken to prevent cheating (Segall et al., 2005), and it is suggested that only assessments with low stakes are delivered (van't Hooft et al., 2004). Moreover, administrative time spent in designing and managing the PDA-based exams are is also an issue to be addressed (Segall et al., 2005). According to Patten et al. (2006) a managing application does not scaffold or support knowledge construction. Instead, this application can be implemented more effectively using desktop or laptop computers.

## MULTIMEDIA ACCESS APPLICATIONS AND DISCUSSION

### Multimedia Access Application

This type of application refers to the information or resources that are downloaded and stored in handheld devices, or can be accessed via the mobile Internet for learners to seek useful information and utilize it (Churchill & Churchill, in press). It includes both stored multimedia access and online multimedia access. Handheld devices can be used for stored multimedia access to reference or revision materials. Such learning experiences often take place in the field of medicine.

In some projects, handheld devices with downloaded software containing electronic reference guides in nursing/medicine, and drug books are distributed to nursing students, so they can look up information as they work (Colevins, Bond, & Clark, 2006). Likewise, handheld devices were used to help health students in the community access learning resources provided by the library (Walton, Childs, & Blenkinsopp, 2005). The downloaded podcasting and videocasting files are considered significant resources for health professionals. A faculty at the University of South Dakota provided students with past quizzes of a psychology course downloaded to their handheld devices for them to review in preparing for the coming course exam (Hackemer & Peterson, 2005). In another project, handheld devices with stored information as guide were used by young learners to look for certain plants in an outdoor garden (Lai et al., 2007). Questions were posed for the learners, and answers could be found using the handheld guide. The mobile version material could be reviewed whenever and wherever a desktop was unavailable. Therefore, the students could get access to additional individualized mobile revision material for reviewing anytime, anywhere they wanted.

Students consider searching learning materials from the Internet via handheld devices useful for their learning (e.g., Corlett et al., 2005). Using the handheld devices, students can get access to a course Web site containing all the lecture notes, slides, references and Web links (e.g., Luckin, Brewster, Pearce, et al., 2005). Chen, Kao, and Sheu (2003) described a bird-watching learning system, a mobile learning environment developed on handheld devices with wireless connectivity. The school students could seek the "just-in-need" information using the handheld devices for outdoor bird watch activities, and deal with the information required. Based on this research, a butterfly-watching mobile learning system was developed and tested by school students in similar ways (Chen et al., 2005). In the *multimedia access* category, instead of *pushing* certain learning resources to student handheld devices, students go to seek the resources which have been downloaded or stored in the handheld devices or can be accessed via the Internet for revision and reference. This allows learners more flexibility in choosing just-in-need learning content, and dealing with information without time and place constraints.

### Discussion

This *multimedia access application* is another example that replicates the practices of using desktop or laptop computers. Using handheld devices to present materials for reference and revision is a means for practicing old information seeking skills for anytime, anywhere information access. However, few students have used the mobile devices to search for information available on the Web, and reading e-books via handheld devices has been considered troublesome due to the constraints of handheld technologies such as small screen size, connectivity, processing speed, and so forth. (Smordal & Gregory, 2003). Therefore, basically, anywhere, any time educational practices have not yet been fully achieved.

## GAMES AND SIMULATIONS APPLICATIONS AND DISCUSSION

### Games and Simulations Applications

*Games and simulations* in handheld devices provide models of real world settings for learners to construct knowledge through active participation in learning activities (Patten et al., 2006). The main characteristics required of simulations are: (1) an adequate model of the complex real-world situation with which the student interacts; (2) a defined role for each participant, with responsibilities and constraints; (3) a data-rich environment that permits students to execute a range of strategies; and (4) feedback for participant actions in the form of changes in the problem or situation (Gredler, 2004, p. 571).

Several simulations applications have been developed on handheld devices for educational use (Roschelle, 2003). A good example is participatory simulations, which were used to help school students to learn science in the classroom (Klopfer, Yoon, & Perry, 2005). This case study showed that these simulations motivated students to engage in self-directed learning and problem solving in science learning. Some games have been developed for collaborative and constructive learning in schools. Lim and Wang (2005) reported a study on a multi-player handheld gaming environment in which students could debate open-ended problems from a variety of perspectives using SMS messages via handheld devices in class. A pair of students could choose their own topics for debate based on the information provided in the virtual environment, and reach conclusions. In this game-based activity, students were provided the opportunity to practice the skills of 'structured academic controversy' (p. 354). *Cooties*, a simulation program that can be operated on Palm OS, lets students work collaboratively to examine the spread of disease. The teacher can determine the variables such as incubation time, individual immunity levels, etc., to start the infection. Students work together and may infect each other by beaming among their handheld devices. Horton and Wiegert (2002) described the Clemson Geometry Project, in which billiards games were created using Geometer's Sketchpad on the handheld devices to help secondary school students explore simple geometric concepts. The students could develop their own games based on their own rules and constructions in class, and made math learning active. These applications have been mostly applied in classroom settings and have provided opportunities for students to engage in a learning experience in a relatively novel manner. However, it seems that these micro-worlds may have been better presented and practiced using desktops or laptops with bigger screens and greater computational power.

## Discussion

*Games and simulations* provide models of real world domains for the learners to construct their knowledge through active participation (Patten et al., 2006). By providing tasks within real world settings, learners are able to engage with topics in a novel and innovative manner" (Patten et al., 2006, p. 298). However, these handheld applications have been practiced mostly in classrooms. Though these applications have provide opportunities for the learners to engage in the learning experience in a relatively novel manner, it seems that it may be neither innovative nor effective enough using small-screen handheld devices in classrooms. Rather, these microworlds may have been better presented and practiced using desktops or laptops with bigger screens and larger computing power. However, this does not mean that mobile games and simulations have no educational implications. On the contrary, educational games and simulations may be more useful for learning activities beyond physical settings. On such example is the Ambient Wood, in which children can explore the authentic woodland habitats themselves by observing, touching, hearing, and smelling different things around them in the real world with the help of a probing tool attached to handheld devices (Price & Roger, 2004). This encourages the children to integrate their physical discoveries of the environment (e.g., spotting a caterpillar on a thistle) with the more abstract knowledge they have learned in the classroom, and make reflections. This study provides good pedagogic underpinning as it aims at engaging learners with exploring their surroundings using a contextual learning approach that facilitates constructive learning (Patten et al., 2006). This educational game augments the reality, and allow learners to explore and learn collaboratively in a contextualized environment, which is better categorized under the context-aware applications that are elaborated in later in this paper.

## DATA COLLECTION APPLICATIONS AND DISCUSSION

### Data Collection Applications

Within this category, handheld devices are used to record data and information, usually in the field or in certain environments (Patten et al., 2006). Handheld devices have been considered powerful, versatile, and rugged tools for data collection in the field (Green, 2001; Vogel, Kennedy, Kuan, et al., 2007). For instance, Lai et al. (2007) conducted an experimental field study. In the study, the students could use handheld devices to collect data about the plants in a garden. They could use the camera-equipped handheld devices to take pictures of the plants as the pictures are assumed to be able to provide richer information than paper-based resources (Whitelock, Domano, Jelfs, & Brna, 2000). The students could also read the descriptions of certain plants via the handheld devices, compare these descriptions with the results observed from the actual experience in the garden, and find out the differences. All these data collected could be used for preparing the introduction to their experiential learning in class. In another project, medical students used the recording function of handheld devices to collect patient data (Cacace, Cinque, Crudele et al., 2004). Note-taking has been found to have the potential for data collection in the field to facilitate the development of field notes during observations (Green, 2001; Lai et al., 2007). These notes have been readily communicated from the students to the teacher by different means. Using PDAs to collect data in research projects has been on the rise. Parr et al. (2004) reported an evaluation of customized handheld data collection software for student use. The results showed that the software helped students expand the types of data that they were aware of, and supported accurate data collection.

### Discussion

*Data collection applications* create learning experiences "that would be either unfeasible, or at least problematic without handheld computers" (Patten et al., 2006, p. 298). Such applications help students collect useful data, and support constructive and collaborative learning in context both during and after a field trip (e.g., Lai et al., 2007). Students can share the pictures taken using the handheld device, with any annotations or voice tags added to the pictures during the field trip, and upload them to a server (Vogel et al., 2007). Without the handheld device, it would be hard for students to collect useful data during the trip, and carry out a series of learning tasks. Using handheld devices to collect data during field trips is a widespread practice in education. More research is needed to explore how handheld devices can be used in other situations like lab experiments, in which students can use the technology to capture the lab results and do reflective and collaborative learning after the experiments.

## CONTEXT-AWARE APPLICATIONS AND DISCUSSION

### Context-Aware Applications

*Context-aware* computing is a relatively new area of research, and is applied more often in ubiquitous computing (Gay & Hembrooke., 2004; Naismith et al., 2004). A system is considered context-aware "if the system uses context to provide relevant information and/or services to the user, where relevancy depends on the user's task" (Dey, 2001, p. 5). Handheld devices are well suited to context-aware applications due to their sensitivity in gathering and responding to real or simulated data unique to the a particular location, environment and time (Klopfer et al., 2005). A context-aware application maybe categorized under three levels of interactivity: personalization, passive context-aware application and active context-aware application (Barkhuus & Dey, 2003), which are illustrated in the following text.

### Personalization

This feature places emphasis on the specific needs and location of individual learners when designing learning contents on handheld devices (Cui & Bull, 2005). An individualized mobile intelligent tutoring system is a good example of this application (Cui & Bull, 2005). The system allows the mobile language learner to make use of individualized grammar learning in terms of tense according to his/her own knowledge and location. Cui and Bull (2005) posit that such learning packages should be: (1) individualized according to the learner's knowledge, (2) individualized according to the learner's location and needs in that location, and (3) portable. Another good example of this is Ogata and Yano's (2004) system, which modeled the learner's knowledge of the usage of Japanese polite expressions depending on locations, and personal attributes (e.g., age).

### Active Context-Aware Application

An active context-aware application refers to a handheld application that automatically changes the content of a service for a user based on sensed information (Barkhuus & Dey, 2003; Dey, 2001). One example is the use of visual codes attached to locations and objects within a museum or park to enable learners to retrieve Web-based information according to their own interest (Mitchell & Race, 2005). Each learner could independently study then make his or her own decisions about the changed context or sensor information. Such applications encourage decision-making and reflection.

### Passive Context-Aware Application

A passive context-aware application presents updated context or sensor information to the user but allows the user to decide how to use the application behavior (Barkhuus & Dey, 2003). The functions of GPS or other positioning engines are used in this application, in which the learners' positions are shown automatically when they move from one place to another. Therefore, this positioning system presents updated context to the students, and gives the students the opportunities to execute the application's features. A game-based activity reported by Facer et al. (2004) is an excellent example. In the game, students pretended to be lions outside in a playing field. They interacted within a virtual Savannah using the global positioning system (GPS) via handheld devices, exploring the opportunities and risks to lions in that space in which they could "see, hear and smell" the world of the Savannah. MobileGame created a prototype for handheld devices; a mixed reality orientation game that allowed university students to explore collaboratively the opportunities for learning support (Schwabe & Goth, 2005). The "physical space is augmented with contextual information" by using this type of tool (Roschelle et al., 2005, p. 1). A passive context-aware application augments physical spaces and encourages the learners to construct knowledge via active participation in designed "learning spaces" in appropriate locations. This creates new opportunities for students to participate in social and collaborative activities, and provides rich instant feedback.

## Discussion

This *context-aware application* has tested educational practices that have been hardly explored using other technologies such as pencil and paper, desktops, or laptops. These practices allow learners to use handheld applications according to contextual factors such as time, location and/or needs on their own (Chen et al., 2003; Colevins et al., 2006), allow learners to make decisions on what applications to use based on the updated context information provided by the positioning systems (Facer et al., 2005), or autonomously change the content of a service for a learner according to the sensed information (Mitchell & Race, 2005). In these applications, the learning environments are adapted and augmented, and learning practices are changed. Learners have the opportunities to execute their own application behavior for their own needs. These educational uses have brought about new learning environments, and changed educational practices. In Sproull and Kiesler's (1995) words, and with Lankshear, Green, and Snyder's (2000) interpretation, these practices have achieved "second-level effects." Nevertheless, this does not mean that there are no problems to be tackled.

There are some significant problems that need to be addressed in designing context-aware learning components or systems. Individual learner goals and needs have to be acknowledged and investigated (Tolmie, 2001), and this is challenging as these needs and goals are diverse. Also, the limited instances of handheld educational uses are insufficient to be used to draw out the general principles to design these context-aware systems (Henricksen & Indulska, 2006). In addition, wireless connectivity is limited to certain contexts, which is a barrier for anytime and anywhere learning (Corlett et al., 2005). Therefore, it takes time, energy and money to design context-aware learning components or environments. Even with this investment the technological complications in context-aware design may sometimes result in limited educational benefits (Patten et al., 2006). Finally, lack of standard platforms prevents replications of the learning outcomes from being achieved via other devices (Patten et al., 2006). As technical breakdowns and platform incompatibility still exist, "It is expensive and financially risky for software authors to write software for many different varied platforms" (Roschelle, 2003, p. 268). Therefore, designing context-aware learning environments or components remains a major challenge.

## Constraints

Handheld devices are different from desktop computers. They have their own unique characteristics as well as constraints. The screen size of handheld devices, especially mobile phones, has constantly been regarded as one of the main issues limiting educational applications (e.g., Chen & Kinshuk, 2005; Corlett et al., 2005; Crowe & van't Hooft, 2006; Kadyte, 2004; Ketamo, 2003; van't Hooft et al., 2004). Because of the limitations of the screen size, resources with longer texts (e-books for example) have been considered inappropriate for viewing on mobile devices. Designers have to condense the learning content to suit the technological environment (e.g., Bull & Reid, 2004). "To be effective, teaching materials and Web sites will need to be designed for display on small screens" (Corlett et al., 2005, p. 170). However, Web pages adapted to the PDA or mobile phone screen size are still limited (Bull & Reid, 2004). Handheld devices are quite restricted on the format and length of the browsed content. They typically display less than 20 lines of text on the screen; and may run different operating systems and support different mark-up languages (Zhang, 2007). Therefore, a variety of approaches have been proposed for implementing adaptation of multimedia Web content such as multiple encoding, transcoding, layered encoding (transcaling and multicasting) and rate shaping (for details of these techniques, please refer to Zhang, 2007).

Though SMS appears to be a popular feature of mobile technologies, it lacks interactivity and space (e.g., Kadyte, 2004), but is useful for organizing social activities (Smordal & Gregory, 2003). In addition, inputting text data to the handheld device is considered troublesome by users (Corlett et al., 2005; Markett et al., 2006; van't Hooft et al., 2004). Students do not think that a keyboard will help them in data entry (Cacace et al., 2004).

Therefore, even if students are provided with handheld devices, they still prefer to use desktop computers to write text (Markett et al., 2006), especially longer texts (Ketamo, 2003). Also, it was reported that students using handheld devices to input or read texts during lectures both distracted the lecturer and were distracted themselves from the lecture (Markett et al., 2006), and trying to write while commuting is difficult.

Wireless connectivity is also not without problems. Learners can only get access to the Internet, e-mail friends, and transmit data within the areas with wireless network coverage (Corlett et al., 2005). As General Packet Radio Service (GPRS) is still relatively expensive and slow for Internet access, Wi-Fi (wireless LAN) is an alternative infrastructure for wireless connections. Recently, there has been an initiative of universities around the world to participate in eduroam, or education roaming, a Wi-Fi mutual access system, which will enable cross-institutional Wi-Fi access for staff and students in local and overseas universities that are eduroam members. The University of Hong Kong has also collaborated with PCCW, a telecommunication service provider and Y5ZONE, to provide staff and students with free Wi-Fi access to the Internet at the PCCW and Y5ZONE Wi-Fi hotspots throughout Hong Kong, including entertainment and leisure stores, coffee shops, fast food chains, the airport, restaurants, shopping malls and MTR stations in Hong Kong. This will greatly enhance the availability of Internet access for using mobile devices for educational purposes. Also more and more 3G services have been promoted by delivering some content in terms of news, games, and so forth. to the 3G users' devices. These high bandwidth networks can also support educational uses.

In addition, because many handheld devices only support IEEE 802.11b (with 11mbps data transfer rate) networks, the speed of data transmission and Internet access is considered slow, especially slow in processing multimedia objects (Smordal & Gregory, 2003). However, with the arrival of the Opera mini and Deepfish dedicated mobile browsers, the speed of accessing and browsing the Internet is closer to that of a wired computer (Godwin-Jones, 2007), and mobile browsing is becoming more and more powerful. The release of the Apple iPhone for the US market adopts a version of the Safari browser that features faithful rendering of original page design (Godwin-Jones, 2007).

There are several other technical issues that impact on using handheld devices, and lack of technical support may be a constraint on their widespread educational use (Hackemer & Peterson, 2005; Walton et al., 2005). Both researchers and students experience data loss due to battery related problems (e.g., Corlett et al., 2005; van't Hooft et al., 2004). There are other constraints such as inconvenience in mobile printing, limited storage (Bradley, Haynes, & Boyle, 2005), slow processing speed (Sharples, Corlett, & Westmancott, 2002), low resolution (Crowe & van't Hooft, 2006), security and cost (Bradley, Haynes, & Boyle, 2005). Even in more recent research studies, technical issues have been encountered (e.g., Crowe & van't Hooft, 2006; Lai & Wu, 2006; Swan et al., 2005). Because of these limitations and others, though handheld devices have the potential to benefit learning and teaching in a range of aspects, it is not advisable to design complicated learning environments or components on unproven technology at this stage. With the rapid development of technology, mobile devices are becoming more and more powerful and useable. Touch phones have been on the market for users to input words by just typing on the virtual keyboard; the storage of the many handheld devices has been expanded by miniSD or SD cards, the processing speed and screen resolution has been increased, and the battery life has become becoming longer. Therefore, the future of the handheld device for educational use is promising.

## SUMMARY

There is no denying that handheld device applications do provide students with more opportunities for their learning. A few innovative uses such as *data collection, context-aware applications*, and *uncontrolled uses* exist. These practices take advantage of the unique characteristics of handheld portability and connectivity (Corlett et al., 2005; Norris & Soloway, 2004), and allow learners some degrees of freedom to take advantage of the handheld devices for their learning practices. These practices using handheld devices are associated closely with the context in which these applications take place. Nevertheless, it seems that the majority of handheld applications have been developed from the perspective of researchers and used by students. These handheld uses have been practiced in a planned or controlled manner (Lankshear et al., 2000). The way that students learn using handheld devices, in many cases, has duplicated the traditional way of learning. Jonassen et al. (2003) point out that "technologies have traditionally been used to support teacher goals, but not those of learners" (p. 8). To date, new information technologies continue to be used this way. Students have limited opportunities to explore how the handheld device can be used for other educational practices. Yet, student use of handheld devices is driven "more by personal experience than by any university-wide promotion" (Hackemer & Peterson, 2005, p. 161). Draper (1998) asserts that most ICT programs in education do not have significant impact on educational practices because only a very limited proportion of their potential is used. This also seems to be the case in handheld educational uses. The uses of handheld devices in education remain largely add-on experiences without bringing about much impact on social practices. In other words, the consequence of handheld educational practices has not achieved "second-level effects." Sproull and Kiesler (1995, p. 4) assert that new communication technology at its early stage tends to be used for improving efficiency, or achieving "first-level effects," but is often to accompanied by negative consequences. Though many research results have been generally positive, which may partly be owing to "strong novelty effect" (Thornton & Houser, 2005, p. 224) or "Hawthorne effect" (Swan et al., 2005, p. 110). "It seems to us that many of the tests and pilots themselves rest on a 'common sense' view of learning," so the credibility of outcomes is challenged (Traxler & Kukulska-Hulme, 2005, p. 7).

## FUTURE RESEARCH DIRECTIONS

With the advent of handheld technology, the context of the technology use in education becomes more and more emergent, dynamic and complex. This calls for new thinking about context and settings in the handheld educational uses as these uses are not triggered by constant proprieties of contexts such as fixed settings and scheduled time, but by a mixture of diverse contextual components. In addition, Jonassen et al. (2003) point out that technologies should be used to support students to articulate their learning goals in contexts so that they can understand more and are better able to use what they have learnt in new situations. Therefore, future research should focus more on how students explore and use handheld technology for their learning needs in varied contexts through a longitudinal empirical study, and less on the technical design or development of complicated, unsustainable software or learning systems. Such a research approach intends to understand the process of how actions or events take place, lead to the research outcomes, and are shaped by the unique context in which they occur (Maxwell, 2004). It should also aim to investigate what educational uses of handheld devices students prefer for their educational practices, what the constraints of handheld devices are, what contextual factors (both broader factors such as institution, family and specific contextual factors such as time, place, users, needs) are critical in influencing educational uses of the handheld device, and how these educational uses result in changes to the environments in which educational practices take place. This requires that handheld devices be used in an uncontrolled and unpredicted manner. These uncontrolled and unpredicted uses should result from the students' own experiences in using the handheld devices. It is expected that such research results will help researchers and teachers acknowledge that educational uses of handheld devices vary from situation to situation, and from student to student. Therefore, when developing learning resources, researchers and teachers need to take these varied situations into consideration and avoid developing "one size fits all" learning resources. In addition, such research results will help researchers, teachers and students establish a clearer understanding of the educational uses and constraints of handheld devices for diversified practices, and make them more aware of the properties of different devices and resources. This, in turn, will help them with their decision making processes regarding using handheld devices for their educational practices. It is hoped that handheld educational uses will bring about changes in the environments in which educational practices take place, and in the practices themselves progressively, hence achieve "second-level effects" in the near future.

## REFERENCES


Abowd, G. D., & Mynatt, E. D. (2000). Charting past, present, and future research in ubiquitous computing. *ACM Transactions on Computer-Human Interaction, 7*(1), 29-58.

Ally, M., Lin, F., McGreal, R., & Woo, B. (2005). *An intelligent agent for adapting and delivering electronic course materials to mobile learners.* Paper presented at the mLearn 2005, Cape Town, South Africa.

Barkhuus, L., & Dey, A. (2003). Is context-aware computing taking control away from the user? Three levels of interactivity examined. *Lecture Notes in Computer Science, 2864*, 149-156.

Belanger, Y. (2005). *Duke University iPod first year experience final evaluation report.* Retrieved June 2, 2006, from http://cit.duke.edu/pdf/ipod_initiative_04_05.pdf

Bradley, C., Haynes, R., & Boyle, T. (2005). *Adult multimedia learning with PDAs -The user experience.* Paper presented at the mLearn 2005, Cape Town, South Africa.

Bull, S., & Reid, E. (2004). Individualised revision material for use on a handheld computer. In J. Attewell & C. Savill-Smith (Eds.), *Mobile learning anytime everywhere* (pp. 35-42). London: Learning and Skills Development Agency.

Cacace, F., Cinque, M., Crudele, M., Iannello, G., & Venditti, M. (2004). The impact of innovation in medical and nursing training: a Hospital Information System for Students (HISS) accessible through mobile devices. In J. Attewell & C. Savill-Smith (Eds.), *Mobile learning anytime everywhere* (pp. 47-52): Published by the Learning and Skills Development Agency.

Chen, J., & Kinshuk. (2005). Mobile technology in educational services. *Journal of Educational Multimedia and Hypermedia, 14*(1), 91-109.

Chen, Y. S., Kao, T. C., & Sheu, J. P. (2003). A mobile learning system for scaffolding bird watching learning. *Journal of Computer Assisted Learning, 19*( 3), 347-359.

Churchill, D., & Churchill, N. (in press). Educational affordances of PDAs: A study of a teacher's exploration of this technology. *Computers & Education.*

Colevins, H., Bond, D., & Clark, K. (2006). Nurses get a hand from handhelds. *Computers in Libraries, 26*(4), 6-48.

Corlett, D., Sharples, M., Bull, S., & Chan, T. (2005). Evaluation of a mobile learning organiser for university students. *Journal of Computer Assisted Learning, 21*(3), 162-170.

Crowe, A., & van 't Hooft, M. (2006). Technology and the prospective teacher: Exploring the use of the TI-83 handheld devices in social studies education. *Contemporary Issues in Technology and Teacher Education, 6*(1), 99-119.

Cui, Y., & Bull, S. (2005). Context and learner modelling for the mobile foreign language learner. *System, 33*(2), 353-367.

Davis, S. (2003). Observations in classrooms using a network of handheld devices. *Journal of Computer Assisted Learning, 19*(3), 298-307.

Dey, A. K. (2001). Understanding and using context. *Personal and Ubiquitous Computing, 5*, 4-7.

Draper, S. W. (1998). Niche-based success in CAL. *Computers & Education, 30*(1-2), pp. 5-8.

Facer, K., Joiner, R., Stanton, D., Reid, J., Hull, R., & Kirk, D. (2004). Savannah: mobile gaming and learning? *Journal of Computer Assisted Learning, 20*(6), 399-409.

Fallahkhair, S., Pemberton, L., & Griffiths, R. (2007). Development of a cross-platform ubiquitous language learning service via mobile phone and interactive television. *Journal of Computer Assisted Learning, 23*(4), 312-325.

Gay, G., & Hembrooke, H. (2004). *Activity-centered design: An ecological approach to designing smart tools and usable systems*. Cambridge, Mass. ; London: MIT Press.

Gredler, M. E. (2004). Games and simulations and their relationships to learning. In D. H. Jonassen (Ed.), *Handbook of research on educational communications and technology* (2nd ed., pp. 571-581). Mahwah, N.J.: Lawrence Erlbaum.

Green, P. D. (2001). Handheld computers as tools for writing and managing field data. *Field Methods, 13*(2), 181-197.

Godwin-Jones, R. (2007). Emerging technologies: E-texts, mobile browsing, and rich internet applications. *Language Learning & Technology, 11*(3), 8-13.

Hackemer, K., & Peterson, D. (2005). University of South Dakota palm initiative. In A. Kukulska-Hulme & J. Traxler (Eds.), *Mobile learning: A handbook for educators and trainers* (pp. 157-163). London: Routledge.

Henricksen, K., & Indulska, J. (2006). Developing contextaware pervasive computing applications: Models and approach. *Pervasive and Mobile Computing, 2*(2), 37-64.

Horton, B., & Wiegert, E. (2002). Using handhelds and billiards to teach reflection. *Learning & Leading With Technology, 29*(8), 32-35.

Jonassen, D. H., Howland, J., Moore, J., & Marra, R. M. (2003). *Learning to solve problems with technology: a constructivist perspective* (2nd ed.). Upper Saddle River, N.J.: Merrill Prentice Hall.

Kadyte, V. (2004). Learning can happen anywhere: a mobile system for language learning. In J. Attewell & C. E. Savill-Smith (Eds.), *Learning with Mobile Devices: Research and Development* (pp. 73-78). London: Learning and Skills Development Agency.

Keegan, D. (2002). *The future of learning: From eLearning to mLearning*: Hagen: Fern Universitat, Gesamthochschule, Germany.

Ketamo, H. (2003). xTask- an adaptable learning environment. *Journal of Computer Assisted Learning, 19*(3), 360-370.

Klopfer, E., Yoon, S., & Perry, J. (2005). Using palm technology in participatory simulations of complex systems: A new take on ubiquitous and accessible mobile computing. *Journal of Science Education & Technology, 14*(3), 285-297.

Lai, C.-H., Yang, J.-C., Chen, F.-C., Ho, C.-W., & Chan, T.-W. (2007). Affordances of mobile technologies for experiential learning: the interplay of technology and pedagogical practices. *Journal of Computer Assisted Learning, 23*(4), 326-377.

Lai, C.-Y., & Wu, C.-C. (2006). Using handhelds in a Jigsaw cooperative learning environment. *Journal of Computer Assisted Learning, 22*(4), 284-297.

Lankshear, C., Green, B., & Snyder, I. (2000). *Teachers and technoliteracy : managing literacy, technology and learning in schools.* St Leonards, N.S.W.: Allen & Unwin.

Laurillard, D., Stratfold, M., Luckin, R., Plowman, L., & Taylor, J. (2000). Affordances for learning in a non-linear narrative medium. *Journal of Interactive Media in Education, 2.*

Levy, M., & Kennedy, C. (2005). Learning Italian via mobile SMS. In A. Kukulska-Hulme & J. Traxler (Eds.), *Mobile learning: A handbook for educators and trainers* (pp. 76-83). London: Routledge.

Liang, J.-K., Liu, T.-C., Wang, H.-Y., Chang, B., Deng, Y.-C., Yang, J.-C., et al., (2005). A few design perspectives on one-on-one digital classroom environment. *Journal of Computer Assisted Learning, 21*(3), 181-189.

Lim, K. Y. T., & Wang, J. Y. Z. (2005). Collaborative handheld gaming in education. *Educational Media International, 42*(4), 351-359.

Liu, C.-C., Tao, S.-Y., Nee, J.-N., Liu, B.-J., Chen, G.-D., Hsu, C.-C., et al., (2005). Supporting Activity Awareness for Teams-Games-Tournaments with GSM Network. In H. Ogata, M. Sharples, Kinshuk & Y. Yano (Eds.), *Proceedings of the Third IEEE International Workshop on Wireless and Mobile Technologies in Education 2005* (pp. 238-242). Tokushima, Japan: The IEEE computer society.

Luckin, R., Brewster, D., Pearce, D., du Boulay, B., & Siddons-Corby, R. (2005). Whether it's m-learning or e-learning, it must be ME learning. In A. Kukulska-Hulme & J. Traxler (Eds.), *Mobile learning: A handbook for educators and trainers* (pp. 116-124). London: Routledge.

Lund, A. (2006). The multiple contexts of online language teaching. *Language Teaching Research, 10*(2), pp. 181-204.

Markett, C., Arnedillo Sanchez, I., Weber, S., & Tangney, B. (2006). Using short message service to encourage interactivity in the classroom. *Computers & Education, 46*(3), 280-293.

Maxwell, J. A. (2004). Causal explanation, qualitative research, and scientific inquiry in education. *Educational Researcher, 33*(2), pp. 3-11.

Mitchell, K., & Race, N. J. P. (2005). uLearn: Facilitating ubiquitous learning through camera equipped mobile phones. In H. Ogata, M. Sharples, Kinshuk & Y. Yano (Eds.), *Proceedings of the Third IEEE International Workshop on Wireless and Mobile Technologies in Education 2005* (pp. 274-281). Tokushima, Japan: The IEEE computer society.

Naismith, L., Lonsdale, P., Vavoula, G., & Sharples, M. (2004). *Literature review in mobile technologies and learning* (No. No. REPORT 11): NESTA Futurelab.

Nicol, D. J., & MacLeod, L. A. (2005). Using a shared workspace and wireless laptops to improve collaborative project learning in an engineering design class. *Computers & Education, 44*(4), 459-475.

Norris, C., & Soloway E. (2004). Envisioning the handheld-centric classroom. *Journal of Educational Computing Research, 30*(4), 281-294.

Ogata, H., & Yano, Y. (2004). *Context-aware support for computer-supported ubiquitous learning.* Paper presented at the 2nd IEEE International Workshop on Wireless and Mobile Technologies in Education, Taiwan.

Parr, C. S., Jones, T., & Songer, N. B. (2004). Evaluation of a handheld data collection interface for science learning. *Journal of Science Education and Technology, 13*(2), 233-242.

Patten, B., Arnedillo-Sanchez, I., & Tangney, B. (2006). Designing collaborative, constructionist and contextual applications for handheld devices. *Computers & Education, 46*(3), 294-308.

Pfeifer, R. S., & Robb, R. (2001). Beaming your school into the 21st century. *Principal Leadership, 1*(9), 30-34.

Pinkwart, N., Hoppe, H. U., Milrad, M., & Perez, J. (2003). Educational scenarios for cooperative use of Personal Digital Assistants. *Journal of computer assisted learning, 19*(3), 383-391.

Price, S., & Rogers, Y. (2004). Let's get physical: The learning benefits of interacting in digitally augmented physical spaces. *Computers and Education, 43*(1-2), 137-151.

Roschelle, J. (2003). Keynote paper: Unlocking the learning value of wireless mobile devices. *Journal of Computer Assisted Learning, 19*(3), 260-272.

Roschelle, J., Sharples, M., & Chan, T. W. (2005). Introduction to the special issue on wireless and mobile technologies in education. *Journal of computer assisted learning, 21*(3), 159-161.

Schwabe, G., & Goth, C. (2005). Mobile learning with a mobile game: Design and motivational effects. *Journal of Computer Assisted Learning, 21*(3), 204-216.

Segall, N., Doolen, T. L., & Porter, J. D. (2005). A usability comparison of PDA-based quizzes and paper-and-pencil quizzes. *Computers & Education, 45*(4), 417-432.

Seppälä, P., & Alamäki, H. (2003). Mobile learning in teacher training. *Journal of Computer Assisted Learning, 19*(3), 330-335.

Sharples, M., & Beale, R. (2003). A technical review of mobile computational devices. *Journal of Computer Assisted Learning, 19*(3), 392-395.

Sharples, M., Corlett, D., & Westmancott, O. (2002). The design and implementation of a mobile learning resource. *Personal and Ubiquitous Computing, 6*, 220 - 234.

Shih, Y. E., & Mills, D. (2007). Setting the new standard with mobile computing in online learning. *International Review of Research in Open and Distance Learning, 8*(2).

Smordal, O., & Gregory, J. (2003). Personal digital assistants in medical education and practice. *Journal of Computer Assisted Learning, 19*(3), 320-329.

Song, Y. (in press). SMS enhanced vocabulary learning for adult learners mobile audiences. *International Journal of Mobile Learning and Organisation.*

Song, Y., & Fox, R. (2007). A review of affordances and constraints of handheld devices in higher education. In P. Tsang, R. Kwan & R. Fox (Eds.), *Learning through technology* (pp. 229-239). Singapore: World Scientific Publishing.

Sproull, L., & Kiesler, S. B. (1995). *Connections: new ways of working in the networked organization.* Cambridge, Mass.: MIT Press.

Swan, K., van't Hooft, M., Kratcoski, A., & Unger, D. (2005). Uses and effects of mobile computing devices in K-8 classrooms. *Journal of Research on Technology in Education, 38*(1), 99-112.

Thornton, P., & Houser, C. (2005). Using mobile phones in English education in Japan. *Journal of Computer Assisted Learning, 21*(3), 217-228.

Tolmie, A. (2001). Examining learning in relation to the contexts of use of ICT. *Journal of Computer Assisted Learning, 17*(3), 235-241.

Traxler, J., & Kukulska-Hulme, A. (2005). *Evaluating mobile learning: Reflections on current practice.* Paper presented at the mLearn 2005, Cape Town, South Africa.

Uden, L. (2007). Activity theory for designing mobile learning. *International Journal of Mobile Learning and Organisation, 1*(1), 81-102.

Vahey, P., & Crawford, V. (2002). *Palm education pioneers program: Final evaluation report.* Menlo Park, CA: SRI International.

van't Hooft, M., Diaz, S., & Swan, K. (2004). Examining the potential of handheld computers: Findings from the Ohio PEP project. *Journal of Educational Computing Research, 30*(4), 295-311.

Vogel, D., Kennedy, D. M., Kuan, K., Kwok, R., & Lai, J. (2007). Do mobile device applications affect learning? In R. H. Sprague (Ed.), *Proceedings of the Hawaii International Conference on System Sciences (HICSS-40).* Hawai'i:: Institute of Electrical and Electronics Engineers.

Walton, G., Childs, S., & Blenkinsopp, E. (2005). Using mobile technologies to give health students access to learning resources in the UK community setting. *Health Information & Libraries Journal, 22*(Supplement2), 51-65.

Wei, F.-H., Chen, G.-D., Wang, C.-Y., & Li, L.-Y. (2007). Ubiquitous discussion forum: Introducing mobile phones and voice discussion into a Web discussion forum. *Journal of Educational Multimedia and Hypermedia, 16*(2), 125-140.

West, P., Foster, G. G., & Clayton, P. G. (2005). *Content exposure of slide show presentations for selective download and annotation via mobile devices.* Paper presented at the mLearn 2005, Cape Town, South Africa.

Whitelock, D., Domano, D., Jelfs, A., & Brna, P. (2000). Perfect presence: What does this mean for the design of virtual learning environments? *Education and Information Technologies, 5*(4), 471-488.

Zhang, D. (2007). Web content adaptation for mobile handheld devices. *Communications of the ACM, 50*(2), 75-79.

## ADDITIONAL READING

Aderinoye, R. A., Ojokheta, K. O., & Olojede, A. A. (2007). Integrating mobile learning into nomadic education programmes in Nigeria: Issues and perspectives. *The International Review of Research in Open and Distance Learning, 8*(2).

Alexander, B. (2004). Going nomadic: Mobile learning in higher education. *EDUCAUSE Review, 39*(5), 28-35.

Attewell, J., & Savill-Smith, C. (2004). Mobile learning and social inclusion: focusing on learners and learning. In J. Attewell & C. Savill-Smith (Eds.), *Learning with mobile devices: Research and development* (pp. 3-12). London: Learning and Skills Development Agency.

Baber, C., Sharples, M., Vavoula, G., & Glew, P. (2004). A "learning space" model to examine the suitability 21 of different technologies for mobile learning. In J. Attewell & C. E. Savill-Smith (Eds.), *Learning with mobile devices: Research and development* (pp. 21-26). London: Learning and Skills Development Agency.

Carlson, S. (2002). Are personal digital assistants the next must-have tool? *The Chronicle of Higher Education, 49*(3), A.33.

Crippen, K., & Brooks, D. (2000). Using personal digital assistants in clinical supervision of student teachers. *Journal of Science Education and Technology, 9*(3), 207-211.

Dede, C. (2002). Augmented reality through ubiquitous computing. *Learning & Leading With Technology, 29*(8), 13.

Eschenbrenner, B., & Nah, F. F.-H. (2007). Mobile technology in education: uses and benefits. *International Journal of Mobile Learning and Organisation, 1*(2), 159-183.

Gay, G., Stefanone, M., Grace-Martin, M., & Hembrooke, H. (2001). The effects of wireless computing in collaborative learning environments. *International journal of Human-Computer Interaction, 13*(2), 257-276.

Hoppe, H. U., Joiner, R., Milrad, M., & Sharples, M. (2003). Guest editorial: Wireless and mobile technologies in education. *Journal of Computer Assisted Learning, 19*(3), 255-259.

Klopfer, E., & Squire, K. (In press). Case study analysis of augmented reality simulations on handheld computers. *The Journal of the Learning Sciences*.

Ling, R. S. (2004). *The mobile connection: The cell phone's impact on society*. San Francisco: Morgan Kaufmann.

Loh, C. (2001). Learning tools for knowledge nomads: using personal digital assistants (PDAs) in Web-based learning environments. *Educational Technology, 41*(6), 5-13.

Maney, K. (2006). 10 years ago Palm Pilot got started on a bluff by inventor. *USA Today*, 03b.

McGuire, L. (2005). Assessment using new technology. *Innovations in Education & Teaching International, 42*(3), 265-276.

Newhouse, C. P., Williams, P. J., & Pearson, J. (2006). Supporting mobile education for preservice teachers. *Australasian Journal of Educational Technology, 22*(3), 289-311.

Paulson, L. D. (2006). TV comes to the mobile phone. *Computer, 39*(4), 13-18.

Penuel, W. R., Tatar, D. G., & Roschelle, J. (2004). The role of research on contexts of teaching practice in informing the design of handheld learning techologies. *Journal of Educational Computing Research, 30*(4), 353-370.

Pulman, & Andy. (2007). Can a handheld gaming device be used as an effective assistive technology tool? *British Journal of Educational Technology, 38*(3), 532-534.

Rau, P.-L. P., Gao, Q., & Wu, L.-M. (2008). Using mobile communication technology in high school education: Motivation, pressure, and learning performance. *Computers & Education, 50*(1), *1-22.*

Roschelle, J., Patton, C., & Tatar, D. (2007). Designing networked handheld devices to enhance school learning. In M. Zelkowitz (Ed.), *Advances in computers, 70*, 1-60.

Sharples, M., Milrad, M., Arnedillo, S. I., & Vavoula, G. (2007). Mobile learning: Small devices, big Issues. In *Technology Enhanced Learning: Principles and Products* (pp. 20 pages, chapter 14).

Staudt, C. (2005). *Changing how we teach and learn with handheld computers*. Thousand Oaks, CA: Corwin Press.

Waycott, J., Jonesb, A., & Scanlonb, E. (2005). PDAs as lifelong learning tools: an activity theory based analysis. *Learning, Media and Technology, 30*(2), 107-130.

Zurita, G., & Nussbaum, M. (2007). A conceptual framework based on activity theory for mobile CSCL. *British Journal of Educational Technology, 38*(2), 211-235.

# Chapter XVI
# Assessing the Benefits of AJAX in Mobile Learning Systems Design

**Feng Xie**
*Massey University, New Zealand*

**David Parsons**
*Massey University, New Zealand*

## ABSTRACT

*Mobile technologies are rapidly changing our lives with increasing numbers of services supported by mobile devices, including Web-based learning applications, providing opportunities for people to study anytime and anywhere. However, using Web-based mobile applications to present learning resources is a challenge for developers because the performance of the mobile Internet over GPRS networks is often unacceptably slow. A new Web development model, Ajax, may help to address this problem. Ajax (asynchronous JavaScript and XML), is an approach to Web application development that uses client-side scripting to reduce traffic between client and server and provide a seamless user application experience. In this chapter, we address the question of whether mobile Ajax provides measurable performance advantages over non-Ajax mobile learning applications. An empirical study was undertaken to measure mobile learning application performance over a GPRS network, comparing an Ajax application and an active server pages (ASP) application with identical functionality. Our results suggest that mobile Ajax can reduce the bandwidth requirement by around 70 percent, and cut the server's response time in half. In addition, these performance improvements were noticed by users in our small group usability test.*

## INTRODUCTION

We live in an information society where learning becomes ever more important, and not all of this learning can take place in a static environment. The mobile revolution is changing our lives and can also facilitate new learning processes, but learners need educational services with a fast response speed and good user interaction if mobile learning systems are to be readily adopted.

An increasing number of people have been using mobile Internet access through wireless networks (Church, Smyth, Cotter, & Bradley, 2007), but due to the limitations of the mobile communications infrastructure and hardware this can still be very problematic. There are a number of reasons, for example: small device screens, high network latency, low bandwidth and interaction complexity (Chakravorty & Pratt, 2002). Such weaknesses can impede the mobile learning process.

The poor performance of commonly used wireless networks such as GSM (global system for mobile communications) and its associated digital packet switched data service, GPRS (general packet radio service) is a major problem for mobile learning systems. There are various reasons for this poor performance, for example high and variable latency, fluctuating bandwidth, occasional link 'blackouts' (Chakravorty, Cartwright & Pratt, 2002), packet loss, and link outages. Sometimes, even simple requests can lead to long delays (Stuckmann, Ehlers & Wouters, 2002). However, despite these problems we should recognise that one distinct feature of mobile learning over other learning activities is *mobility* (Leung & Chan, 2003). The GPRS mobile phone network is the most commonly used network in the world, with the widest coverage, considerably larger globally than 3G (third generation) wireless networks. Using this network for mobile learning can provide services anytime almost anywhere around the world, with extensive international roaming. In addition, although 3G was intended to resolve technological fragmentation in the wireless communications market, this has not happened in practice and there are several competing 3G technologies. Cost is also an important factor, with a large number of low cost GSM/GPRS devices on the market, and in many territories the GSM/GPRS service fee is cheaper than 3G. Therefore, despite their limitations, we will need to continue to work with GPRS systems for some time to come.

### Mobile Learning Technical Challenges

The combination of wireless telecommunications and mobile computing is resulting in a transformation of the educational landscape (Alexander, 2004). The growth and rapid evolution of wireless technology have created new opportunities for the 'anytime and anywhere' learning paradigm (Seong, 2006) that is mobile learning.

Various researchers have defined mobile learning (m-learning) in different ways. Pinkert et al. (2003) define it as e-learning that uses mobile devices and wireless transmission. Polsani (2003) defines it as a form of education whose site of production, circulation, and consumption is the network. Traxler (2005) defines m-learning as any educational provision where the sole or dominant technologies are handheld or palmtop devices. Sharples (2005) defines it as a process of coming to know, by which learners in cooperation with their peers and teachers, construct transiently stable interpretations of their world.

Regardless, mobile learning is different from our traditional learning experience; it has its own problems and limitations. Mobile learners may feel uncomfortable because they cannot have face-to-face interaction with teachers or other students (Stodel, Thompson, & MacDonald, 2006). There are also limitations on what it can deliver. Berri, Benlamri & Atif (2006) describe it as mainly a time-constrained exercise with lightweight content-oriented instruction.

Traxler (2007) categorizes previous mobile learning studies into 6 aspects:

- Technology-driven
- Miniature but portable e-Learning
- Connected classroom learning
- Informal, personalized, situated mobile learning
- Mobile training / performance support
- Remote / rural / development mobile learning

From a design process point of view, these aspects fall into three major categories: pedagogical learning models, adapting to mobile environments, and technical challenges (Berri et al., 2006). Currently, most mobile learning studies focus on pedagogical learning models, or mobile environments. In contrast, this study focuses on technical challenges.

Despite extensive research into mobile learning systems, technical support for Web-based mobile learning has been explored in a relatively small number of studies. On example is a mobile learning system using Instant Messaging and wireless networks (GSM/GPRS and 802.11 Wi-Fi), developed by Kadirie (2007). This system is Web-based, and the kernel of the instant messaging server is written in Java. Learners use a mobile Web browser to login to this learning system to chat with other learners or view their messages. Nakahara, Hisamatsu, Yaegashi, and Yamauchi (2005) describe a Web-based collaborative learning site with a bulletin board system (BBS) that allows learners to interact, exchange information, engage in discussion, and collaborate on projects. Another technical paper is by Seong (2006). He developed a mobile learning course manager portal to demonstrate and exemplify the usability guidelines proposed. Kukka and Ojala (2006) developed a Java and XML based learning system for both desktop and mobile devices. Other technical examples include Cao, Tin, McGreal, et al. (2006), Lee and Lu (2003) and Theng (2007). These examples all use Web-based technologies plus a mobile Web browser or mobile application as the mobile learning client.

These Web-based mobile learning systems use a traditional synchronous client-server application architecture, typically using the GSM/GPRS network, but in most cases Web application performance over a high latency GPRS network is poor (Chakravorty et al., 2002; Chakravorty & Pratt, 2002; Stuckmann et al., 2002). One issue is that the TCP/IP Internet protocol needs to initialize before it starts to transfer actual data; this process takes over 7 seconds to enable the connection when the initial request is made (Chakravorty, Clark, & Pratt, 2003). In addition all wireless networks have high latency; GPRS link latency is 600ms-3,000ms for the downlink and 400ms-1,300ms on the uplink. Round-trip latencies are therefore are least 1,000ms (Chakravorty et al., 2002). In a lab environment, download times for CNN's Web site over GPRS was between 125 and 170 seconds (Chakravorty & Pratt, 2002). This latency does not improve with increased bandwidth, and deploying a 3G mobile phone network is not necessarily helpful (Hunaiti, Garaj, Balachandran & Cecelja, 2005), for example the TCP initialization time on GPRS is 5-7 seconds, but the TCP initialization time on 3GSM is up to 12-15 seconds (Nortel, 2007).

This poor network performance is compounded by the synchronous request/response model of traditional Web applications. Therefore to improve mobile learning system performance we might use a higher performance network, where one is available, or perhaps apply some new approach to the architecture of the Web-based mobile learning system itself.

## Ajax Web Applications

To address performance issues in Web-based mobile learning applications, we may look to recent developments in desktop browser technologies that might be implemented in the mobile environ-

*Figure 1. The pioneers of the Ajax approach (Google Suggest & Google Maps)*

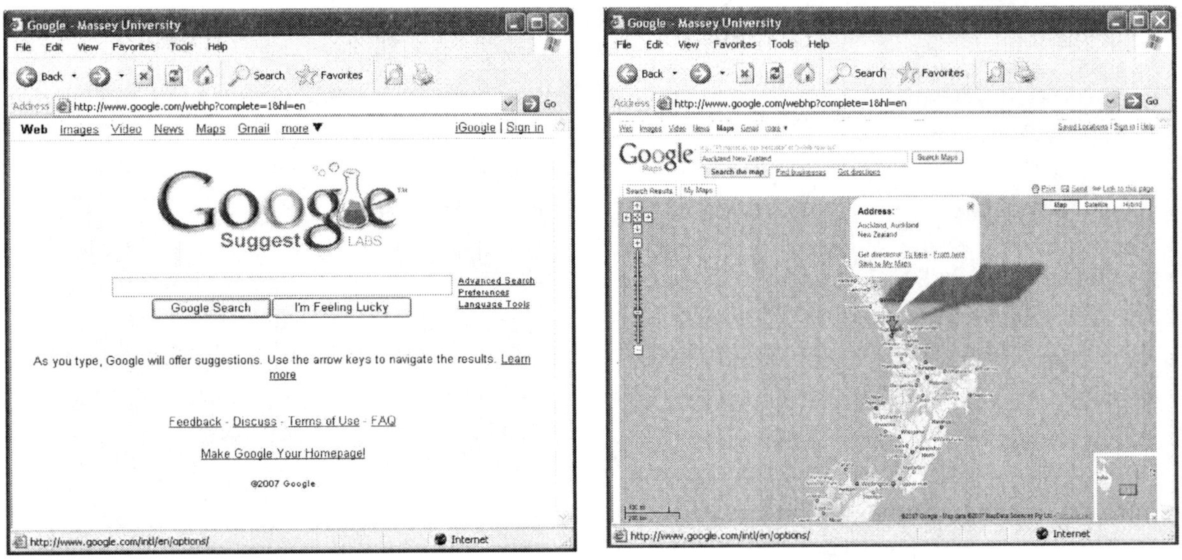


ment. In 2005, Garrett introduced the concept of Ajax (Asynchronous JavaScript and XML) (Garrett, 2005). Since then, an increasing number of Web applications have used the Ajax approach (including learning systems—see chapter VIII of this book). Although Garrett's article was the first to use the term 'Ajax,' the technique he described was already being used by many Web applications. The pioneer developers of this technique were Google, with Google Suggest and Google Maps being two early examples of this approach (Figure 1), though some of the key technologies (such as the XMLHttpRequest object) were first developed by Microsoft.

Ajax was originally introduced as a desktop Web programming model, not a mobile environment solution, but because of its key characteristics (small transmission volumes, asynchronous communication and partial Web page updates), it can actually be very useful in the mobile environment, especially in high latency networks where Ajax can reduce the frequency and volume of data transfer.

The traditional Web application allows users to submit a request, which the server will process and respond to. Then the client browser will refresh the whole Web page, even if only a small part of the content is changed. In most cases, however, the new Web page will be very similar to the old one, which means that during these transmissions some duplicate content is transferred. This can waste both network resources and users' time.

Ajax uses a browser hosted 'Ajax engine' to handle both data transmission and partial updates to the Web page. The Ajax engine only requests new content from the server, reducing unnecessary data transmission. It can also partially update the current Web page when the response is received. Because the Ajax engine runs locally in the client browser, the Web application's response speed is faster and the user's experience is improved. In addition, client server communication in Ajax can be carried out asynchronously, enabling the user to continue interacting with the system even while the browser is waiting for data from the server. Compared with the traditional Web application model, Ajax can significantly increase a Web application's performance in the desktop browser environment (Smullen & Smullen, 2006, 2007; White, 2005).

### Applying Ajax to a Mobile Learning Application

Previous studies have shown that Ajax reduces the data transmission volumes between the server and client device, and improves the user experience on the desktop. These are particularly relevant issues for Web based mobile learning systems, which tend to be content rich and require extensive client server interaction. Such systems can surely benefit from a strategy to reduce transmission volumes in the context of expensive and low speed connections, and provide a better user experience in browsers with limited screen real estate and navigation tools. With the increasing sophistication of mobile browsers, there are many mobile devices that can easily support Ajax applications. Therefore, we might ask the question, if we apply Ajax to a mobile learning environment, what might be the result? Does an Ajax mobile learning application provide measurable advantages over a non-Ajax mobile learning application?

In this chapter, we describe the development and analysis of a mobile learning system built using Ajax and tested in a GPRS mobile environment. We evaluate the system in terms of both measured performance against a non-Ajax system and usability testing, and assess what benefits we might gain from this approach.

We developed two mobile learning systems with identical functionality, one using Ajax and the other using a more traditional active server pages (ASP) architecture. These two Web applications have the same user interface and can both be accessed by commonly available mobile browsers. We measured the performance of these two applications over a GPRS network based on the data collected from the Web server's log files. We also carried out a small group usability test to assess the perceived benefits of Ajax over a non-Ajax system.

To perform our experiments we needed to implement a representative mobile learning system that was structured enough to enable effective performance measurement. To meet these requirements we chose to implement a mobile quiz as the learning content of the applications. Quizzes are popular components of learning systems, which can help the learner to improve their personal knowledge and problem solving ability (Yokomoto, 2000). They force the learner to think about every question, answer them all carefully, and review all answers at the end. A good quiz-based learning system needs a responsive user interface to give rapid feedback. This requirement can be easily met in a desktop environment, but it is not so easy in a mobile Web-based system. Nevertheless there have been some examples of using quizzes in mobile learning systems (Black & Hawkes 2006; Bar et al., 2007; Seong, 2006). Therefore we concluded that a mobile quiz was both an appropriate and effective example to use for our experiments.

## WEB APPLICATIONS, AJAX AND MOBILITY

Since the first Web page built by Tim Berners-Lee in 1991 (Lee, 1992; Watson, Rainer & Koh, 1991), a synchronous request and response cycle has been used for Web access, with a 'click, wait and refresh' action approach (Figure 2) where the user needs to click to send a request and wait until the browser refreshes with a new page (Wei, 2005).

In this Web interaction model the user must wait for a request to be converted into a data stream, sent by HTTP over the Internet, processed on the server, and have its response returned by HTTP to be displayed in the browser (Crane, Pascarello & James, 2005). The user's waiting

*Figure 2. Classic Web application model: Full page refresh and synchronous communication (adapted from Wei, 2005)*

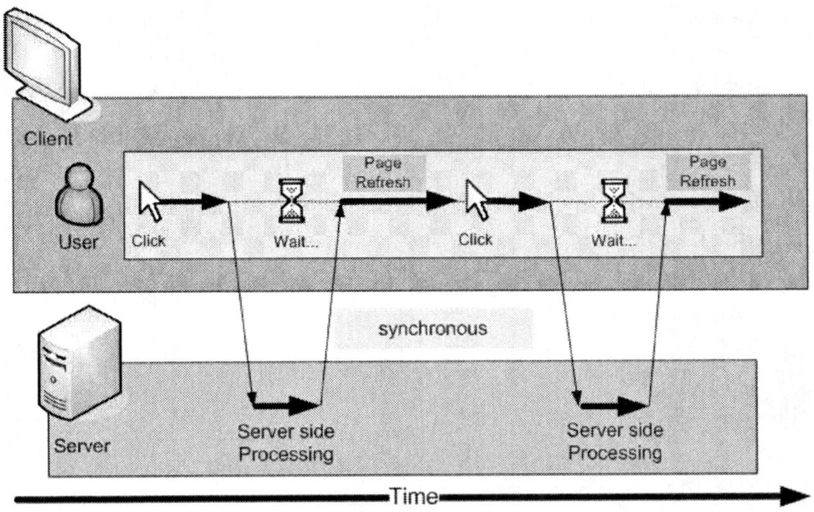


*Figure 3. The synchronous request and response model*

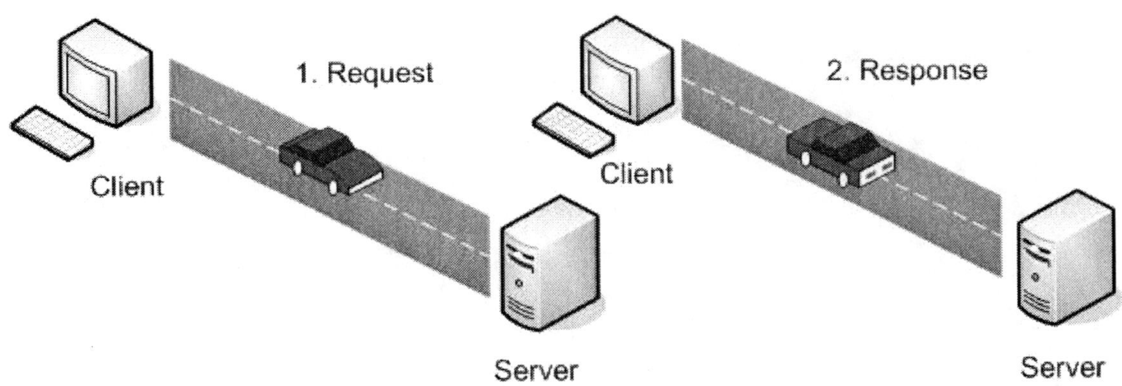


time includes network transition time, server response time and the browser's display time. The problem is that users cannot do anything when they are waiting, and this waiting time can be up to minutes depending on the network status. The 'click, wait and refresh' approach makes a lot of technical sense, because it makes it easy for Web developers to build applications, but it is not efficient or good for the user's experience. It forces the user to act like a machine; input, wait for the process to finish, and finally get the output.

The traditional Web application communication model is based on synchronous requests and responses. The client initiates a request and the server responds to it. The communication is always initiated one way, from the client to the server. During communication, users cannot do anything; they have to wait until the server responds to the request. This approach is like a car running on a freeway connected between the client and the server. The freeway allows two lines of traffic to travel at the same time, but due to the approach we are using, at any point in time only one car is actually running on the freeway (Figure 3). The traditional Web application cannot therefore provide a seamless user experience.

According to Watson et al. (1991), most users will run out of patience after six seconds, so we need some way of avoiding such delays. Under the limitations of network bandwidth and latency, the only way to improve Web application performance is to break down the synchronous model, and jump off the 'click, wait and refresh' cycle. One approach to this problem is Ajax (asynchronous JavaScript and XML).

## Asynchronous JavaScript and XML (Ajax)

Ajax is a Web development technique for creating interactive Web applications. It is not a single new technology but combines a set of powerful, widely-used, well-known and mature technologies, mostly hosted by the client browser and largely independent of the server. The intent of Ajax is to make Web pages feel more responsive by exchanging small amounts of data with the server behind the scenes, so that the entire Web page does not have to be reloaded each time the user makes small changes. This increases the Web page's interactivity, speed, and usability.

Garrett (2005) first defined Ajax as a combination of:

*Figure 4. Asynchronous communication and partial user interface updates (adapted from Wei, 2005)*

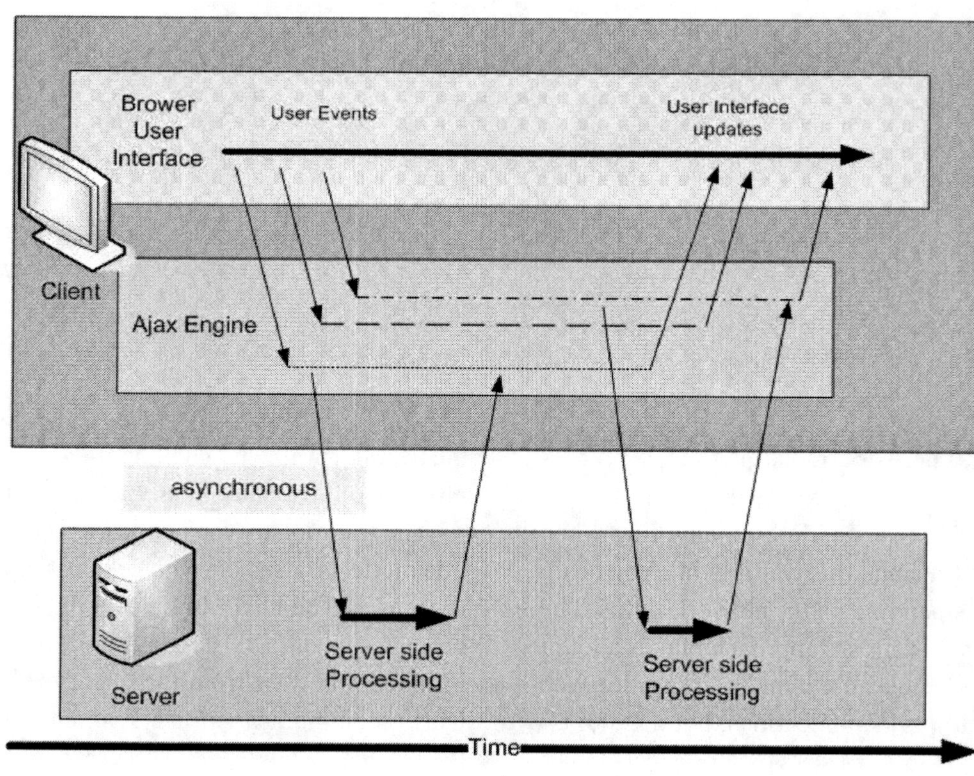

- Standards-based presentation using extensible hypertext markup language (XHTML) and cascading style sheets (CSS);
- Dynamic display and interaction using the document object model (DOM);
- Data interchange and manipulation using the XML and extensible stylesheet language transformations (XSLT);
- Asynchronous data retrieval using the XMLHttpRequest object;
- JavaScript binding everything together.

By controlling the submission and properties of XMLHttpRequest objects using JavaScript, partial screen updates replace the click, wait, and refresh user interaction model, and the synchronous request/response model is replaced by asynchronous communication (Wei, 2005). Asynchronous communication allows users to continue to interact with Web applications without waiting for server responses. The Ajax engine will automatically request information from the server in the background. Meanwhile, users can keep using other functions of the Web application. When the new information arrives, the Ajax engine will partially update the user interface (Figure 4).

With this interaction model, the user can enjoy a seamless browsing experience. Compared to the analogy of freeway with a single vehicle, it is like a busy and efficient freeway with multiple vehicles travelling in both directions (Figure 5).

The weaknesses of the traditional Web interaction paradigm and communication model are the strengths of the Ajax model. The Ajax approach for the first time introduces some common features of desktop applications to Web-based applications. The partial update and asynchronous features provide a number of advantages.

- Instead of submitting form data a page at a time when explicitly requested by the user, Ajax does submissions automatically, when the user triggers some event.
- Requests can be sent asynchronously, with the browser receiving results continually.

*Figure 5. The asynchronous request and response model*

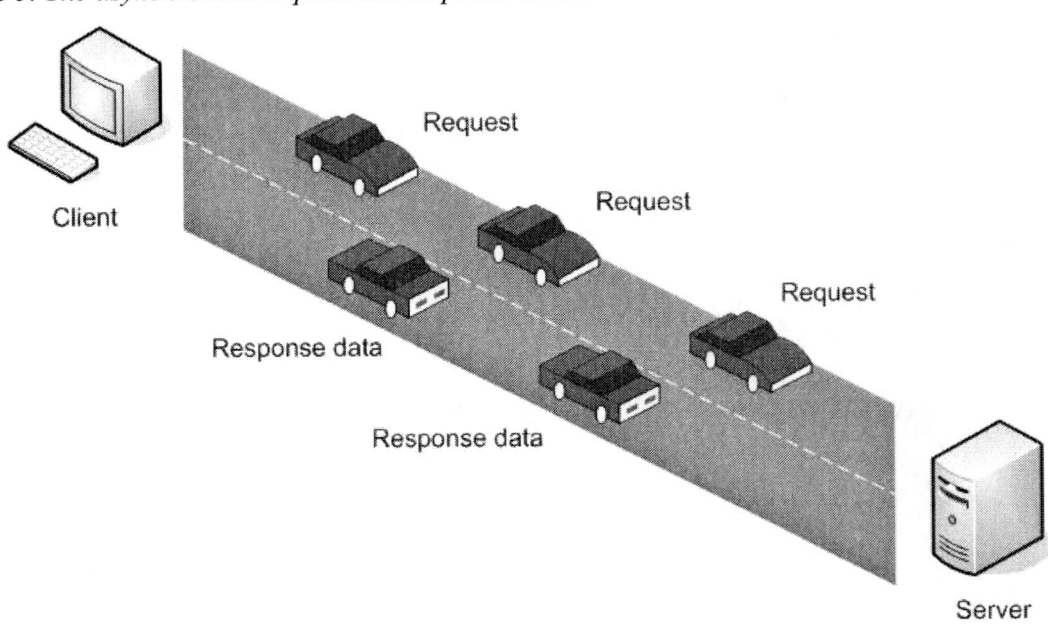

- Smaller, incremental in-place partial Web page updates can be made, instead of full page rebuilds.
- The size of data transmission can be reduced as the Ajax engine only downloads new content, not new pages.

In summary, the key characteristics of Ajax are asynchronous requests to the server, retrieving data continually, and partial page updates. Theoretically, these features can significantly improve the performance and user experience of Web applications.

*Figure 6. Classic Web application architecture (adapted from Wei, 2005)*

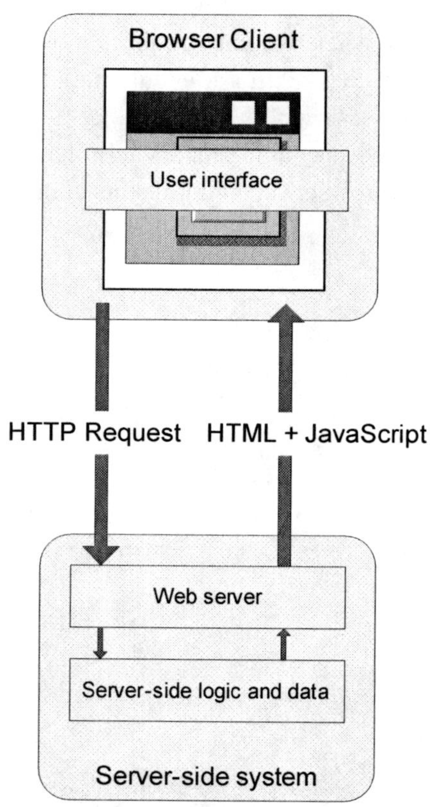


## The Ajax Engine

Not only is the Ajax communication model different from the traditional Web application, but the structural layers of Ajax are different. A traditional Web application architecture has two layers; user interface and server (Figure 6) while Ajax has three layers; user interface, Ajax engine, and server (Figure 7).

The Ajax engine plays the role of middleware sitting between the user interface and the server. It controls the user interface and the data trans-

*Figure 7. Ajax architecture (adapted from Wei, 2005)*

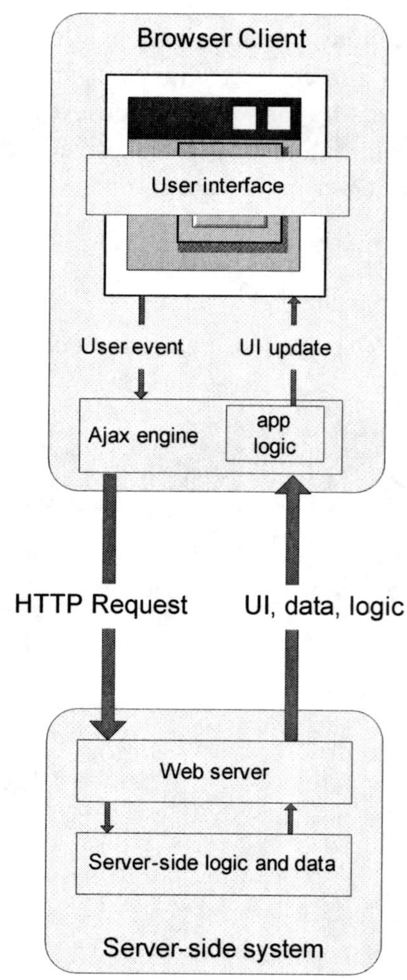

*Figure 8. Ajax engine*

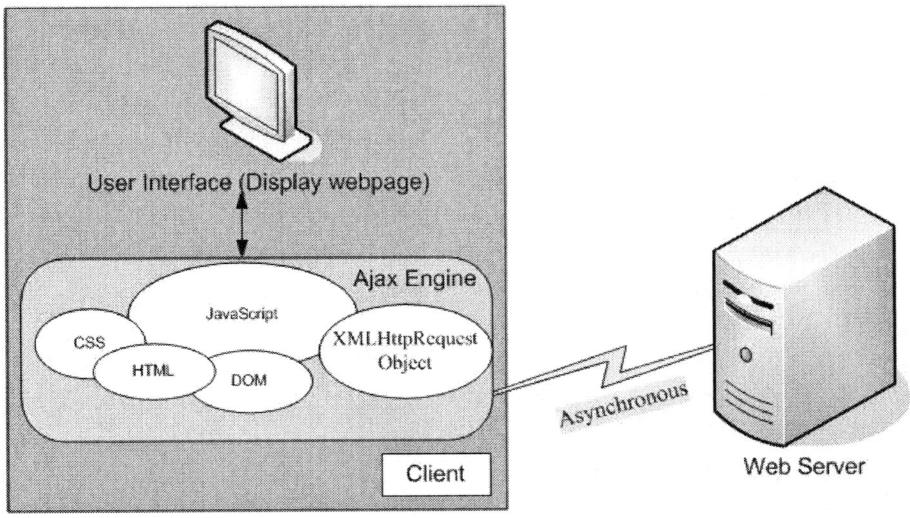


mission between client and server, and also some logical functions. When the user initiates his or her request, this request will pass to the Ajax engine first, instead of the server directly as it does in a traditional client-server model. Then the Ajax engine will combine the data from the Web browser cache and data requested from the server to update the page. The main advantage of this combining data process is that it can reduce the size of data downloaded from the server, and increase the response speed.

Data transfer between the server and the Ajax engine is via XML documents (or character streams), not entire Web pages. These XML documents only contain the new content required by the client; they do not include any Web page structure or presentation information. This can also reduce the client-server data transfer volume and time.

The Ajax engine is written in JavaScript and uses the document object model (DOM), cascading style sheets (CSS), HTML, and the XMLHttpRequest object to control the Web application (Figure 8). The Ajax engine can be loaded into the client browser at any time, depending on the application design.

The XMLHttpRequest object is a JavaScript object. It supports a set of operations that that allows JavaScript to perform HTTP client functionality for transferring data between a client and a server, such as submitting form data or loading data from a server (W3C, 2007). By using the XMLHttpRequest object, a Web developer can partially update the current page with data from the server after the page has loaded using asynchronous client-server communication.

XMLHttpRequest objects can be used by browser scripting languages to transfer XML and other text data to and from a Web server using HTTP, by establishing an independent communication channel between client and server. The user's requests are received by the Ajax engine, the engine analyses the request and sends out a new request to the server; when the client sends the request to the server via an XMLHttpRequest object, the server will respond to the client using an XML document or character stream rather than a new page of markup. The Ajax engine will use this content to partially update the user interface using the DOM.

In traditional HTML Web applications the 'submit' action is manually controlled by the user's clicking a button following the 'click, wait, and refresh' interaction model, but with Ajax, the JavaScript can directly communicate with the server through the XMLHttpRequest object. It means that not only can 'click events' activate submit actions, but also other user activities, such as mouse or keyboard events. XMLHttpRequest objects allow asynchronous requests to be sent to the server, so requests can be sent continuously every time a user triggers an event, and one does not need to wait until the previous request returns before sending out another request. When the new content arrives, JavaScript uses the received XML document to partially rebuild the Web page using the DOM and CSS without disrupting the user's activities.

## Ajax Performance

Previous research suggests that Ajax can significantly improve Web application performance on the desktop. One commonly cited Ajax performance evaluation is the one reported by White (2005). In this study, the Ajax application transferred on average just 27 percent of the bytes that were transferred by a traditional HTML application. Not only was there an improvement in the transferred byte volume, but there was also an improvement in performance, a 68 percent overall improvement in data transfer time. The two applications used in this study, however, did not have the same user interface. The users' skill levels and training were also not assessed in the report, and these factors may have affected the experimental outcomes.

A more controlled Ajax performance measurement comes from Smullen and Smullen (2006). They compared the client-side performance of a real-life HTML application and an Ajax application that implemented the same user interface. In this case, the variable of training and users' skill level can be ignored. Experimental data was collected on the performance of each when presented with the same set of tasks. Performance measures were computed for the HTML application and for the Ajax application.

Later they extended their study by collecting data on a statistically significant sample size and included server performance results. Response size and service time performance measures computed for the applications provided significant performance improvements in response size for the Ajax application (56%), thereby reducing bandwidth requirements. Ajax provided a mean service time improvement of approximately 16 percent (Smullen & Smullen, 2007).

Another Ajax performance study used a Web-chat application, and compared a traditional client-server Web chat model with an asynchronous Ajax model (Angelaccio & Buttarazzi, 2006). Their results showed better performance with the Ajax system, measured by the ratio of chat messages to the second.

Although Ajax performance has been measured on the desktop, there is no currently available performance measurement research about Ajax on mobile devices. Therefore we do not know if it can improve mobile Web-based application performance, or indeed if Ajax is effective in the mobile environment. Ajax is a client-side technology, which means when the client platform moves from the desktop to a mobile device the results may be different. Therefore this study focuses on measuring Ajax performance in the context of mobile learning.

## Mobile Browser Support for Ajax

Nokia studies of S60 Smartphone users show browsing is generating over 60 percent of data traffic (Nokia, 2005). Today, hundreds of mobile phone models are available on the market, and more and more mobile phones are offering mobile browsers. A mobile Web browser is designed for the constrained environment of mobile devices, such as mobile phones, PDAs, or PPC (pocket PCs). These mobile browsers are device-oriented. They need to be optimized or modified for the devices they are going to run on, because different vendors' mobile phones have different kinds of hardware and embedded operating systems and applications.

There is no common standard for mobile client development; each mobile phone or mobile device is unique. To illustrate, the size of the screen can vary greatly. The Nokia N70 uses 176 x 208 pixels, 35 x 41 mm TFT and a 256K color screen, but the Motorola RAZR2 V8 uses a 2.2" QVGA (320 x 240 pixels) internal display with a 262K color screen. Their display sizes and resolutions are totally different. Not only is the hardware different for each mobile phone, but operating systems also vary. For example the Nokia N70 uses the Symbian operating system while the Motorola RAZR2 V8 uses Linux. Because of these variations in both hardware and software, a mobile application cannot run on both mobile phones without some customization. However, an increasing number of mobile Web browsers are being developed that can overcome these hardware and operating system issues and provide a common platform for mobile Ajax. To support Ajax, the mobile Web browser must be JavaScript enabled and support the XMLHttpRequest object. It also needs to support DOM, CSS and XML. Opera Mobile, Nokia's S60 Browser and Microsoft Internet Explorer Mobile all claim to support JavaScript and Ajax, and more mobile browsers will support Ajax in the future, which should help the acceptability of Ajax as a pervasive mobile learning platform. For our experiments we used Opera Mobile, a stand-alone Web browser that can run on a number of different mobile devices. Opera Mobile supports all of the necessary desktop browser technologies that enable Ajax.

## METHODOLOGY

The methodology used in this study was design science, which has five basic steps: awareness of the problem, suggestion, development, evaluation, and conclusion. Design science research is sometimes called "improvement research," emphasizing the problem-solving/performance-improving nature of the activity. Suggestions for a problem solution are abductively drawn from the existing knowledge/theory base for the problem area. An attempt to build an artefact that implements the solution follows (development). Partially or fully successful implementations are then evaluated. This cycle of development, evaluation and further suggestions is frequently performed iteratively. The basis of the iteration, from partial completion of the cycle back to awareness of the problem, is *circumscription*. Conclusion indicates the termination of a specific design project (Vaishnavi & Kuechler, 2007).

Our study consisted of a single iteration with two evaluations (technical performance measures and usability testing). The results of these measures can be used in further research. In terms of awareness of the problem, we know that mobile Web access is hampered by issues like small screens, high network latency, small bandwidth and interface complexity. One suggestion for addressing this problem is to develop a mobile Ajax system, but so far Ajax application performance over wireless networks has remained untested. To address this issue, two mobile quiz Web applications were developed. The first mobile quiz system is an Ajax enabled Web application, while the second uses a more traditional ASP (active server pages) Web technology. Although these two Web applications use different technologies, they have the same user interface. Because we were interested in comparing the performance of these two applications we needed to exclude other aspects that might affect system performance, such as the user interface. Also, the learning process had to be the same, including very similar questions and answers. Finally, both systems had to be accessed from the same mobile device over the same GPRS network. We evaluated the performance of the two Web-based mobile learning applications from three aspects: theoretical evaluation, real-life performance evaluation, and usability testing.

We began with a theoretical evaluation (quantitative study) where we modeled the mobile learning process and calculated all the data on paper to find out how different system implementations mapped to our design. First, we assume there is a user visiting the Web application using a Web browser and predict the data transfer usage between client and server. Then we compare the ASP Web application's predicted data transfer volumes with the Ajax enabled Web application's data. According to this data, we might consider modifying the Web application's design. This can be considered a tentative evaluation.

The second part of the study was a real-life computer logging evaluation (quantitative study). Here, we set up an experimental test system (using a mobile device, a GPRS network and a Web server) for physical interaction, and deployed the Web applications onto a server. We tested the Web applications over a GPRS network and collected all the data transfer information between the Web server and the client using the Web server's log. Finally, we analyzed the log data to evaluate the performance of the Ajax and non-Ajax Web applications over the GPRS network.

The final part of the study was user evaluation (qualitative study). A small group of users was invited to participate in an observational usability test. Cooperative evaluation, a variation of the think-aloud observational technique, was used in this test. Participants were asked to use both the Ajax and non-Ajax mobile quiz sites using a single mobile device (an i-mate™ SP5 mobile phone). Their comments were recorded and analyzed, with particular emphasis on comments relating to Web application performance. This user evaluation attempted to find out whether the users thought that the Ajax application's performance was noticeably better than the non-Ajax version.

According to March and Smith (1995), there are four outputs from the design research method: constructs, models, methods, and instantiations. Later, Rossi and Sein (2003) and Purao (2002) have set forth their own list of design research outputs. However, they can be mapped directly to March and Smith's list as a fifth output, better theories (Table 1).

In this research, the main constructs are Ajax and non-Ajax Web application technologies using mobile devices over the TCP/IP GPRS network. Our models describe the traditional Web application interaction model and the Ajax interaction model. Our methods are based on software engineering approaches to developing Web-based applications while our instantiations are an Ajax enabled Web application and a non-Ajax application. Measuring the comparative performance of these applications gives us better theories about how to optimize the performance of mobile learning applications.

## User Evaluation Design

To evaluate the two systems from the user perspective we used within-subject design, using coopera-

*Table 1. The outputs of design research (March & Smith, 1995; Purao, 2002; Rossi & Sein, 2003)*

| | Output | Description |
|---|---|---|
| 1 | Constructs | The conceptual vocabulary of a domain |
| 2 | Models | A set of propositions or statements expressing relationships between constructs |
| 3 | Methods | A set of steps used to perform a task – how-to knowledge |
| 4 | Instantiations | The operationalization of constructs, models and methods. |
| 5 | Better theories | Artefact construction as analogous to experimental natural science |

tive evaluation and the post-task walkthrough as our observational techniques.

Because the goal of this usability test was to find out if end users gain any benefits from mobile Ajax, we first measured task completion performance. We also gathered preference data to see if users preferred using Ajax over the ASP application. In a within-subject design, the values of the dependent variable for a task or a set of tasks are compared with the values for another task or another set of tasks within one participant's data. In other words, the same variable is measured repeatedly on the same participant under different task conditions (Ergosoft Laboratories, 2003). Applying within-subject design in this usability test, participants needed to complete the same task using both Web applications. The task was to finish all five multi-choice questions in the quiz (all questions are of equal difficulty). Since the users will have had experience of both versions of the system, they could compare them and make some statements about them. At the same time we could collect some performance data (task completion time). To avoid users transferring learning from the first application to the second, we provided the users with a demonstration Web site with the same interface but different questions, to help users get familiar with the test device and the task process.

## Observational Techniques

A popular way to gather information about actual use of a system is to observe users interacting with it. The think-aloud method (Lewis & Reiman, 1993) is used to gather data in usability testing in product design and development, in psychology and a range of social sciences. Think-aloud protocols involve participants thinking aloud as they are performing a set of specified tasks. Cooperative evaluation is a variation of think-aloud, encouraging the user to be a collaborator in the evaluation. It allows the user to ask the evaluator any questions if a problem arises and the evaluator can encourage them to express themselves. The protocol for think-aloud sessions can include paper and pencil, audio recording, video recording, computer logging, and user notebooks (Dix, Finlay, Abowd, & Beale, 1998). Paper and pencil plus video recording were applied in this usability test. A post-task walkthrough was also used to reflect their actions back to the subjects after the event. The advantage of a delayed walkthrough is that the analyst has time to frame suitable questions and focus on specific incidents (Dix et al., 1998). In our case, the post-task walkthrough helped users become more aware of some performance problems.

# IMPLEMENTATION

This study is based on a controlled measurement of data transfer volumes and times, comparing an Ajax Web application with a similar non-Ajax implementation. First, we predicted system performance based on paper calculations, theoretically analyzing the data volumes and response times. Second, we analyzed real-life data collected from the system log of the server including bandwidth usage, response times, and mobile unit storage requirements. Finally, small group usability tests were conducted to find out if the users noticed any difference between the Ajax and non-Ajax Web applications. In order to compare the actual performance of a traditional Web application and Ajax, two mobile learning quiz systems were created: an Ajax Web application and a similar non-Ajax implementation (ASP). They have the same user interface and functionality (Figure 9).

The mobile learning Web application was based on a three-layer architecture: the mobile phone user interface (Web browser), server pages and a database server. The Web server was Microsoft Internet Information Services (IIS), the server pages were written using ASP 2.0 and the test database was stored using Microsoft Access. All dynamically generated Web pages were valid

*Figure 9. Mobile learning Web application interface (in the Opera Mobile browser)*

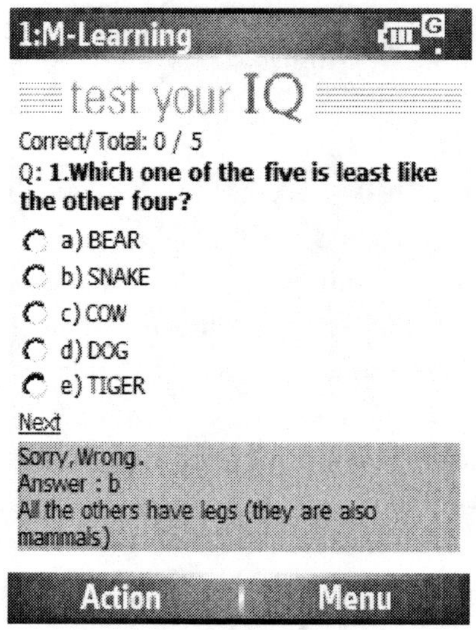


XHTML 1.1, compatible with the mobile browser (Opera Mobile) used in the test. All data transfer was through the Vodafone New Zealand GPRS network by Class 2 '2+1' multi-slots (2 downlink, 1 uplink channels). The system logs were activated in the IIS server, so all traffic going to and from the server was recorded. The collected data was used to analyze and measure the Web applications' performance over the GPRS network.

## The ASP Mobile Learning System

The ASP Web application only contains three server pages: a start page, a question display page, and a finish page. These pages interact with the database to provide page content. Other supporting files are an image file (in GIF format) and a JavaScript file to manage the question fetching process. When the user clicks on the submit button on the screen, it activates a JavaScript function that sends a request to the server to fetch a new question or answer. The storage sizes of all the files on the Web server are shown in Table 2.

The ASP Web application is developed in the traditional way. An ASP file includes CSS and JavaScript file paths in its header (the CSS is used to manage the page's presentation). It also contains the Web page layout information, Web content, and ASP code to dynamically generate the HTML pages. The ASP code also includes a function to verify the user's answers as they are submitted.

The JavaScript file ('asp_m.js') only has two main functions. The first fetches the next question for the client, using a question number attached to the URL. The second validates the user's submission to avoid empty or error submissions being sent to the server.

The database file is used to store all the mobile learning multi-choice questions. In the question database, there is only a single table. All questions, answers and explanations are included in this table.

## The Ajax Mobile Learning System

The Ajax implementation is similar to the ASP Web application. It has the same user interface, the same image file ('iq_header.gif'), the same CSS file ('hello.css') and shares the same database. However on the client side it includes three HTML pages and has a different JavaScript file. The ASP file used in the Ajax application is also different from those in the ASP application. It is used as both a database connector and XML file generator, because the communication between client and server is via XML. It is not used to generate HTML pages. The storage sizes of all the files on the Web server for the Ajax application are shown in Table 3.

The total size of the ASP Web application is about 2Kb bytes smaller than the Ajax version. The major difference between them is the size of the JavaScript file. The ASP Web application's JavaScript file is only 1.5Kb but the Ajax Web application's JavaScript file is almost 5Kb. That is because the Ajax JavaScript file not only contains the same functions as the ASP version, but also

*Table 2. ASP Web application file sizes (server side)*

| File Name | Size(bytes) | Detail |
|---|---|---|
| index.asp | 818 | Start Web page |
| mlearning.asp | 3,010 | Display questions |
| iq_header.gif | 1,470 | Header image file |
| asp_m.js | 1,530 | JavaScript file |
| hello.css | 45 | CSS file |
| learning.mdb | 496,000 | Question database |
| end.asp | 744 | Finish Web page |
| Total size | 504,309 bytes | |

*Table 3. Ajax Web application file sizes (server side)*

| File Name | File Size (bytes) | Detail |
|---|---|---|
| majax.js | 4,750 | JavaScript file(Ajax Engine) |
| Hello.css | 45 | CSS file |
| 2.asp | 1,040 | Database connection & XML generator |
| end.htm | 1,530 | Finish Web page |
| index.htm | 526 | Start Web page |
| iq_header.gif | 1,470 | Header image file |
| learning.mdb | 496,000 | Questions database(same as asp version) |
| mlearning.htm | 1520 | Display questions |
| Total size | 506,836 bytes | |

has some logical functions, such as verifying the user's answer.

Because Ajax uses a different Web development approach, the purpose of some its files are different from those in the ASP Web application. In the Ajax application, the HTML file size is very small because it is only used to lay out the main structure for the Web page, such as where to display the questions and where to display the answers and explanation. The most important part of the HTML file is that it provides an entry page for the m-learning Web application and the JavaScript file path. The HTML can be considered as a content template.

The JavaScript file, which is acting as an Ajax engine, is very important for the Ajax Web application and provides most of its functions. There are two main aspects and five basic functions provided by this JavaScript file (Ajax engine). The first part of the Ajax engine is used to asynchronously transfer the data between the client browser and server. The second part is used to manage the client side user interface, with a logical function used to verify the user's answer. When the Ajax engine submits a request to the server, the ASP file will dynamically generate an XML document, containing the data returned from a database query. In this case, just a few characters need to be returned to the client browser. The response only includes content data; it does not include any page mark-up information. Back on the client side, when the Ajax engine receives the XML response from the server it will partially update the user interface based on the latest information.

## Using the Mobile Learning System

Because both mobile learning quizzes are Web-based, users do not need to install or setup anything as long as their mobile device supports an Ajax enabled browser. Because the Ajax and ASP mobile learning systems are almost identical, both systems are used in the same way. Figure 9 shows some screenshots from the mobile quiz. It begins with a welcome page (1) with a hyperlink to start the quiz. Each answer is selected using a radio button (2). The 'Answer' hyperlink will provide the result. The correct answer will appear under the questions, with the explanation for the answer. Meanwhile, the 'Answer' link will change to a 'Next' link for the next question (3). When all questions have been completed, a finish page is displayed with a link back to the start page (4).

## Theoretical ASP System Performance

In this section, we predict the Web applications' performance based on application analysis and paper calculations, theoretically analyzing the data volumes and response times.

For the ASP mobile learning Web application, when the browser requests an ASP file, the server passes the request to the ASP engine. The ASP engine reads the ASP file and executes the scripts in the file to dynamically generate a client Web page. Finally, the generated page is returned to the browser as plain XHTML. These XHTML files also contain references to other required resources such as stylesheets and script files, so we can predict the total file load size based on the source code (Table 4).

*Figure 10. Using the mobile learning quiz Web application (screen captures from Opera Mobile browser)*

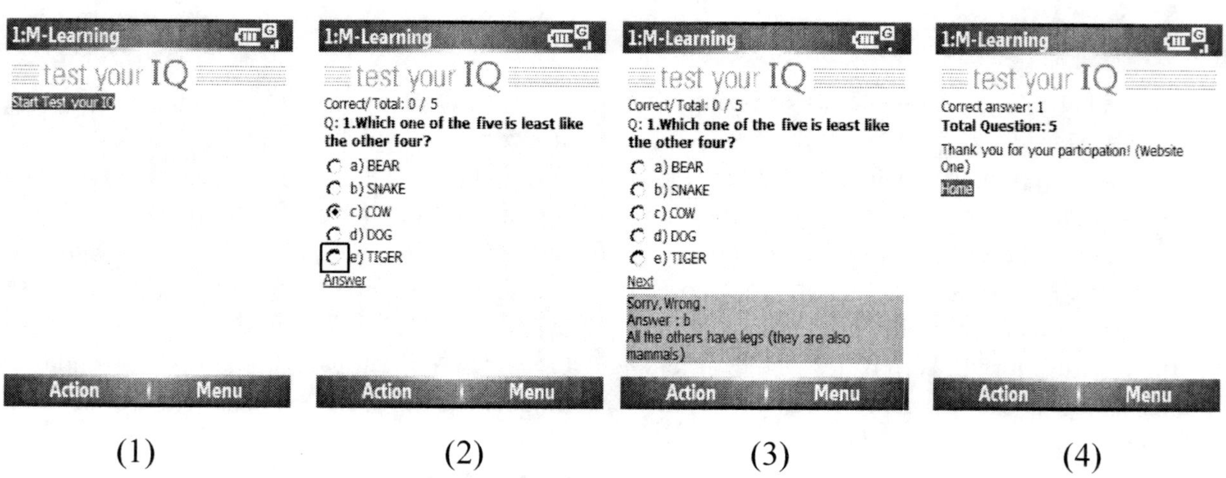

*Table 4. ASP application initial load file sizes*

| File Name | Load Size (bytes) | Detail |
|---|---|---|
| mlearning.asp | 2K × 2 | Loads questions, and refreshes the whole page after submitting answers. |
| iq_header.gif | 2K | Header image file |
| asp_m.js | 2K | JavaScript file |
| hello.css | 45 | CSS file |
| Initial Load Size | 8K | |

When the ASP application is running, a welcome page and a finish page are used to start and end the application; 'index.asp' (load size: 1K) and 'end.asp' (load size: 1K). These two files are not included in the calculated load size of the application.

The initial load size of the ASP application is about 8K, including the main ASP application file ('mlearning.asp') and a header image file. The JavaScript file controls the downloading of questions, verifies the submitted form and submits the answer back to the server. The initial file size includes the load size for the first question.

The header image file, JavaScript file and the CSS file need to be loaded only once if the browser supports caching. In our measurements, all browsers support caching.

The size of the XHTML file generated by 'mlearning.asp' is 2K, smaller than the ASP file size of 3K (Table 1). This is because the code generated by the scripts in the page is smaller than the source code. Because of the nature of the ASP file, the ASP scripts need to be executed on the server and rebuild the whole page each time. Since the question and answer pages are generated separately, the client browser will download 2K × 2 = 4K bytes for each question.

The mobile learning application contains 10 multi-choice questions in total, but when the application initially loads, only one question has been loaded. Subsequently the 'mlearning.asp' file needs to generate the other nine questions for the user. Additionally, it sends back nine answer pages, so the ASP file will generate 18 pages in total comprising about 2K × 18 = 36K bytes of XHTML mark-up. The formula to calculate the total download size of a Web application is:

Initial Load Size + Process Load Size = Total Size

Therefore the total download size for the ASP mobile learning Web application is:

8K + 36K= 44K

As a result, we expect that about 44K of data needs to be sent back to the browser from the m-learning Web server.

## Theoretical Ajax System Performance

When an Ajax application is requested by a browser, an XHTML page is loaded into the browser. This XHTML page contains the JavaScript needed to run the user interface and to issue XMLHttpRequest calls, the XML handler to format and present the results, and any other elements needed for the user interface, such as XHTML mark-up and CSS. The client interacts with the user interface presented on the browser

*Table 5. Ajax application initial load file sizes*

| File Name | Load Size (bytes) | Detail |
|---|---|---|
| 2.asp | 500 | Send XML back to browser (XML file size) |
| majax.js | 5K | JavaScript file (Ajax Engine) |
| hello.css | 45 | CSS file |
| iq_header.gif | 2K | Header image file |
| mlearning.htm | 2K | Load first questions |
| Initial Load Size | 10K | |

page (e.g., clicking hyperlinks, submitting forms, or triggering an event). The Ajax engine handles all interaction events between user and server by generating an XMLHttpRequest message to the server. The server returns new content in XML file format. This XML data is handled by the Ajax engine and presented to the user by partially updating the current page in the browser.

As with the ASP application, the Ajax system displays a welcome page at the beginning (index.htm) and a finish page at the end (end.htm) which are both about 1K. These two files are again excluded from the download calculations.

The main application file (mlearning.htm) and associated resources (e.g., JavaScript file, CSS file, and header image file) will be loaded into the Web browser for the first question. This initial load of the Ajax application requires 10K of files, but unlike the ASP Web application, only a small XML file needs to be downloaded for subsequent questions. The file sizes for the Ajax application are shown in Table 5.

As with the ASP version, the header image file, JavaScript file and CSS file only need to be loaded once if the browser supports caching.

In the Ajax version of the m-learning application, the initial load size is more than the ASP version, but the advantages of the Ajax application show up when the user is using the application. No further pages need to be downloaded from the Web server, only the question's content and answers in XML file format. All the answers from the user can be verified locally by the Ajax engine; the browser does not need to submit any data back to the server. The size of the XML file sent back from the Web server is only about 500 bytes. With this data the Ajax engine can partially update the Web page to tell the user the correct answer. Therefore the main part of the Ajax application only needs to transfer about $9 \times 500 = 4.5K$ of data during the quiz process. So, according to our formula the total download size for the Ajax application is:

$$10K + 4.5K = 14.5K$$

In summary, according to our calculations, the ASP m-learning system needs to transfer 44K of data; Ajax can provide the same interface by only transferring 14.5K of data. That is a big theoretical improvement, but can it be replicated using empirical testing? To find out, a practical performance measurement was carried out on an experimental test system.

## Experimental Performance Measures

The experimental system setup used to measure the m-learning system performance is shown in Figure 11. We used a laptop connected to an

*Figure 11. Experimental setup for performance testing*

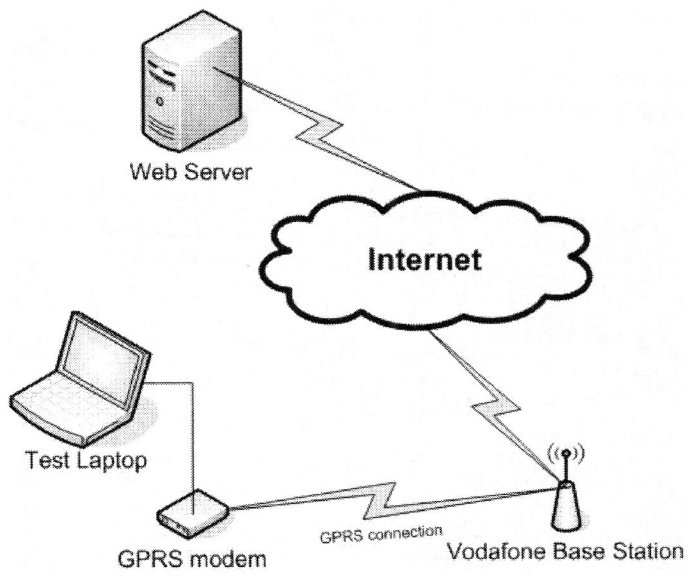


iTegno 3000 GPRS modem by USB 1.1 through a serial PPP (point-to-point) link to act as a GPRS mobile terminal to emulate the GPRS connection from mobile devices to the Web server in the Vodafone New Zealand (GPRS/GSM) network. The laptop used the Windows XP operating system and the Web browser was Opera Mobile 8.65 for Windows mobile 5/6 PPC running in a Windows Mobile 6 emulator. The Vodafone New Zealand GPRS network (two download channels and one upload channel, Coding Scheme 4) was used. Since wireless network signal levels may vary across different locations, all access to the Web server from the test laptop was via the Albany Stadium base station adjacent to Massey University's Auckland campus.

The Web server was running Internet Information services 6.0 (IIS 6.0); system logs were active in IIS so all traffic going to and from the Web server was recorded in the system log, monitoring all data transfer between the Web server and mobile terminal. These logs, comprising data from over 200 requests, were analyzed using Microsoft Log Parser 2.2. (Microsoft TechNet 2007). All log data was collected from the dedicated test domain.

Table 6 shows a summary of part of the server log for the ASP Web application. 'cs-method' means the HTTP request type, 'sc-bytes' means the number of bytes that the server sent and 'cs-bytes' means the number of bytes that the server received. 'Time-taken' means the length of time that the action took in milliseconds. This table illustrates one of the most important aspects of the ASP Web application, which is that the 'mlearning.asp' file needs to be executed twice for each question; the first time is the 'GET' request for the question, and the second time is the 'POST' request for the answer, which refreshes the whole page.

Table 7 shows an example of the IIS log for the Ajax application. This uses only 'GET' requests.

## ASP Application Response Size

Response size means the number of bytes sent by the server, which is one of the most useful

*Table 6. ASP application log example*

| cs-method | cs-uri-stem | sc-bytes | cs-bytes | time-taken |
|---|---|---|---|---|
| GET | /asp1/index.asp | 1,117 | 466 | 468 |
| GET | /asp1/hello.css | 346 | 454 | 156 |
| GET | /asp1/iq_header.gif | 1,815 | 459 | 296 |
| GET | /asp1/asp_m.js | 1,893 | 469 | 187 |
| GET | /asp1/index.asp | 1,117 | 407 | 140 |
| GET | /asp1/iq_header.gif | 1,815 | 352 | 156 |
| GET | /asp1/mlearning.asp | 2,181 | 493 | 265 |
| POST | /asp1/mlearning.asp | 2,274 | 686 | 499 |
| .<br>.<br>. | | | | |
| GET | /asp1/mlearning.asp | 2,184 | 495 | 203 |
| POST | /asp1/mlearning.asp | 2,280 | 690 | 343 |
| GET | /asp1/end.asp | 963 | 460 | 156 |

*Table 7. Ajax application log example*

| cs-method | cs-uri-stem | sc-bytes | cs-bytes | time-taken |
|---|---|---|---|---|
| GET | /ajax1/index.htm | 851 | 508 | 171 |
| GET | /ajax1/hello.css | 346 | 350 | 140 |
| GET | /ajax1/iq_header.gif | 1,814 | 354 | 281 |
| GET | /ajax1/mlearning.htm | 2,166 | 515 | 374 |
| GET | /ajax1/majax.js | 5,190 | 353 | 296 |
| GET | /ajax1/2.asp | 518 | 377 | 296 |
| GET | /ajax1/2.asp | 514 | 377 | 203 |
| GET | /ajax1/2.asp | 500 | 376 | 187 |
| GET | /ajax1/2.asp | 621 | 377 | 187 |
| GET | /ajax1/2.asp | 659 | 377 | 187 |
| GET | /ajax1/2.asp | 528 | 376 | 234 |
| GET | /ajax1/2.asp | 533 | 377 | 203 |
| GET | /ajax1/2.asp | 571 | 378 | 249 |
| GET | /ajax1/2.asp | 583 | 377 | 203 |
| GET | /ajax1/2.asp | 524 | 379 | 187 |
| GET | /ajax1/end.htm | 1,951 | 465 | 156 |

*Table 8. ASP application 'sc-bytes' statistics*

| cs-uri-stem | sc-bytes | | |
|---|---|---|---|
| ASP application | AVG | MIN | MAX |
| /asp1/mlearning.asp | 2 x 2,305 | 2,256 | 2,406 |
| /asp1/asp_m.js | 1,893 | 1,833 | 1,897 |
| /asp1/iq_header.gif | 1,815 | 1,800 | 1,880 |
| /asp1/hello.css | 346 | 356 | 396 |
| | | | |
| Initial load size. | ≈ 8.7Kb | | |

*Table 9. Ajax application 'sc-bytes' statistics*

| cs-uri-stem | sc-bytes | | |
|---|---|---|---|
| Ajax application | AVG | MIN | MAX |
| /ajax1/majax.js | 5,190 | 5,100 | 5,280 |
| /ajax1/mlearning.htm | 2,166 | 2,007 | 2,356 |
| /ajax1/iq_header.gif | 1,814 | 1,754 | 2,014 |
| /ajax1/hello.css | 346 | 326 | 386 |
| /ajax1/2.asp | 555 | 500 | 614 |
| | | | |
| Initial load size. | ≈ 10kb | | |

measurable features of application performance. In the IIS log, its identifier is 'sc-bytes.'

According to the IIS system log (Table 8), the average ASP application initial load size is about 8.7K (including the first question). As expected, the main ASP application file needs to load 10 times to display the questions and load another 10 times after users submit the answer. The average load size of the file generated by 'mlearning.asp' is 2,305K, so the total bytes needed to be sent back from the Web server are calculated as follows:

$$8.7K + 9 \times 2 \times 2305 \approx 50K$$

So, in real-life performance tests, the ASP M-learning application needs to load about 50K of data from the server during the whole application process, somewhat larger than our estimate of 44K.

## Ajax Application Response Size

The number of bytes returned for Ajax application is different from the ASP application. First, the load size is very different from the ASP version, about 10K of initial load data. The major file is the JavaScript (Ajax engine) file which is about 5K, and the other one is an HTML layout file that is about 2K (Table 9).

However, while the Ajax application is running, just a few bytes need to be transferred between the client browser and the server. All

*Table 10. ASP application 'time-taken' statistics*

| cs-uri-stem | Time-taken | | |
|---|---|---|---|
| ASP Application | AVG(ms) | MIN(ms) | MAX(ms) |
| /asp1/mlearning.asp | 285×2 | 187 | 499 |
| /asp1/iq_header.gif | 207 | 156 | 296 |
| /asp1/asp_m.js | 202 | 187 | 218 |
| /asp1/hello.css | 163 | 156 | 171 |
| Initial Time taken(ms) | 1,142 | | |

*Table 11. Ajax application 'time-taken' statistics*

| cs-uri-stem | Time-taken | | |
|---|---|---|---|
| Ajax Application | AVG (ms) | MIN (ms) | MAX (ms) |
| /ajax1/hello.css | 319 | 140 | 499 |
| /ajax1/majax.js | 296 | 296 | 296 |
| /ajax1/iq_header.gif | 281 | 201 | 381 |
| /ajax1/mlearning.htm | 272 | 171 | 374 |
| /ajax1/2.asp | 210 | 171 | 343 |
| Initial Time taken(ms) | 1,378 | | |

this data is XML and the average size is only about 555 bytes (generated by the '2.asp' file), including the question and the answer. It allows the Ajax engine to verify the answer locally, and provides users with a very fast response time. Also, because the question and answer load at the same time, each question only needs one post back from the server. The total load size is calculated as follows:

$$10K + (9 \times 555K) \approx 15K$$

This is only slightly larger than our estimate of 14.5K. The Ajax m-learning application only needs to transfer about 15K of data between the server and the client, whereas the ASP Web application needs the server to transfer 50K of data. There is therefore a significant difference between the two applications in terms of data transfer from server to client.

## ASP Application Response Time

As well as measuring the size of downloaded data, the server's response time is another important feature that can be used to measure a Web application's performance. In the IIS system log, there is a field called 'time-taken,' which is the length of time that the action took in milliseconds on the server, and can be considered the server's action response time.

The measurement formula is similar to measuring the bytes sent back from the server. According to the IIS system log (Table 10), the initial response time for the ASP m-learning application is about 1,142ms (initial load time, including the first question load time), and the main application file needs to load twice the number of times as there are questions in the quiz. The average response time for the 'mlearning.asp' file is 285ms; the 'mlearning.asp' page needs to process twice for each question, so for a total of 10 questions, minus the response time for first question, the process response time is

$9 \times 2 \times 285 = 5{,}130$ms

Similar to the total load size formula, the total response time for the whole application formula is:

Initial Response Time + Process Response Time = Total Response Time

So, the total Web server response time for the ASP mobile learning Web application is:

1,142ms + 5,130ms = 6,272ms

## Ajax Application Response Time

For the Ajax m-Learning application, the IIS system log shows the initial response time is about 1,378ms (including the first question), and the average server response time for the XML generator '2.asp' file is 210ms (Table 11). However, '2.asp' only runs once for each question on the server. Therefore for a total of 10 questions, minus the response time for the first question, the process response time is

$9 \times 210 = 1{,}890$ms.

According to the formula, the total server response time for the whole ASP mobile learning Web application is:

1,378ms + 1,890ms = 3,268ms

In summary, for the whole application, the ASP system takes about 6,272ms of server response time, but the Ajax application takes only 3,268ms.

## ASP Application Request Size

The response size from the Web server and the server's task time show us the Ajax m-learning application transfers smaller volumes of data between the client and server and has a better response time than the ASP m-learning application. However, what about the data submitted from the client? Does it reveal a similar story?

The client's request size can affect the transfer time; if the request size is big, not only does it need more bandwidth to transfer the data, but also it takes a long time to transfer before the server can actually process the request. Then, from the user's point of view, the Web application's response time is slow. Therefore, we should also include the client's request sizes in our evaluation.

The 'cs-bytes' field in the IIS system log shows us the number of bytes that the server received, so this data can be regarded as a measure of the data sent from the client browser to request new data from server. Table 12 shows query size of submitted data from the client to the server. For the 'mlearning.asp' file there are both 'GET' and 'POST' methods, and the query size differs for these two methods.

These queries are used to request the next question for the user. The average query size for the main 'mlearning.asp' file is about 496 bytes (Table 12, 'GET' method). Of course, in the ASP version, users need to submit their answers back to the server to have them verified. The average transfer size is 686 bytes (Table 12, 'POST' method) when users submit their answers. Therefore the initial request size (including the first question) formula is:

mlearning.asp (GET) + mlearning.asp (POST) + other files = Initial Request Size

*Table 12. ASP application 'cs-bytes' statistics*

| cs-uri-stem | cs-bytes (bytes) | | |
|---|---|---|---|
| ASP Application | AVG | MIN | MAX |
| /asp1/mlearning.asp("GET") | 496 | 492 | 566 |
| /asp1/mlearning.asp("POST") | 686 | 600 | 750 |
| /asp1/iq_header.gif | 398 | 352 | 459 |
| /asp1/asp_m.js | 423 | 378 | 469 |
| /asp1/hello.css | 416 | 379 | 454 |
| Initial Query Size | 2,428 (including first question) | | |

*Table 13. Ajax application 'cs-bytes' statistics*

| cs-uri-stem | cs-bytes | | |
|---|---|---|---|
| Ajax Application | AVG (bytes) | MIN (bytes) | MAX (bytes) |
| /ajax1/2.asp | 377 | 373 | 379 |
| /ajax1/iq_header.gif | 354 | 350 | 374 |
| /ajax1/majax.js | 353 | 351 | 363 |
| /ajax1/hello.css | 350 | 350 | 370 |
| /ajax1/mlearning.htm | 515 | 514 | 519 |
| Initial Request Size | 1,949 (including first question) | | |

According the formula, the initial request size is therefore calculated as:

$496 + 686 + 398 + 423 + 416 = 2{,}428$ bytes

The total request size calculation formula is:

Initial Request Size + Process Request Size = Total Request Size

Which gives us the following result:

$2{,}428 + (686 \times 9 + 496 \times 9) = 13{,}066$ bytes $\approx$ 13K

In total, therefore, the client needs to submit about 13K of data to the Web server.

## Ajax Application Request Size

Table 13 shows the size of data submitted from the Ajax client to the server.

The average request size is 377 bytes. These submissions only request the next question: one question, one submission. Therefore it is very easy to calculate the initial request size (including the first question) for the Ajax application; simply add all the request sizes together:

$377 + 354 + 353 + 350 + 515 = 1{,}949$ bytes

The total query size is therefore calculated as:

$1{,}949 + 9 \times 377 = 5{,}342$ bytes $\approx$ 5.3K

*Table 14. Performance improvement*

| | Response size | Response seconds | Request size |
|---|---|---|---|
| ASP | 50K | 6,272ms | 13K |
| Ajax | 15K | 3,268ms | 5.3K |
| Improvement | 71.00% | 48.00% | 59.23% |

*Figure 12. Experimental setup for usability testing*

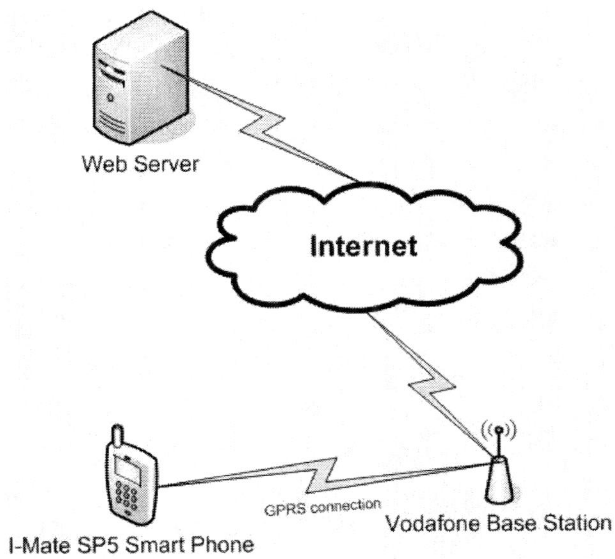


The overall result is that the ASP Web application submitted around 13K of request data to the server; while the Ajax Web application sent only about 5.3K of data.

## Overall Performance Measures

When measuring a Web application's performance, the data can be separated into three parts: data transfer time (in two ways, from client to server and from server to client) and server process time. Table 6 summarizes these performance measures for the two Web applications. The average data volume sent from the Web server is 50K for the ASP version and 15K for the Ajax version during the whole in the m-Learning Web application process. The average number of bytes received by the server is about 13K for the ASP version and about 5.3K for the Ajax version. It takes about 6,272ms for the server to process all requests for the ASP version and about 3,268ms for the Ajax version. White (2005) and Smullen and Smullen (2006) use the following algorithm to define the percentage of an Ajax application's performance improvement:

(HTML-Ajax)/HTML

This formula has been used to calculate the 'improvement' data in Table 14 (in our case 'HTML' would represent the ASP application results).

The Ajax approach in our experiment provides a significant performance improvement across all three of our measures. We can therefore confidently state that our Ajax M-learning Web application has much better performance than the non-Ajax m-learning Web application when tested over a GPRS mobile phone network.

## USABILITY EVALUATION

The goal of this study was to evaluate the Ajax enabled mobile learning system's performance over a GPRS network both by quantitative measurement statistics and user feedback. This was necessary because although we could demonstrate that the Ajax system performed better in quantitative terms, we also needed to find out if these differences were significant enough to be noticed by end users. The goal of the usability test was therefore to find out if users noticed any differences between the Ajax and the non-Ajax mobile learning applications, and if these differences were meaningful from an end-user perspective.

In order to simulate a real-life mobile learning environment as realistically as possible, the usability testing experimental setup (Figure 12) was somewhat different from the performance measurement setup. In this system, we used an i-mate™ SP5 SmartPhone instead of the laptop. The mobile Web browser was Opera Mobile, version 8.65 for Windows Mobile 5/6 SmartPhone operating system. The same Vodafone New Zealand GPRS network (two download channels and one upload channel, Coding Scheme 4) was used. All participants used the same mobile phone, again connected via the Albany Stadium base station.

In the evaluation, users were asked to complete the quiz using the mobile device, think aloud about their feelings and answer a questionnaire. In order to reduce the interview time, we reduced the number of questions in the quiz from 10 down to five for each system. The test Web applications were again hosted at http://www.ajaxnz.com.

There were two tasks in this usability test. Task one required users to finish all the multi-choice test questions in the Ajax mobile learning system, and task two required users to finish a similar set of multi-choice test questions using the ASP mobile learning system. A training session was provided before users started their mobile learning activity. Finally, the post-walkthrough method was applied to wrap up the test. The questionnaire was completed and the Opera Mobile cache was cleared ready for the next test.

In this usability test, there were two variables we collected from the users to represent both quantitative data and qualitative data. The first quantitative variable is the task completion time, which was used to measure the two different performances (within-subject design). The second variable is a qualitative variable, the preferred data, which is used to find out if users notice any difference between the two Web applications.

The questionnaire included some user information, such as gender, the type of mobile phone they have, if they have ever used their mobile phone to access the Internet and if so, how often. Participants were also asked if they noticed any differences between the two Web applications, and if so, if they preferred to use one over the other.

### Usability Test Results

Twenty people attended this small group usability test, totaling 19 valid examples; one participant did not finish the test. In all of the valid examples, 18 of the participants had mobile phones, and only one participant did not have a mobile phone. There were six female participants and 13 male participants. Approximately 40 percent of the participants had previous experience of using a mobile phone to access the Internet. Twenty percent had accessed the Internet a few times, while 20 percent have used the mobile Internet just once. Only 10 percent of the participants used the mobile Internet at least once a week.

During the usability test, all participants were comfortable about the mobile learning system itself, but had some problems with the mobile device. The first complaint was about the navigation button being hard to use and not functioning very well. Sometimes users had to keep pressing it until it worked (i.e., an action occurred on the screen). One participant aborted the test because of these problems with the navigation button. The second problem was about the small screen and font size. Observations about the device tended to overshadow observations about the application itself. However one participant observed that the loading order of questions and answer options was different between the two applications (a consequence of Ajax updating the page asynchronously).

Based on the 19 test examples, applying pairwise T-Testing to the task completion time, our participants performed significantly better with the Ajax mobile learning system.

$t(18) = 11.71, p<0.05.$

The mean value of the task completion time for the first Web application was 119.26 seconds and the task completion time for the second Web application was 64.11 seconds. This means that users finished their task on the second application much faster than the first, about 50 percent faster. The Ajax mobile learning system therefore enabled much better performance than the ASP mobile learning system when assessed by the quantitative variable.

In terms of frequency to their preference, we can use a chi-square test, but to some extent it does not seem to meet its assumption.

Chi-square $(2) = 12.41, p<0.05$

The expected sample is too small, so it would be unwise to use this statistical analysis for sure. However, analytically (not statistically) speaking, users noticed there were some differences between the two applications, even though they have the same interface and the same learning process. Twelve out of the 13 participants who noticed a difference preferred the Ajax mobile learning system. These two results from the usability test indicate that an Ajax mobile learning system can provide better performance in the mobile environment on the user experience level.

## SUMMARY AND CONCLUSION

In our tests we applied the Ajax approach to a mobile learning system and compared it with a traditional ASP mobile learning system. After the evaluation of these systems in terms of their performance and usability, our results indicate that an Ajax mobile learning system can reduce data transmission volumes and the server's response time, and is preferred by users. Moreover, applying an Ajax approach to Web-based mobile applications is a much more practical way to improve system performance than updating the mobile hardware or the wireless network.

The advantages of Ajax are; first, it allows users to access to the system as long as their mobile browser supports Ajax and has connectivity to a GSM/GPRS network, meaning that it is a widely available option. Second, system performance is much faster than a traditional Web application. Third, reduced data traffic can save money for mobile learners in territories where mobile telecommunications companies charge for the amount of data sent and received. In addition, although this is not a major aspect of our research, Ajax systems also can be accessed from a desktop browser environment without modification.

In our experimental measurements, Ajax reduced network transfer traffic from the server to the client by 71 percent, saved 48 percent of the Web server's processing time and reduced submission data from the client to the server by 59 percent. Furthermore, on the user experience level, the task completion time statistics show users finished the same m-learning multi-choice questions on the Ajax system 50 percent faster than they did on the ASP system. Finally, more than 60 percent (12 out of 19) of users noticed the difference between these systems and preferred to use the Ajax system.

The results of our experiments demonstrate that the Ajax approach can significantly reduce both the data transmission size and the server's response time. Meanwhile, reducing the bandwidth required, and speeding up the user interface on the mobile device provides the user with a better mobile Internet experience. In addition, the Ajax approach has little infrastructure dependence, because it is a software technology and does not require any hardware or environment updates in order to be implemented.

## FUTURE RESEARCH DIRECTIONS

There were a number of limitations to this study. Our experimental system was only tested on one GPRS network, with two download channels and one upload channel, and only one mobile device (i-mate™ SP5). There are some other GPRS network configurations that could be used as an evaluation network and may give different results, as well as a number of other types of mobile data network such as 3G, WiMAX and WiFi. Other mobile devices may also perform differently from those used in our experiments. Another limitation of our study is that the Web application's functionality is very limited, only measuring the performance of a multi-choice quiz. Whilst such quizzes are an important component of mobile learning systems, other types of mobile learning interaction may well give different profiles of data transfer. In addition, the test group used in our usability experiment was too small and narrow (most of the participants were students or teachers) to enable us to draw more generalized conclusions.

There are a number of avenues for further research that could be explored. In this study, both the ASP and Ajax Web applications had the same user interface to enable us to evaluate performance statistics. However, one consequence of this is that we are not using the full potential of Ajax, simply mimicking a more traditional UI. To properly evaluate the benefits of Ajax in mobile learning systems it would be helpful to undertake both quantitative and qualitative evaluations of systems that leverage all the features of Ajax. Further study might also be undertaken using other types of wireless network, to see if the kinds of results measured here over GPRS are replicated under different network conditions. Finally it would be useful to do performance testing with a more realistic, larger scale mobile learning system than the multiple choice quizzes used in this study.

## REFERENCES

Alexander, B. (2004). Going nomadic: Mobile learning in higher education. *EDUCAUSE Review, 39*(5), 28-35.

Angelaccio, M., & Buttarazzi, B. (2006, December 4-7). *A Performance Evaluation of Asynchronous Web Interfaces for Collaborative Web Services.* Paper presented at the Frontiers of High Performance Computing and Networking – ISPA 2006 Workshops (pp. 864-872). Sorrento, Italy.

Bar, H., Häussge, G., & Röβling, G. (2007, June 23–27). *An integrated system for interaction support in lectures.* Paper presented at the 12th annual SIGCSE conference on Innovation and technology in computer science education (ITiCSE'07) (pp. 281-285). Dundee, Scotland.

Berri, J., Benlamri, R., & Atif, Y. (2006, July 3-6). *Ontology-based framework for context-aware mobile learning.* Paper presented at the International Conference on Wireless Communications and Mobile Computing (pp. 1307-1310). Vancouver, Canada.

Black, J. T., & Hawkes, L. W. (2006, July 3-6). *A prototype interface for collaborative mobile learning.* Paper presented at the International Conference on Wireless Communications and

Mobile Computing (pp. 1277-1282). Vancouver, Canada.

Cao, Y., Tin, T., McGreal, R., Ally, M., & Coffey, S. (2006, July 3-6). *The Athabasca University mobile library project: increasing the boundaries of anytime and anywhere learning for students.* Paper presented at the International Conference on Wireless Communications and Mobile Computing (pp. 1289-1294). Vancouver, Canada.

Chakravorty, R., Cartwright, J., & Pratt, I. (2002, November 17-21). *Practical experience with TCP over GPRS.* Paper presented at the IEEE Global Telecommunications Conference (GLOBECOM '02) (pp. 1678-1682). Taipei, Taiwan.

Chakravorty, R., Clark, A., & Pratt, I. (2003, May 5-8). *GPRSWeb: Optimizing the Web for GPRS Links.* Paper presented at the MobiSys 2003: The 1st International Conference on Mobile Systems, Applications, and Services, San Francisco.

Chakravorty, R., & Pratt, I. (2002, September 9-11). *WWW performance over GPRS.* Paper presented at the 4th International Workshop on Mobile and Wireless Communications Network, Stockholm, Sweden.

Church, K., Smyth, B., Cotter, P., & Bradley, K. (2007). Mobile information access: A study of emerging search behavior on the mobile internet. *ACM Transactions on the Web, 1*(1), Article 4.

Crane, D., Pascarello, E., & James, D. (2005). *Ajax in action.* Manning Publications Co.

Dix, A. J., Finlay, J. E., Abowd, G. D., & Beale, R. (1998). *Human-computer interaction* (2nd ed.). Hertfordshire: Prentice Hall Europe.

Ergosoft Laboratories. (2003). What is a within-subjects design? Retrieved January 3, 2008, from http://www.ergolabs.com/within_subjects.htm

Garrett, J. J. (2005). Ajax: A New Approach to Web Applications. Retrieved January 3, 2008, from http://adaptivepath.com/publications/essays/archives/000385.php

Kadirire, J. (2007). Instant messaging for creating interactive and ollaborative m-learning environments. *The International Review of Research in Open and Distance Learning, 8*(2) 1-14.

Kukka, H., & Ojala, T. (2006, October 25–27). *mobileDOK: Culture in your pocket.* Paper presented at the 3rd international conference on mobile technology, applications & systems (Mobility 06), Bangkok, Thailand .

Lee, T. B. (1992). What's new in '92. Retrieved January 3, 2008, from http://www.w3.org/History/19921103-hypertext/hypertext/WWW/News/9201.html

Lee, W., & Lu, C. (2003). Customising WAP-based information services on mobile networks. *Personal and Ubiquitous Computing, 7*(6), 321-330.

Leung, C.-H., & Chan, Y.-Y. (2003, July 9-11). *Mobile learning: a new paradigm in electronic learning.* Paper presented at the 3rd IEEE International Conference on Advanced Learning Technologies (pp. 76-80). Athens, Greece.

Lewis, C., & Rieman, J. (1993). Task-centered user interface design: A practical introduction. *Shareware.* Retrieved January 3rd, 2008, from http://www.hcibib.org/tcuid/

March, S., & Smith, G. (1995). Design and natural science research on information technology. *Decision Support Systems 15*(4), 251-266.

Microsoft TechNet. (2007). W3C Extended Log File Format (IIS 6.0). Retrieved January 3, 2008, from http://www.microsoft.com/technet/prodtechnol/WindowsServer2003/Library/IIS/bea506fd-38bc-4850-a4fb-e3a0379d321f.mspx?mfr=true

Nakahara, J., Hisamatsu, S., Yaegashi, K., & Yamauchi, Y. (2005, May 30 - June 4). *iTree:*

*Does the mobile phone encourage learners to be more involved in collaborative learning?* Paper presented at the 2005 conference on Computer support for collaborative learning (pp. 470–478). Taipei, Taiwan.

Nokia. (2005). Nokia introduces a new Web browser for S60 3rd Edition. Retrieved January 3, 2008, from http://press.nokia.com/PR/200511/1019239_5.html

Nortel. (2007). Long-Term Evolution (LTE): The vision beyond 3G. Retrieved January 3, 2008, from http://www.nortel.com/solutions/wireless/collateral/nn114882.pdf

Pinkwart, N., Hoppe, H. U., Milrad, M., & Perez, J. (2003). Educational scenarios for the cooperative use of personal digital assistants. *Journal of Computer Assisted Learning, 19*(3), 383-391.

Polsani P (2003). Network learning. In K. Nyiri (Ed.), *Mobile learning. Essays on philosophy, psychology and education* (pp. 139–150). Vienna: Passagen Verlag.

Purao, S. (2002). Design Research in the Technology of Information Systems: Truth or Dare. *GSU Department of CIS* Working Paper, Atlanta, Retrieved January 3, 2008, from http://purao.ist.psu.edu/working-papers/dare-purao.pdf

Rossi, M., & Sein, M. (2003, August 9-12). *Design Research Workshop: A Proactive Research Approach.* Presentation delivered at the 26th Information Systems Research Seminar in Scandinavia (IRIS 26), Porvoo, Finland.

Sayar, A., Pierce, M., & Fox, G. (2006, February 19-25). *Integrating AJAX Approach into GIS Visualization Web Services.* Paper presented at the International Conference on Internet and Web Applications and Services (AICT-ICIW '06), Guadeloupe, French Caribbean.

Seong, D. S. K. (2006, October 25–27). *Usability guidelines for designing mobile learning portals.* Paper presented at the 3rd international conference on mobile technology, applications & systems (Mobility 06), Bangkok, Thailand.

Sharples, M., Corlett, D., & Westmancott, O. (2002). The design and implementation of a mobile learning resource. *Personal and Ubiquitous Computing, 6*(3), 220-234.

Smullen, C., & Smullen, S. (2006, July 17-19). *Modelling AJAX Application performance.* Paper presented at the Conference on Web Technologies, Applications, and Services (WTAS 2006), Calgary, Canada.

Smullen, C., & Smullen, S. (2007, March 22-25). *AJAX application server performance.* Paper presented at the IEEE SoutheastCon (pp. 154-158). Richmond, USA.

Stodel, E. J., Thompson, T. L., & MacDonald, C. J. (2006). Learners' perspectives on what is missing from online learning: Interpretations through the community of inquiry framework. *The International Review of Research in Open and Distance Learning, 7*(3).

Stuckmann, P., Ehlers, N., & Wouters, B. (2002, September 24-28). *GPRS traffic performance measurements.* Paper presented at the 56th Vehicular Technology Conference (VTC 2002), Vancouver, Canada.

Theng, Y, Tan, K, Lim, E, Zhang, J., Goh, D., Chatterjea, K. et al. (2007, June 18-23). *Mobile G-Portal supporting collaborative sharing and learning in geography fieldwork: an empirical study.* Paper presented at the International Conference on Digital libraries (pp. 462–471). Vancouver, Canada.

Traxler, J. (2007). Defining discussing and evaluating mobile learning: The moving finger writes and having writ... *International Review of Research in Open and Distance Learning. 8*(2) unpaged.

Vaishnavi, V., & Kuechler, B. (2007). Design Research in Information Systems. Retrieved

January 3, 2008 from http://www.isworld.org/Researchdesign/drisISworld.htm

W3C.org. (2007). The XMLHttpRequest Object. Retrieved January 3, 2008 from http://www.w3.org/TR/XMLHttpRequest/

Watson, H., Rainer, R., & Koh, C. (1991). Executive information systems: A framework for development and a survey of current practices. *MIS Quarterly, 15*(1), 13-30.

Wei, C. K. (2005). AJAX: Asynchronous Java + XML? Retrieved March 4, 2007, from http://www.developer.com/design/article.php/3526681

White, A. (2005). Measuring the Benefits of Ajax. Retrieved August 15, 2006, from www.developer.com/java/other/article.php/3554271

Yokomoto, C. F. (2000, October 18-21). *Promoting depth of knowledge through a new quiz format to improve problem solving abilities.* Paper presented at the 30th ASEE/IEEE Annual Frontiers in Education Conference (FIE 2000), Kansas City, USA.

# Chapter XVII
# Recommended Readings and Resources

**Hokyoung Ryu**
*Massey University, New Zealand*

**David Parsons**
*Massey University, New Zealand*

## ABSTRACT

*We have found several books, journals, articles, and newspaper clips to be particularly helpful and inspiring in thinking about mobile learning systems design. Included in this list of good reads are the following.*

## LEARNING EXPERIENCE DESIGN AND ITS THEORETICAL FOUNDATIONS

Bellamy, R.K.E., (1996). Designing educational technology: Computer-mediated change. In B. A. Nardi (Ed.), *Context and Consciousness: Activity Theory and Human-Computer-Interaction*, pp. 123-146, Cambridge, MA: MIT Press.

Bendy, G. Z., Seglin, M. H., & Meister, D. (2000). Activity theory: history, research and application. *Theoretical Issues in Ergonomic Sciences*, 1(2), 168-206.

Brown, J. S., Collins, A., & Duguid, P. (1989). Situated cognition and culture of learning. *Educational Technology*, 18(1), 32-42.

Brown, J.S., and Duguid, P. (2000). *The Social Life of Information*. Harvard Business School, Cambridge, MA.

Carlen, U., & Jobring, O. (2005). The rationale of online learning communities. *International Journal of Web Based Communities*, 1(3), 272-295.

Collier, K. G. (1980). Peer-group learning in higher education: The development of higher-order skills. *Studies in Higher Education*, 5(1), 55-62.

Csikszentmihalyi, M. (1996). *Creativity: Flow and the Psychology of Discovery and Invention*. New York: Harper Collins.

Csikszentmihayli, M.,(1992), *Flow: the Psychology of Happiness*. London, U.K: Rider.

Engeström, Y. (1987). *Learning by Expanding*. Helsinki: Orienta-Konsultit.

Laurillard, D. (2002). *Rethinking University Teaching: A Conversational Framework for the Effective Use of Learning Technologies* (2nd Ed.). London, U.K.: Routledge

Lave, J., & Wenger, E. (1991). *Situated Learning: Legitimate Peripheral Participation*, Cambridge, U.K.: Cambridge University Press

Notterman, J. M. & Drewry, H. N. (1993). *Psychology and Education, Parallel and Interactive Approaches*. New York: Plenum Press.

Price, S., Rogers, Y., Stanton, D., and Smith, H. (2003). A new conceptual framework for CSCL: Supporting diverse forms of reflection through multiple interactions', Paper presented at the *International Conference on Computer Supported Collaborative Learning 2003*, Bergen Norway

Rogoff, B., & Lave, J. (1984). *Everyday Cognition: Its Development in Social Context*. Cambridge, MA: Harvard University Press.

Roschelle, J., Rosas, R., and Nussbaum, M. (2005). Towards a design framework for mobile computer-supported collaborative learning. Paper presented at *the International Conference on Computer Supported Collaborative Learning*, Taipei, Taiwan

Ryu, H., & Parsons, D. (2008). A learner-centred design of a location-aware learning reminder. *International Journal of Mobile Learning and Organisation*, *2*(2), 187-200.

Sharples, M., Taylor, J., & Vavoula, G. (2005). Towards a theory of mobile learning. Paper presented at *mLearn 2005* Conference, Cape Town, South Africa

Uden, L. (2007). Activity theory for designing mobile learning. *International Journal of Mobile Learning and Organisation*, Vol. 1, No. 1, pp.81–102.


## DESIGNING THEMES IN MOBILE LEARNING


Alexander, B. (2004). Going nomadic: Mobile learning in higher education. *EDUCAUSE Review*, 39(5), 28-35.

Attewell, J. (2005). *Mobile technologies and learning: A technology update and m-learning project summary*. London, UK: Learning and Skills Development Agency.

Benford, P., Magerkurth, C., & Ljungstrand, P. (2005). Bridging the physical and digital in pervasive gaming. *Communications of the ACM*, 48(3), 54-57.

Chen, Y.S., Kao, T.C., & Sheu, J.P. (2003). A mobile learning system for scaffolding bird watching learning. *Journal of Computer Assisted Learning*, 19, 347-359

Garris, R., Ahlers, R. & Driskell, J. E. (2002). Games, motivation, and learning: A research and practice model. *Simulation and Gaming*, 33, 441 - 467.

Goh, T., & Kinshuk. (2006). Getting ready for mobile learning – Adaptation perspective. *Journal of Educational Multimedia and Hypermedia*, 15(2), 175-198.

Holzinger, A., Nischelwitzer, A., & Meisenberger, M. (2005). Lifelong-learning support by m-learning: example scenarios. *ACM e-Learn*, 11, November, 2.

Jegers, K. & Wiberg, M. (2006). Pervasive gaming in the everyday world. *Pervasive Computing*, 5(1), 78-85.

Johanson, M. & Törlind, P. (2004). Mobility support for distributed collaborative teamwork. *ITcon*, 9, 356-366

Jonassen, D. H., Peck, K. L., & Wilson, B. G. (2000). *Learning with Technology: A Constructivist Approach*. Upper Saddle River, NJ: Prentice Hall.

Kirkpatrick, D., & McLaughlan, R. (2000). Flexible lifelong learning in professional education. *Education and Information Technologies*, 3(1), 24-31.

Kukulska-Hulme, A., & Traxler, J. (2005). *Mobile Learning: A Handbook for Educators and Trainers*. London, U.K: Routledge.

Massey, A. (2006). Design, development, and assessment of mobile applications: The case for problem-based learning. *IEEE Transactions on Education*, 49(2), 183-192.

Pasnik, S. (2007). *iPod in Education: The Potential for Teaching and Learning*. Cupertino, CA: Apple

Perry, D. (2003). *Handheld Computers (PDAs) in Schools*. Coventry, UK: British Educational Communications and Technology Agency.

Prensky, M. (2001). *Digital Game-Based Learning*. NY: McGraw-Hill

Roschelle, J. (2003). Unlocking the Learning Value of Wireless Mobile Devices. *Journal of Computer Assisted Learning*. 19, 260-272.

Salen, K. & Zimmermann, E. (2003). *Rules of Play. Game Design Fundamentals*. Cambridge, MA: The MIT Press.

Schwabe, G., & Göth, C. (2005). Mobile learning with a mobile game: Design and motivational effects. *Journal of Computer Assisted Learning*, 21(3), 204-216.

Seppälä, P., & Alamäki, H. (2003). Mobile learning in teacher training. *Journal of Computer Assisted Learning*, 19, 330-335.

Soloway, E., Grant, W., Tinker, R., Roschelle, J., Mills, M., Resnick, M., Berg, R., & Eisenberg, M. (1999). Science in the palms of their hands. *Communications of the ACM*, 42(8), 21-27.

Thomas, P. (2006). Pervasive learning games, explorations of hybrid educational gamescapes. *Simulation and Gaming*, 37 (1), 41-55.

Walther, Bo Kampmann. (2005). Atomic Actions – Molecular Experience, Theory ·of Pervasive Gaming. *ACM Computers in Entertainment*, 3(2). 1-13

Weiser, M. (1998). The future of ubiquitous computing on campus. *Communications of the ACM*, 41(1), 42-43.

Zurtia, G., & Nussbaum, M. (2004). Computer supported collaborative learning using wirelessly interconnected handheld computers. *Computers and Education*, 2004, 42(3), 289-314.

## DESIGN GUIDELINES FOR MOBILE LEARNING

Resources listed below often have sections or parts containing designing mobile learning systems.

Alavi, M. (1994). Computer-mediated collaborative learning: An empirical evaluation. *MIS Quarterly*. 18, 159-174.

Alavi, M., Marakas, G.M., & Yoo, Y. (2002). A comparative study of distributed learning environments on learning outcomes. *Information Systems Research*. 13(4), 404-415.

Ballard, B. (2007). *Designing the Mobile User Experience*, Chichester, U.K., John Wiely

Churchill, D., & Churchill, N. (2008). Educational affordances of PDAs: A study of a teacher's exploration of this technology. *Computers and Education*, *50*(4), 1439-1450.

Corlett, D., Sharples, M., Bull, S., & Chan, T. (2005). Evaluation of a mobile learning organiser for university students. *Journal of Computer Assisted Learning*, 21, 162-170.

Crippen, K., & Brooks, D. (2000). Using personal digital assistants in clinical supervision of student teachers. *Journal of Science Education and Technology*, 9(3), 207-211.

Davis, S. (2003). Observations in classrooms using a network of handheld devices. *Journal of Computer Assisted Learning*, 19(3), 298-307.

Eschenbrenner, B., & Nah, F. F. (2007). Mobile technology in education: Uses and benefits. *International Journal of Mobile Learning and Organisation*, 1(2), 159-183.

Gay, G., Stefanone, M., Grace-Martin, M., & Hembrooke, H. (2001). The effects of wireless computing in collaborative learning environments. *International Journal of Human-Computer Interaction*, 13(2), 257-276.

Johnson, M., & Marsden, G. (2006). *Mobile Interaction Design*, Chichester, U.K., John Wiely

Lindholm, C., Keinonen, T., & Kiljander, H. (2003). *Mobile Usability :How Nokia Changed the Face of the Mobile Phone*, New York, McGraw-Hill

Lonsdale, P., Baber, C., & Sharples, M. (2004) A context awareness architecture for facilitating mobile learning. In J. Attewell & C. Savill-Smith (Eds.) *Learning with Mobile Devices: Research and Development*. London. Learning and Skills Development Agency

Luppicini R.J., *Online learning communities*. Charlotte, NC: Information Age Publishing.

Lyytinen, K., & Yoo, Y. (2003). Issues and challenges in ubiquitous computing. *Communications of ACM*, 45(12), 62-65.

Norris, C., & Soloway E. (2004). Envisioning the handheld-centric classroom. *Journal of Educational Computing Research*, 30(4), 281-294.

Parr, C. S., Jones, T., & Songer, N. B. (2004). Evaluation of a handheld data collection interface for science learning. *Journal of Science Education and Technology*, 13(2), 233-242.

Patten, B., Arnedillo-Sanchez, I., & Tangney, B. (2006). Designing collaborative, constructionist and contextual applications for handheld devices. *Computers and Education*, 46(3), 294-308.

Rodden, T., Cheverst, K., Davies, N. and Dix A. (1998), Exploiting Context in HCI design for Mobile Systems, Paper presented at *the 1st Workshop on HCI with Mobile Devices*, Glasgow, U.K.

Segall, N., Doolen, T. L., & Porter, J. D. (2005). A usability comparison of PDA-based quizzes and paper-and-pencil quizzes. *Computers and Education*, 45(4), 417-432.

Sharples, M., & Beale, R. (2003). A technical review of mobile computational devices. *Journal of Computer Assisted Learning*, 19(3), 392-395.

Sharples, M., Corlett, D., & Westmancott, O. (2002). The design and implementation of a mobile learning resource. *Personal and Ubiquitous Computing*, 6, 220 - 234.

Traxler, J., & Kukulska-Hulme, A. (2005). Evaluating mobile learning: Reflections on current practice. Paper presented at the *mLearn 2005*, Cape Town, South Africa.

Waycott, J.,Scanlon, E., & Jones, A. (2002). Evaluating the use of PDAs as learning and workplace tools: An Activity Theory perspective. Paper presented at the *European Workshop on Mobile and Contextual Learning*, Birmingham, U.K

## INDIVIDUAL LEARNING EXPERIENCE AND ITS MOBILE APPLICATIONS

Benford, P., Rowland, D., Flintham, M., Hull, R., Reid, J., Morrison, J., Facer, K. & Clayton,

B. (2004). Savannah, Designing a location-based game simulating lion behaviour. Paper presented at the *Conference on Advances in Computer Entertainment Technology*, Singapore.

Cebeci, Z., & Tekdal, M. (2006). Using podcasts as audio learning objects. *Interdisciplinary Journal of Knowledge and Learning Objects*, 2, 47-57.

Chen, Y.S., Kao, T.C., & Sheu, J.P. (2003). A mobile learning system for scaffolding bird watching learning. *Journal of Computer Assisted Learning*, 19, 347-359

Hogan, K., & Pressley, M. (1997). *Scaffolding Student Learning: Instructional Approaches and Issues*. Cambridge, MA: Brookline Books.

Jonassen, D., Peck, K. & Wilson, B. (2000). *Learning with Technology: A Constructivist Approach*. Upper Saddle River, NJ: Prentice Hall.

Lindquist, D., Denning, T., Kelly, M., Malani, R., Griswold, W. G., & Simon, B. (2007). Exploring the potential of mobile phones for active learning in the classroom. Paper presented at the *38th ACM Technical Symposium on Computer Science Education (SIGCSE'07)*, Covington, Kentucky

Scanlon, E., Jones, A., & Waycott, J. (2005). Mobile technologies: prospects for their use in learning in informal science settings. *Journal of Interactive Media in Education,* 25, 1-17

Soloway, E., Grant, W., Tinker, R., Roschelle, J., Mills, M., Resnick, M., Berg, R., & Eisenberg, M. (1999). Science in the palms of their hands. *Communications of the ACM*, 42(8), 1999, 21-27.

Soloway, E., Norris, C., Blumenfeld, P., Fishman, B., Krajcik, J., & Marx, R. (2001). Log on education: Handheld devices are ready-at-hand. *Communications of the ACM*, 44(6), 15-20.

## COLLABORATIVE LEARNING EXPERIENCE AND ITS MOBILE APPLICATIONS

Alavi, M. (1994). Computer-mediated collaborative learning: An empirical evaluation. *MIS Quarterly*. 18, 159-174.

Alavi, M., Marakas, G.M., & Yoo, Y. (2002). A comparative study of distributed learning environments on learning outcomes. *Information Systems Research*. 13(4), 404-415.

Benford, P., Rowland, D., Flintham, M., Hull, R., Reid, J., Morrison, J., Facer, K. & Clayton, B. (2004). Savannah, Designing a location-based game simulating lion behaviour. Paper presented at the *Conference on Advances in Computer Entertainment Technology*, Singapore.

Collella. V. (2001). Participatory simulations: Building collaborative understanding through immersive dynamic modeling. *Journal of the Learning Science*, 9(4), 471-500.

Gay, G., Stefanone, M., Grace-Martin, M., & Hembrooke, H. (2001). The effects of wireless computing in collaborative learning environments. *International Journal of Human-Computer Interaction*, 13(2), 257-276.

Hoppe, U., Lingnau, A., Machado, I., Paiva, A., Prada, R., & Tewissen, F. (2000). Supporting collaborative activities in computer integrated classrooms – the NIMIS approach. Paper presented at *the 6th International Workshop on Groupware (CRIWG 2000)*. Madeira, Portugal. IEEE Computer Society.

Lai, C. Y., & Wu, C. C. (2006). Using handhelds in a Jigsaw cooperative learning environment. *Journal of Computer Assisted Learning*, 22(4), 284-297.

Laine, T. H., & Suhonen, J. (2008). Establishing a mobile blog system in a distance education environment. *International Journal of Mobile Learning and Organisation*, 2(2), 149-165.

Lim, K. Y. T., & Wang, J. Y. Z. (2005). Collaborative handheld gaming in education. *Educational Media International*, 42(4), 351-359.

Luppicini R.J. (2007). *Online learning communities*. Charlotte, NC: Information Age Publishing.

Nova N., Girardin F. and Dillenbourg P. (2005). 'Location is not enough!': An empirical study of location-awareness in mobile collaboration', Paper presented at the 3rd *IEEE International Workshop on Wireless and Mobile Technologies in Education*, Tokushima, Japan.

Patten, B., Arnedillo-Sanchez, I., & Tangney, B. (2006). Designing collaborative, constructionist and contextual applications for handheld devices: Virtual learning. *Computers and Education*, 46(3), 294-308.

Rogers Y., Price S., Randel C., Frases D.S, Weal M. and Fitzpatrick G. (2005). Ubi-Learning Integrates Indoor and Outdoor Experiences. *Communications of the ACM*, 18(1), 55-59.

Slavin, R. E. (1983). Cooperative learning. *Review of Educational Research*, 50(2), 315-342.

Wilensky, U. (1999). Learning through participatory simulations: Network-based design for systems learning in classrooms, Paper presented at the *Computer Support for Collaborative Learning 1999*, Mahwah, NJ

Zurtia, G., & Nussbaum, M. (2004). Computer supported collaborative learning using wirelessly interconnected handheld computers. *Computers and Education*, 2004, 42(3), 289-314.

## SITUATED LEARNING EXPERIENCE AND ITS MOBILE APPLICATIONS

Beale, R., & Lonsdale, P. (2004). Mobile context aware systems: the intelligence to support tasks and effectively utilise resources. Paper presented at *Mobile HCI 2004*, Glasgow, UK.

Benford, S., Rowland, D., Flintham, M., Drozd, A., Hull, R., Reid, J., Morrison, J., & Facer, K. (2005). Life on the edge: Supporting collaboration in location-based experiences, Paper presented at *CHI 2005*, Portland, OR.

Benford S. (2005). *Future Location-Based Experiences*, JISC Tech Report (TSW0501).

Benford, P., Rowland, D., Flintham, M., Hull, R., Reid, J., Morrison, J., Facer, K. & Clayton, B. (2004). Savannah, Designing a location-based game simulating lion behaviour. Paper presented at the *Conference on Advances in Computer Entertainment Technology*, Singapore.

Brown, J. S., Collins, A., & Duguid, P. (1989). Situated cognition and culture of learning. *Educational Technology*, 18(1), 32-42.

Brown, J.S., and Duguid, P. (2000). *The Social Life of Information*. Harvard Business School, Cambridge, MA.

Dourish, P. (2004). What we talk about when we talk about context. *Journal of Personal and Ubiquitous Computing*, 8(1), 19-30

Lave, J., & Wenger, E. (1991). *Situated Learning: Legitimate Peripheral Participation*, Cambridge, U.K.: Cambridge University Press

Lonsdale, P., Baber, C., & Sharples, M. (2004) A context awareness architecture for facilitating mobile learning. In J. Attewell & C. Savill-Smith (Eds.) *Learning with Mobile Devices: Research and Development*. London. Learning and Skills Development Agency

Nardi, B. A. (1996). *Context and Consciousness*. Cambridge, MA: The MIT Press

Nova N., Girardin F. and Dillenbourg P. (2005). 'Location is not enough!': An empirical study of location-awareness in mobile collaboration', Paper presented at the 3rd *IEEE International*

*Workshop on Wireless and Mobile Technologies in Education*, Tokushima, Japan.

Rogers Y., Price S., Randel C., Frases D.S, Weal M. and Fitzpatrick G. (2005). Ubi-Learning Integrates Indoor and Outdoor Experiences. *Communications of the ACM,* 18(1), 55-59.

Rogoff, B., & Lave, J. (1984). *Everyday Cognition: Its Development in Social Context.* Cambridge, MA: Harvard University Press.

Smailagic, A. and Siewiorek, D. (2002), Application design for wearable and context-aware computers, *IEEE Pervasive Computing,* 1(4), October-December 2002, 20-29.

Stone, A., & Livingstone, D. (2004). Don't lose the learning from learning objects! A request for care in the adaptive context community. *International Journal of Continuing Engineering Education and Lifelong Learning,* 14(4/5), 318-330.

Waycott, J.,Scanlon, E., & Jones, A. (2002). Evaluating the use of PDAs as learning and workplace tools: An Activity Theory perspective. Paper presented at the *European Workshop on Mobile and Contextual Learning*, Birmingham, U.K

Vygotsky, L. S. (1926). *Pedagogical Psychology.* Moscow: Educational Worker Publisher.

## CHALLENGES FOR DESIGNING MOBILE LEARNING

Davis, S. (2003). Observations in classrooms using a network of handheld devices. *Journal of Computer Assisted Learning,*19, 298-307.

Holzinger, A., Nischelwitzer, A., & Meisenberger, M. (2005). Lifelong-learning support by m-learning: Example scenarios. ACM eLearn 2005, 11, November, 2.

Lankshear, C., and Knobel, M. (2006). *New Literacies: Everyday Practices and Classroom Learning.* Berkshire, U.K: Open University Press.

Roschelle, J. (2003). Unlocking the learning value of wireless mobile devices. *Journal of Computer Assisted Learning.* 19, 260-272.

Ryu, H. (2007). Is mobile learning a necessary evil?: The goodness-of-fit of mobile learning situations. Paper presented at *the Workshop of Pervasive Learning 2007*: Design Challenges and Requirements. Toronto, Ontario, Canada.

Sharples, M. (2006). *Big Issues in Mobile Learning.* Nottingham, U.K: Learning Sciences Research Institute.

Swan, K., Kratcoski, A., & van 't Hooft, M. (2007). Highly mobile devices, pedagogical possibilities, and how teaching needs to be reconceptualized to realize them. *Education Technology,* 3, 10-12

## REVIEWING THE LANDSCAPE OF MOBILE LEARNING

Cobcroft, R. (2006) Literature Review into Mobile Learning in the University Context. Queensland University of Technology. Retrieved 21 August, 2008, from http://eprints.qut.edu.au/archive/00004805/

Kukulska-Hulme, A., Sharples, M., Milrad, M., Arnedillo-Sánchez, I. & Vavoula, G. Massey, A. (2008). Innovation in Mobile Learning: a European Perspective. International Journal of Mobile and Blended Learning 1(1).

Naismith, L., Lonsdale, P., Vavoula, G. & Sharples, M. (2004). Literature Review in Mobile Technologies and Learning, Report 11 for NESTA Futurelab, University of Birmingham. Retrieved 21 August, 2008, from http://elearning.typepad.com/thelearnedman/mobile_learning/reports/futurelab_review_11.pdf

Pachler N. (ed.) (2007). Mobile Learning: Towards a Research Agenda. London: WLE Centre.

Traxler, J. (2008). Learning in a Mobile Age. International Journal of Mobile and Blended Learning 1(1).

Trifonova, A. (2003). Mobile Learning – Review of the Literature. Technical Report DIT-03-009, Informatica e Telecomunicazioni, University of Trento

## RELEVANT SCHOLARLY JOURNALS

Mobile learning researchers then to introduce their latest findings in the following journals:

ACM Computers in Entertainment

ACM E-Learn

Communications of the ACM

Computers and Education

Educational Technology

IEEE Pervasive Computing

IEEE Transactions on Education

International Journal of Human-Computer Interaction

International Journal of Human-Computer Studies

International Journal of Mobile and Blended Learning

International Journal of Mobile Learning and Organisation

Journal of Computer Assisted Learning

Journal of Educational Multimedia and Hypermedia

Journal of Personal and Ubiquitous Computing

Journal of the Learning Science

Personal and Ubiquitous Computing

Review of Educational Research

## CONFERENCES

Mobile learning practitioners and academics meet and present the findings of their latest work at a number of international conferences, including the following:

Handheld Learning

IADIS Mobile Learning

International Conference on Interactive Mobile and Computer Aided Learning

IEEE Workshop on Wireless, Mobile and Ubiquitous Technologies in Education

International Conference on Advanced Learning Technologies

International Conference on Computers in Education

Mobile Learning Technologies and Applications

mLearn

Wireless and Mobile Technologies in Education

# Glossary

**3G:** The third generation of mobile phone standards and technology. The most significant feature of 3G mobile technology is that it supports greater numbers of voice and data customers—especially in urban areas—and higher data rates at lower incremental cost than 2G.

**Advanced Audio Coding (AAC):** A standardized encoding scheme for digital audio, which is used as the default audio format of Apple's iPhone, iPod, iTunes.

**Activity Theory:** Activity theory originated as a dialectical materialist psychology developed by Vygotsky and his students in the Soviet Union in the beginning of the twentieth century. It rejects the isolated human being as an adequate unit of analysis, insisting on cultural and technical mediation of human activity. The unit of analysis accordingly includes technical artifacts and the cultural organization that the human being is both determined by and actively creating (Bertelsen & Bodker, 2003, p. 298)

**Ajax:** Ajax (asynchronous JavaScript and XML) is a group of inter-related Web development techniques used for creating interactive Web applications. A primary characteristic is the increased responsiveness and interactiveness of Web pages achieved by exchanging small amounts of data with the server "behind the scenes" so that the entire Web page does not have to be reloaded each time the user performs an action. This is intended to increase the Web page's interactivity, speed, functionality, and usability, which seem to be considered as essential qualities in designing mobile learning contents.

**Augmented Reality:** Augmented reality, the reality overlaid with virtual reality, and virtual reality, in which representations of the real world have been embedded and contextualized, is enabling interactions both in real and virtual spaces. For instance, Lonsing (2004) suggests that an augmented reality system generates a composite view in real time, a combination of a real scene viewed by a user and a virtual scene generated by a computer, where the real scene is submerged with additional information in order to enhance the perception of the user.

**Behaviorism:** A philosophy of psychology based on the proposition that all things which organisms do—including acting, thinking and feeling—can and should be regarded as behaviors, that is, it is the term that explains human behaviors with simple responses of an organism to stimuli. For instance, Skinner believed that behavior is sustained by reinforcements or rewards, not by free will (Skinner, 1976). Therefore, traditional classroom learning environments are designed on the principles of the behaviorist view of learning and knowing. They are organized with the goal of students acquiring a maximum accumulation of organized information and procedural knowledge (Greeno, Collins, & Resnick, 1996) and Noam Chomsky (1967) criticized behaviorism for giving a description of the stimuli and responses solely in terms of observable behaviors.

**Blended Learning:** The term used to describe learning or training events or activities where e-learning, in its various forms, is combined with more traditional forms of training such as "classroom" training. It is now encompassing mobile learning technologies in its broadest definition, for instance, as a way to deliver learning contents prior to actual face-to-face lectures at a course or program.

**Camera Phone:** A mobile phone that has a camera built-in, so MMS (see multimedia messaging service) may be enabled.

**Collaboration:** A structured, recursive process where two or more people work together toward a common goal—typically an intellectual endeavor that is creative in nature—by sharing knowledge, learning and building consensus (*copied from Wikipedia*). With respect to education, it is widely believed that, regardless of the subject matter, students working in groups tend to learn more of what is taught and retain it longer than when the same content is presented in other instructional formats. That is, learning can be even more effective when learners can converse with each other, by interrogating and sharing their descriptions of the world

**Constructivism:** Developed during the 1960s and 70s and inspired by the rise in cognitive theories of learning (Bruner, 1966, 1986, 1996). It depicts that learners are constructing their own knowledge by testing ideas and approaches based on their prior knowledge and experiences, actively applying these to a new situation and constructively integrating the new knowledge gained with pre-existing intellectual constructs. Therefore, the commitments and practices that the teacher or other educational resources should support would be the creation of suitable learning situations and settings, rather than the organized (or pre-defined) contents to deliver.

**Context Awareness:** A term which seeks to deal with linking changes in the circumstance or environment with computer systems. Context-aware learning systems adapt according to the location of use, the collection of nearby people, hosts, and accessible devices, as well as to changes to such things over time. For this reason, mobile devices have been paid much attention for sensing and collecting their physical information in the context of use.

**Contextual Learning:** Occurs in close relationship with actual experience, allowing, for instance, students to test theories or contents delivered in classroom through real world applications (real problem-solving). This approach to learning and teaching assumes that the mind naturally seeks meaning in context—that is, in relation to the person's current environment—and that it does so by searching for relationships that make sense and appear useful. In conjunction with various context awareness technologies, it is highly possible to anchor teaching in students diverse life-contexts.

**Distributed Learning Environment:** Distributed learning is an instructional model that allows instructor, students, and content to be located in different, non-centralized locations so that both instruction and learning occur independent of time and place. The distributed learning environment can be designed in combination with traditional classroom-based courses, with traditional distance learning courses, or it can be used to create wholly virtual classrooms, so it allows one to learn something at a distance from the traditional education providers (institutes and universities).

**Edutainment:** A form of entertainment designed to educate as well as to amuse. Edutainment typically seeks to instruct or socialize its audience by embedding lessons in some familiar form of entertainment: television programs, computer and video games, films, music, Web sites, multimedia software, and etc. It is widely believed to give a basis for teaching in a more efficient and faster way (edited from Wikipedia).

**Electronic Learning (E-Learning):** Another way of teaching and learning. In its broadest definition, e-learning includes instruction delivered via all electronic media including the Internet, intranets, extranets, satellite broadcasts, audio/video tape, interactive TV, and CD-ROM (Govindasamy, 2002, p. 291).

**Engagement (Participatory Engagement):** Refers to a conceptual approach to engage students in the construction of knowledge that embodies complex concepts, necessitates collaboration, and contextualizes learning within contexts. Rather than directly presenting instructional treatments, the goal from this perspective is to establish rich environments that encourage explanation and discovery, nurture reflection, and support students in the carrying out of practices that embody personally meaningful and practically functional representations (Barab, Hay, Barnett, & Squire, 2001 p. 48; Lave & Wenger, 1991).

**Field Trip:** A group excursion for the purpose of firsthand observation, as to a museum, the woods, or a historic place. Recently, practitioners in education technology have developed various types of *virtual field trips* that are designed to be more entertaining and educational. These include the simple, such as a photo tour of a famous museum, to extremely detailed and high-tech field trips that offer video and audio segments to make the visit more interactive.

**Flow Experience:** The mental state of operation in which the person is fully immersed in what he or she is doing, characterized by a feeling of energized focus, full involvement, and success in the process of the activity, proposed by psychologist Mihaly Csikszentmihalyi (1990). In education, there is the concept of overlearning which seems to be an important factor in this technique, in that Csikszentmihalyi states that overlearning enables the mind to concentrate on visualizing the desired performance as a singular, integrated action instead of a set of actions (copied from Wikipedia)

**Framework (Theory and Architecture):** The specification of a few broad principles of design from a general perspective, where particular features are left. In contrast, a 'theory' develops a framework, adding some data structures or mechanisms for explaining and understanding the nature of the phenomenon. Furthermore, an "architecture" encompasses both theory and framework.

**Game-Based Learning:** An approach to use games (especially, video games) as learning tools. With the widespread interest in games by teenagers or adults, it employs games as an engaging and effective instructional medium to integrate them into the learning process. Prensky (2000) explains how those principles can be leveraged in learning activities.

**Global System for Mobile Communications (GSM):** A digital mobile telephone system that is widely used in Europe and other parts of the world. GSM uses a variation of time division multiple access (TDMA) and is the most widely used of the three digital wireless telephone technologies (TDMA, GSM, and CDMA).

**Global Positioning Systems (GPS):** A U.S. space-based radionavigation system that provides reliable positioning, navigation, and timing services to civilian users on a continuous worldwide basis, freely available to all. For anyone with a GPS receiver, the system will provide location and time. GPS provides accurate location and time information for an unlimited number of people in all weather, day and night, anywhere in the world.

**Hand-Held Devices:** Pocket-sized computing devices, typically comprising a small visual display screen for user output and a miniature keyboard or touch screen for user input. In the case of the personal digital assistant (PDA) the input and output are combined into a touch-screen interface. Smartphones and PDAs are popular amongst those who require the assistance and convenience of a conventional computer, in environments where carrying one would not be practical (edited from Wikipedia.org).

**High Speed Downlink Packet Access (HSDPA):** Anew protocol for mobile telephone data transmission, especially for 3G mobile network.

**Interaction:** In this book interaction means any mutual or reciprocal communication between a user (learners or educators) and technology (computers or mobile devices), whether it is direct or indirect. In particular, *interactive technology* has greatly changed in meaning with the advent of extended computer or mobile device use in the education domain in the last decades. For example, WebCT™ provides an online proprietary virtual interactive learning environment for e-learning.

**iPods (podcasting):** A portable digital media player designed by Apple Computer™. iPods have a reputation for being user-friendly and quality design features, such as a "touch wheel" which is a centrally-placed circular disk designed for one-hand operation. Many versions of new iPods include features like a calendar, address book, to-do list, alarm clock with sleep timer, games, and text reader. Several universities have employed iPods as an instructional medium using the 'Podcasting' technology which prepares and distribute audio files using RSS (see real simple syndication) to subscribed users.

**Learning Activity:** The essence of a learning activity is that it must have one or more learning outcomes associated with it. Learning outcomes are what the learners should know, or be able to do, after completing the learning activity; for example, understand, demonstrate, design, produce, appraise. For instance, pop quizzes in classroom are to present students with a learning activity to test the knowledge of the contents delivered by posing questions.

**Learning Experience:** Creating of the practical knowledge, feeling, understanding and incidents which impact a learner's perception of all the learning activities that are being performed by the learner. To harness effective learning experiences, education providers should consider learner's needs, desires, beliefs, knowledge, skills, and perceptions.

**Learning Management Systems (LMS):** A term used to describe software tools designed to manage user learning interventions, for example, WebCT™ or Blackboard™.

**Learning Outcomes:** Take the measure of the knowledge, skills, and abilities the individual student possesses and can demonstrate upon completion of a learning experience or sequence of learning experiences (e.g., course, program, and degree).

**Life-Long Learning:** Allows people to continually enhance their knowledge to address immediate problems and to participate in the process of professional development, facilitating learning at different times and places. Mobile learning appears to be a promising approach to enable this life-long learning in context and analysis of real world problems with the use of mobile devices and applications to better support the learning process.

**Media Richness Theory:** Explains the relationship between characteristics of a communication medium and communication activities (Daft & Lengel, 1986). According to Daft and Lengel's theory, media richness is a function of (1) the medium's capacity for immediate feedback, (2) the number of cues and channels available, (3) language variety; and (4) the degree to which intent is focused on the recipient. The greater social presence of a medium creates a greater immediacy and warmth of the communication, because of the greater number of channels.

**Mobile Phone (or Cellular Phone):** Uses a network of short-range transmitters, often in an automobile, located in overlapping cells throughout a region, with a central station making connections to regular telephone lines. It is also called cellular phone (edited from The American Heritage Dictionary of the English Language)

**MMS (Multimedia Messaging Service):** A mobile telephone messaging system that allows sending multimedia messages such as images, audio, video and rich text.

**Mobile Learning:** Wikipedia puts it that "Learning that happens across locations, or that takes advantage of learning opportunities offered by portable technologies."

**Participatory Simulation:** Allows the learner to understand the course content better through participation/practice. Through this active participation, the learner can discuss and get the correct answer, consequently understand what they have learned better, for example (Colella, 2000)

**Personal Digital Assistants (PDAs):** Handheld computers, which have been used for organizing personal schedules or an address book, and running office programs (e.g., spreadsheets, word processor). Newer PDAs are similar to Smartphones in that they allow one to access the Internet via Wi-Fi or carrier mobile phone networks, but tend to have a larger form factor.

**Really Simple Syndication (RSS):** A family of Web feed formats used to publish frequently updated content such as blog entries, news headlines or podcasts (*edited from Wikipedia*)

**Scaffolding:** An instructional method or process by which a teacher provides temporary support to learners to help bridge the gap between what the learner knows and can do and what he or she needs to accomplish in order to succeed at a particular learning task, like training wheels when learning to ride a bicycle, that scaffolding is gradually taken away over time until the learner is able to perform the skill on their own, with no support.

**Semacode:** Use two dimensional barcodes that support and complement development of the concept of using mobile phones as devices for seamless information gathering and exchange. Its tagging system is an open system with the SDK software tools available for educational purposes.

**Short Text Messaging Services (SMS):** A feature available in most mobile phones, that lets users receive and send short text messages from 150 to 160 characters to other mobile phones. On Nokia™ phones, users can even receive new ringing tones or background logos in an SMS text message.

**Situated Learning:** A general theory of knowledge acquisition that is based on the notion that learning occurs in the context of authentic activities (Lave & Wenger 1991). Other researchers have further developed the theory of situated learning. Brown, Collins & Duguid (1989), for instance, emphasize the idea of cognitive apprenticeship: "Cognitive apprenticeship supports learning in a domain by enabling students to acquire, develop and use cognitive tools in authentic domain activity. Learning, both outside and inside school, advances through collaborative social interaction and the social construction of knowledge." Brown et al. also emphasize the need for a new epistemology for learning—one that emphasizes active perception over concepts and representation.

**Virtual Learning Environment:** A software system designed to help teachers by facilitating the management of educational courses for their students, especially by helping teachers and learners with course administration. The system can often track the learners' progress, which can be monitored by both teachers and learners. While frequently thought of as primarily tools for distance education, they are most often used to supplement the face-to-face classroom. Recently, learning inside a 3D virtual immersive workspace (e.g., Second Life™) has been paid more attention for a future VLE, in conjunction with its virtual reality, environment and course management systems.

**Wi-Fi:** An acronym for *wi*reless *fi*delity and is meant to be used generically when referring of any type of 802.11 network, whether 802.11b, 802.11a, dual-band, and so forth.

**Working Memory:** A system for temporarily storing and managing the information required to carry out complex cognitive tasks such as learning, reasoning, and comprehension. It is considered as the bottleneck of human information processing, thereby making a stronger implication on learning activities and outcomes.

**Ubiquitous (Pervasive) Computing:** A post-desktop model of human-computer interaction in which information processing has been thoroughly integrated into everyday objects and activities. As opposed to the desktop paradigm, in which a single user consciously engages a single device for a specialized purpose, someone "using" ubiquitous computing engages many computational devices and systems simultaneously, in the course of ordinary activities, and may not necessarily even be aware that they are doing so. Therefore, it seeks to embed computers into our everyday lives in such ways as to render them invisible and allow them to be taken for granted.

## REFERENCES

Barab, S. A., Hay, K. E., Barnett, M., & Squire, K. (2001). Constructing virtual worlds: Tracing the historical development of learner practices. *Cognition and Instruction, 19*(1), 47-94.

Bertelsen, O. W., & Bodker, S. (2003). Activity theory. In J. Carroll (Ed.), *HCI models, theories, and frameworks* (pp. 292-324). San Francisco: Morgan Kaufmann Publishers.

Brown, J. S., Collins, A., & Duguid, P. (1989). Situated cognition and culture of learning. *Educational Technology, 18*(1), 32-42.

Bruner, J. (1966). *Toward a theory of instruction.* Cambridge, MA: Harvard University Press.

Bruner, J. (1986). *Actual minds, possible worlds.* Cambridge, MA: Harvard University Press.

Bruner, J. (1996). *The culture of education.* Cambridge, MA: Harvard University Press.

Chomsky, N. (1967). A review of B. F. Skinner's verbal behavior. In L. A. Jakobovits & M. S. Miron (Eds.), *Readings in the psychology of language* (pp. 142-143). New York: Prentice-Hall.

Colella, V. (2000). Participatory simulations: Building collaborative understanding through immersive dynamic modeling. *The Journal of the Learning Science, 9*(4), 471-500.

Csikszentmihalyi, M. (1990). *Flow: The psychology of optimal experience.* New York: HarperCollins.

Daft, R. L., & Lengel, R. H. (1986). Organizational information requirements, media richness and structural design. *Management Science, 32*(5), 554-571.

Govindasamy, T. (2002). Successful implementation of e-learning pedagogical consideration. *The Internet and Higher Education, 4*(3-4), 287-299.

Greeno, J. G., Collins, A. M., & Resnick, L. B. (1996). Cognition and learning. In D. Berliner & R. Calfee (Eds.), *Handbook of educational psychology* (pp. 15-41). New York: Macmillan.

Lave, J., & Wenger, E. (1991). *Situated learning: Legitimate peripheral participation.* Cambridge: University of Cambridge Press.

Lonsing, W. (2004). Augmented reality, augmented reality as tool in architecture. Paper presented at the *22nd Conference on Education and Research in Computer Aided Architectural Design in Europe*, Copenhagen, Denmark.

Prensky, M. (2000). *Digital game-based learning.* McGraw-Hill.

Skinner, B. F. (1976). *About behaviorism: Why do people behave as they do?* New York: Vintage.

# Compilation of References


Aalto, L., Göthlin, N., Korhonen, J., & Ojala, T. (2004). Bluetooth and WAP Push based location-aware mobile advertising system. In *Proc. Second International Conference on Mobile Systems, Applications and Services*, (pp. 49 - 58). Boston, MA.

Abowd, G. D. (1999). Classroom 2000: An experiment with the instrumentation of a living educational environment. *IBM Systems Journal., 38*(4), 508-530.

Abowd, G. D., & Mynatt, E. D. (2000). Charting past, present, and future research in ubiquitous computing. *ACM Transactions on Computer-Human Interaction, 7*(1), 29-58.

Ackerman, M. S., & Cranor, L. (1999). Privacy critics: UI components to safeguard users' privacy. In *Proc. Computer Human Interaction Conference (CHI '99) Extended Abstracts on Human Factors in Computing Systems*, (pp 258-259). Pittsburgh, Pennsylvania

Ackermann, E. (1996). Perspective-taking and object construction. In Y. Kafai, & M. Resnick (Ed.), *Construction in practice: Designing, thinking, and learning in a digital world* (pp. 25–35). Marwah, NJ: Lawrence Erlbaum.

Ackermann, E., Chabert, A., Decortis, F., Magli, R., Napoletano, L., & Owen, M. (2006). "La piazza": Convivial spaces for intergenerational learning: What places for the digital age? In A. Mendez-Vilas, A. Solano Martin, J. Mesa-Gonzalez, & J.A. Mesa Gonzalez (Eds.) *Current developments in computer-assisted education, (*pp. 1359-1363), Badajoz, Spain: FORMATEX.

Adidam, P., & Bingi, R. (2000). The importance of decision confidence to strategy outcomes. *The Journal of Applied Business Research, 16*(2), 35-50.

Alavi, M. (1994). Computer-mediated collaborative learning: An empirical evaluation. *MIS Quarterly, 18*, 159-174.

Alavi, M., Marakas, G.M., & Yoo, Y. (2002). A comparative study of distributed learning environments on learning outcomes. *Information Systems Research, 13*(4), 404-415.

Aldrich, C. (2005). *Learning by doing: A comprehensive guide to simulations, computer games, and pedagogy in e-learning and other educational experiences.* San Francisco: John Wiley & Sons.

Al-Dujaily, A. (2007). *Personality effect in the design of adaptive e-learning systems.* Unpublished thesis. Massey University, Auckland.

Alexander, B. (2004). Going nomadic: Mobile learning in higher education. *EDUCAUSE Review, 39*(5), 28-35.

Ally, M., Lin, F., McGreal, R., & Woo, B. (2005). *An intelligent agent for adapting and delivering electronic course materials to mobile learners.* Paper presented at the mLearn 2005, Cape Town, South Africa.

Althaus, S.L. (1997). Computer-mediated communication in a university classroom: An experiment with on-line discussion. *Communication Education, July,* 158-174.

Alvarez-Cavazos, F., Garcia-Sanchez, R., Garza-Salazar, D.A., Lavariega, J.C., Gomez-Martinez, L.G., & Sordia, M. (2005, October 17-20). Universal access architecture for digital libraries. In *Proceedings of the 2005 conference of the Centre for Advanced Studies on Collaborative Research*, (pp. 12-28). Toronto, Ontario, Canada

Amab, D., Mehmet, P., & Schnitker, T. (2001). Could mobile banking go global? *The McKinsey Quarterly, 4*, 71-80.

Amory, A., Naicker, K., Vincent, J., & Adams, C. (1999). The use of computer games as an educational tool: identification of appropriate game types and game elements. *British Journal of Educational Technology, 30*(4), 311–321.

Anderson, T. (2004). Teaching in an online learning context. In T. Anderson & F. Elloumi (Eds.), *Theory and practice of online learning* (pp. 273-294). Athabasca, AB: Athabasca University.

Angelaccio, M., & Buttarazzi, B. (2006, December 4-7). *A Performance Evaluation of Asynchronous Web Interfaces for Collaborative Web Services.* Paper presented at the Frontiers of High Performance Computing and Networking – ISPA 2006 Workshops (pp. 864-872). Sorrento, Italy.

Angelo, T. A., & Cross, K. P. (1993). *Classroom assessment techniques: A handbook for college teachers* (2nd ed.). San Francisco: Jossey-Bass Publishers.

Apple Education. (2007a). Elementary school students become podcasting pros. Retrieved August 7, 2007, from www.apple.com/education/profiles/wells

Apple Education. (2007b). Podcasts: Earmarked for success. Retrieved August 7, 2007, from www.apple.com/education/profiles/wisconsin-madison

Applefield, J. M., Huber, R., & Mahnaz, M. (2000). Constructivism in theory and practice: Toward a better understanding. *High School Journal., 84*(2), 35-59.

Aronson, E., Blaney, N., Stephin, C., Sikes, J., & Snapp, M. (1978). *The jigsaw classroom.* Beverly Hills, CA: Sage Publishing Company.

Atkins, C. (2005). *Adult education and social reform.* Lecture notes in the distant lifelong education of the University of Nottingham. Hong Kong.

Avouris, N., Margaritis, M., Komis, V., Saez, A., & Meléndez, R. (2003). Modellingspace: Interaction design and architecture of a collaborative modelling environment. In *Proceedings of the Computer Based Learning in Science 2003*

Baajour H., Magoulas G. D., & Poulovassilis A. (2007, June 25-29). Designing services-enabled personalisation for planning of lifelong learning based on individual and group characteristics. *In* P. Brusilovsky, M. Grigoriadou, K. Papanikolaou (eds.) *Proc. of Workshop on Personalisation in E-Learning Environments at Individual and Group Level,* 11th International Conference on User Modeling (UM 2007), (pp. 8-15). Corfu, Greece.

Baluja, S. (2006, May 23 - 26). Browsing on small screens: recasting Web-page segmentation into an efficient machine learning framework. In *Proceedings of the 15th International Conference on World Wide Web,* (pp. 33-42), Edinburgh, Scotland WWW '06. New York, NY: ACM Press.

Bar, H., Häussge, G., & Rößling, G. (2007, June 23–27). *An integrated system for interaction support in lectures.* Paper presented at the 12th annual SIGCSE conference on Innovation and technology in computer science education (ITiCSE'07) (pp. 281-285). Dundee, Scotland.

Barak, M., Harward, J., & Lerman, S. (2007). *Studio-based learning via wireless notebooks: a case of a Java programming course. International Journal Mobile Learning and Organisation, 1(1), 15–29.*

Barkhuus, L., & Dey, A. (2003). Is context-aware computing taking control away from the user? Three levels of interactivity examined. *Lecture Notes in Computer Science, 2864*, 149-156.

Barnes, S. J., & Scornavacca, E. (2005). *Strategic Impact Of Wireless Applications In New Zealand Business.* Paper presented at the Hong Kong Mobility Roundtable, Hong Kong.

Barth, B., & Cerny, P. (2006). *Kommunikation und IT in Österreich.* Wien: Integral Markt-und Meinungsforschungsgep.m.b.H.

Bealle, R., & Lonsdale, P. (2004). Mobile context aware systems: the intelligence to support tasks and effectively utilise resources. In *Proceedings of Mobile HCI 2004*, (pp. 240-251), Glasgow, UK.

Becerra-Fernandez, I., Cousins, K. C., & Weber, R. O. (2007). Nomadic context-aware knowledge management systems: Applications, challenges and research problems. *International Journal of Mobile Learning and Organization, 1*(2), 103-121.

Beckman, M. (1990). Collaborative learning: Preparation for the workplace and democracy. *College Teaching, 38*(4), 128-133.

Belanger, Y. (2005). *Duke University iPod first year experience final evaluation report.* Retrieved June 2, 2006, from http://cit.duke.edu/pdf/ipod_initiative_04_05.pdf

Belanger, Y. (2005, June). *Duke University iPod fist year experience final evaluation report.* Retrieved July 1, 2007, from http://cit.duke.edu/pdf/ipod_initiative_04_05.pdf

Benbunan-Fich, R., & Hiltz, S.R. (1999). Educational applications of CMCS: Solving case studies through asynchronous learning networks. *Journal of Computer-Mediated Communication,* 4(3), March, http://www.ascusc.org/jcmc/vol4/issue3/benbunan-fich.html

Bendy, G. Z., Seglin, M. H., & Meister, D. (2000). Activity theory: history, research and application. *Theoretical Issues in Ergonomic Sciences, 1*(2), 168-206.

Benford, P., Magerkurth, C., & Ljungstrand, P. (2005). Bridging the physical and digital in pervasive gaming. *Communications of the ACM, 48*(3), 54-57.

Benford, P., Rowland, D., Flintham, M., Hull, R., Reid, J., Morrison, J. et al. (2004). *Savannah, designing a location-based game simulating lion behaviour.* Paper presented at the Conference on Advances in Computer Entertainment.

Benford, S., Rowland, D., Flintham, M., Drozd, A., Hull, R., Reid, J. et al. (2005). Life on the edge: Supporting collaboration in location-based experiences. In *Proceedings of CHI2005* (pp. 721–730), Portland, OR.

Berri, J., Benlamri, R., & Atif, Y. (2006, July 03 - 06). Ontology-based framework for context-aware mobile learning. In *Proceedings of the 2006 International Conference on Communications and Mobile Computing,* (pp. 1307-1310), Vancouver, British Columbia, Canada.

Berri, J., Benlamri, R., & Atif, Y. (2006, July 3-6). *Ontology-based framework for context-aware mobile learning.* Paper presented at the International Conference on Wireless Communications and Mobile Computing (pp. 1307-1310). Vancouver, Canada.

Bertelsen, O. W., & Bødker, S. (2003). Activity theory. In J. M. Carroll (Ed.), *HCI models, theories, and frameworks: Toward a multidisciplinary science* (pp. 291-324). San Francisco, CA: Morgan Kaufmann.

Bhakhtin, M. (1981). *The dialogical imagination: four essays.* trans. Emerson C. and Holquist M. Austin: University of Texas Press.

Biros, D., Daly, M., & Gunsch, G. (2004). The influence of task load and automation trust on deception detection. *Group Decision and Negotiation, 13,* 173-189.

Bishop, A. L., Dinkins, R. K., & Dominick, J. L. (2003). Programming handheld devices to enhance learning. *EDUCAUSE Quarterly, 26,* 50-53.

Black, J. T., & Hawkes, L. W. (2006, July 3-6). *A prototype interface for collaborative mobile learning.* Paper presented at the International Conference on Wireless Communications and Mobile Computing (pp. 1277-1282). Vancouver, Canada.

Blekas, A., Garofalakis, J., & Stefanis, V. (2006, May 22 - 22). Use of RSS feeds for content adaptation in mobile Web browsing. In *Proceedings of the*

*2006 international Cross-Disciplinary Workshop on Web Accessibility (W4a): Building the Mobile Web: Rediscovering Accessibility?* (vol. 134) (pp. 79-85), Edinburgh, U.K, W4A. New York, NY: ACM Press.

Blodgett, D. L. (2006). *The Effects of Implementing an Interactive Student Response System in a College Algebra Classroom.* Unpublished Masters Thesis, the University of Maine, USA.

Bloom, B. S. (Ed.). (1956). *Taxonomy of educational objectives: The classification of educational goals. Handbook 1: Cognitive domain* (1st ed.). New York: Longmans Green.

Bollen, L., Eimler, S., & Hoppe, H. U. (2004). SMS-Based Discussions - Technology Enhanced Collaboration for a Literature Course. *Proceedings of the 2nd IEEE International Workshop on Wireless and Mobile Technologies in Education (WMTE?'04).*

Bonk, C.J., & Kim, K.A. (1998). Extending sociocultural processes to adult learning. In M.C. Smith & T. Poiurchot (Eds.), *Adult learning and development: Perspectives from educational psychology* (pp. 67-88). Mahwah, NJ: Lawrence Erlbaum Press,

Bonoma, T. (1985). Case research in marketing: Opportunities, problems and a process. *Journal of Marketing Research, 12*, 199-208.

*Bonwell, C., & Eison, J. (1991). Active Learning: Creating Excitement in the Classroom AEHE-ERIC Higher Education Report No.1. Washington, DC: Jossey-Bass.*

Bote, M. L., Vaquero, L. M., Vega, G., Dimitriadis, Y., Asensio, J. I., Gómez, E., & Hernández, D. (2004). Tailorable collaborative learning system that combines OGSA grid services and IMS – LD scripting. In *Proceedings on the X international workshop on groupware, CRIWG 2004*, Springer-Verlag, LNCS 3198, (pp. 205-321). San Carlos, Costa Rica.

Bradley, C., Haynes, R., & Boyle, T. (2005). *Adult multimedia learning with PDAs -The user experience.* Paper presented at the mLearn 2005, Cape Town, South Africa.

Branch, J., & Oberg, D. (2004). *Focus on inquiry: A teacher's guide to implementing inquiry-based learning.* Edmonton, AB: Alberta Learning Teaching Resources Branch.

Brar, A., & Kay, J. (2005). *Privacy and security in ubiquitous personalized applications* (Tech. Rep. No. 561), School of Information Technologies, University of Sydney.

Brittain, S., Glowacki, P., Van Ittersum, J., & Johnson, L. (2006). Formative evaluation strategies helped identify a solution to a learning dilemma. *EDUCAUSE Quarterly, 29*(3). Retrieved August 6, 2007, from http://www.educause.edu/apps/eq/eqm06/ eqm0634.asp?bhcp=1

Brodersen, C., Christensen, B. G., Grnbk, K., Dindler, C., & Sundararajah, B. (2005). Ebag: a ubiquitous Web infrastructure for nomadic learning. In *Proceedings of the 14th international conference on World Wide Web (WWW '05)*, (pp. 298-306), New York, NY:ACM Press.

Brown, A. L., & Campione, J. C. (1994). Guided discovery in a community of learners. In K. McGilly (Ed.), *Classroom lessons: Integrating cognitive theory and classroom practice* (pp. 229-272). Cambridge, MA: MIT Press/Bradford Books.

Brown, J. S., & Duguid, P. (2000). *The social life of information.* Cambridge, MA: Harvard Business School.

Brown, J. S., Collins, A., & Duguid, P. (1989). Situated cognition and culture of learning. *Educational Technology, 18*(1), 32-42.

Brown, J.S., & Duguid, P. (2000). *The social life of information.* Cambridge, MA: Harvard Business School.

Bruce, B., & Hogan, M. (1998). The disappearance of technology: Toward an ecological model of literacy. In D. Reinking & M. McKenna & L. Labbo & R. Kieffer (Eds.), *Handbook of literacy and technology.* Mahwah, NJ: Lawrence Erlbaum Associates.

Brufee, K.A. (1993). *Collaborative learning: Higher education, interdependence, and the authority of knowledge.* Baltimore: Johns Hopkin University Press.

Bruhn, J., Gräsel, C., Mandl, H., & Fischer, F. (1998). Befunde und Perspektiven des Lernens in Computernetzen. In F. Scheuermann, F. Schwab & H. Augenstein (Eds), *Studieren und weiterbildung mit mulitmedia* (p. 390). Nürnberg.

Bruner, J. (1966). *Toward a theory of instruction.* Cambridge, MA: Harvard University Press.

Bruner, J. (1986). *Actual minds, possible worlds.* Cambridge, MA: Harvard University Press.

Bruner, J. S. (1961). *The act of discovery. Harvard Educational Review, 31(1), 21–32.*

Brusilovsky, P. (2001). Adaptive hypermedia systems. *User Modeling and User-Adapted Interaction, 11*(1-2), 87-110.

Brusilovsky, P., & Rizzo, R. (2002). Map-Based Access to Multiple Educational On-Line Resources from Mobile Wireless Devices. In *Proceedings of the 4th international Symposium on Mobile Human-Computer interaction,* (pp. 404-408), Pisa, Italy

Brusilovsky, P., Chavan, G., & Farzan, R. (2004, August 23-26) Social Adaptive Navigation Support for Open Corpus Electronic Textbooks. In: W. Nejdl, & P. De Bra, (Eds.), *Proceedings of the Third International Conference on Adaptive Hypermedia and Adaptive Web-Based Systems,* (pp. 24-33), Eindhoven, the Netherlands.

Brusilovsky, P., Schwarz, E., & Weber, G. (1996). A tool for developing adaptive electronic textbooks on www., *The Webnet 96 Conference of the Web Society.* San Francisco, CA.

Brynat, J. (2003) *Map data on the move.* Retrieved November 20th 2007 http://www.ordnancesurvey.co.uk/oswebsite/business/sectors/wireless/news/articles/pdf/Map%20data%20on%20the%20move.pdf

Buchanan G, Jones M., Thimbleby H., Farrant S., & Pazzani M. (2001, May) Improving mobile Internet usability. In *Proceedings of the 10th International Conference on World Wide Web,* (pp. 673-680), Hong Kong: ACM Press.

Buehner, M., Krumm, S., & Pick, M. (2005). Reasoning = working memory ≠ attention. *Intelligence, 33,* 251-272.

Bull, S., & Reid, E. (2004). Individualised revision material for use on a handheld computer. In J. Attewell & C. Savill-Smith (Eds.), *Mobile learning anytime everywhere* (pp. 35-42). London: Learning and Skills Development Agency.

Bunce, D., & Sisa, L. (2002). Age differences in perceived workload across a short vigil. *Ergonomics,* 45(13), 949-960.

Cacace, F., Cinque, M., Crudele, M., Iannello, G., & Venditti, M. (2004). The impact of innovation in medical and nursing training: a Hospital Information System for Students (HISS) accessible through mobile devices. In J. Attewell & C. Savill-Smith (Eds.), *Mobile learning anytime everywhere* (pp. 47-52): Published by the Learning and Skills Development Agency.

Caillois, R. (2001). *Man, play and games.* Urbana: The University of Illinois Press.

Cantor, J., & Engle, R. W. (1993). Working-memory capacity as long-term memory activation: An individual-differences approach. *Journal of Experimental Psychology: Learning, Memory, and Cognition, 19*(5), 1101-1114.

Cao, Y., Tin, T., McGreal, R., Ally, M., & Coffey, S. (2006, July 3-6). *The Athabasca University mobile library project: increasing the boundaries of anytime and anywhere learning for students.* Paper presented at the International Conference on Wireless Communications and Mobile Computing (pp. 1289-1294). Vancouver, Canada.

Carlen, U., & Jobring, O. (2005). The rationale of online learning communities. *International Journal of Web Based Communities, 1*(3), 272-295.

Carlson, J.R., & Zmud, R.W. (1999). Channel expansion theory and the experiential nature of media richness perceptions. *Joural of the Academy of Management, 42*(2), 153-170.

Carnaghan, C., & Webb, A. (2005). Investigating the effects of group response systems on learning outcomes and satisfaction in accounting education. Retrieved 16 June, 2006, from http://www.learning.uwaterloo.ca/LIF/responsepad_june20051.pdf

Carroll, A., Barnes, S. J., & Scornavacca, E. (2005). Consumers Perceptions and Attitudes towards SMS Mobile Marketing in New Zealand. In *Proceedings of the Fourth International Conference on Mobile Business* (pp. 434-440). Sydney, Australia: IEEE Computer Society.

Carroll, J., & Carrithers, C. (1984). Training wheels in a user interface. *Communications of the ACM, 27*(8), 800-806.

Casey, E. (1979). *The fate of place: a philosophical history.* Berkeley, The University of California Press.

Cebeci, Z., & Tekdal, M. (2006). Using podcasts as audio learning objects. *Interdisciplinary Journal of Knowledge and Learning Objects, 2*, 47-57.

Celorrio, C., & Verdejo, M. F. (2007). Adapted Activity Deployment and Configuration in a Pervasive Learning Framework. In *Proceedings of the Workshop of Pervasive Learning 2007: Design Challenges and Requirements.* Toronto, Ontario, Canada.

Celorrio, C., Verdejo, M. F., & Barros, B. (2006). Un Modelo de Distribución de Repositorios para un Portal de Comunidades de Aprendizaje Colaborativo. *IE Comunicaciones, Revista Iberoamericana de Informática Educativa, 3*, 27-42.

Chakravorty, R., & Pratt, I. (2002, September 9-11). *WWW performance over GPRS.* Paper presented at the 4th International Workshop on Mobile and Wireless Communications Network, Stockholm, Sweden.

Chakravorty, R., Cartwright, J., & Pratt, I. (2002, November 17-21). *Practical experience with TCP over GPRS.* Paper presented at the IEEE Global Telecommunications Conference (GLOBECOM '02) (pp. 1678-1682). Taipei, Taiwan.

Chakravorty, R., Clark, A., & Pratt, I. (2003, May 5-8). *GPRSWeb: Optimizing the Web for GPRS Links.* Paper presented at the MobiSys 2003: The 1st International Conference on Mobile Systems, Applications, and Services, San Francisco.

Chan, A., Lee, M. J. W., & McLoughin, C. (2006). Everyone's learning with podcasting: A Charles Sturt University experience. *Proceedings of the 23rd Annual Ascilite Conference: Who's Learning? Whose technology?* Retrieved August 1, 2007, from http://www.ascilite.org.au/conferences/sydney06/proceeding/pdf_papers/p171.pdf

Chan, T., & Sharples, M. (2002). A Concept Mapping Tool for Pocket PC Computers. *In Proceedings of IEEE WMTE2002* (pp. 163–164), Vaxjo, Sweden.

Chang, C., & Sheu. J. (2002). Design and Implementation of Ad Hoc Classroom and eSchoolbag Systems for Ubiquitous Learning. In *Proceedings of IEEE WMTE2002* (pp. 8–14), Vaxjo, Sweden.

Chang, R., Chang, Y., & Paper, D. (2003). The effect of task uncertainty, decentralization and AIS characteristics on performance of AIS: An empirical case in Taiwan. *Information & Management,40*, 691-703.

Chang, S.F., Zhong, D., & Kumar, R. (2001, December). Real-Time Content-Based Adaptive Streaming of Sports Video. In *Proceedings of the IEEE Workshop on Content-Based Access to Video/Image Library,* (p.139), Hawaii, Los Alamitos.

Chatfield, C., Carmichael, D., Hexel, R., Kay, J., & Kummerfeld, B. (2005, November 21 - 25). Personalisation in intelligent environments: managing the information flow. In *Proceedings of the 19th Conference of the Computer-Human Interaction Special Interest Group (CHISIG) of Australia on Computer-Human interaction: Citizens online: Considerations For Today and the Future,* (pp. 1-10), Canberra, Australia.ACM International Conference Proceeding Series, (vol. 122).

Narrabundah, Australia: Computer-Human Interaction Special Interest Group (CHISIG) of Australia.

Chen, J., & Kinshuk. (2005). Mobile technology in educational services. *Journal of Educational Multimedia and Hypermedia, 14*(1), 91-109.

Chen, J., & Lee, S. (2003). An exploratory cognitive DSS for strategic decision making. *Decision Support Systems, 36*, 147-160.

Chen, Y. S., Kao, T. C., & Sheu, J. P. (2003). A mobile learning system for scaffolding bird watching learning. *Journal of Computer Assisted Learning, 19*( 3), 347-359.

Chen, Y., Kao, Y., Yu, G., & Sheu, J. (2004). A Mobile Butterfly-Watching Learning System for Supporting Independent Learning. In *Proceedings of IEEE WMTE2004* (pp. 11–18), Taoyuan, Taiwan.

Chen, Y.S., Kao, T.C., & Sheu J.P. (2003). A mobile learning system for scaffolding bird watching learning. *Journal of Computer Assisted Learning, 19*, 347-359.

Chen, Y.S., Kao, T.C., Sheu, J.P., & Chiang, C.Y. (2002). A mobile scaffolding-aid-based bird -watching learning system, Paper presented at *the IEEE International Workshop on Wireless and Mobile Technologies in Education* (WMTE'02) (pp. 15-22). Växjö, Sweden

Chenoweth, T., Dowling, K., & St. Louis, R. (2004). Convincing DSS users that complex models are worth the effort. *Decision Support Systems, 37*, 71-82.

Church, K., Smyth, B., Cotter, P., & Bradley, K. (2007). Mobile information access: A study of emerging search behavior on the mobile internet. *ACM Transactions on the Web, 1*(1), Article 4.

Churchill, D., & Churchill, N. (in press). Educational affordances of PDAs: A study of a teacher's exploration of this technology. *Computers & Education.*

Cisco. E-Learning. (2003). Retrieved 5 April 2003 from http://business.cisco.com/prod/tree.taf%3Fasset_id-44748&public_ view=true&kbns=1.html

Clark, J. (April 10, 2007). School bus becomes mobile classroom thanks to iPods. Retrieved August 7, 2007, from http://www.todaysthv.com/news/news.aspx?storyid=44254

Cohen, D., Herscovici, M., Petruschka, Y., Maarek, Y. S., & Soffer, A. (2002, May 07 - 11). Personalized pocket directories for mobile devices. In *Proceedings of the 11th international Conference on World Wide Web*, (pp. 627-638), Honolulu, Hawaii.

Colazzo, L., Molinari, A., Ronchetti, M., & Trifonova, A. (2003). Towards a Multi-Vendor Mobile Learning Management System. Paper presented at the World Conference on E-Learning.

Colella. (2000). Participatory simulations: Building collaborative understanding through immersive dynamic modeling. *Journal of the Learning Science, 9*(4), 471-500.

Colevins, H., Bond, D., & Clark, K. (2006). Nurses get a hand from handhelds. *Computers in Libraries, 26*(4), 6-48.

Collier, K. G. (1980). Peer-group learning in higher education: The development of higher-order skills. *Studies in Higher Education, 5*(1), 55-62.

Conati, C. (2002). Probabilistic assessment of user's emotions in educational games. *Applied Artificial Intelligence, 16*(7-8), 555-575.

Constant, D., Kiesler, S.B., & Sproull, L.S. (1996). The kindness of strangers: The usefulness of weak electronic ties for technical advice. *Organization Science, 7*(2), 119-135.

Conway, A. R., Cowan, N., Bunting, M. F., Therriault, D. J., & Minkoff, S. R. (2002). A latent variable analysis of working memory capacity, short-term memory capacity, processing speed, and general fluid intelligence. *Intelligence, 30*, 163-183.

Copley, J. (1992). The integration of teacher education and technology: A constructivist model. In D. Carey, R. Carey, D. Willis & J. Willis (Eds.), *Technology and teacher education annual – 1992* (pp. 617-622). Charlottesville, VA: Association for the Advancement of Computing in Education.

Corlett, D., Sharples, M., Bull, S., & Chan, T. (2005). Evaluation of a mobile learning organizer for university students. *Journal of Computer Assisted Learning, 21*, 162-170.

Cowley, E. (2004). Recognition confidence, recognition accuracy and choice. *Journal of Business Research, 57*, 641-646.

Cramton, C. (2001). The mutual knowledge problem and its consequences for dispersed collaboration. *Organization Science, 12,* 346-371.

Crane, D., Pascarello, E., & James, D. (2005). *Ajax in action.* Manning Publications Co.

Crawford, C. (1981). *The art of computer game design.* New York, McGraw Hill

Crook, C. (1994). *Computers and the collaborative experience of learning.* London: Routledge International Library of Psychology.

Crowe, A., & van 't Hooft, M. (2006). Technology and the prospective teacher: Exploring the use of the TI-83 handheld devices in social studies education. *Contemporary Issues in Technology and Teacher Education, 6*(1), 99-119.

Csikszentmihalyi, M. (1992). *Flow: The psychology of happiness.* London, Rider.

Csikszentmihalyi, M. (2000). *Das Flow-Erlebnis, jenseits von Angst und Langeweile - im Tun aufgehen.* Stuttgart: Klett-Cotta.

Cuartielles, D., Malmborg, L., & Schlaucher, P. (2003). Components for a Ubiquitous Learning Environment. In *Proceedings of HCII2003*, Crete, Greece.

Cui, Y., & Bull, S. (2005). Context and learner modelling for the mobile foreign language learner. *System, 33*(2), 353-367.

Curtis, D.D., & Lawson, M.J. (2001). Exploring collaborative on line learning. *Journal of Asynchronous Learning Networks*, 5(1), 21-34.

CyberTracker (2000). *CyberTracker.* Retrieved June 2007, from http://www.cybertracker.co.za/

Daft, R., & Lengel, R. (1986). Organizational information requirements, media richness and structural design. *Management Science, 32*(5), 554-571.

Danaher, P. A. (2006, November). The social control–social capital debate about distance and online teacher education: Critical reflections on two courses at the University of Southern Queensland, Australia. *Journal of Open Learning and Teacher Education, 1*(3), 81-93.

Danaher, P. A., Hickey, A., Brown, A., & Conway, J. M. (2007). Exploring elements for creating an online community of learners within a distance education course at the University of Southern Queensland. In R. J. Luppicini (Ed.), *Online learning communities* (pp. 219-240). Charlotte, NC: Information Age Publishing.

Daneman M., & Green, I. (1986). Individual differences in comprehending and producing words in context. *Journal of Verbal Learning and Verbal Behavior, 19*, 450–466.

Daneman, M., & Carpenter, P. A. (1980). Individual differences in working memory and reading. *Journal of Verbal Learning and Verbal Behavior, 19*, 450-466.

Davis, F. (1986). *A technology acceptance model for empirically testing new end-user information systems: theory and results.* Unpublished Doctoral dissertation, MIT.

Davis, F. (1989). Perceived usefulness, perceived ease of use, and user acceptance of information technology. *MIS Quarterly, 13*(3), 319-340.

Davis, S. (2003). Observations in classrooms using a network of handheld devices. *Journal of Computer Assisted Learning, 19*(3), 298-307.

Davis, S. M., (2002, August). Research to industry four years of observations in classrooms using a network of handheld devices. In *Proc. of IEEE International Workshop on Wireless and Mobile Technologies in Education (WMTE 2002)*, (pp. 31-38), Vaxjo, Sweden.

De Crom, E. P., & De Jager, A. (2005). The "ME"-learning experience: PDA technology and e-learning in ecotourism at the Tshwane University of Technology (TUT). In *Proceedings of the 4th World Conference in M-Learn 2005*. Cape Town, South Africa

De Croock, M., Van Merrienboer, J., & Paas, F. (1998). High versus low contextual interference in simulation-based training of troubleshooting skills: Effects on transfer performance and invested mental effort. *Computers in Human Behavior, 14*(2), 249-267.

De Jong, T. (2001). Knowledge construction and sharing with media based applications. *Journal of Educational Research, 21*(1).

Dede, C.J. (1996). The evolution of distance education: Emerging technologies and distributed learning. *American Journal of Distance Education, 10*(2), 4-36.

Deguchi, A, Yamaguchi, E., Inagaki, S., Sugimoto, M., & Kusunoki, F. (2006). CarettaKids: A System for Supporting Children's Face-to-Face Collaborative Learning by Integrating Personal and Shared Spaces. In *Proceedings of IDC2006* (pp. 45–48), Tampere, Finland.

Demirjian, A., & David, B. (1995). Learning medical and dental sciences through interactive multi-media. *Medinfo, 8*, 1705.

Denning, P. (2004, April 21-27). Building a culture of innovation. *Ubiquity, 5*(8), 1-13. Retrieved May 31, 2004, from http://www.acm.org/ubiquity/issues5.html

DeSanctis, G., & Gallupe, R.B. (1987). A foundation for the study of group decision support systems. *Management Science, 33*(5), 589-609.

Design-Based Research Collective. (2003). Design-based research: An emerging paradigm for educational inquiry. *Educational Researcher, 32*(1), 5-8.

Dey, A. K. (2001). Understanding and using context. *Personal and Ubiquitous Computing, 5*, 4-7.

Dias, E., Beinat, E., Rhin, C., & Scholten, H. (2004). Location aware ICT in addressing protected areas' goals. In *Research on computing Science 11, Special Edition on e-Environment*, 273–289.

Dimakopoulos, D.N., & Magoulas, G.D. (2006, October 5-8). A personalised mobile environment for lifelong learners. In *Proc. of IADIS International Conference on WWW/Internet 2006*, (pp. 31-38). Murcia, Spain.

Dimitrova, M., Sadler, C., Hatzipanagos, S., & Murphy, A. (2003). Addressing learner diversity by promoting flexibility in e-learning environments. In *Proceedings of the 14th international workshop on database and expert systems applications.*

Dimoco. *Zukunft für Mobile-Marketing. SMS-Werbung auf dem Vormarsch.* Retrieved March 16, 2007, from http,//www.report.at/artikel.asp?kid=&mid=1&aid=6005.

Diws. (2004). Building Smart Client using .NET. Retrieved 17 January, 2008, from http://www.codeproject.com/dotnet/DotNetBuildSmClnts.asp

Dix, A. J., Finlay, J. E., Abowd, G. D., & Beale, R. (1998). *Human-computer interaction* (2nd ed.). Hertfordshire: Prentice Hall Europe.

Dix, A., Finlay, J., Abowd, G., & Beale, R. (2005). *Human-computer interaction* (3rd ed.). London: Prentice Hall.

Dix, A., J. Finlay, et al. (2004). *Human-computer interaction.* Harlow, Pearson Education Limited.

Doolittle, P. (2007). *The Effect of Working Memory Capacity on Multimedia Learning: Does Attentional Control Result in Improved Performance?* Manuscript submitted for publication.

Döring, S. (1997). *Lernen durch Spielen - Spielpädagogische Perspektiven institutionellen Lernens.* Weinheim:Beltz.

Doswell, J. (2006). Context-aware mobile augmented reality architecture for lifelong learning. Paper presented at the sixth International Conference on

Advanced Learning Technologies, Kerkrade, The Netherlands.

Dourish, P. (2004). What we talk about when we talk about context. *Journal of Personal & Ubiquitous Computing, 8*(1), 19-30.

Draper, S. W. (1998). Niche-based success in CAL. *Computers & Education, 30*(1-2), pp. 5-8.

Draper, S. W., & Brown, M. I. (2004). Increasing interactivity in lectures using an electronic voting system. *Journal of Computer Assisted Learning 20*, 81-94.

Drengner, J., & Zanger, C. (2003). Die Eignung des Flow-Ansatzes zur Wirkungsanalyse von Marketing-Events, *Marketing ZFP, 25*(1), 25-40.

Druin, A. (1998). *The design of children's technology.* San Francisco, CA: Morgan Kaufmann Publishers.

Dufresne, R., Geracce, W., Leonard, W., Mestre, J., & Wenk, L. (1996). Classtalk: A classroom communications system for active learning. *Journal of Computing in Higher Education, 7*, 3–47.

Duke University. (2005). iPod first year experience final evaluation report. Retrieved August 1, 2007, from http://cit.duke.edu/pdf/reports/ipod_initiative_04_05.pdf

Duke University. (2006). Duke Digital Initiative end of year report. Retrieved August 10, 2007, from http://www.duke.edu/ddi/pdf/ddi_exec_report_05_06.pdf

Duncan, D. (2005). *Clickers in the classroom: How to enhance science teaching using classroom response systems.* San Francisco: Pearson.

Edelmann, W. (2000). *Lernpsychologie.* Weinheim: Psychologie Verlags Union.

Edelson D. (2002). Design research: What we learn when we engage in design. *The Journal of the Learning Sciences, 11*(1), 105-121.

Edelstein, W., & de Haan, G. (2004). Empfehlung 5, Lernkonzepte für eine zukunftsfähige Schule - von Schlüsselkompetenzen zum Curriculum. In Heinrich-Böll-Stiftung und Bildungskommission der Heinrich-Böll-Stiftung (Ed.), *Selbstständig lernen. Bildung stärkt Zivilgesellschaft* (pp. 130-188). Beltz: Weinheim.

Edirisingha, P., Rizzi, C., Nie, M., & Rothwell, L. (2007). Podcasting to provide teaching and learning support for an undergraduate module on English language and communication. *Turkish Online Journal of Distance Education, 8*(3). Retrieved July 22, 2007, from http://tojde.anadolu.edu.tr/tojde27/articles/article_6.htm

Edirisingha, P., Salmon, G., & Fothergill, J. (2007). Profcasting – A pilot study and guidelines for integrating podcasts in a blended learning environment. In U. Bernath & A. Sangrà (Eds.) *Research on competence development in online distance education and e-learning.* Oldenburg: BIS-Verlag. Retrieved July 20, 2007, from http://hdl.handle.net/2381/404

Effects of portable music-player use on driver performance. *Proceedings of the SIGCHI Conference on Human Factors in Computer Systems, 1*, 243-250.

Einsiedler, W. (1999). *Das Spiel der Kinder.* Bad Heilbrunn; Klinkhardt.

Elbel, G., & Laake, A. (2002). *Handlungsorientierter Unterricht . Qualifikationen für den Arbeitsmarkt – Qualifikationen für das Leben.* Göttingen: Institut für berufliche Bildung und Weiterbildung.

El-Bishouty, M., Ogata, H., &Yano, Y. (2006). Personalized knowledge awareness map in computer supported ubiquitous learning, *Paper presented at the International Conference on Advanced Learning Technologies (ICALT2006)*, Amsterdam, The Netherlands.

Elgort, I., Marshall, S., & Mitchell, G. (2003). *NESB student perceptions and attitudes to a new online learning environment.*

Engestrom, Y. (1987). *Learning by expanding.* Helsinki: Orienta-Konsultit.

Engeström, Y. (1999). Activity theory and individual and social transformation. In Y. Engestrom & R.

Miettinen & R.-L. Punamaki (Eds.), *Perspectives on activity theory* (pp. 19-38). New York: Cambridge University Press.

Enjoji, H., Ravasio, P., & Sugimoto, M. (2006). A Collaboration Support Tool for Cartography Learning in the Open Air. In *Proceedings of CollabTech2006* (pp.90–95), Tsukuba, Japan.

Ergosoft Laboratories. (2003). What is a within-subjects design? Retrieved January 3, 2008, from http://www.ergolabs.com/within_subjects.htm

Erikson, E.H. (1963). *Childhood and society* (2nd ed). New York: Norton.

Eschenbrenner, B., & Nah, F. F. (2007). Mobile technology in education: Uses and benefits. *International Journal of Mobile Learning and Organization, 1*(2), 159-183.

Euler, D. (2005). Didaktische Gestaltung von E-Learning-unterstützten Lernumgebungen. In Euler, D. & Seufert, C. (Eds): *E-learning in Hochschulen und Bildungszentren - Gestaltungshinweise für pädagogische Innovationen* (pp. 227–242). München: Oldenburg.

Facer, K., Joiner, R., Stanton, D., Reid, J., Hull, R., & Kirk, D. (2004). Savannah: mobile gaming and learning, *Journal of Computer Assisted Learning, 20*, 399-409.

Facer, K., Joiner, R., Stanton, D., Reid, J., Hull, R., & Kirk, D. (2004). Savannah: mobile gaming and learning? *Journal of Computer Assisted Learning, 20*(6), 399-409.

Fallahkhair, S., Pemberton, L., & Griffiths, R. (2007). Development of a cross-platform ubiquitous language learning service via mobile phone and interactive television. *Journal of Computer Assisted Learning, 23*(4), 312-325.

Faux, F., McFarlane A., Roche, N., & Facer, K. (2006). *Learning with handheld technologies.* Futurelab Handbook. December 2006. Retrieved July 18, 2007, from http://www.futurelab.org.uk/resources/publications_reports_articles/handbooks/Handbook198

Faux, F., McFarlane, A., Roche J., & Facer, K. (2006). *Learning with handheld technologies.* Bristol: Futurelab.

Feierabend, P., & Rathgeb, T. (2006). *JIM – Studie. Jugend, Information, (Multi-) Medi.* Stuttgart: Medienpädagogischer Forschungsverband Südwest.

Feldman Barrett, L., Tugade, M. M., & Engle, R. A. (2004). Individual differences n working memory capacity and dual-process theories of the mind. *Psychological Bulletin, 130*(4), 553-573.

Feldman, S. (2003). *Building scaffolds in your classroom.* Teaching Pre K-8, pp. 6.

Flannigan, J. C. (1954). The critical incident technique. *Psychological Bulletin, 51*(4), 327-358.

Flannigan, J. C. (1962). *Measuring human performance.* Pittsburgh, PA: American Institutes for Research.

Fleiss, L.F., Levin, B., & Myunghee, C.P. (2003). *Statistical methods for rates and proportions* (3rd ed.). New Jersey: John Wiley & Sons, Inc.

Flintham, M., Anastasi, R., Benford, S.D., Hemmings, T., Crabtree, A., Greenhalgh, C.M. et al. (2003). Where on-line meets on-the-streets: Experiences with mobile mixed reality games. *Paper presented at the Conference on Human Factors in Computing Systems CHI 2003,* Fort. Lauderdale, FL.

Flitner, A.(2004). *Spielen - Lernen. Praxis und Deutung des Kinderspiels.* München: Beltz.

Forum Mobilkommunikation. (2007). *Klares Votum für das Handy.* Retrieved March 16, 2007, from http,//www.fmk.at/mobilkom/dl/FMK_4-6.pdf.

Freeman, M., & Blayney, P. (2005). Promoting interactive in-class learning environments: A comparison of an electronic response system with a traditional alternative. *Proceedings of the 11th Australasian Teaching Economics Conference.*

Gaeta, M., Ritrovato, P., & Salerno, S. (2005). Making e-learning a service oriented utility: the European learning grid infrastructure project. In P. Ritrovato, C.Allison, S.A. Cerri, T. Dimitrakos, M. Gaeta & S. Salerno (Eds.), *Towards the learning grid: advances in human learning services, frontiers in artificial intelligence and applications,* (vol. 127) pp. 63-76.

Galegher, J., & Kraut, R.E. (1992). Computer-mediated communication and collaborative writing: Media influence and adaptation to communication constraints. *Proceedings of the 1992 ACM Conference on Computer-Supported Cooperative Work.* Toronto, Ontario, Canada. Pp. 155-162. Retrieved 16 November 2004 from http://delivery.acm.org/10.1145/150000/143474/p155-galegher.pdf?key1=143474&key2=6761560011&coll=GUIDE&dl=ACM&CFID=31360932&CFTOKEN=55657576,.

Garrett, J. J. (2005). Ajax: A New Approach to Web Applications. Retrieved January 3, 2008, from http://adaptivepath.com/publications/essays/archives/000385.php

Garris, R., Ahlers, R., & Driskell, J. E. (2002). Games, motivation, and learning: A research and practice model. *Simulation & Gaming, 33*, 441 - 467.

Gaultney, J. F., Kipp, K., & Kirk, G. (2005). Utilization deficiency and working memory capacity in adult memory performance: Not just for children anymore. *Cognitive Development, 20,* 205-213.

Gay, G., & Hembrooke, H. (2004). *Activity-centered design: An ecological approach to designing smart tools and usable systems.* Cambridge, Mass. ; London: MIT Press.

Gay, G., Stefanone, M., Grace-Martin, M., & Hembrooke, H. (2001). The effects of wireless computing in collaborative learning environments. *International Journal of Human–Computer Interactions, 13*(2), 257-276.

Gee, J. P. (2003). *What video games have to teach us about learning and literacy.* New York: Palgrave.

Gee, J. P. (2003). *What video games have to teach us about learning and literacy.* New York: Palgrave/Macmillan.

Georgia College and State University. (2006). The GCSU iPod story. Retrieved August 7, 2007, from http://ipod.gcsu.edu/GCSU%20iPod%20Story/index.html

Gerhard, F., & Masanori, S. (2006). Supporting self-directed learners and learning communities with socio technical environments. *Research and Practice in Technology Enhanced Leaning, 1*(1), 31-64.

Gibson, J.J. (1977). The theory of affordances. In R. Shaw & J. Bransford (Eds.), *Perceiving, acting and knowing.* Hillsdale, NJ: Erlbaum.

Gilbert, A. (2005). New for back-to-school: 'Clickers.' CNet news.com.

Ginns, P. (2005). Meta-analysis of the modality effect. *Learning and Instruction, 15*, 313-331.

Glaser, R., & Bassok, M. (1989). Learning theory and the study of instruction. *Annals of Psychology, 40*, 631-666.

Glu Mobile. (2007). *Challenges Mobile Myths With New Consumer Survey.* Retrieved March 13, 2007, from http,//www.glu.com/about/presp.php?press=100106b

Godwin-Jones, R. (2007). Emerging technologies: E-texts, mobile browsing, and rich internet applications. *Language Learning & Technology, 11*(3), 8-13.

Goggin, G. (2005, March). Mobile learning. *Professional Educator, 4*(1), 35-37.

Goh, T., & Kinshuk. (2006). Getting ready for mobile learning – adaptation perspective. *Journal of Educational Multimedia and Hypermedia, 15*(2), 175-198.

Goldman, P., & Kaufman, B. J. (2001). How to push an elephant through a straw: Using wireless technology in a Web-enhanced skills program. *International Review of Law, Computers & Technology, 15*(3), 281-299.

Goode, J. (2003). Are pilots at risk of accidents due to fatigue? *Journal of Safety Research, 34*, 309-313.

Goodison, T. A. (2001). The implementation of e-learning in UK higher education. In *Proceedings of world conference on educational media, hypermedia and telecommunications 2001* (pp. 613-618). Chesapeake, VA: AACE.

Gopal, R. D., & Tripathi, A. K. (2006). Advertising via wireless networks. *International Journal of Mobile Communications, 4*(1), 1 - 16.

Gosh, D., & Ray, M. (1997). Risk, ambiguity, and decision choice: Some additional evidence. *Decision Sciences, 28*(1), 81-104.

Gräsel, C. (1997). Wir können auch anders: Problemorientiertes Lernen an der Hochschule. In Gruber, H. & Renkl, A. (Eds), *Wege zum Können: Determinanten des Kompentenzerwerbs.* Bern: Verlag Hans Huber.

Grasso, A., & Roselli, T. (2005, November). Guidelines for Designing and Developing Contents for Mobile Learning. *3rd IEEE International Workshop on Wireless and Mobile Technologies in Education* (WMTE'05), (pp. 28-30). Tokushima, Japan.

Gredler, M. E. (2004). Games and simulations and their relationships to learning. In D. H. Jonassen (Ed.), *Handbook of research on educational communications and technology* (2nd ed., pp. 571-581). Mahwah, N.J.: Lawrence Erlbaum.

Green, P. D. (2001). Handheld computers as tools for writing and managing field data. *Field Methods, 13*(2), 181-197.

Greeno, J. G. (1997). On claims that answer the wrong questions. *Educational Researcher, 26*(1), 5-17.

Greer, L., & Heaney, P. J. (2004). Real-time analysis of student comprehension: An assessment of electronic student response technology in an introductory earth science course. *Journal of Geoscience Education, 52*, 345-351.

Grohmann, G., Hofer, A., & Martin, G. (2005, July 11). *ARIS MOBILE: Helping to define the future of mobile learning.* Paper presented at the 4th international conference on mobile business, University of Technology Sydney, Sydney, NSW.

Gururajan, R., Kerr, D., Moloney, C., & Houghton, L. (2005). An investigation into the factors of adoption of wireless applications for data management by nurses. In L. Eldon (Ed.), *Collaborative decision making in the internet era* (pp. 1-5). Asia-Pacific Decision Sciences Institute.

Hackemer, K., & Peterson, D. (2005). University of South Dakota palm initiative. In A. Kukulska-Hulme & J. Traxler (Eds.), *Mobile learning: A handbook for educators and trainers* (pp. 157-163). London: Routledge.

Hafeez-Baig, A., & Danaher, P. A. (2007, February). Future possibilities for mobile learning technologies and applications at the University of Southern Queensland, Australia: Lessons from an academic focus group. In D. Parsons & H. Ryu (Eds.), *Mobile learning technologies and applications (MoLTA) 2007: Proceedings of the conference on mobile learning technologies and applications (MoLTA) Massey University, Auckland, New Zealand February 19th 2007* (pp. 16-22). Auckland, New Zealand: Institute of Information and Mathematical Sciences, Massey University.

Haghirian, P., Madlberger, M., & Tanuskova, A. (2005). *Increasing advertising value of mobile marketing - An empirical study of antecedents.* Paper presented at the 38th Hawaii International Conference on System Sciences, Big Island, Hawaii.

Hake, R. (1998). Interactive-engagement vs. traditional methods: A six-thousand-student survey of mechanics test data for introductory physics courses. *American Journal of Physics, 66*, 64-74.

Hall, R. H., Collier, H. L., Thomas, M. L., & Hilgers, M. G. (2005). A Student Response System for Increasing Engagement, Motivation, and Learning in High Enrolment Lectures. *Proceedings of the Eleventh Americas Conference on Information Systems, Omaha, NE, USA August 11-14, 2005.*

Hancock, J.T., Thom-Santelli, J., & Ritchie, T. (2004). Deception and design: The impact of communication technology on lying behavior. *Proceedings of*

*CHI2004.* Vienna, Austria: The Association for Computing Machinery, 24-29 April. Retrieved 16 November 2004 from http://delivery.acm.org/10.1145/990000/985709/p129-hancock.pdf?key1=985709&key2=11205600&coll=GUIDE&dl=ACM&CFID=31360932&CFTOKEN=55657576,

Harasim, L., Hiltz, S.R., Teles, L., & Turoff, M. (1995). *Learning networks: A field guide to teaching and learning on line.* Cambridge, MA: The MIT Press.

Harreveld, R. E., & Danaher, P. A. (2004, November). Innovation in open and distance learning and teacher education: The case of pre-service secondary vocational education and training at an Australian regional university. *Journal of Open Learning and Teacher Education, 1*(1), 1-13.

Haythornthwaite, C., Kazmer, M.M., & Robins, J. (2000). Community development among distance learners: Temporal and technological dimensions. *Journal of Computer-Mediated Communication,* 6(1), September. Retrieved 10 April 2003 from http://www.ascusc.org/jcmc/vol6/issue1/ hathornthwaite.html

Henricksen, K., & Indulska, J. (2006). Developing contextaware pervasive computing applications: Models and approach. *Pervasive and Mobile Computing, 2*(2), 37-64.

Hevner, A., March, S., Park. J., & Ram, S. (2004). Design science in information systems research. *MIS Quarterly, 28*(1), 75-105.

Hignite, K. (2007). The word on generational diversity. *HR Horizons, 2*(3), July 2007.

Hiltz, S. R., Coppola, N., Rotter, N., & Turoff, M. (2000, September). Measuring the importance of collaborative learning for the effectiveness of ALN: A multi-measure, multi-method approach. *Journal of Asynchronous Learning Networks, 4*(2). Retrieved June 25, 2006, from http://www.sloan-c.org/publications/jaln/v4n2/v4n2_hiltz.asp

Hiltz, S.R., & Wellman, B. (1997). Asynchronous learning networks as a virtual classroom. *Communications of the ACM, 40*(9), 44-49.

Hirschfelder, D. (2006). Der „Nürnberger Trichter" – Ein Allheilmittel gegen die Dummheit?.In Großmann, U. (Ed), *KulturGUT - Aus der Forschung des Germanischen Nationalmuseums* (pp. 3-5)

Hoadley, C. (2004). Methodological alignment in design-based research. *Educational Psychologist, 39*(4), 203-212.

Hodas, S. (1993, September 14). Technology refusal and the organizational culture of schools. *Education Policy Analysis Archives, 1*(10). Retrieved April 9, 2005, from http://epaa.asu.edu/epaav1n10.html

Hogan, K., & Pressley, M. (Eds.). (1997). *Scaffolding student learning: instructional approaches and issues.* Cambridge, MA: Brookline Books.

Holtzblatt, K. (2005). Customer-centered design for mobile applications. *Personal Ubiquitous Computer, 9*(4), 227-237.

Holzinger, A., Nischelwitzer, A., & Meisenberger, M. (2005). Lifelong-learning support by m-learning: example scenarios. *eLearn 2005, 11*(Nov. 2005), 2.

Hoppe, H.U., Pinkwart, N., Oelinger, M., Zeini, S., Verdejo, F., Barros, B. et al. (2005). Building Bridges within Learning Communities through Ontologies and Thematic Objects. Proceedings of the *International Conference on Computer Supported Collaborative Learning* (CSCL2005). Taiwan.

Hoppe, U., Lingnau, A., Machado, I., Paiva, A., Prada, R., & Tewissen, F. (2000). Supporting collaborative activities in computer integrated classrooms – the NIMIS approach. In P. Antunes & C. Salgado (Eds.), *Proceedings of CRIWG 2000* (pp. 94-101). Los Alamitos, CA: IEEE Computer Society Publication.

Hoppe, U., Verdejo, M.F., Pinkwart, N., Barros, B., Oelinger, M, Mayorga, J.I., & Zeini, S. (2005). Building bridges within learning communities through ontologies and "thematic bjects." Computer Supported Collaborative Learning, 2005

Horton, B., & Wiegert, E. (2002). Using handhelds and billiards to teach reflection. *Learning & Leading With Technology, 29*(8), 32-35.

Horwood, A. (2003, 01.03.2003). Txt 2 help u quit smoking. New Zealand Herald.

Howard, A., Hafeez-Baig, A., Howard, S., & Gururajan, R. (2006). A framework for the adoption of wireless technology in healthcare: An Indian study. In S. Spencer & A. Jenkins (Eds.), *Proceedings of the 17th Australasian conference on information systems (ACIS 2006).* Australasian Association for Information Systems.

Hürst, W., Lauer, T., & Nold, E. (2007). A study of algorithm animations on mobile devices. *Proceedings of the ACM Technical Symposium on Computer Science Education* (SIGCSE), *39*(1), 160-164.

IMS Tools Interoperability Guidelines. (2006). Retrieved July 18, 2007, from http://www.imsglobal.org/ti

InfoSynch World. (2004). 1.87 billion mobile users by 2007. Retrieved 2006, from http://www.infosyncworld.com/news/n/5048.html

Ishizuka, T., Horita, T., Takada, K., Ishihara, K., Ogawa, M., Moriya, K. et al. (2004). Development and Practical Study on Mobile Learning Environment using Personal Digital Assistants for Elementary School Children. In *Proceedings of ICCE 2004* (pp. 969–981), Melbourne, Australia.

IVES Training (2007). What's Wrong With This? Retrieved 17 January, 2008, from http://www.ivestraining.com/forklift_training_aids/training-videos.php

Izard, C. E. (1977). *Human emotions.* New York: Plenum Press.

Jackson, S.L., Stratford, S.J., Krajcik, J., & Soloway, E. (1996). *Model-It: A Case Study of Learner-centered Software Design for Supporting Model Building.* Report No. MVK50700. Michigan University, Ann Arbor, MI. (ERIC Document Reproduction Service No. ED 446903)

Japanese Ministry of Education, Culture, Sports, Science and Technology (2006a). Retrieved September 2006, from http://www.mext.go.jp/b_menu/houdou/17/12/05120900.htm (in Japanese)

Japanese Ministry of Education, Culture, Sports, Science and Technology (2006b). Retrieved September 2006, from http://www.mext.go.jp/b_menu/houdou/17/12/05120900/007/{005,020,021,024}(in Japanese)

Jay, E. F. (2007). iPods viable for military education. *Military Advanced Education Online Edition.* Retrieved August 7, 2007, from http://www.military-advanced-education.com/print_article.cfm?DocID=2043

Jegers, K. (2006). *Pervasive GameFlow, Understanding Player enjoyment in Pervasive Gaming.* Paper presented at the 3rd International Workshop on Pervasive Gaming Applications.

Jegers, K., & Wiberg, M. (2006). Pervasive gaming in the everyday world. *Pervasive Computing, 5*(1), 78-85.

Johanson, M., & Torlind, P. (2004). Mobility support for distributed collaborative teamwork. *Proceedings of the 2004 Information Technology Conference (ITCON),* August. Retrieved 23 January 2006 from http://www.itcon.org/data/works/att/2004_25.content.00610.pdf

Johnson, D.W., Johnson, R.T., & Smith, K. (1991). Creative learning: Increasing college faculty productivity. Washington, DC: ASHE-ERIC Higher Education Report (4), Clearinghouse on Higher Education. The George Washington University.

Johnston, R. (1999). Beyond flexibility: Issues and implications for higher education. *Higher Education Review, 32*, 55-67.

Jonassen, D. H., Howland, J., Moore, J., & Marra, R. M. (2003). *Learning to solve problems with technology: a constructivist perspective* (2nd ed.). Upper Saddle River, N.J.: Merrill Prentice Hall.

Jonassen, D. H., Peck, K. L., & Wilson, B. G. (1999). *Learning with technology: A constructivist perspective.* Upper Saddle River, NJ: Merrill.

Jonassen, D., Peck, K., & Wilson, B. (2000). *Learning with technology: A constructivist approach.* Upper Saddle River, NJ: Prentice Hall.

Jones, A., Blake, C., Davies, C., & Scanlon, E. (2004). Digital maps for learning: review and prospects. *Computers and Education, 43*, 91–107.

Judson, E., & Sawada, D. (2002). Learning from past and present: Electronic response systems in college lecture halls. *Journal of Computers in Mathematics and Science Teaching 21*, 167-181.

Just, M., & Carpenter, P. A. (1992). A capacity theory of comprehension: Individual differences in working memory. *Psychological Review, 99*, 122-149.

Kadirire, J. (2007). Instant messaging for creating interactive and ollaborative m-learning environments. *The International Review of Research in Open and Distance Learning, 8*(2) 1-14.

Kadyte, V. (2004). Learning can happen anywhere: a mobile system for language learning. In J. Attewell & C. E. Savill-Smith (Eds.), *Learning with Mobile Devices: Research and Development* (pp. 73-78). London: Learning and Skills Development Agency.

Kane, M. J., & Engle, R. W. (2000). Working-memory capacity, proactive interference, and divided attention: Limits on long-term memory retrieval. *Journal of Experimental Psychology: Learning , Memory, and Cognition, 26*(2), 336-358.

Kane, M. J., & Engle, R. W., (2003). Working-memory capacity and the control of attention: The contributions of goal neglect, response competition, and task set to Stroop interference. *Journal of Experimental Psychology: General, 132*(1), 47-70.

Kane, M. J., Bleckley, M. K., Conway, A. R., & Engle, R. W., (2001). A controlled-attention view of working memory capacity. *Journal of Experimental Psychology: General, 130*(2), 169-183.

Kane, M. J., Hambrick, D. Z., Tuhoski, S. W., Wilhelm, O., Payne, T. W., & Engle, R. W. (2004). The generality of working memory capacity: A latent-variable approach to verbal and visuospatial memory span and reasoning, *Journal of Experimental Psychology: General, 133*(2), 189-217.

Keegan D. (2001). *The future of learning: From eLearning to m Learning.* Retrieved July 31, 2007, from http://learning.ericsson.net/mlearning2/project_one/book.html

Keegan, D. (2002). *The future of learning: From eLearning to mLearning*: Hagen: Fern Universitat, Gesamthochschule, Germany.

Kehrwald, B. A. (2006, June). *Inhabiting online spaces: Reconsidering social presence in contemporary online learning environments.* Paper presented at the 4th international lifelong learning conference, Yeppoon, Qld.

Ketamo, H. (2003). xTask- an adaptable learning environment. *Journal of Computer Assisted Learning, 19*(3), 360-370.

Kiesler, S., & Sproull, L. (1992). Group decision making and communication technology. *Organization Behavior and Human Decision Processes, 52*, 96-123.

Kiewra, K. A., & Benton, S. L. (1988). The relationship between information-processing ability and notetaking. *Contemporary Educational Psychology, 13,* 33-44.

Kim, K.J., & Bonk, C.J. (2002). Cross-cultural comparisons in on-line collaboration. *Journal of Computer-Mediated Communication,* 8(1) http://www.ascusc.org/jcmc/vol8/issue1/ kim.html

Kirriemuir, J., & McFarlane, A. (2006). *Literature review in games and learning.* London: Futurelab.

Klopfer, E., Squire, K., & Jenkins, H. (2002). Environmental detectives: PDAs as a window into a virtual simulated world. *In the proceedings of the International Workshop on Wireless and Mobile Technologies in Education.* Växjö, Sweden, (pp. 95-98).

Klopfer, E., Yoon, S., & Perry, J. (2005). Using palm technology in participatory simulations of complex systems: A new take on ubiquitous and accessible mobile computing. *Journal of Science Education & Technology, 14*(3), 285-297.

Kopf, S., Scheele, N., Winschel, L., & Effelsberg, W. (2005). *Improving Activity and Motivation of Students with Innovative Teaching and Learning Technologies*, Methods and Technologies for Learning, pp. 551-556.

Koschembahr, C. (2005a). Optimizing Your Sales Workforce through Mobile Learning. Learning Circuits. Retrieved 17 January 2008 from http://www.learningcircuits.org/2005/apr2005/vonKoschembahr.htm

Koschembahr, C. (2005b). Mobile Learning: The Next Evolution in Education. Retrieved 17 January 2008 from http://www.clomedia.com/content/templates/clo_article.asp?articleid=849&zoneid=71

Koschembahr, C., & Sagrott S. (2005). The future of learning at IBM: Empowering employees through mobile learning. In A. Kukulska-Hulme & J. Traxler (Eds.), *Mobile learning: A handbook for educators and trainers* (pp. 164-173). Routledge.

Koschmann, T. (1996). *Paradigm shifts and instructional technology: an introduction.* In CSCL: Theory and practice, (pp. 1-23). Lawrence: Erlbaum Associates.

Krikos, V., Stamou, S., Kokosis, P., Ntoulas, A., & Christodoulakis, D. (2005). DirectoryRank: ordering pages in Web directories. In *Proceedings of the 7th Annual ACM international Workshop on Web information and Data Management*, (pp. 17-22), Bremen, Germany.

Kristensen, B., Laursen, B., Pilegaard, M., & Jensen, B. (2004). Physical workload during use of speech recognition and traditional computer input devices. *Ergonomics*, *47*(2), 119-133.

Kroeber-Riel, W., & Weinberg, P. (2003). *Konsumentenverhalten.* München: Vahlen.

Kukka, H., & Ojala, T. (2006, October 25–27). *mobileDOK: Culture in your pocket.* Paper presented at the 3rd international conference on mobile technology, applications & systems (Mobility 06), Bangkok, Thailand .

Kukulska-Hulme A. (2002, June). Cognitive, Ergonomic and Affective Aspects of PDA Use for Learning. In *Proceedings of the European Workshop on Mobile and Contextual Learning*, (pp. 32-33), Birmingham, UK.

Kukulska-Hulme, A. (2007). Hooked into learning: Designing for the Unexpected. In *W. o. P. L.* 2007 (Ed.). Toronto, Canada.

Kukulska-Hulme, A., & Traxler, J. (2005). Mobile teaching and learning. In A. Kukulska-Hulme & J. Traxler (Eds.), *Mobile learning: A handbook for educators and trainers* (pp. 25-44). Routledge.

Kukulska-Hulme, A., & Traxler, J. (Eds.). (2005). *Mobile learning: A handbook for educators and trainers.* London: Routledge.

Kules, B., Kustanowitz, J., & Shneiderman, B. (2006). Categorizing Web search results into meaningful and stable categories using fast-feature techniques. In *Proceedings of the 6th ACM/IEEE-CS Joint Conference on Digital Libraries*, (pp. 210-219). Chapel Hill, USA.

Kumar A. (2005). How Mobile Web Applications Work. *The Code Project.* Retrieved 17 January 2008 from http://www.codeproject.com/dotnet/mobilewebappsecret.asp

Kuo, F., Chu, T., Hsu, M., & Hsieh, H. (2004). An investigation of effort-accuracy trade-off and the impact of self-efficacy on web searching behaviors. *Decision Support Systems*, *37*, 331-342.

Kurti, A., Milrad, M., & Spikol, D. (2007, July 18-20). Designing Innovative Learning Activities Using Ubiquitous Computing. *Proceedings of the 7th IEEE International Conference on Advanced Learning Technologies* on, 2007 in Niigata, Japan, (pp. 386-390).

Kurti, A., Spikol, D., & Milrad, M. (2007). Bridging outdoors and indoors educational activities in schools with the support of mobile and positioning technologies. *International Journal of Mobile Learning and Organization.*

Kurti, A., Spikol, D., & Milrad, M. (2008). Bridging Outdoors and Indoors Educational Activities in Schools with the Support of Mobile and Positioning Technologies. Accepted for publication at the *International Journal of Mobile Learning and Organization,2*(2), 166-168.

Kurti, A., Spikol, D., Milrad, M., & Svensson, M. (2007). *Exploring how pervasive computing can support situated learning, Pervasive Learning.* Toronto, Canada: Centre for Mobile Computing, Massey University.

Kyllonen, P. C., & Christal, R. E. (1990). Reasoning ability is (little more than) working-memory capacity?! *Intelligence, 14*, 389-433.

La Pointe, L. B., & Engle, R. W. (1990). Simple and complex word spans as measures of working memory capacity. *Journal ofExperimental Psychology: Learning, Memory and Cognition, 16,* 1118-1133.

Lai, C.-H., Yang, J.-C., Chen, F.-C., Ho, C.-W., & Chan, T.-W. (2007). Affordances of mobile technologies for experiential learning: the interplay of technology and pedagogical practices. *Journal of Computer Assisted Learning, 23*(4), 326-377.

Lai, C.-Y., & Wu, C.-C. (2006). Using handhelds in a Jigsaw cooperative learning environment. *Journal of Computer Assisted Learning, 22*(4), 284-297.

Laine, T. H., & Suhonen, J. (2008). Establishing a mobile blog system in a distance education environment. *International Journal of Mobile Learning and Organisation, 2*(2), 149-165.

Lankshear, C., & Knobel, M. (2006). *New literacies: Everyday practices and classroom learning.* Berkshire, UK: Open University Press.

Lankshear, C., Green, B., & Snyder, I. (2000). *Teachers and technoliteracy : managing literacy, technology and learning in schools.* St Leonards, N.S.W.: Allen & Unwin.

Laouris, Y., & Eteokleous, N. (2005). *We need an educationally relevant definition of mobile learning.* Paper presented at the mLearn2005: 4th World Conference on mLearning, Cape Town, South Africa.

Larrañaga, M., Rueda, U., Elorriaga, J.A., & Arruarte A. (2002). Using CM-ED for the Generation of Graphical Exercises Based on Concept Maps. In Kinshuk, R. Lewis, K. Akahori, R. Kemp, T. Okamoto, L. Henderson & C.-H. Lee, (Eds.), *Proceedings of ICCE'2002*, (pp. 173-177).

Laurillard, D. (2002). *Rethinking university teaching: A conversational framework for the effective use of learning technologies* (2nd ed.). London, U.K.: Routledge Falmer.

Laurillard, D. (2002). *Rethinking university teaching: A conversational framework for the effective use of learning technologies* (2nd ed.). London: Routledge.

Laurillard, D., Stratfold, M., Luckin, R., Plowman, L., & Taylor, J. (2000). Affordances for learning in a non-linear narrative medium. *Journal of Interactive Media in Education, 2.*

Lave J., & Wenger E. (1991). *Situated learning: Legitimate peripheral participation,* Cambridge, UK: University of Cambridge Press.

Lave, J. (1988). *Cognition in practice.* Cambridge: Cambridge University Press.

Lave, J., & Wenger, E. (1991). *Situated learning: Legitimate peripheral participation.* Cambridge: University of Cambridge Press.

Leah, A. R. (2005). Who are the millennials? *Deloitte Consulting.* Retrieved 17 January, 2008, from http://www.deloitte.com/dtt/cda/doc/content/us_consulting_millennialfactsheet_080606.pdf

Lee, T. B. (1992). What's new in '92. Retrieved January 3, 2008, from http://www.w3.org/History/19921103-hypertext/hypertext/WWW/News/9201.html

Lee, W., & Lu, C. (2003). Customising WAP-based information services on mobile networks. *Personal and Ubiquitous Computing, 7*(6), 321-330.

Lee, Y. E., & Benbasat, I. (2003). Interface design for mobile commerce. *Communications of the ACM, 46(12)*, 48-52, 2003.

Leedy, P. D., & Ormrod, J. E. (2004). *Practical research: Planning and design.* Upper Saddle River, NJ: Pearson Education.

Leem, C. S., Suh, H. S., & Kim, D. S. (2004). A classification of mobile business models and its applications. *Industrial Management & Data Systems, 104*, 78-87.

Leont'ev, A. N. (1978). *Activity, consciousness, personality.* Englewood Cliffs, NJ: Prentice Hall.

Leung, C.-H., & Chan, Y.-Y. (2003, July 9-11). *Mobile learning: a new paradigm in electronic learning.* Paper presented at the 3rd IEEE International Conference on Advanced Learning Technologies (pp. 76-80). Athens, Greece.

Levy, M., & Kennedy, C. (2005). Learning Italian via mobile SMS. In A. Kukulska-Hulme & J. Traxler (Eds.), *Mobile learning: A handbook for educators and trainers* (pp. 76-83). London: Routledge.

Lewis, C., & Rieman, J. (1993). Task-centered user interface design: A practical introduction. *Shareware.* Retrieved January 3rd, 2008, from http://www.hcibib.org/tcuid/

Liang, J.-K., Liu, T.-C., Wang, H.-Y., Chang, B., Deng, Y.-C., Yang, J.-C., et al., (2005). A few design perspectives on one-on-one digital classroom environment. *Journal of Computer Assisted Learning, 21*(3), 181-189.

Liebowitz, J. (2007). Developing knowledge and learning strategies in mobile organisations. *International Journal of Mobile Learning and Organisation, 1*(1), 5–14

Lim, K. Y. T., & Wang, J. Y. Z. (2005). Collaborative handheld gaming in education. *Educational Media International, 42*(4), 351-359.

Lindquist, D., Denning, T., Kelly, M., Malani, R., Griswold, W. G., & Simon, B. (2007, March 7-10). Exploring the potential of mobile phones for active learning in the classroom. In *Proc. of SIGCSE'07*, (pp. 384-388). Covington, Kentucky.

Littleton, K., & Hakkinen, P. (1999). Learning together: Understanding the processes of computer-based collaborative learning. In P. Dillenbourg (Ed.) *Collaborative learning* (pp. 20-30). Oxford: Pergamon.

Liu, C.-C., Tao, S.-Y., Nee, J.-N., Liu, B.-J., Chen, G.-D., Hsu, C.-C., et al., (2005). Supporting Activity Awareness for Teams-Games-Tournaments with GSM Network. In H. Ogata, M. Sharples, Kinshuk & Y. Yano (Eds.), *Proceedings of the Third IEEE International Workshop on Wireless and Mobile Technologies in Education 2005* (pp. 238-242). Tokushima, Japan: The IEEE computer society.

Liu, T. C., Wang, H. Y, Liang, J. K., Chan, T. W., Ko, H. W., & Yang, J. C. (2003). Wireless and mobile technologies to enhance teaching and learning. *Journal of Computer Assisted Learning, 19*, 371-382.

Liu, T., & Choudary, C. (2004). Real-time content analysis and adaptive transmission of lecture videos for mobile applications. In *Proc. of the 12th Annual ACM international Conference on Multimedia*, (pp. 400-403), New York.

Liu, X. (2002). Using concept mapping for assessing and promoting relational conceptual change in science. *Science Education, 88*(3), 373-396

LLN (2007). Lifelong Learning Networks, Higher Education Funding Council for England (HEFCE). Retrieved from http://www.hefce.ac.uk/widen/lln/

Lonsdale, P., Baber, C., & Sharples, M. (2004) A Context Awareness Architecture for Facilitating Mobile Learning. In J. Attewell & C. Savill-Smith (Eds.) *Learning with Mobile Devices: Research and Development. London. Learning and Skills Development Agency*, (pp. 79-85).

Looi, C.-K. (2005). Exploring the affordances of online chat for learning. *International Journal of Learning Technology, 1*(3), 322-338.

Luchini, K., Quintana, C., & Soloway, E. (2004). Design guidelines for learner-centered handheld tools. In *Proc. of the SIGCHI Conference on Human Factors in Computing Systems*, (pp. 135-142), Vienna, Austria.

Luchini, K., Quintana, C., & Soloway, E. (2004, 24-29 April). *Design Guidelines for Learner-Centered Handheld Tools.* Paper presented at the Human Factors in Computing Systems, Vienna, Austria.

Luckin, R., Brewster, D., Pearce, D., du Boulay, B., & Siddons-Corby, R. (2005). Whether it's m-learning or e-learning, it must be ME learning. In A. Kukulska-Hulme & J. Traxler (Eds.), *Mobile learning: A handbook for educators and trainers* (pp. 116-124). London: Routledge.

Lund, A. (2006). The multiple contexts of online language teaching. *Language Teaching Research, 10*(2), pp. 181-204.

Lustig, C., May, C. P., & Hasher, L. (2001). Working memory span and the role of proactive interference. *Journal of Experimental Psychology: General, 130*(2), 199-207.

Lyytinen, K., & Yoo, Y. (2003). Issues and challenges in ubiquitous computing. *Communications of ACM, 45*(12), 62-65.

Madden, K. M. (2000). The role of communication in building customer relationships in the Australian financial services industry: An in-depth study. Unpublished Doctor of Philosophy thesis, Faculty of Business, University of Southern Queensland, Toowoomba, Qld.

Majchrzak, A., Malholtra, A., & John, R. (2005). Perceived individual collaboration know-how development through information technology-enabled contextualization: Evidence from distributed teams. *Information Systems Research, 16*(1), 9-27.

Malone, T.W. (1981). What makes computer games fun? *Byte*, 6, 258-277

Man, L. (2007). Software lets students use iPods to cram for SATs. *The Kansas City Star.* Retrieved August 14, 2007, from http://www.technewsworld.com/rsstory/58464.html

Mandl, H., & Gräsel, C. (1997). Multimediales und problemorientiertes Lernen, Thyroidea – ein Lernprogramm für das Medizinstudium. In Hamm, I. & Müller-Bölling, D. (Eds), *Hochschulentwicklung durch neue Medien* (p.175). Gütersloh.

March, S., & Smith, G. (1995). Design and natural science research on information technology. *Decision Support Systems 15*(4), 251-266.

Mark, G. (2002). Conventions for coordinating electronic distributed work: A longitudinal study of groupware use. In P.J. Hinds and S. Keisler (Eds.), *Distributed work.* Cambridge, MA: MIT Press.

Markett, C., Arnedillo Sanchez, I., Weber, S., & Tangney, B. (2006). Using short message service to encourage interactivity in the classroom. *Computers & Education, 46*(3), 280-293.

Markett, C., Sanchez, I. A., Weber, S., & Tangney, B. (2006). Using short message service to encourage interactivity in the classroom. *Computers and Education, 46*, 280-293.

Markus, M.L. (1994). Electronic mail as the medium of managerial choice. *Organization Science, 5*(4), 502-527.

Mathieson, K. (1991). Predicting user intentions: comparing the technology acceptance model with the theory of planned behaviour. *Information Systems Research, 2*(3), 173-191.

Mautone, P. D., & Mayer, R. E. (2001). Signaling as a cognitive guide in multimedia learning. *Journal of Educational Psychology, 93*(2), 377-389.

Maxwell, J. A. (2004). Causal explanation, qualitative research, and scientific inquiry in education. *Educational Researcher, 33*(2), pp. 3-11.

Mayer, R. E. (2005b). Principles of multimedia learning based on social cues: Personalization, voice, and image principles. In R. E. Mayer (Ed.), *The cambridge handbook of multimedia learning* (pp. 201-212). New York: Cambridge.

Mayer, R. E. (Ed.) (2005a). *The cambridge handbook of multimedia learning.* Cambridge: Cambridge Press.

Mayer, R. E., & Anderson, R. B. (1992). The instructive animation: Helping students build connections between words and pictures in multimedia learning. *Journal of Educational Psychology, 84*(4), 444-452.

Mayer, R. E., & Moreno, R. (1998). A split-attention effect in multimedia learning: Evidence for dual processing systems in working memory. *Journal of Educational Psychology, 90*(2), 312-320.

Mayorga, J.I., Barros, B., Celorrio, C., & Verdejo, M. F. (2006). Accessing a learning object repository through a semantic layer. In *Workshop on 'Learning Object Repositories as Digital Libraries: Current challenges'2006.*

Mazur, J. E. (1998). *Learning and behavior* (4th ed.). Upper Saddle River, NJ: Prentice Hall.

McConnell, D. A., Steer, D. N., Owens, K. D., Knott, J. R., Van Horn, S., Borowski, W., et al. (2006). Using conceptests to assess and improve student conceptual understanding in introductory geoscience courses. *Journal of Geoscience Education 54*, 61-68.

McGee, M. K. (2003). E-learning on the fly. *Information Week*, November 3rd 2003. Retrieved 17 January 2008 from http://www.informationweek.com/story/showArticle.jhtml?articleID=15800505

McGrath, J.E. (1984). *Groups, interaction and performance.* Englewood Cliffs, NJ: Prentice-Hall.

McKenzie, J. (1999). Scaffolding for success. From now on. *The Educational Technology Journal, 9*(4), Retrieved August 27, 2005, from http://www.fno.org/dec99/scaffold.html

Meisenberger, M., & Nischelwitzer, A. (2005). The mobile learning engine (MLE) – a mobile, computer-aided, multimedia based learning application. Retrieved 20 April 2007 from http://drei.fh-joanneum.at/mle/docs/Matthias_Meisenberger_MApEC_Paper_mLearning.pdf

Merriam, S. B. (1988). *Case study research in education: A qualitative approach.* San Francisco: Jossey-Bass Publications.

Merten, J., & Krause, R. (1993). *DAS (Differentielle Affekt Skala).* Arbeiten der Fachrichtung Psychologie, University of Saarland.

Microsoft (2007). ASP.NET Mobile Controls Overview. Retrieved 17 January 2008 from http://www.asp.net/mobile/intro.aspx?tabindex=3&tabID=44

Microsoft TechNet. (2007). W3C Extended Log File Format (IIS 6.0). Retrieved January 3, 2008, from http://www.microsoft.com/technet/prodtechnol/WindowsServer2003/Library/IIS/bea506fd-38bc-4850-a4fb-e3a0379d321f.mspx?mfr=true

Miertschin, S. L., & Willis, C. L. (2004). Mobile computing in the freshman computer literacy course what impact? In *Proc. of the 5th Conference on information Technology Education*, (pp. 149-152), Salt Lake City.

Mikleia, E. (2006, June 27 - July 01). Mobile devices to be applied as supporting tools in Research Method class for undergraduate students. In *Proc. of the 7th international Conference on Learning Sciences*, (pp. 918-919), Bloomington, IN: International Society of the Learning Sciences.

Miles, M. B., & Huberman, A. M. (1994). *Qualitative data analysis: An expanded sourcebook* (2nd ed.). Thousand Oaks, CA: Sage Publications.

Miller, G. (1956). The magical number seven, plus or minus two: Some limits on our capacity for processing information. *Psychological Review, 63*, 81-97.

Milrad, M., Chan, T-W., Roschelle, J., Hsi, S. et al. (2006). One-to-one technology-enhanced learning: An opportunity for global research collaboration. *Research and Practice in Technology Enhanced Learning Journal, 1*(1), 3-29.

Mitchell, K., & Race, N. J. P. (2005). uLearn: Facilitating ubiquitous learning through camera equipped mobile phones. In H. Ogata, M. Sharples, Kinshuk & Y. Yano (Eds.), *Proceedings of the Third IEEE International Workshop on Wireless and Mobile Technologies in Education 2005* (pp. 274-281). Tokushima, Japan: The IEEE computer society.

Moher, T., Hussain, S., Halter, T., & Kilb, D. (2005). *RoomQuake: Embedding dynamic phenomena within the physical space of an elementary school classroom.* Paper presented at the Conference on Human Factors in Computing Systems: ACM Press.

Mor, Y., & Winters N. (2007). Design approaches in technology-enhanced learning. *Interactive Learning Environments, 15*(1), 61 – 75.

Moreno, R., & Mayer, R. E. (2000). A coherence effect in multimedia learning: The case for minimizing irrelevant sounds in the design of multimedia instructional messages. *Journal of Educational Psychology, 92*(1), 117-125.

Morrison, G. R., & Lowther, D. L. (2005). *Integrating computer technology into the classroom* (3rd ed.). Upper Saddle River, NJ: Prentice Hall.

Morton, A., Ackerman, F., & Belton, V. (2003). Technology-driven and model-driven approaches to group decision support: Focus, research philosophy, and key concepts. *European Journal of Information Systems*, 12.

Muylle, S., Moenaert, R., & Despontin, M. (2004). The conceptualization and empirical validation of web site user satisfaction. *Information & Management, 41*, 543-560.

Naismith, L., Lonsdale, P., Vavoula, G., & Sharples, M. (2004). *Literature review in mobile technologies and learning* (No. No. REPORT 11): NESTA Futurelab.

Naismith, L., Lonsdale, P., Vavoula, G., Sharples, M. (2004). *Mobile technologies and learning*, Bristol: Futurelab.

Nakahara, J., Hisamatsu, S., Yaegashi, K., & Yamauchi, Y. (2005, May 30 - June 4). *iTree: Does the mobile phone encourage learners to be more involved in collaborative learning?* Paper presented at the 2005 conference on Computer support for collaborative learning (pp. 470–478). Taipei, Taiwan.

Nardi, B. A. (1996). *Studying context: A comparison of activity theory, situated action models, and distributed cognition. Context and consciousness.* MA: The MIT Press.

Neuman, W. L. (2003). *Social research methods: Qualitative and quantitative approaches* (5th ed.). Boston, MA: Allyn & Bacon.

Newcomb, E., Pashley, T., & Stasko, J. (2003). Mobile computing in the retail arena. In *Proc. of the SIGCHI Conference on Human Factors in Computing Systems*, (pp. 337-344). Ft. Lauderdale, USA.

Nicol, D. J., & MacLeod, L. A. (2005). Using a shared workspace and wireless laptops to improve collaborative project learning in an engineering design class. *Computers & Education, 44*(4), 459-475.

Nokia. (2005). Nokia introduces a new Web browser for S60 3rd Edition. Retrieved January 3, 2008, from http://press.nokia.com/PR/200511/1019239_5.html

Nonaka, I., & Takeuchi, H. (1995). *The knowledge crating company.* New York: Oxford University Press,

Norris, C., & Soloway E. (2004). Envisioning the handheld-centric classroom. *Journal of Educational Computing Research, 30*(4), 281-294.

Nortel. (2007). Long-Term Evolution (LTE): The vision beyond 3G. Retrieved January 3, 2008, from http://www.nortel.com/solutions/wireless/collateral/nn114882.pdf

Notterman, J. M., & Drewry, H. N. (1993). *Psychology and education, parallel and interactive approaches.* New York and London: Plenum press.

Novak , J. D. (1977). *A theory of education.* Ithaca, NY: Cornell University Press.

O'Hara, K., Mitchell, A. S., & Vorbau, A. (2007). Consuming video on mobile devices. *Proceedings of the SIGCHI Conference on Human Factors in Computer Systems, 1*, (pp. 858-866).

O'Malley, C., & Stanton Fraser, D. (2004). *Learning with tangible technologies.* Futurelab Bristol.

O'Malley, C., Vavoula, G., Glew, J. P., Taylor, J., Sharples, M., Lefrere, P. et al. (2005). *WP4 – Pedagogical Methodologies and Paradigms, Guidelines for learning/teaching/tutoring in a mobile environment.* MOBIlearn. Retrieved 18 September, 2006 from www.mobilearn.org/download/results/public_deliverables/

Oberauer, K., Süß, H. M., Schulze, R., Wilhelm, O., & Wittmann, W. W. (2000) Working memory capacity—facets of a cognitive ability construct. *Personality and Individual Differences, 29*, 1017-1045.

Ogata, H., & Yano, Y. (2004). *Context-Aware Support for Computer-Supported Ubiquitous Learning.* Proc. of WMTE 2004, IEEE Computer Society Press, (pp.27-34).

Ogata, H., & Yano, Y. (2004). Context-Aware Support for Computer-Supported Ubiquitous Learning. In *Proceedings of IEEE WMTE2004* (pp. 27–34). Taoyuan, Taiwan.

Ogata, H., &Yano, Y. (2003a). *How ubiquitous computing can support language learning.*Proc. of KEST 2003, (pp.1–6).

Ogata, H., &Yano, Y. (2003b). *Supporting knowledge awareness for a ubiquitous CSCL*, Proc. of E-Learn 2003, (pp.2362–2369).

Okada, M., Yamada, A, Tarumi, H., Yoshida, M. & Moriya, K (2003). DigitalEE II: RV-Augmented Interface Design for Networked Collaborative Environmental Learning. In *Designing for Change in Networked Learning Environments* (pp. 265–274). Kluwer Academic Publishers.

Okada, M., Yamada, A., Tarumi, H., Yoshida, M., & Morita, K. (2003). *Digital-EE II: RV-augumented interface design for networked collaborative environmental learning*, Proc. of CSCL 2003, (pp. 265-274).

Osgood, C.E., Suci, G. J., & Tannenbaum, P. H. (1975). *The measurement of meaning.* Urbana: University of Illinois Press.

Owen, M. (2005). Driving demand for broadband. *ITSE, 3*(1)

Owen, M. (2006). *Social software and learning.* Futurelab, Bristol

Papert, S., & Harel, I. (1991). *Constructionism, chapter Situating Constructionism.* Ablex Publishing Corporation, Norwood, NJ.

Parr, C. S., Jones, T., & Songer, N. B. (2002). CyberTracker in BioKIDS: Customization of a PDA-Based Scientific Data Collection Application for Inquiry Learning. In *Proceedings of ICLS2002* (pp. 574–575), Seattle WA.

Parr, C. S., Jones, T., & Songer, N. B. (2004). Evaluation of a handheld data collection interface for science learning. *Journal of Science Education and Technology, 13*(2), 233-242.

Parsons D., Ryu H., & Cranshaw M. (2006, July 5-7). A Study of Design Requirements for Mobile Learning Environments. Sixth IEEE International Conference on Advanced Learning Technologies, 2006, Kerkrade, The Netherlands.

Parsons, D., Ryu, H., & Cranshaw, M. (2006). A Study of Design Requirements for Mobile Learning Environments. In *Proc. of the 6th IEEE International Conference on Advanced Learning Technologies*, (pp. 96-100). Washington, DC: IEEE Computer Society.

Parsons, D., Ryu, H., & Cranshaw, M. (2007). A design requirements framework for mobile learning environments.*Journal of Computers,* **2**(4), 1-8.

Patten, B., Arnedillo-Sanchez, I., & Tangney, B. (2006). Designing collaborative, constructionist and contextual applications for handheld devices: Virtual learning. *Computers and Education, 46*(3), 294-308.

Patten, B., Arnedillo-Sanchez, I., & Tangney, B. (2006). Designing collaborative, constructionist and contextual applications for handheld devices. *Computers & Education, 46*(3), 294-308.

Patton, M. (1990). *Qualitative methods.* Newbury Park, CA: Sage Publications.

Paul, R. (April 17, 2007). Rural Arkansas school turns school buses into classrooms. *Ars Technica.* Retrieved August 7, 2007, from http://arstechnica.com/news.ars/post/20070417-rural-arkansas-school-uses-ipods-to-turn-schoolbuses-into-classrooms.html

Payne, F., Ball, J., & Snow, R. (2000). A corporate approach to the introduction of flexible delivery education. In *Proceedings of the 30th ASEE/IEEE conference: Frontiers in education.* Kansas City, MO.

Pea, R., and Maldonado, H. (2006). WILD for learning: Interacting through new computing devices anytime, anywhere. In K. Sawyer (Eds.), *Cambridge handbook of the learning sciences* (pp. 427 - 442). New York: Cambridge University Press.

Penuel, W. R., Roschelle, J., & Abrahamson, L. (2005). Research on Classroom Networks for Whole-Class Activities. *IEEE International Workshop on Wireless and Mobile Teachnologies in Education (WMTE'05),* (pp. 222-229).

Perry, D. (2003). *Handheld computers (PDAs) in schools.* Coventry, UK: British Educational Communications and Technology Agency.

Pfeifer, R. S., & Robb, R. (2001). Beaming your school into the 21st century. *Principal Leadership, 1*(9), 30-34.

Pinkwart, N., Hoppe, H. U., Milrad, M., & Perez, J. (2003). Educational scenarios for cooperative use of Personal Digital Assistants. *Journal of computer assisted learning, 19*(3), 383-391.

Pinkwart, N., Hoppe, H. U., Milrad, M., & Perez, J. (2003). Educational scenarios for the cooperative use of personal digital assistants. *Journal of Computer Assisted Learning, 19*(3), 383-391.

Poddar, S. (2003). Introduction to .NET Web Services. Retrieved 10 October, 2007, from http://www.codeproject.com/dotnet/intro2websvc.asp

Polsani P (2003). Network learning. In K. Nyiri (Ed.), *Mobile learning. Essays on philosophy, psychology and education* (pp. 139–150). Vienna: Passagen Verlag.

Pratt, M.W., Boyes, C., Robins, S., & Manchester, J. (1989). Telling tales: Aging, working memory, and the narrative cohesion of story retellings. *Developmental Psychology, 25,* 628-663.

Prensky, M. (2000). *Digital game-based learning.* New York: McGraw-Hill.

Prensky, M. (2004) *What can you learn from a cell phone?* Retrieved November 20th 2007. http://www.marcprensky.com/writing/Prensky What_Can_You_Learn_From_a_Cell_Phone-FINAL.pdf

Price, S., & Rogers, Y. (2004). Let's get physical: The learning benefits of interacting in digitally augmented physical spaces. *Computers and Education, 43*(1-2), 137-151.

Price, S., Rogers, Y., Stanton, D., & Smith, H. (2003, June 14-18). A new conceptual framework for CSCL: Supporting diverse forms of reflection through multiple interactions. *In Proceedings of the International Conference on CSCL'03* (pp. 513-522). Bergen, Norway. Kluwer Academic Publishers.

Purao, S. (2002). Design Research in the Technology of Information Systems: Truth or Dare. *GSU Department of CIS* Working Paper, Atlanta, Retrieved January 3, 2008, from http://purao.ist.psu.edu/working-papers/dare-purao.pdf

Quinn, C. (2000). mLearning: mobile, wireless, in-your-pocket learning. *LiNE Zine* Retrieved 16 June, 2006, from http://www.linezine.com/2.1/features/cqmmwiyp.htm

Ractham, P., & Zhang, X. (2006). Podcasting in academia: a new knowledge management paradigm within academic settings. In *Proc. of the 2006 ACM*

*SIGMIS CPR Conference on Computer Personnel Research*, (pp. 314-317), Claremont, USA.

Randel, J.M., Morris, B.A., Wetzel, D.C., & Whitehill, B.V. (1992). The effectiveness of games for educational purposes: A review of recent research. *Simulation & Gaming, 23*, 261 - 276.

Ravasio, P., Tscherter, V., & Sugimoto, M. (2006). SketchMap: A Collaborative Tool to Learn Basic Cartographical Concepts. In *Proceedings of COOP '06* (pp.115–122), Carry-le-Rouet, Province, France.

Research, Research and Consultancy Outsourcing Services (2007). *Europe Mobile Handset Market Analysis*. Retrieved March 16, 2007, from http://www.researchandmarkets.com/reportinfo.asp?report_id=337216

Revell, S. (2004). From Hide and Seek to ARQuake: Considering the Challenges of Location-Based Gaming. *Proceedings of GI Days 2004*, Muenster, Germany.

Rheinberg, F., & Vollmeyer, R. (2003). Flow-Erleben in einem Computerspiel unter experimentell variierten Bedingungen. *Zeitschrift für Psychologie, 211*(4), 161–170.

Rheinberg, F., Vollmeyer, R., & Engeser, P. H. (2002). Die Erfassung des Flow-Erlebens. In Stiensmeier-Pelster, J., & Rheinberg, F. (Eds), *Diagnostik von Selbstkonzept, Lernmotivation und Selbstregulation* (pp. 261-279). Göttingen: Hogrefe.

Riedl, A., & Schelten, A. (2000). Handlungsorientiertes Lernen in technischen Lernfeldern. In Bader, R. & Sloane, P. F. (Eds), *Lernen in Lernfeldern. Theoretische Analysen und Gestaltungsansätze zum Lernfeldkonzept* (pp.155 – 164). Eusl: Markt Schwaben.

Riege, A. (1996). Marketing communication of international travel and tourism: A study of UK and German markets for Australia and New Zealand. Unpublished Doctor of Philosophy thesis, Queensland University of Technology, Brisbane, Qld.

Roberts, R. J., Hager, L. D., & Heron, C. (1994). Prefrontal cognitive processes: Working memory and inhibition in the antisaccade task. *Journal of Experimental Psychology: General, 123*, 374-393.

Robson, S. (1993, Spring). The qualitative survey. *Survey*, 13-14.

Rogers, Y., Price, S., Fitzpatrick, G., Fleck, R., Harris, E., Smith, H. et al. (2004, June 1-3, 2004). *Ambient Wood: Designing New Forms of Digital Augmentation for Learning Outdoors.* Paper presented at the IDC 2004, College Park, Maryland, USA.

Rogers, Y., Price, S., Randell, C., Fraser, D. S., Weal, M., & Fitzpatrick, G. (2004). Ubi-learning Integrates Indoor and Outdoor Experiences. *Communications of the ACM, 48*(1), 55–59.

Roschelle, J. (2003). Keynote paper: Unlocking the learning value of wireless mobile devices. *Journal of Computer Assisted Learning, 19*(3), 260-272.

Roschelle, J. (2003). Unlocking the learning value of wireless mobile devices. *Journal of Computer-Assisted Learning, 19*, 260-272.

Roschelle, J., Rosas, R., & Nussbaum, M. (2005, May 30 - June 04). Towards a design framework for mobile computer-supported collaborative learning. *In Proceedings of The 2005 Conference on Computer Support For Collaborative Learning: Learning 2005: the Next 10 Years!* (pp. 520-524). Taipei, Taiwan: Computer Support for Collaborative Learning, International Society of the Learning Sciences.

Roschelle, J., Sharples, M., & Chan, T. W. (2005). Introduction to the special issue on wireless and mobile technologies in education. *Journal of computer assisted learning, 21*(3), 159-161.

Rosen, V. M., & Engle, R. W. (1997). The role of working memory capacity in retrieval. *Journal of Experimental Psychology: General, 126*(3), 211-227.

Rossi, M., & Sein, M. (2003, August 9-12). *Design Research Workshop: A Proactive Research Approach.* Presentation delivered at the 26th Information Systems Research Seminar in Scandinavia (IRIS 26), Porvoo, Finland.

Rushby, N. J. (1998). Learning without strings: Exploiting mobile technologies for flexible learning. *ACM-SIGCUE Outlook, 26*(2). Retrieved April 29, 2007, from http://www.eee.bham.ac.uk/handler/ePapers/rushby_1998.pdf

Ryu, H., & Parsons, D. (2007). A learner-centred design of a location-aware learning reminder. *International Journal of Mobile Learning and Organization.*

Ryu, H., & Wong, L. (2008). Empirical studies on generic mobile environment issues that may affect mobile quiz implementation. *IEEE Wireless, Mobile, Ubiquitous Technologies in Education.* Beijing, China: IEEE Press.

SAIL. (2007). Scalable Architecture for Inquiry Learning. *SAIL home page.* http://www.telscenter.org/confluence/display/SAIL/

Saito, N., Ogata, H., Paredes, R., & Yano, Y., & Ayala, G. (2005). *Supporting Classroom Activities with the BSUL Environment.* Proc. of WMTE 2005, IEEE Computer Society Press, (pp. 243-250).

Salvucci, D. D., Markley, D., Zuber, M., & Brumby, D. P. (2007). iPod distraction:

Sauer, J., Wastell, D., Hockey, G., Crawshaw, C., Ishak, M., & Downing, J. (2002). Effects of display design on performance in a simulated ship navigation environment. *Ergonomics, 45*(5), 329-347.

Saunders, C. (2000). Virtual teams: Piecing together the puzzle. In R. Zmud (Ed.), *Framing the domains of IT management.* Cincinnati, OH: Pinnaflex Publishing.

Sayar, A., Pierce, M., & Fox, G. (2006, February 19-25). *Integrating AJAX Approach into GIS Visualization Web Services.* Paper presented at the International Conference on Internet and Web Applications and Services (AICT-ICIW '06), Guadeloupe, French Caribbean.

Scanlon, E., Jones, A., & Waycott, J. (2005). Mobile technologies: prospects for their use in learning in informal science settings. *Journal of interactive media in education (25)*

Schellekens, A., Paas, F., & van Merriënboer, J. J. G. (2003, April). Flexibility in higher professional education: A survey in business administration programmes in the Netherlands. *Higher Education, 45*(3), 281-305.

Schiefele, U., & Roussakis, E. (2006). Die Bedingungen des Flow-Erlebens in einer experimentellen Spielsituation. *Zeitschrift für Psychologie, 214*(4), 207–219.

Schwabe, G., & Goth, C. (2005). Mobile learning with a mobile game: Design and motivational effects. *Journal of Computer Assisted Learning, 21*, 204-216.

Schwabe, G., Göth, C., & Frohberg, D. (2005). Does Team Size Matter in Mobile Learning? In *Proceedings of the Fourth International Conference on Mobile Business* (pp. 227-234). Sydney, Australia: IEEE Computer Society.

Scornavacca, E., & McKenzie, J. (2006). *Perceived Critical Success Factors of SMS Based Campaigns: a managerial perspective.* Paper presented at the European Conference on Information Systems, Gotheburg, Schweden.

Scornavacca, E., Barnes, S. J., & Huff, S. (2006). Mobile business research published in 200-2004: Emergence, current status, and future opportunities. *Communications of the Association for Information Systems, 17*, 635 - 646.

Scornavacca, E., Huff, S. L., & Marshall, S. (2007). *Developing a SMS-based classroom interaction, mobile learning technologies and applications.* Auckland, New Zealand.

Scornavacca, E., Huff, S., & Marshall, S. (2007). *Developing a SMS-based classroom interaction system, mobile learning technologies and applications.* Paper presented at the Conference on Mobile Learning Technologies and Applications.

Sebok, A. (2000). Team performance in process control: Influences of interface design and staffing levels. *Ergonomics, 43*(8), 1210-1236.

Segall, N., Doolen, T. L., & Porter, J. D. (2005). A usability comparison of PDA-based quizzes and paper-and-pencil quizzes. *Computers & Education, 45*(4), 417-432.

Seong, D. S. K. (2006, October 25–27). *Usability guidelines for designing mobile learning portals.* Paper presented at the 3rd international conference on mobile technology, applications & systems (Mobility 06), Bangkok, Thailand.

Seppala, P., & Alamaki, H. (2003). Mobile learning in teacher education. *Journal of Computer Assisted Learning, 19*, 330-335.

Shaffer, D. W. (2007). *How computer games help children learn.* New York: Palgrave Macmillan.

Sharples, M. (2000). The design of personal mobile technologies for lifelong learning. *Comput. Educ., 34(3-4)*, 177-193.

Sharples, M., & Beale, R. (2003). A technical review of mobile computational devices. *Journal of Computer Assisted Learning, 19*(3), 392-395.

Sharples, M., Corlett, D., & Westmancott, O. (2002). The design and implementation of a mobile learning resource. *Personal and Ubiquitous Computing, 6*, 220 - 234.

Sharples, M., Corlett, D., Bull, S., Chan, T., & Rudman, P. (2005). The student learning organizer. In A. Kukulska-Hulme & J. Traxler (Eds.), *Mobile learning: A handbook for educators and trainers.* London: Routledge.

Sharples, M., Taylor, J., & Vavoula, G. (2005). *Theory of learning for the Mobile age.* Paper presented at MLEARN.

Sharples, M., Taylor, J., & Vavoula, G. (2005). Towards a theory of mobile learning. In *Prococeedings of mLearn 2005 Conference.*

Shefton-Green (2004) *Literature review in Informal learning with technology outside school.* Report NESTA Future Labs. (7).

Sheng, H., Nah, F. F.-H., & Siau, K. (2005). *Using Wireless Technology to Facilitate Learning: A Grounded Theory Approach.* Paper presented at the 11th Americas Conference on Information Systems, Omaha.

Shih, Y. E., & Mills, D. (2007). Setting the new standard with mobile computing in online learning. *International Review of Research in Open and Distance Learning, 8*(2).

Shusterman, R. (2005) Somaesthetics and social theory. *Human Affairs, 2*, 105-115,

Shute, V. J. (1991). Who is likely to acquire programming skills? *Journal of Educational Computing Research, 7*, 1-24.

Singer, D., Golinkoff, R. M., & Hirsh-Pasek, K. (2006). *Play=learning: How play motivates and enhances children's cognitive and social-emotional growth.* New York, NY: Oxford University Press.

Slain, D., Abate, M., Hodges, B. M., Stamatakis, M. K., & Wolak, S. (2004). An interactive response system to promote active learning in the doctor of pharmacy curriculum. *American Journal of Pharmaceutical Education, 68*, 1-5.

Slavin, R. E. (1983a). Cooperative learning. *Review of Educational Research, 50*(2), 315-342.

Slavin, R. E. (1983b). When does cooperative learning increase student achievement? *Psychological Bulletin, 94*(3), 429-445.

Smailagic, A., & Kogan, D. (2002). Location sensing and privacy in a context-aware computing environment. *IEEE Wireless Communications, 9*(5), 10-17.

Smith, C., Johnston, J., & Paris, C. (2004). Decision support for air warfare: Detection of deceptive threats. *Group Decision and Negotiation, 13,* 129-148.

Smordal, O., & Gregory, J. (2003). Personal digital assistants in medical education and practice. *Journal of Computer Assisted Learning, 19*(3), 320-329.

Smullen, C., & Smullen, S. (2006, July 17-19). *Modelling AJAX Application performance.* Paper presented at the Conference on Web Technologies, Applications, and Services (WTAS 2006), Calgary, Canada.

Smullen, C., & Smullen, S. (2007, March 22-25). *AJAX application server performance.* Paper presented at the IEEE SoutheastCon (pp. 154-158). Richmond, USA.

Smyth, B., & Cotter, P. (2002). Personalized adaptive navigation for mobile portals. In *Proc. of the 15th European Conference on Artificial Intelligence*, (pp. 608-612), Lyon, France.

Sohn, S., & Jo, Y. (2003). A study on the student pilot's mental workload due to personality types of both instructor and student. *Ergonomics, 46*(15), 1566-1577.

Soloway, E., Grant, W., Tinker, R., Roschelle, J., Mills, M., Resnick, M. et al. (1999). Science in the palms of their hands. *Communications of the ACM, 42(8)*, 21-27.

Soloway, E., Norris, C., Blumenfeld, P., Fishman, B., Krajcik, J., & Marx, R. (2001). Log on education: Handheld devices are ready-at-hand. *Communications of the ACM, 44(6)*, 15-20.

Soloway, E., Norris, C., Blumenfeld, P., Fishman, B., Krajcik, J. & Marx, R. (2001). Handheld devices are ready-at-hand. *Communications of the ACM, 44*(6), 15-20.

Sommerville, I. (1996). *Software engineering.* London: Addison-Wesley.

Song, Y. (in press). SMS enhanced vocabulary learning for adult learners mobile audiences. *International Journal of Mobile Learning and Organisation.*

Song, Y., & Fox, R. (2007). A review of affordances and constraints of handheld devices in higher education. In P. Tsang, R. Kwan & R. Fox (Eds.), *Learning through technology* (pp. 229-239). Singapore: World Scientific Publishing.

Spohrer, J. (1999). Information in places. *IBM Systems Journal, 38*(4), 602-628.

Sproull, L., & Kiesler, S. (1986). Reducing social context cues: Electronic mail in organizational computing. *Management Science, 32*(11), 1492-1512.

Sproull, L., & Kiesler, S. B. (1995). *Connections: new ways of working in the networked organization.* Cambridge, Mass.: MIT Press.

Squire, K., & Klopfer, E. (2007). Augmented reality simulations on handheld computers. *The Journal of the Learning Sciences, 16*(3), 371–413.

Stodel, E. J., Thompson, T. L., & MacDonald, C. J. (2006). Learners' perspectives on what is missing from online learning: Interpretations through the community of inquiry framework. *The International Review of Research in Open and Distance Learning, 7*(3).

Stone, A., & Livingstone, D. (2004). Don't lose the learning from learning objects! A request for care in the adaptive context community. *International Journal of Continuing Engineering Education and Lifelong Learning, 14*(4/5), 318-330.

Stone, A., Briggs, J., & Smith, C. (2002, August 29-30). *SMS and Interactivity - Some Results from the Field, and its Implications on Effective Uses of Mobile Technologies in Education.* Paper presented at the IEEE International Workshop on Wireless and Mobile Technologies in Education (WMTE'02), Vaxjo, Sweden.

Storck, J., & Sproull, L. (1995). Through a glass darkedly: What do people learn in videoconferencing? *Human Communication Research, 22*, 197-219.

Stringer, M., Fitzpatrick, G., Halloran, J., & Hornecker, E. (2005). *Moving Beyond the Application: Design Challenges For Ubiquitous Computing.* Position paper presented at Aarhus2005 workshop 'Ambient Computing in a Critical, Quality of Life Perspective'. Aarhus, Denmark, 21st August .

Stuckmann, P., Ehlers, N., & Wouters, B. (2002, September 24-28). *GPRS traffic performance measurements.* Paper presented at the 56th Vehicular Technology Conference (VTC 2002), Vancouver, Canada.

Suchman, L. A. (1987). *Plans and situated actions: The problem of human machine communication.* Cambridge, U.K.: Cambridge University Press.

Sugimoto, M. (2005). A support for learning by enhancing interactions with physical worlds. *Journal of Japanese Society for Information and Systems in Education, 22*(4), 161–168 (in Japanese).

Sugimoto, M., Ravasio, P., & Enjoji, H. (2006). SketchMap: A System for Supporting Outdoor Collaborative Learning by Enhancing and Sharing Learners' Experiences, In *Proceedings of ICCE Workshop on Design and Experiments of Mobile and Ubiquitous Learning Environments* (pp. 1– 8), Beijing, China.

Swan, K., van't Hooft, M., Kratcoski, A., & Unger, D. (2005). Uses and effects of mobile computing devices in K-8 classrooms. *Journal of Research on Technology in Education, 38*(1), 99-112.

Syuvanen, A., Beale, R., Sharples, M., Ahonen, M., & Lonsdale, P. (2005, November 28-30). *Supporting Pervasive Learning Environments: Adaptability and Context Awareness in Mobile Learning.* In proceedings of the 3rd IEEE International Workshop on Wireless and Mobile Technologies in Education (WMTE'05). Tokushima, Japan.

Tabard, A., Mackay, W., Roussel, N., & Letondal, C. (2007). PageLinker: Integrating contextual bookmarks within a browser. In *Proc. of the SIGCHI conference on Human factors in computing systems (CHI '07)*, (pp. 337-346), San Jose, California, USA.

Takenaka, M., Inagaki, S., Kuroda, H., Deguchi, A., & Ohkubo, M. (2006). Fieldwork Support System Using Mobile Phones: Evaluations of Information Sharing in the Second Grade's Life Environment Study. In *Proceedings of ED-MEDIA2006* (pp. 1325–1331), Chesapeake, VA.

Talkline, *Talkline - 98 Prozent der jugendlichen Handybesitzer nutzen SMS.* Retrieved March 16, 2007, from http,//www.dsltarife.net/news/900.html

Tarasewich, P. (2003). Designing mobile commerce applications. *Communications of the ACM, 46*, 57-60.

Tatar, D., Roschelle, J., Vahey, P., & Penuel, W. R. (2003). Handhelds go to school: Lessons Learned. *SRI International. Journal Computer, September*, 30-37

Taylor, J. (2003). A Task-Centred Approach to Evaluating a Mobile Learning Environment for Pedagogical Soundness. In *Presentation at the Mlearn 2004, Birmingham.*

Taylor, J. (2004). A task-centered approach to evaluating a mobile learning environment for pedagogical soundness. *Learning with mobile devices: research and development* (pp. 167-171). London, UK: Learning and Skills Development Agency.

Taylor, S., & Todd, P. (1995). Assessing IT usage: The role of prior experience. *MIS Quarterly, 19*(4), 561-570.

Telephia: *European Subscriber and Device Report, Q3 2006*, Retrieved March 12, 2007, from http://telephia.com/html/Smartphonepress_release_template.html

Tewissen, F., Lingnau, A., Hoppe, U., Mannhaupt, G., & Nischk, D. (2001). Collaborative writing in a computer-integrated classroom for early learning. In P. Dillenbourg, A., Eurelings & K. Hakkarainen (Eds.), *Proceedings of Euro-CSCL 2001* (pp. 593-600). Mahwah, NJ: Lawrence Erlbaum Associates.

Thayer, R.E. (1978). Toward a psychological theory of multidimensional activation (arousal). *Motivation and Emotion, 2*(1), 1-34.

Thayer, R.E. (1989). *The biopsychology of mood and arousal.* New York: Oxford University Press.

Theng, Y, Tan, K, Lim, E, Zhang, J., Goh, D., Chatterjea, K. et al. (2007, June 18-23). *Mobile G-Portal supporting collaborative sharing and learning in geography fieldwork: an empirical study.* Paper presented at the International Conference on Digital libraries (pp. 462–471). Vancouver, Canada.

Thissen, F. (1997). Das Lernen neu erfinden? Konstruktivistische Grundlagen einer Multimedia-Didaktik. In Beck, U. & Sommer, W. (Eds.), *LEARNTEC 97 Europäischer Kongreß für Bildungstechnologie und betriebliche Bildung* (pp. 69 – 79). Karlsruhe.

Thomas, P. (2005). *Pervasive, persuasive e-learning, modeling the pervasive learning space.* Paper presented at the 3rd International Conference on Pervasive Computing and Communications Workshops.

Thomas, P. (2006). *From theory to practice, on designing a pervasive learning game.* Paper presented at the 4th Annual IEEE International Conference on Pervasive Computing and Communications Workshops.

Thomas, P. (2006). Pervasive learning games, explorations of hybrid educational gamescapes. *Simulation & Gaming, 37*(1), 41-55.

Thompson, T., Heer, D., Brown, S., Traylor, R., & Fiez, T.S. (2004 October 20–23). *Educational Design, Evaluation & Development of Platforms for Learning.* Proceedings of the 34th ASEE/IEEE Frontiers in Education Conference, Savannah, GA, USA.

Thornton, P., & Houser, C. (2005). Using mobile phones in English education in Japan. *Journal of Computer Assisted Learning, 21*(3), 217-228.

Todd, P., & Benbasat, I. (1999). Evaluating the impact of DSS, cognitive effort, and incentives on strategy selection. *Information Systems Research, 10*(4), 357-374.

Tolmie, A. (2001). Examining learning in relation to the contexts of use of ICT. *Journal of Computer Assisted Learning, 17*(3), 235-241.

Traxler, J. (2007). Defining discussing and evaluating mobile learning: The moving finger writes and having writ... *International Review of Research in Open and Distance Learning. 8*(2) unpaged.

Traxler, J., & Kukulska-Hulme, A. (2005). *Evaluating mobile learning: Reflections on current practice.* Paper presented at the mLearn 2005, Cape Town, South Africa.

Traxler, J., & Kukulska-Hulme, A. (2005, October 25-28). Evaluating Mobile Learning: Reflections on Current Practice. In *Proceedings of MLEARN 2005.* Cape Town, South Africa.

Trelease, R. (2006). Diffusion of innovations: Anatomical informatics and iPods. *The Anatomical Record (Part B: New Anatomy), 289B*, 160-168.

Trifonova, A. (2003). *Mobile Learning: Review of the literature.* (Technical Report #DIT-03-009).

Trommsdorff, V. (2004). *Konsumentenverhalten.* Stuttgart: Kohlhammer.

Turner, M. L., & Engle, R. W. (1989). Is working memory capacity task dependent? *Journal of Memory and Language, 28*, 127-154.

Uden, L. (2007). Activity theory for designing mobile learning. *International Journal of Mobile Learning and Organization, 1*(1), 81-102.

Uden, L. (2007). Activity theory for designing mobile learning. *International Journal of Mobile Learning and Organisation, 1*(1), 81-102.

University of Minnesota School of Public Health. (2007a). Public health matters. Retrieved August 7, 2007 from http://blog.lib.umn.edu/sphpod/sphprograms/

University of Minnesota School of Public Health. (2007b). Tune in to our podcasts and blogs. Retrieved August 7, 2007 from http://www.sph.umn.edu/podcast/home.html

University of Utah. (2007). Podcasting from the U. Retrieved August 7, 2007 from http://www2.utah.edu/podcast/info.php

University of Virginia. (2007). University of Virginia podcasts and Webcasts. Retrieved August 7, 2007, from http://www.virginia.edu/uvapodcast/

Unsworth, N., & Engle, R. W. (2007). The nature of individual difference n working memory capacity: Active maintenance in primary memory and controlled search from secondary memory. *Psychological Review, 114*(1), 104-132.

Vahey, P., & Crawford, V. (2002). Palm Education Pioneers Program: Final Evaluation Report, Menlo Park, CA: *SRI International.*

Vaishnavi, V., & Kuechler, B. (2007). Design Research in Information Systems. Retrieved January 3, 2008 from http://www.isworld.org/Researchdesign/drisISworld.htm

Valiquette, P., Williams, M., Kutner, N., & Seeger, E. (2005). Rensselaer's mobile computing program a whale of an idea! In *Proc. of the 33rd Annual ACM SIGUCCS Conference on User Services*, (pp. 405-410), Monterey, USA.

Van den Akker, J. (1999). Principles and methods of development research. In J. Van den Akker et al. (Eds.), *Design approaches and tools in education and training* (pp. 1-14). Kluwer Academic Publishers

Van Meel, R. M. (1992). *Perspectives on organization and management of higher education (OTIC research report 49).* Heerlen, the Netherlands: Open University of the Netherlands.

van't Hooft, M., Diaz, S., & Swan, K. (2004). Examining the potential of handheld computers: Findings from the Ohio PEP project. *Journal of Educational Computing Research, 30*(4), 295-311.

Vanderbilt University. (2007). Aspirnaut Initiative: A pilot educational initiative to elevate the mathematics and science achievement of rural community K-12 students who commute on long bus rides to school. Retrieved August 7, 2007, from http://www.aspirnaut.org/

Vavoula, G., & Sharples, M. (2002). *KLeOS: A personal, mobile, knowledge and learning organisation system.* Paper presented at the IEEE International Workshop on Mobile and Wireless Technologies in Education, Vaxjo, Sweden.

Vavoula, G., Sharples, M., & Rudman, P. (2002, Aug 28–29). Developing the 'Future Technology Workshop' method. In M. Bekker, P. Markopoulos & M. Kersten-Tsikalkina (Eds.), *Proceedings of the International Workshop on Interaction Design and Children (IDC2002)* (pp. 65–72). Eindhoven, The Netherlands: Shaker Publishing.

Vélez. J., Barros, B., & Verdejo, M. F. (2006). A framework to define web based communities. *International Journal of Web Based Communities (IJWBC), 2(*3).

Venegas, V. (January 27, 2007). Teaching the troops: NMSU offers distance education through "Sociology in a Sack". Retrieved August 7, 2007, from http://www.nmsu.edu/ ~ucomm/Releases/2007/january/sociologysack.htm

Verdejo, M. F., Celorrio, C., & Lorenzo, E. (2006b). Improving Learning Object Description Mechanisms to Support an Integrated Framework for Ubiquitous Learning Scenarios. In *Proceedings of the 4th IEEE International Workshop on Wireless, Mobile and Ubiquitous Technologies in Education (WMUTE06).* Athens, Greece.

Verdejo, M.F., Celorrio, C., Lorenzo, E., & Sastre, T. (2006a). An Educational Networking Infrastructure Supporting Ubiquitous Learning for School Students. In *Proceedings of 6th IEEE International Conference on Advanced Learning Technologies (ICALT06)*, Kerkrade, The Netherlands.

Vogel, D., Kennedy, D. M., Kuan, K., Kwok, R., & Lai, J. (2007). Do mobile device applications affect learning? In R. H. Sprague (Ed.), *Proceedings of the Hawaii International Conference on System Sciences (HICSS-40).* Hawai'i:: Institute of Electrical and Electronics Engineers.

Vygostsky, L. (1933). *Play and its role in the mental development of the child.* First Published: 1933, Source: Voprosy psikhologii, 1966, No. 6; Translated: Catherine Mulholland; Transcription/Markup: Nate Schmolze; Retrieved November 20th 2007, http://www.marxists.org/archive/vygotsky/works/1933/play.htm

Vygotsky, L. S. (1926). *Pedagogical psychology.* Moscow: Educational Worker Publisher.

W3C, A P3P Preference Exchange Language 1.0 (APPEL1.0). (2006). Retrieved July 31, 2007, from http://www.w3.org/TR/P3P-preferences/

W3C, Platform for Privacy Preferences (P3P) Project, (2006), Retrieved July 31, 2007, from http://www.w3.org/P3P/

W3C.org. (2007). The XMLHttpRequest Object. Retrieved January 3, 2008 from http://www.w3.org/TR/XMLHttpRequest/

Walker, K. (2004, October15). *Learning on Location with Cinematic Narratives.* Proceedings of the 1st ACM Workshop on Story Representation, Mechanism and Context (pp. 55-58). New York.

Walther, B.K. (2005). Atomic actions—molecular experience, theory of pervasive gaming. *Computers in Entertainment, 3*(3).

Walton, G., Childs, S., & Blenkinsopp, E. (2005). Using mobile technologies to give health students access to learning resources in the UK community setting. *Health Information & Libraries Journal, 22*(Supplement2), 51-65.

Wang, Y. (2003). Assessment of learner satisfaction with asynchronous electronic learning systems. *Information & Management, 41*, 75-86.

Wang, Y. (2004). *Context Awareness and Adaptation in Mobile Learning.* Paper presented at the 2nd International Workshop on Wireless and Mobile Technologies in Education.

Ward, C. R., Reeves, J. H., & Heath, B. P. (2003). Encouraging active student participation in chemistry classes with a Web-based, instant feedback, student response system. *CONFCHEM: Non-Traditional Teaching Methods* Retrieved 16 June, 2006, from http://aa.uncw.edu/chemed/papers/srs/confchem/confchem_srs.htm

Watson, H., Rainer, R., & Koh, C. (1991). Executive information systems: A framework for development and a survey of current practices. *MIS Quarterly, 15*(1), 13-30.

Waycott, J.,Scanlon, E., & Jones, A. (2002, June). Evaluating the Use of PDAs as Learning and Workplace Tools: An Activity Theory Perspective. In *Proc. of the European Workshop on Mobile and Contextual Learning*, (pp. 34-35), Birmingham, UK.

Wei, C. K. (2005). AJAX: Asynchronous Java + XML? Retrieved March 4, 2007, from http://www.developer.com/design/article.php/3526681

Wei, F.-H., Chen, G.-D., Wang, C.-Y., & Li, L.-Y. (2007). Ubiquitous discussion forum: Introducing mobile phones and voice discussion into a Web discussion forum. *Journal of Educational Multimedia and Hypermedia, 16*(2), 125-140.

Weiser, M. (1998). The future of ubiquitous computing on campus. *Communications of the ACM, 41*(1), 42-43.

Weiβenberg, N., Voisard, A., & Gartmann, R. (2004). Using ontologies in personalized mobile applications. In *Proc. of the 12th Annual ACM international Workshop on Geographic information Systems*, (pp. 2-11), Washington DC.

Wellmann, B. (1999). The network community: An introduction to networks in the global village. In B. Wellmann (Ed.). *Networks in the global village* (pp. 1-48). Boulder, CO: Westview Press

Wellmann, G., & Guila, M. (1999). The network basis of social support: A network is more than the sum of its ties. In B. Wellmann (Ed.). *Networks in the global village* (pp. 83-118). Boulder, CO: Westview Press

West, P., Foster, G. G., & Clayton, P. G. (2005). *Content exposure of slide show presentations for selective download and annotation via mobile devices.* Paper presented at the mLearn 2005, Cape Town, South Africa.

White, A. (2005). Measuring the Benefits of Ajax. Retrieved August 15, 2006, from www.developer.com/java/other/article.php/3554271

Whitelock, D., Domano, D., Jelfs, A., & Brna, P. (2000). Perfect presence: What does this mean for the design of virtual learning environments? *Education and Information Technologies, 5*(4), 471-488.

Wilensky, U. (1999). *Learning through participatory simulations: Network-based design for systems learning in classrooms.* Computer Support for Collaborative Learning, (pp. 667-676).

Wilensky, U., & Stroup, W. (1999). Learning Through Participatory Simulations: Network-based Design for Systems Learning in Classrooms. In *Proceedings of CSCL'99* (pp. 667–676), Stanford, CA.

Wilensky, U., & Stroup, W. (2000). *Networked gridlock: Students enacting complex dynamic phenomena with the HubNet architecture.* Proc. ICLS 2000, Lawrence Erlbaum Associates, (pp. 282-289).

Winters, N., & Price, S. (2005, September 19-22). Mobile HCI and the learning context. *The 7th International Conference on Human Computer Interaction with Mobile Devices and Services.* Austria: University of Salzburg.

Wiredu, G. O. (2006). Perception, mobility and the affordances of portable computers: An historical epistemology of mobile computing. Retrieved July 29, 2007, from http://www.carstensorensen.com/download/Wiredu2006.pdf

Wixom, B.H., & Todd, P.A. (2005). A theoretical integration of user satisfaction and technology acceptance. *Information Systems Research, 16*(1), 85-102.

Wood, D., Bruner, J.S., & Ross, G. (1976). The role of tutoring in problem solving. *Journal of Child Psychology and Psychiatry, 17*, 89-100.

Woodroof, J., & Burg, W. (2003). Satisfaction/dissatisfaction: Are users predisposed? *Information & Management, 40*, 317-324.

Wright, N., & McGown, A. (2001). Vigilance on the civil flight deck: Incidence of sleepiness and sleep during long-haul flights and associated changes in physiological parameters. *Ergonomics, 44*(1), 82-106.

Xie, B., & Salvendy, G. (2000). Review and reappraisal of modeling and predicting mental workload in single-and multi-task environments. *Work & Stress, 14*(1), 74-99.

Yatani, K., Sugimoto, M., & Kusonoki, F. (2004). Musex: A System for Supporting Children's Collaborative Learning in a Museum with PDAs. In *Proceedings of IEEE WMTE2004* (pp. 109–113), Taoyuan, Taiwan.

Yin, C., Ogata, H., & Yano, Y. (2004). Supporting Japanese polite expressions learning using PDA towards ubiquitous learning. *The Journal of Information and Systems in Education, 3*(1), 33-39.

Yin, R. K. (2002). *Case study research: Design and methods* (3rd ed.). Newbury Park, CA: Sage Publications.

Yokomoto, C. F. (2000, October 18-21). *Promoting depth of knowledge through a new quiz format to improve problem solving abilities.* Paper presented at the 30th ASEE/IEEE Annual Frontiers in Education Conference (FIE 2000), Kansas City, USA.

Young, M. F., De Palma, A., & Garrett, S. (2002). An ecological psychology perspective on situations, interactions, process and affordances. *Instructional Science, 30*, 47-63.

Zhang, D. (2007). Web content adaptation for mobile handheld devices. *Communications of the ACM, 50*(2), 75-79.

Ziefle, M., Schroeder, U., Strenk, J., & Michel, T. (2007, April 28 - May 03). How younger and older adults master the usage of hyperlinks in small screen devices. In *Proc. of the SIGCHI Conference on Human Factors in Computing Systems*, (pp. 307-316), San Jose, California.

# About the Contributors

**Hokyoung Ryu** holds a master's degree in industrial engineering from the Korea Advanced Institute of Science and Technology, Korea (1999) and a PhD in psychology from the University of York, U.K. (2004). His current role is as a lecturer in information technology, at Massey University, Auckland, New Zealand. His research interests include human-computer interaction (HCI), mobile usability and mobile learning, publishing more than hundred peer-reviewed articles in conferences, journals, and book chapters. He is currently preparing to edit another mobile learning book tentatively entitled *Designing Themes in Mobile Learning.* He is also authoring two books, *Mobile User Interface Analysis and Design* and *Affordance Design*, and is currently an associate editor of the *International Journal of Mobile Learning and Organisation.* Dr. Ryu is a member of the British Computer Society's HCI group and the Centre for Mobile Computing at Massey University.

**David Parsons** holds an MPhil in electronics and computer science from the University of Southampton, UK (1996) and a PhD in information technology from Nottingham Trent University, UK (1999). He has worked both in the university sector and commercial software development, including academic posts at Southampton Solent University, U.K., and professional roles at The Object People, BEA Systems and Valtech. He is currently a senior lecturer in information technology, at Massey University, Auckland, New Zealand. He has published books on C++ and Java programming and Web application development. His areas of research interest include mobile learning and agile software development. He is currently editor-in-chief of the *International Journal of Mobile and Blended Learning,* and the *Journal of Principles and Applications in Information Science and Technology* (PAIST). Dr. Parsons is a professional member of the British Computer Society.

* * *

**C. Noel Byrd** is a doctoral student in the educational psychology program at Virginia Tech. Her academic background includes a bachelor's degree in nursing, and a master's degree in education (SPED). Her research interests include conceptions of intelligence, motivation, self-efficacy, and teacher practices.

**Carlos Celorrio** is a researcher and a PhD student of the learning technologies and collaborative systems (LTCS) research group at the National Distance Learning University (UNED), in Madrid, Spain. He has a computer science background and is interested in modeling and implementing computer supported collaborative learning environments based on a socio-constructivist approach. Currently, he is researching the design of an agent-based platform for supporting students and teachers in formal and informal settings that would be able to configure, personalize, and deploy tools and data for mobile learning scenarios, and dynamically adapt them according to the situated requirements.

**Patrick A. Danaher** is an associate professor (education research) in the faculty of education at the Toowoomba campus of the University of Southern Queensland, Australia. He is the co-author of *Teaching Traveller Children: Maximizing Learning Outcomes* (Trentham Books, 2007) and the sole/co-editor of *Changing University Learning and Teaching: Engaging and Mobilizing Leadership, Quality and Technology* (Post Pressed, in preparation), *Doctrina Perpetua: Brokering Change, Promoting Innovation and Transforming Marginalization in University Learning and Teaching* (Post Pressed, 2006), *Strategic Uncertainties: Ethics, Politics and Risk in Contemporary Educational Research* (Post Pressed, 2004) and *Beyond the Ferris Wheel: Educating Queensland Show Children* (Central Queensland University Press, 1998).

**Dionisios N. Dimakopoulos** has a degree in mechanical engineering and an MSc in e-commerce technologies (distinction). He is currently writing up his PhD at Birkbeck College. He has worked in several projects in industry and academia, involving combinations of Internet and mobile technologies. His research interests include user modeling, usability, semantic Web, application of AI to learning environments and advanced Web-based applications. He is currently the vice president of the Java Hellenic User Group, where his role is to expand the use of Java among beginners, as an easy to use and learn programming language suitable for current development needs. His webpage is at www.dcs.bbk.ac.uk/~dionisis/.

**Peter E. Doolittle** is currently the director of the educational psychology research program in the department of learning sciences and technology at Virginia Tech, Blacksburg, VA. He is also the executive editor of the *International Journal of Teaching and Learning in Higher Education* (IJTLHE). His academic background includes 19 years teaching primary, secondary, undergraduate, and graduate students, in public schools and private schools, using traditional and online formats, across several subject areas including mathematics, computer science, statistics, and educational psychology. His current research focus includes the investigation of learning efficacy in multimedia instructional environments.

**Ana Dzartevska** was born in Macedonia and now lives in New Zealand. She holds an honors degree in information sciences from Massey University, majoring in computer science and information systems. She works as a professional systems developer in Auckland city.

**Francika Edegger**, born in Bosnia and Herzegovina, studied economics at the Karl-Franzens University in Graz from 2001 to 2004 and graduated as a master of finance with a diploma thesis on the "Business Location Croatia." She subsequently started a PhD in economics dealing with information systems and their application in learning situations. Supported by the evolaris research lab, Ms. Edegger performed extensive empirical studies on mobile game-based learning and pervasive learning games. The results arising from these research activities also form a core part of her PhD thesis, which she completed in summer 2007 and which is planned to be published by the "Deutscher Universitätsverlag."

**Martha Grabowski** is a professor and the director of the information systems program at LeMoyne College in Syracuse, New York and research professor of decision sciences and engineering systems at Rensselaer Polytechnic Institute in Troy, New York. Her research interests include the impact of technology in distributed large-scale systems, human and organizational error, and risk analysis and management in safety-critical settings.

**Raj Gururajan** is an associate professor in information systems in the faculty of business at the University of Southern Queensland, Australia. His main area of research is user behavior issues in using wireless technology, the management of technology and the diffusion of technology at enterprise level. Currently he is applying these interests to health settings. He has published over 100 refereed articles, won over $1 million in grant income, managed six funded projects and supervised over 10 research students. He is also the leader of the technology management cluster in the faculty of business at the University of Southern Queensland.

**Abdul Hafeez-Baig** is a lecturer in information systems in the faculty of business at the University of Southern Queensland (USQ), Australia. Since joining USQ in 2004, he has published more than a dozen refereed publications in the domains of health and education with technology as the area of research. He is currently studying for his PhD with Raj Gururajan in establishing factors that determine the adoption of wireless technology in healthcare settings. Abdul is very conversant with wireless technology as well as emerging learning management systems such as Moodle. He teaches information systems concepts to both undergraduate and postgraduate students, including MBA students.

**Sid Huff** is a professor of information systems and head of the school of information management at Victoria University of Wellington, New Zealand. His research focuses on IS strategy, IT governance, senior management roles in IT, and IT management. He has published extensively in both academic and practitioner journals, and has also written over 50 teaching cases for educational use. His most recent book is *Managing IT Professionals in the Internet Age*, co-authored with Dr. Pak Yoong. He holds degrees in applied mathematics, electrical engineering and business administration, from Queens University. He received his PhD in information systems from the M.I.T. Sloan School of Management.

**Kinshuk** is professor and director in the school of computing and information systems at Athabasca University, Canada, since August, 2006. Before moving to Canada, Dr. Kinshuk worked at the German National Research Centre for Information Technology as a postdoctoral fellow, and at Massey University, New Zealand as an associate professor of information systems and director of the Advanced Learning Technology Research Centre. He also holds a Docent position with the University of Joensuu, Finland. He has been involved in large-scale research projects for adaptive and mobile learning environments and by 2007, he had published over 200 research papers in international refereed journals, conferences and book chapters. He is chair of the IEEE Technical Committee on Learning Technology and editor of the *SSCI indexed Journal of Educational Technology & Society* (ISSN 1436-4522).

**Christian Kittl** holds masters degrees in telematics from the Technical University of Graz, Austria, and in business administration from Karl-Franzens University, Graz. He worked at the Institute for Innovation Transfer as EU Interreg IIc project coordinator from 1998 to 2000 and then joined the evolaris research lab, the Austrian Competence Centre for Mobile and Electronic Business, where he is currently head of international programs. Mr. Kittl is a member of the steering committee of the "Mobile Game-Based Learning" project within the European Research Framework Program 6 led by evolaris, and a lecturer at the department of information science and information systems at the Karl-Franzens-University.

**George Kulick** is the MBA director and an associate professor of business in the applied management analysis program at LeMoyne College in Syracuse, New York. His research interests include the use of quantitative methods in pedagogy and statistical methods.

**Arianit Kurti** is a PhD student in computer science with an orientation in Informatics at Växjö University, Sweden. His fields of research include mobile systems, and pervasive and ubiquitous computing, especially in exploring the use of contextual information for designing innovative services and applications in specific domains. His current research focus is on collaborative learning and collaborative knowledge building activities where he explores the role of mobile and positioning technologies to support these processes. He also holds a lecturer position at South East European University in Macedonia and at the University of Prishtina in Kosova.

**Greg Lepak** is a professor of business in the applied management analysis program at LeMoyne College in Syracuse, New York. His research interests include statistical methods, non parametric analysis, and the use of quantitative methods for decision making.

**Emilio Julio Lorenzo** is a teaching assistant and PhD student of the learning technologies and collaborative systems (LTCS) research group at the National Distance Learning University (UNED), in Madrid, Spain. He has a MS degree in computer engineering and his research is related to computer supported collaborative learning in secondary school scenarios, developing tools to gather, store and model information in ubiquitous environments. He is now very interested in modeling the learning products that the students produce in different activities using concept mapping.

**Danielle L. Lusk** is a doctoral student in the educational psychology program at Virginia Tech. She is an associate editor of the *International Journal of Teaching and Learning in Higher Education* (IJTLHE). Her academic background includes bachelor's and master's degrees in English. Her research interests include motivation in multimedia environments, conceptions of intelligence, interdisciplinary learning (particularly in STEM fields), and grading anxiety.

**Dr George D. Magoulas** is a reader in the school of computer science and information systems at Birkbeck College, University of London. He has edited two volumes, *Adaptable and Adaptive Hypermedia Systems* (IRM Press, 2005) and *Advances in Web-based Education: Personalized Learning Environments* (Information Science Publishing, 2006), and has published extensively in the area of computational intelligence and adaptive systems. He has received awards from the IEEE (2000), the EUNITE (2001 and 2004) and the IADIS (2006), and has secured research grants from the EPSRC, the AHRC, the ESRC and the JISC, UK. His research interests are in modeling, learning, adaptation and evolution in complex environments with applications in user modeling, sequence processing, adaptive Web-based systems and personalized learning environments. His webpage is at www.dcs.bbk.ac.uk/~gmagoulas.

**Gina J. Mariano** is currently a graduate student in the educational psychology program in the department of learning sciences and technology at Virginia Tech, Blacksburg, VA. She is an associate editor of the *International Journal of Teaching and Learning in Higher Education* (IJTLHE). Her academic background includes a bachelor's degree in psychology and gerontology, and a master's degree in counseling psychology. Her research focus includes the investigation of knowledge transfer, knowledge application, working memory, and long-term memory retrieval.

**Stephen Marshall** is a senior lecturer in educational technology and acting director at the VUW University Teaching Development Centre. Mr. Marshall is a member of the New Zealand Ministry of Education Tertiary E-Learning Reference Group and Chair of the E-Learning Directors of the New Zealand Universities Group. Stephen researches in the areas of e-learning benchmarking, plagiarism and academic integrity, mobile e-learning, intellectual property and that of development of policy and strategy to support and encourage the effective use of technology.

**Marta Millán** is a secondary school teacher working in the science department at Instituto Diego Vélazquez de Torrelodones, in Madrid, Spain. She has a degree in physics by the Universidad Complutense de Madrid, and has participated in several pedagogical projects about significant learning construction and the use of new technologies in the classroom.

**Marcelo Milrad** is an associate professor in computer science at the school of mathematics and systems engineering, at Växjö University (VXU), Sweden. His current research interest focuses on the design and development of mobile and wireless applications to support collaborative learning. Dr. Milrad has published numerous articles in international journals, refereed conferences, books and technical reports. He has also been presenting and giving lectures about his work in more than 25 countries worldwide. Recently, he has been serving as a program committee member in a number of international scientific conferences such as CSCL, ICALT, WMTE, ICCE and CELDA. Dr. Milrad is one of the initiators of the IEEE international conference on wireless and mobile technologies in education (WMTE).

**Sosuke Miura** received his BA from the school of informatics and sciences from Nagoya University in 2003, and a MA from the graduate school of arts and sciences at the University of Tokyo in 2005. Currently, he is a PhD student at the graduate school of frontier sciences, University of Tokyo. His research concern includes user interface, interaction design and computer-supported learning. He is also interested in music, and investigating new interactive technologies for music education.

**Hiroaki Ogata** is an associate professor in the faculty of engineering at Tokushima University. He received BE, ME, and PhD degrees from the department of information science and intelligent systems from the Tokushima University, Japan, in 1992, 1994, and 1998, respectively. He was a visiting researcher of Center of Lifelong Learning and Design, the University of Colorado at Boulder, USA from 2001 to 2003. His current interests are in computer-supported ubiquitous learning. Also he is a member of IPSJ, JSiSE, IEEE, ACM, and AIED. He received the best paper award from JSiSE in 1998, from WebNet 1999, and from ICALT 2006.

**Martin Owen** is at present developing unique playful, tangible devices for early learning with his own company Smalti. He is also working as an independent researcher in EU projects in technology enhanced learning with particular emphasis on games, mobility and augmented reality. He was until recently the director of learning and head of concept development at Futurelab. At Futurelab he was responsible for the key concepts in the mobile game Savannah and the physics game Racing Academy. Previously he was an academic researcher and teacher trainer at University of Wales, Bangor. He has also been a school teacher.

**Otto Petrovic** is a professor at the University of Graz in the department of information science and information systems, and chairman of the board of directors of evolaris. He is active in numerous projects and technical committees relating to commerce and government in the USA, Germany and Austria. He has held research assignments and visiting professorships in Tucson Arizona, at IBM Maryland, at the University of Central Florida, and at the Technical University of Eindhoven. He is also a member of the task force for "e-austria" under the Federal Ministry of Commerce and Employment and of the "media@kom" advisory committee of the German Ministry of Commerce and Technology.

**Sergio Prades** is a PhD student and a researcher in the learning technologies collaborative systems (LTCS) research group at the National Distance Learning University (UNED), in Madrid, Spain. He has a degree in computer science and is interested in improving the performance and increasing the participation, integration and motivation of the low level students through the use of collaborative learning systems in secondary school scenarios. Currently he works in Web-based collaborative tools for synchronous activities that reuse the learning products previously produced.

**Pamela Ravasio** earned her MSc degrees in computer science (1999) and computer science education (2005), and her PhD in human-computer interaction (2004) from the Swiss Federal Institute of Technology (ETHZ), Zurich, Switzerland. From 2001 to 2005 she was researcher and project manager for the Man-Machine Interaction Group, Institute of Occupational Health and Medicine, ETHZ. She was a visiting researcher with Professor Sugimoto's Interaction Technology Laboratory from 2005 to 2006. As of mid 2006, she joined Siemens-Asahi Medical Technologies Ltd., Tokyo, Japan as a project manger in the area of hospital information systems (HIS).

**Eusebio Scornavacca** is a senior lecturer in the school of information management at Victoria University of Wellington, New Zealand. Dr. Scornavacca is interested in the business and end-user aspects of mobile and wireless technologies. He has a particular interest on mobility, user acceptance, strategy, quality and organizational/individual impact of mobile systems. Before moving to Wellington, Eusebio spent two years as a researcher at Yokohama National University, Japan. He has published and presented more than 60 papers in conferences and academic journals including *ICIS, ECIS, HICSS*, *Communications of AIS,* and *Communications of ACM.* Dr. Scornavacca is currently on the editorial boards of the *International Journal of Mobile Communications*, the *International Journal of Electronic Finance* and he is the editor and founder of first online database dedicated to m-business.

**Yanjie Song** is a PhD student at the division of information technology studies in the faculty of education at the University of Hong Kong. She has been interested in exploring the potential of mobile technologies in education since she pursued her postgraduate study in the UK in the year 2002. She sustains her interest in her current PhD research study, exploring educational uses of PDAs through empirical studies.

**Daniel Spikol** is a PhD student in computer science with an orientation in informatics at the school of mathematics and systems engineering at Växjö University (VXU), Sweden. His current research interests include the design of mobile learning environments that explore novel modes of collaboration. He is presently involved in a number of European and national projects exploring how mobile and wireless technologies can be used to support new ways of learning and how these technologies can support groups of learners when they, collectively, share their understanding in these learning environments. Previously he has worked for the Interactive Institute, in Sweden and the LEGO group in Denmark.

**Masanori Sugimoto** received BEng, MEng, and DrEng degrees from the University of Tokyo, Japan, in 1990, 1992, and 1995, respectively. In 1995, he joined the research and development department at the National Center for Science Information Systems (currently National Institute of Informatics), Tokyo, Japan. He became an associate professor at the Information Technology Center, University of Tokyo, in August 1999. Currently, he is an associate professor in the department of frontier informatics at the graduate school of frontier sciences at the University of Tokyo. He was a visiting scientist in the department of computer science at the University of Colorado, Boulder, USA (from January 1997 to November 1997). His research concern is related to human-computer interaction, especially mixed reality, mobile computing, computer supported collaborative work/learning, intelligent systems, interactive robotics, and so on. Dr. Sugimoto is a member of ACM, IEEE, IEICE, JSAI, IPSJ, VRSJ, JSSE, and JCSS.

**Javier Vélez** is a computer engineer and currently PhD student working at Universidad Complutense de Madrid as a teaching assistant. His research interests lay in the pedagogical technologies and computer-supported collaborative learning fields. A main part of his on-going PhD work consists in developing a platform to support collaborative learning scenarios and to integrate several external tools within it.

**Dr. M. Felisa Verdejo** is a professor (computer systems and languages) and head of the learning technologies and collaborative systems (LTCS) research group and the Department Lenguajes y Sistemas Informáticos (LSI) and the at the National Distance Learning University (UNED), in Madrid, Spain. The main research line of her group concerns modeling and principled design of interactive and collaborative learning environments. Learning design specification based on activity theory, knowledge-based methods to support and analyze computer supported collaborative learning, and semantic approaches to create the infrastructure for virtual learning communities are the themes that her research group currently focus on. Mobile learning has been already explored in this context in previous projects, and currently also for primary school settings.

**Feng Xie** holds an honors degree in information sciences, majoring in computer science, and a masters of information sciences (information systems) from Massey University, New Zealand. His research interests include mobile web application architecture and mobile technologies. He works as a technical professional in the Internet services industry and is currently a resident in Singapore.

**Yoneo Yano** is currently a professor and also the dean at the faculty of engineering at the Tokushima University, Japan. He received his BE, ME, and PhD degrees in communication engineering from the Osaka University, Japan, in 1969, 1971, and 1974, respectively. He was a visiting research associate at the computer-based education research lab at the University of Illinois, USA. He is a member of the

IPSJ, IEICE, and JSiSE, the JAPAN Society for Educational Technology, AACE, and IEEE. Currently, he is the vice president and the vice editor in chief of *JSET.*

**Chengjiu Yin** graduated from the China Jinzhou Normal College Computer Department. He worked as a software engineer for Beijing Automatic Science and Technology Co. Ltd from 2000 to 2002. In 2005, he received his ME degree from the department of information science and intelligent systems from the Tokushima University, Japan. At the moment, he is pursuing his PhD degree in the department of information science and intelligent systems from Tokushima University, Japan. Currently, he is committing himself in mobile learning and ubiquitous computing. He is also a student member of JSiSE and APSCE.

# Index

## G



## H



## I



## J



## L



## M



## O



## P

## R



## S



## T



## U



## V



## W



## X